A Bedside Nature

About the Editor

Walter Gratzer is a member of the Medical Research Council's Muscle and Cell Motility Unit and a Professor at King's College, University of London. He has a long-standing association with *Nature*, for which he regularly writes book reviews. He is the Editor of *The Literary Companion to Science* (Longman/Norton, 1989).

Editor's Note

To preserve the sense of period conveyed by the extracts in this book, they have been reproduced in untouched facsimile form, exactly as they appeared.

Acknowledgements

The essence and form of this book, like so much else that has sprung from *Nature* in the past 30 years, were conceived by John Maddox. That it was brought to completion owes much to the efforts of Tim Lincoln and Peter Tallack, and the arduous task of fitting the extracts and accompanying commentary into the pages was heroically undertaken by Camilla Clanchy, who, with Jane Walker and Steve Sullivan, also chose the illustrations and created the design. The index is the work of Jean Macqueen, and Desmond King-Hele kindly read the proofs and corrected several errors and ambiguities. Eric Freeman, Librarian of the Wellcome Institute for the History of Medicine, and Andrew Melvin, a member of the Institute's staff, were generous in providing help during the early stages of the project.

A BEDSIDE *NATURE*

Genius and Eccentricity in Science 1869-1953

Edited by Walter Gratzer

W. H. Freeman and Company
New York

Cover design by Pixel Press.

Library of Congress Cataloging-in-Publication Data

A bedside Nature : genius and eccentricity in science 1869–1953 / edited by Walter Gratzer.
    p.   cm.
   Includes index.
   ISBN 0-7167-3139-8
    1. Science—History—19th century.   2. Science—History—20th century.
   3. Nature (London, England)   I. Gratzer, W. B. (Walter Bruno). 1932–   .
   Q125.B375, 1997
   509′.034—dc21
                                           97-15964
                                             CIP

Published in the United States in 1997 by W. H. Freeman and Company, 41 Madison Avenue, New York, NY 10010. First published in the United Kingdom by Macmillan Magazines, Ltd., 1996.

Printed in the United States of America
First U.S. printing, 1997

selected contents

foreword

"Art may err", Dryden wrote, "but Nature cannot miss". So how could this collection drawn from nature's finest midwife, interpreter and namesake be anything less than wonderful? Macrocosmic nature is of course not always so marvellous, hence a large part of her fascination. Pliny the Elder called nature "a kind parent or a merciless stepmother" and then died, a few years later, in the same eruption of Mount Vesuvius that buried Pompeii.

Wordsworth contrasted nature's "sweet lore" with our "meddling intellect", but the wonder of this volume lies in their conjunction — the power of science with the humanity (including sometimes the silliness) of interpreters. I intended to skim the volume as I prepared this foreword, but ended up reading every item. The text, aided by a brilliantly discerning editorial hand, is consistently charming and even ennobling. The items may be full of error, tendentiousness, even viciousness on occasion — but one feature stands out through all the variety, one trait that captures the essence of the peculiar institution we call science: the people who wrote these items cared passionately about the content of their contribution, whether this be a French mnemonic for π to 30 decimal places, or James Watson and Francis Crick on the structure of DNA. Every writer knows that science matters in the deepest human way. God bless them all (and nature is, as Chaucer said, "the vicaire of the almyghty lorde").

I was most moved by the interplay of topicality and timelessness in these items of NATURE's history, for science is both a quintessential human activity in the struggle to know (hence its necessary and particular temporal location and change) and an institution dedicated to finding empirical truth with some stability (however unattainable the total goal, for numerous reasons rooted in both nature and science). All items share in this interplay at their various levels. Three categories stand out for me in the amazing pot-pourri that NATURE has displayed for science in all its multifariousness and fluctuating worth (with constant fascination).

In a first category — the "greats" if you will — NATURE has sprinkled its pages, virtually in every issue, with papers that have become classics or have been written by extraordinary scientists with a flair for expression and distinctiveness of ideas. A conventional anthology would have focused on these papers alone — and such a volume would have been a great mistake and a poor indication of NATURE's totality and meaning through the years. But an anthology that did not include some of these conventional items would also have been a poor thing indeed.

This volume strikes just the right balance — enough of the "greats", without any dominating presence. Anchors at both ends fall into this category, as we begin with T. H. Huxley and end — where else and what other conceivable choice? — with Watson and Crick in 1953. In between, Charles Darwin writes his last contribution to NATURE in the year of his death (1882), on a characteristically "small" topic with global implications: "On the dispersal of freshwater bivalves." Huxley, in his magisterial prose, then writes Darwin's obituary:

An intellect which had no superior, and with a character which was even nobler than the intellect.... He found a great truth, trodden under foot, reviled by bigots, and ridiculed by all the world; he lived long enough to see it, chiefly by his own efforts, irrefragably established in science, inseparably incorporated with the common thoughts of men.... What shall a man desire more than this?

Eadweard Muybridge proposes the photo finish to resolve "dead heats" in racing (1882). Alfred Russel Wallace reminisces in old age, showing both Darwin's decency and his own modest generosity: "The one great result which I claim for my paper of 1858 is that it compelled Darwin to publish his 'Origin of Species' without further delay" (1903). Charles's son, the great physicist George Darwin, displays fealty by using newly discovered radioactivity to challenge Lord Kelvin's dangerously low estimates (for evolution) of the age of the Earth and Sun (1903). Francis Galton, bored silly while sitting for portraits, counts brushstrokes on two occasions and learns that very different artistic styles still require about the same number (1905). D'Arcy Thompson comments on Aristotle's remedies for killing fleas by exposure to the Sun (1911).

remedies for killing fleas by exposure to the Sun (1911).

A second category — call it idiosyncrasy, or even crankiness — also embodies timelessness, but from the other end, of items too small or too personal for participation in a *Zeitgeist*. Americans in particular (I am one, and a New Yorker to boot) are amused (just as the British note Americans' boorishness and philistinism) by the British propensity for passionate and endless letter-writing on scholarly minutiae (usually in Greek or Latin), as though such items matched World War II in importance. *NATURE* is a gold mine for such dubiety, and the charm of the journal (and this anthology) lies in the interspersing of such items among the "greats" — both are equally important attributes of human nature. Such idiosyncrasy is truly timeless and not confined to the nineteenth century when males of proper breeding (virtually the entire pool of potential scientists, in a great injustice of the time) all studied the classics. As I write this foreword, the letters column of *NATURE* is incandescent with volleys of contributions on whether or not we should pronounce the second "p" in apoptosis: no, say some, according to the Greek silent source in *ptosis*; yes, say others, with the litany of "helicopter" and "*Archaeopteryx*". (I am, by the way, in no position to complain about letters on classical trivia, for I wrote one myself a few years back to correct Karl Popper on the Latin source and meaning of the Royal Society's motto, *Nullius in verba*.)

On it goes. What meaning, if any, can we draw from the fact that the number 137 crops up in many physical formulae? We may easily remember π to 30 decimal places with nifty mnemonics in French and German, a letter published in 1905 informs us (count the number of letters in each successive word). A few weeks later a correspondent offers an English version:

Sir,—I send a rhyme excelling
In sacred truth and rigid spelling.
Numerical sprites elucidate
For me the lexicon's dull weight.
If "Nature" gain
Not you complain,
Tho' Dr. Johnson fulminate.

But nothing suits this category so well as the plethora of "animal anecdotes" that have always filled the pages of *NATURE*, perhaps most in the nineteenth century, during a more narrative age of natural history, but by no means absent today. Thus Charles Darwin's famous protégé George J. Romanes on whether animals have an explicit sense of humour (1875): "I once had a Skye terrier... he had several tricks... the sole object of which was evidently to excite laughter." A correspondent from North Manchester (Indiana, not England) writes of a spider who called in a compatriot to help to subdue a prey too large for her alone; after securing the fly, the helper retreated, permitting her friend to enjoy the feast alone (1880). My favourite items, however, detail the "suicide" of scorpions and a remarkable case of altruism among sea urchins (as three companions helped an injured individual to turn over). The correspondent, another American, this time from Maine, closed by noting: "This is the best instance of 'giving a lift' I have ever met with among animals of so low a grade" (1890).

Science, largely through the immediacy of technology, is the prime mover of social change, but social trends also influence science in a complex interactivity. In my third category of timely items, I was fascinated most of all by the topical chronology of science's participation in this most tumultuous and transforming of all our centuries. As we read this anthology in temporal order, we note the unique perspective of the scientists' parish on a panoply of history created, in large part, by their own successes as well as their arrogant foibles.

Consider the real scoop on the psychology behind Henry Morton Stanley's restrained greeting to David Livingstone (1872):

... anxious to enter the African town with as much éclat *as possible, [Stanley] disposed his little band in such a manner as to form a somewhat imposing procession. At the head was borne the American flag; next came the armed escort.... In an instant he recognised the European as none other than Dr. Livingstone himself; and he was about to rush forward and embrace him, when the thought occurred he*

was in the presence of Arabs, who, being accustomed to conceal their feelings, were very likely to found their estimate of a man upon the manner in which he conceals his own. A dignified Arab chieftain, moreover, stood by, and this confirmed Mr. Stanley in his resolution to show no symptoms of rejoicing or excitement. Slowly advancing towards the great traveller, he bowed and said, "Dr. Livingstone, I presume?" to which address the latter, who was fully equal to the occasion, simply smiled and replied "Yes."

Paul Broca hopes that "men of eminence" will donate their brains to science so that the correlation of size with intelligence may be affirmed (1876). Results fall short, however, or at least into ambiguity. Ivan Turgenev weighs in at a mammoth size of more than 2,000 grams (1884), but poor Léon Gambetta could only manage a distinctly sub-par 1,100 grams (1883), though *NATURE*'s correspondent reports that the dissector "found the structure of the brain to be very fine, and the third convolution, which M. Broca associates with the speechifying faculty, to be remarkably developed."

Mr William Ewart Gladstone, temporarily out of office as prime minister, publishes a book on Homer, and several correspondents, with copious and untranslated Greek citations, discuss the Grand Old Man's theory that Homer's informant must have been colour-blind (1878). Clever Hans, the counting horse, receives his due (1904) and then his comeuppance in a beautifully crafted methodological note on the necessity of double-blind testing (also 1904, as Hans didn't fool scientists for long). On the ephemeral (but, alas, sometimes enduring) prejudices of our scientific culture, several items extol various forms of biological determinism in racial, sexual or class-biased modes. One correspondent writes in 1895: "The labouring classes are much less sensitive to pain than the non-labouring classes, and the women of the lower classes are much less sensitive to pain than those of the better classes. The general conclusion is that the more developed the nervous system, the more sensitive it is to pain."

A Henry E. Armstrong writes a chillingly patronising obituary notice for a woman scientist he regarded as diligent but necessarily uninspired by nature (1923): "In fine, my conclusion is, that *das ewig Weibliche* was in no way overcome in Mrs. Ayrton: nor could we wish that a thing so infinitely precious should be: she was a good woman, despite of her being tinged with the scientific afflatus." Long after the battle, Hilaire Belloc attacks Darwin (1927); Arthur Conan Doyle defends spiritualism; and *NATURE* gives much space in rebuttal to René-Prosper Blondlot's illusory n-rays, Paul Kammerer's pseudo-Lamarckian midwife toads and T. D. Lysenko's perversion of Soviet genetics.

Contributions of wartime are, to my mind, the most interesting and the most revealing. *NATURE* reports on papers read before the Académie des Sciences during the Prussian siege of Paris in 1871: "M. Payen writes on hippophagy, M. Frémy on the use of osseine as food, M. Richie on the use of black puddings of ox-blood." World War I hit the pages of *NATURE* with all its jingoism and intensity. J. P. Lotsy defended the internationalism of science and regretted the renunciation of British honorary degrees by several German scientists (1914). Ernest Rutherford wrote to bemoan the death of a young physicist (1915): "Our regret for the untimely end of Moseley is all the more poignant that we cannot but recognise that his services would have been far more useful to his country in one of the numerous fields of scientific inquiry rendered necessary by the war than by exposure to the chance of a Turkish bullet." Fritz Haber, in a post-mortem, even defends Germany's use of poison gas (1922). But science marched on regardless; in 1917, Chandrasekhara Venkata Raman wrote from Calcutta about the production of "wolf-tones" by violins and cellos.

The times around World War II became even uglier. The Nobel laureate Johannes Stark wrote an early and vicious piece (1934) on Aryan physics and its pollution by Jews. The Nazis try to purge their language of foreign terms by substituting Germanic neologisms (*Höchstwert* for *Maximum*, for example). *NATURE*'s correspondent describes the re-dedication of "the once illustrious Institute of Physics at Heidelberg" in 1936:

"... an imposing number of German physicists assembled to make public confession of their union against the Jewish evil (*jüdischer Ungeist*) from which German science must be completely freed." Dr Stark again attacks Jewish science for defending internationalism, when Aryan purity must be maintained (1938): "Under cover of the term 'exchange of experience' there lurks the doctrine of the internationality of science which the Jewish spirit has always propagated, because it provides the basis for unlimited self-glorification." (I believe that NATURE proceeded admirably by opening its pages to such Nazi letters, thus allowing the ideologues to hang themselves.) In tough times of pre-war Depression, France issued a series of stamps to raise money "*pour les chomeurs intellectuels*" (for unemployed intellecturals); Pasteur holds a test tube on the first issue. And NATURE carried on during the Blitz (1941): "... so far, the offices of NATURE have suffered nothing but a few broken windows and occasional interruption of work caused by the proximity of delayed-action bombs."

Let me end with two heart-warming stories about the preservation of science and generosity amid the rigours of war. During the Japanese occupation of Malaya, the Raffles' Museum of Singapore issued C. F. Symington's *Foresters' Manual of Dipterocarps*, despite strict prohibitions against such enemy and English publications. The work appeared because the museum's head in occupation, a Professor Tanakadate of Tohoku Imperial University, bravely bent the rules. He even had proofs brought to the author at a prisoner-of-war camp. "It was insisted by Prof. Tanakadate that the book should conform exactly with the previous series of *Malayan Forest Records* of which it is No. 16, so that it should stand the test of time, as a scientific work, regardless of hostilities and racial prejudice. He therefore added a brief preface, as a single page of romanized Japanese, and he issued the Manual from the Museum to give it official standing."

Under the title "'Prison Camp Geology', NATURE reported (1949) a remarkable monograph on local petrology published by several geologists at the "University of Edelbach" founded by French prisoners in Oflag XVIIA:

Not content with lectures alone, the geologists made a thorough investigation of the area — only 400 metres square — enclosed within the barbed wire. No stone was left unturned, and trenches and secret tunnels provided many critical exposures. A microscope was constructed in the camp and equipped with polarizers improvised from piled cover glasses. Thin sections were mounted with a mixture of violin wax and edible fat.

Science will persevere so long as its practitioners retain their zeal and commitment. The continuity of this great culture, celebrated here through a flagship journal, will continue through this millennium and into many others to come if *Homo sapiens* can harness what Abraham Lincoln called the better angels of our nature (and what evolutionary theory might designate an attainable side of our flexible behavioural potentiality) in order to eschew the vicious parochiality of Dr Stark and to emulate the prisoner petrologists of Oflag XVIIA.

Stephen Jay Gould
Harvard University
July 1995

introduction

Dean Inge, the worldly prelate (columnist for the *London Evening Standard* and an occasional contributor to *NATURE*), also known as the Gloomy Dean, related that he had once received a letter from a disaffected reader that concluded: "I am praying for your death: I have been successful on two other occasions." Some 15 years ago the then editor of *NATURE* was favoured with an almost identical missive from a failed author. We know of course from the work of one of the journal's most illustrious contributors, Francis Galton (as you will see if you turn to p.50), that the efficacy of prayer is statistically nugatory, so the threat would have caused less alarm than the assassination attempt by another rejected author on the editor of *Physical Reviews* (which spared him but killed a secretary). It shows nevertheless that the ambition to publish in *NATURE* can be a powerful urge, and there are some at least who would kill to gratify it. For *NATURE* is now, as it instantly became at its inception in 1869, the chic place for scientists to disport themselves.

The idea of a journal dedicated to the reporting of science emerged mainly from a little coterie of intellectual bruisers — a familiar feature of the Victorian scene — that could fairly be described as the scientific establishment. The X Club consisted of nine men and included, besides T. H. Huxley, its founder, Joseph Hooker, John Tyndall, Edward Frankland and John Lubbock. It convened for a monthly dinner and once a year for an excursion, to which wives were admitted ("the x's and yv's"). The group had given strenuous support to a periodical called *The Reader*, a part of which was set aside for science, and indeed the less wealthy among them, such as Huxley and Tyndall, relied on the income afforded by writing for a popular readership. But *The Reader* folded, probably for want of a resolute editor. A year later Norman Lockyer, a young astronomer working for the War Office, became adviser on matters of science to the publisher Alexander Macmillan, and represented to him the view of the X Club and others that science urgently needed a voice. Macmillan accepted the challenge, invited Lockyer (who had just discovered in the solar spectrum a new element, which he called helium) to edit the new journal and through a decade or more of dismaying losses kept faith with his editor, the contributors

and the readers. *NATURE* was the name chosen and some rather sententious lines from a sonnet by Wordsworth became the motto on its masthead. The mathematician J. J. Sylvester, for one, was enchanged: "What a glorious title, *NATURE*, a veritable stroke of genius to have hit upon. It is more than a Cosmos, more than a universe. It includes the seen as well as the unseen, the possible as well as the actual, *NATURE* and Nature's God, mind and matter. I am lost in admiration of the effulgent blaze of the ideas it calls forth."

The first number appeared on 4 November 1869. "I think we will look nice," Macmillan had written to a friend on the eve of the launch. *NATURE* proclaimed as its objectives to inform the reading public of new scientific and technological advances and to promote the interests of science politically, to disseminate new results among scientists and to allow the ventilation of controversies. There was nothing parochial about *NATURE*. From the beginning its outlook was international and each issue carried reports of the proceedings of such bodies as the Astronomical Society of Riga and the Montevideo Natural History Association. Many of the titans of nineteenth-century science contributed leaders and commentary, often anonymously: the file copies of the journal often reveal the identity of the anonymous contributors, for the secretary has written their names in sepia ink across the head of the column, together with a note of the remuneration ("Lord Rayleigh, 2s 6d"). It also reveals, for instance, that the reviewer of one book is its author. (He approved.)

Browsing in these early volumes conveys a vivid impression of collegiality; the contributors and the readers must have felt themselves to be members of a community, even a club, riven from time to time, to be sure, by scholarly disputes, in which everyone seemed to take sides. These were pursued with an uninhibited ferocity unthinkable today, but the standard of debate and especially of literacy is even more striking. (Obituaries, by sharp contrast, and likewise a series of tributes to living scientists, entitled "Scientific Worthies", were disappointingly decorous. Dr Johnson's generous sentiment that in lapidary inscriptions a man is not under oath has sadly deprived us of much good biographical material.)

Adlai Stevenson described an editor as one who separates the wheat from the chaff and prints the chaff; but in *NATURE* there was (and is) remarkably little chaff. Editorial standards were rigorous and the quality of the contributions consistently high. The concern for good English has endured, but it is unlikely that in Lockyer's day there would have been much need for copy-editing. An editor of *Science*, *NATURE*'s transatlantic counterpart, wrote wistfully: "It has been one of the trials of my life that *NATURE* is better than *Science*. I like to blame this on the circumstance that the subscription price of *Science* is only one fourth that of *NATURE*, but I suspect that it is also in large measure due to the greater skill of your people as writers on science. This I attribute to the fact that nearly all your scientific men come from upper-class families where good English is used, while these traditions have been lacking in many of the families from which our scientific men come." He might have added that they were also quick with a quip or a quote in Latin or Greek (with no concessions in the way of translation). The dean of a great American university deplored the introduction of the Bachelor of Science degree on the grounds that, while it would not ensure that the recipients knew any science, it would undoubtedly guarantee that they knew no Latin. Lockyer's contributors shared these worries and displayed a far more enlightened concern for educational values than the Greats men and historians who dominated the universities, made policy and had no time for science.

Lockyer served as editor for half a century, and certainly as time went on he became increasingly opinionated and intolerant of dissenting views. W. T. Thiselton-Dyer, the botanist and a frequent contributor, cites in the jubilee issue the observation by the mathematician H. J. S. Smith to the effect that Lockyer failed at times to recognise the difference between the editor of *NATURE* and the author of Nature.

Between the turn of the century and World War I an unmistakable change came over science as it is reflected in the columns of the journal. Gone for the most part were the leisurely ruminations on phenomena involving rainbows and lightning and the curious behaviour of ants and pet spaniels. New professorial chairs had been founded in the universities and both the physical and

biological sciences were becoming professionalised. The new physics of Einstein, Bohr, Planck and Lorentz was establishing itself and the structures of atoms and molecules were being uncovered by Thomson, Rutherford and the Curies. Genetics was at the advancing front of biology, and chemistry too was being transformed by the structural revelations of X-ray crystallography. There was a new editor, R. L. (already in fact Sir Richard) Gregory, who was to occupy the chair for 20 years.

By the 1920s Gregory's had become the voice of the new scientific mandarins: nor was he prey to self-doubt, for there proceeded from him a stream of pronouncements on how society, industry, the country and the world should be managed. Gregory was much influenced by his close friend H. G. Wells (and probably tried, though unsuccessfully, to help Wells fulfil his gnawing ambition to get into the Royal Society). Wells was allowed liberal access to the columns of *NATURE* and his books received expansive reviews. He repaid the debt by many fulsome allusions to *NATURE* and its editor, most famously in his long novel *The World of William Clissold*, in which the hero refers to *NATURE* as "the ideal newspaper", its influence on society being stronger and far more humane than that of any other journal: it "tells you of things that matter". Since science itself had more influence on man's life in the twentieth century than almost any other factor, *NATURE*'s impact was incalculable. And, Wells added, no English-speaking scientist could go for a week without reading it.

Wells and Gregory, together with other public figures, such as Julian Huxley and the science journalist Ritchie Calder, sought to promote, through the unlikely channel of the British Association for the Advancement of Science, Utopian schemes for a new world order, based on the application of "scientific principles". "We intellectual workers", H. G. declared, "have to decide... whether we will take our place as the servant masters of the world." Harold Nicolson made the following sagacious observation about the growing confidence of the men of science: "In England intellectuals are not expected to exercise any corporate influence at all. It is only our scientists (a fresh breed of men) who suppose that their opinions on subjects outside their own competence are of importance. The illusion will pass." So it came about that the following could find its way into the journal: "...a large proportion of the slum populations consists of ...'morons' — that is of mental defectives of comparatively high grade. These people are lacking not only in intelligence but also in self-control, which is the basis of morality, and they reproduce recklessly." The Fabian socialist G. D. H. Cole summed up this view more succinctly:

The middle-classes — me and you —
Already know a thing or two.
But, oh the poor, they breed like rabbits!
They have the most distressing habits.

To be fair, Gregory was reflecting the views of many politically committed scientists, not excluding communists such as J. D. Bernal.

But more often, and when it mattered most, as when first Stalin and then Hitler rose to power, NATURE was far ahead of the daily press, or indeed most politicians, in grasping the implications and in forthright condemnation. It lent its support to A. V. Hill, Gowland Hopkins and others, who set up an "Academic Assistance Council" to find shelter and laboratories for refugees within days of the announcement of the first dismissals of Jewish scholars from German universities. It was cause for pride that NATURE was proscribed by the German Ministry of Education under Hitler and censored in the Soviet Union.

Although the remarkable proportion of the world's most important scientific results that first saw the light in NATURE never diminished, the two decades following World War II saw some of the sparkle fade from its pages. The third editor (some of the time in double harness with A. J. V. Gale) was L. J. F. Brimble – note the initials: first names in professional circles appeared only much later (except no doubt in the United States, though even there H. L. Mencken defined the first Rotarian as the first man to call John the Baptist Jack). Brimble was a less assertive figure than his two predecessors: one can scarcely imagine that either of them would have uncomplainingly edited for publication, as Brimble did, a highly disobliging review of his own botany text. He was in poor health, overworked and weakened by a wound sustained

during the London Blitz. The typefaces looked wilted, the layout unalluring. A referee would still receive with the manuscript an unsigned letter, which avowed that the editors of NATURE presented their compliments to him and trusted that he would find the enclosed of interest. And what other learned journal would have rejected Hans Krebs's communication on his discovery of the citric-acid cycle (one of the foundations of modern biochemistry) with the apologetic statement that the letters section was for the present sufficiently supplied? All the same, Brimble did later accept for publication a letter describing a highly conjectural structure that had by no means found universal favour but proved to be a turning point in the history of science (see p.248). So his record as an editor could be said to stand well in the black.

With the arrival in 1966 of John Maddox the journal was rejuvenated and it achieved and has maintained under him (and in an intervening period under Dai Davies) a pre-eminence in science, a standing in the world, and a circulation for that matter, far beyond anything Macmillan and Lockyer could have imagined. Karl Pearson's comment on the demise of *The Reader* and the birth of NATURE ran: *Pereat lector, Natura resurget.* And so it still does, a century and a quarter on.

What follows, by way of celebration of the longevity of NATURE and the stewardship of Sir John Maddox, as he relinquishes the chair to his successor, Philip Campbell, is not a collection of Great Papers from NATURE. Such compilations may impress but seldom give much pleasure. Nor do the selections extend all the way to the present day: the line has been drawn roughly where the memory of scientists still at work fades into history. The aim has been to present a panorama of science, seen against the backdrop of nineteenth- and twentieth-century history, with its triumphs, débâcles, surprises and absurdities, all reflected in NATURE's mirror. And if any justification is needed, beyond the hope to entertain the reader, it is always worth remembering that, in science as elsewhere, those who do not learn the lessons of history are for ever destined to repeat them.

Walter Gratzer
University of London
June 1995

chapter one
1869-1871

NATURE PROCLAIMS ITS CREED IN ITS FIRST LEADING ARTICLE, PENNED BY T. H. HUXLEY. THE REDOUBTABLE
EDWIN RAY LANKESTER KICKS OFF WITH A DISSERTATION ON RED EXUDATES FROM VARIOUS FAUNA. PROPOSALS FOR
A CROSS-CHANNEL BRIDGE ARE THE TARGET OF LEARNED MOCKERY. ALFRED RUSSEL WALLACE SCORNS PUBLIC
FUNDING OF SCIENCE. THE EGREGIOUS HENRY CHARLTON BASTIAN WEIGHS IN WITH THE FIRST OF MANY DOGMA-
RIDDEN PHILIPPICS IN FAVOUR OF SPONTANEOUS GENERATION OF LIFE. COCKROACHES NIBBLE TOENAILS. VIEWS ARE
AIRED ON THE PROSPECTS FOR MILITARY BALLOONING. THE CONDUCT OF THE SAVANTS DURING THE SIEGE OF PARIS
IS COMMENDED AND THE ASTRONOMER CHARLES PIAZZI SMYTH DENOUNCES THE REFEREES OF LEARNED JOURNALS.

1869-1871

These were the years of the Franco-Prussian War — the rout of the French army at Sedan, the overthrow of the Third Empire, the siege of Paris and the Commune. Otto von Bismarck became Imperial Chancellor of the new German Empire. The Suez Canal was opened by Empress Eugénie. In Britain, the half-penny postage stamp was issued, the ancient universities opened their doors to non-Anglicans, and Charles Darwin published *The Descent of Man*.

NATURE leaders in the early volumes were written by Norman Lockyer, the editor, T. H. Huxley and other luminaries. Here is one such from one of the first issues, probably by Huxley — a ripe example of the genre, with its stately prose, literary ornamentation and at times ponderous humour, articulating the complaint of the scientist through the years that society does not sufficiently value his calling. Note the echo years later from G. K. Chesterton: "Is ditchwater dull? Naturalists with microscopes have told me that it teems with quiet fun."

THE DULNESS OF SCIENCE

WE have all heard of the fox who, when he had lost his own tail, tried to prevail upon his comrades to dispense with theirs ; and we think it must surely have been in a congress of the blind that the question was first started, " Is it dull to use your eyes and look about you ? "

For, in fact, what is science but this ? We come unexpectedly into a great mansion, of which we know nothing ; and if it be dull to seek out the various inmates of the house, and to ascertain its laws and regulations, then is science dull ; but if this be important and interesting, then so also is science interesting.

But, alas ! the blind in this sense are numbered by myriads ; and as they, for a time, almost threaten to carry their point, a few remarks upon the dulness of science, or rather, perhaps, the dulness of men, may not be out of place.

We have in our mind's eye at the present moment several notable specimens of blind men. One of these lives not very far from where we write—a most hopeless individual ; we had better not inquire too narrowly concerning his occupation ; he will be found somewhere in the purlieus of this great city. His one sense is the sense of gain. We remember once seeing through a microscope the animalcules of a drop of water, and we noticed that one of the largest of these had one end fixed to the side of the vessel, while its arms and mouth were busy gathering up and swallowing its smaller neighbours. Now, the man of whom we speak is only this animalcule magnified without the microscope. Ignorant of all laws, civil, religious, physical, moral, social, sanatory, he rots in his place until Dame Nature, in one of her clearing-out days, fetches at him with her besom the plague ; and he is swept aside and seen no more.

Our country readers are no doubt well acquainted with Farmer Hodge. One day he happened to sit next the poet Coleridge, listening, with that reverence for his betters to which he had been early trained, to the marvellous sayings of the great man, and it was only when the apple dumpling made its appearance that he exclaimed, " Them's the jockeys for me ! " Hodge, we fear, maintains no sort of relations with the universe around him. He farms in the same way in which his grandfather did, and has the most profound aversion for the steam plough.

He told Tennyson—

" But summun' ull come ater meä mayhap wi' 'is kittle o' steäm,
Huzzin' an' maäzin the blessed feälds wi' the Divil's oän teäm.
Gin I mun doy I mun doy, an' loife they says, is sweet ;
But gin I mun doy I mun doy, for I couldn abear to see it."

Nevertheless, Hodge has some sense of his duty to his neighbour. Indeed, we learn from D'Arcy Thompson, that being once asked What is thy duty towards thy neighbour ? he wrote as follows upon a slate :—

" My duty tords my nabers, is to love him as thyself, and to do to all men as I wed thou shall do and to me, to love, onner and suke my farther and mother, to onner and to bay the Queen, and all that are pet in a forty under her, to smit myself to all my gooness, teaches, sportial pastures and marsters, to oughten myself lordly and every to all my betters, to hut nobody by would nor deed, to be trew in jest in all my deelins, to beer no malis nor ated in yours arts, to kep my hands from pecken and steel, my turn from evil speaking, lawing and slanders, not to civet nor desar othermans good, but to laber trewly to git my own leaving, and to my dooty in that state if life, and to each it is please God to call men."

Ascending in the scale, we come next to our friend " Cui Bono ; " a very good sort of man, very fussy, very philanthropic, and very short-sighted,—in fact, he sees nothing distinctly that is more than one inch from his face. He called upon us the other day to give us a little good advice : it was about the time when our astronomers were investigating the chromosphere of the sun. " What," he asked, " is the use of all this ? will it put one penny in your pocket or mine ? will it help to feed, or clothe, or educate your family or mine ? Take my advice, sir, and have nothing to do with it." We did not reply to him ; indeed we learned afterwards that he had just written an article on the subject in one of the journals. Next day he called upon us in a state of high jubilation ; he had just seen a friend of his who had succeeded in making a useful application of some great discovery, which, being within the requisite *inch*, was clearly perceived by " Cui Bono "—" A very useful and practical discovery, sir, which will greatly alleviate human suffering ; none of your hydrogen-in-the-sun business." And so the successful adapter got all the praise, while the wretched man of science who discovered the principle was left out in the cold.

Still ascending the scale, we come to a man of strong mental eyesight, but without leisure to use it ; one that it makes us grieve to see, inasmuch as he is capable of far better things. His ears are not altogether stopped to the mighty utterance that all nature gives, nor yet is he wholly ignorant, when at night he looks upwards, of that which the firmament declares ; but its utterance is drowned in the tumult of a great city, while its starlight is quenched in the smoke. Our sentiment for such a man is that of pity ; for indeed, what with the cares of this world and the deceitfulness of riches, he has a hard battle to fight.

But is it not melancholy to reflect how great a proportion of the energy of this country is devoted to the acquisition of gain, and how small a proportion to the acquisition of knowledge ?

We have now arrived at the ranks of the affluent and the nobly-born, where, if anywhere, we might expect to find " tastes refined by reading and study, and judgments matured by observation and experience ; " but how seldom is this the case ? The mental eyesight is often weak to begin with, and often is it rendered still weaker by poring over classics without end. The unfortunate youth is then sent to make the tour of Europe. He is sent to Switzerland and the Alps to see all that is grand in nature, and to Rome and Paris to see all that is great in art, and he comes home wretched and disgusted, and no wonder. He has been made the unfortunate subject of a senseless experiment—an experiment much the same as that of turning a man with weak eyes into a picture gallery in order to improve them. His friends forget that appreciation of the beautiful and the true is the product of the coming together of two things—eye-sight and nature. In fact, the result is much the same, whether a man with no eyes is carried out into a glorious landscape, or whether a man with good eyes is shut up in a dark room.

It is of this the poet speaks, when he says :—

"O Lady! we receive but what we give,
And in our life alone does Nature live;
Ours is her wedding-garment, ours her shroud!
   And would we aught behold of higher worth,
Than that inanimate cold world allow'd
To the poor, loveless, ever-anxious crowd,—
   Ah! from the soul itself must issue forth
A light, a glory, a fair luminous cloud
   Enveloping the earth;
And from the soul itself must there be sent
   A sweet and potent voice, of its own birth,
Of all sweet sounds the life and element!"

But let us hasten to our friend Philosophus, who is a man of quite a different mould. Once, when he was young, his tutor said to him, "Have the goodness, sir, to solve the following problem : 'A hemispherical bowl is filled with a heavy fluid, the density of which varies as the nth power of the depth below the surface; find the whole pressure and the resultant pressure on the semi-lune of the surface contained between two vertical planes passing through the centre of the bowl, and making with each other an angle 2β.'" But Philosophus thrust the paper violently aside, saying "I will have none of that," and in fact was extremely rude. You may be sure, therefore, that when he came to be a man he had a mind of his own, and carried out his own ideas. He told us lately that he had been studying the laws of energy. It is a mistake, he said, to suppose that these laws are difficult of comprehension; they are merely remote from our ordinary conceptions, and must be patiently pursued until you grasp them. He had studied them, he said, at all times and on all occasions—in the railway carriage, on the thorough-fare, in the study, on his bed, in the night watches; and now that he had come to perceive their exceeding grandeur, and beauty, and simplicity, they were a source of great and continual joy to him, and recompensed him more than a thousandfold for all the trouble he had taken. Philosophus lately told us certain truths which may, perhaps, be of service to the readers of NATURE. He said that, not far from London, there was a place where the spirits and understandings of men were annually ground to pieces in a huge machine made of the very best metal; ay, such is its temper, said he, that were it only made into good broadswords, it might enable us to cleave our way to the very heart of the universe. Again, he said : "No doubt the dulness of science is a cry of the blind; nevertheless, men of science are much to blame. It is their sense of beauty that leads them to Truth, whom they discover by means of the glorious garments which she wears. But she is immediately stripped of these, and dressed in an antiquated mediæval garb, worse than that of any charity-school girl, and equal to that of any Guy

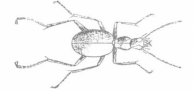

Faux : no wonder that in such guise her beauty is unperceived by those who cannot pierce the veil, and that as a consequence she is slightly esteemed."

There was another thing he told us—a thing of the highest importance. "The priests of Science," he said, "must consent to use the vernacular, before they will ever make a profound impression upon the heart of humanity; and when they have learned to do this, let them not fear the sneers of their deacons who will call their teaching sensational.   F. R. S.

Vol I, 43 1869

NATURE *became immediately a forum for "naturalists" of the time, among whom Charles Darwin numbered himself. Edwin Ray Lankester, a model supposedly for Arthur Conan Doyle's irascible but lovable Professor Challenger, was one of the most prolific, and published several collections of popular essays (prototypes in some sort of Stephen Jay Gould's). Here he is in characteristic vein.*

THE ORIGIN OF BLOOD-LETTING

THE flamingo in the gardens of the Zoological Society has recently been observed to vomit a red-coloured fluid over certain smaller birds kept with it; and it has been shown that this red fluid contains true blood-corpuscles, and inferred that the flamingo is in the habit of feeding its young by this ejection of a blood-stained "pigeon's milk" into their mouths. Further, the habit of the flamingo has been with great probability connected with the story of the pelican, which, as is well known, is stated to wound its own breast in order to feed its young with the blood. It is not at all improbable that birds so alike in their plumage and habitat as the pelican and flamingo should be confused in the way suggested by Mr. Bartlett, who, I believe, first observed the habit of the captive flamingo. The extravasation of *blood corpuscles* normally from the pharynx or œsophagus of such an animal is a matter of great interest. Mr. Lowne has a paper in the Journal of the Queckett Microscopical Club, in which he gives a full account of the case, having examined the bloody exudation microscopically. To this the reader is referred; but I have something to add to it.

The connection of the flamingo with the classical story of the pelican's self sacrifice is increased in interest, since it appears that the red exudation of the hippopotamus is connected with an equally ancient and more important tradition—namely, the history of the origin of blood-letting. Before giving this tradition, I would mention that two years since, by the kindness of Dr. Murie, I obtained some of the red exudation of the hippopotamus on a slip of glass, and on examining it with the spectroscope, I did *not* obtain a blood-spectrum. Mr. Tomes (Proc. Zool. Society, 1857) described the microscopic appearances of the exudation of the hippopotamus, and stated that he found in it remarkable corpuscles with pigmentary granules, but not *blood corpuscles*. The folds of the skin in various parts of the body of the hippopotamus are coloured bright pink by a distinct pigment, and the same tint suffuses the darker parts of the skin. I believe it is this pigmentary matter which causes the red colour of the exudation of the hippopotamus, and that it is not a sweat of blood at all. The case of Mr. Jamrack's rhinoceros mentioned by Mr. Lowne may be otherwise. Mr. Lowne says that cases of blood-stained sweat from the skin of man are, though rare, well authenticated. This is perhaps true; but many apparent cases of such staining are due to the formation of a purpurate in the sweat, from the decomposition of the uric acid which it contains.

Now, with regard to the hippopotamus, it is important to note how popular tradition has attributed the origin of a very valuable medical art to a totally false inference on the part of Egyptian priests.

M. Milne-Edwards, in the 3rd volume of his "Leçons sur la Physiologie" (p. 3), has the following note :—" Homer, whose poems constitute a sort of encyclopædia of the science which the Greeks possessed about the ninth century before Jesus Christ, does not speak of bleeding : but if we are to believe an author of the fifth century, Stephanus of Byzantium, this operation was known to the surgeons of the army of Agamemnon. In fact, he relates that one of them, Podalirius, son of Æsculapius, and brother of Machaon, on the return from the siege of Troy, practised it on a patient whose cure obtained for him the sovereignty of the Chersonese. This would be the first case of blood-letting of which the remembrance has been preserved; and, on consideration of a fable reported by Pliny, I am induced to believe that this practice had taken its rise in Upper Egypt : in fact, this naturalist tells us that the hippopotami, when they become too obese, have the habit of piercing for themselves the vein of the thigh, by pressing against a pointed reed; and that these animals have thus taught physicians to practise analogous operations. Now, this account does not apply to the sea-horse (or *Syngnathus*), as the author of an estimable work on the history of medicine (Leclerc) supposes, but to the great pachyderm which inhabits the rivers of the interior of Africa, and which is found in Upper Egypt. It is evidently a fable : but this fable could only have reached us from Egypt."

M. Milne-Edwards was not aware of, at any rate does not refer to, the red oozing observed on the skin of the hippopotamus sometimes after emerging from his bath, or when enraged, which gives so marked a confirmation to the Egyptian story. We may conclude fairly enough, either that the Egyptian priests saw this red exudation, 'and imitated it with the practice of bleeding, or, as is infinitely more probable, that the Egyptian laity noticed the blood-coloured sweat of the great river-horse, and connected it with the practice of bleeding then in operation, by the interpolation of the sharp reed, and an inability to understand that their wise men could discover a remedy untaught.   E. Ray Lankester

Vol I, 76 1869

The late nineteenth century was of course a period of prodigious growth in scientific activity. But how many men (no women to speak of) were there who could legitimately be regarded as scientists? The census of 1861 (which broke down the population by trades and was supposed to have concluded "Queens — 1; Bumboat-women — 1") recognised several classes of scientists. Here is the voice of Nature *on the subject.*

CAN we reduce to the certainty of numbers the amount of interest taken in England in the advance of science? How many are there devoted to its pursuit?

How many sufficiently concerned in its progress, as to be willing to make some sacrifices for its promotion? Interested in the results of science, ready to grasp its countless benefits, eager to catch its earliest gifts, we all are; but how many love science for its own sake, and are actively engaged in cultivating and promoting it? Are there not very many in this sordid age ready to exclaim with D'Ailly—

Dieu me guarde d'être savant,
D'une science si profonde
Les plus doctes, le plus souvent,
Sont les plus sottes gens du monde !

The Census of 1861 gave the number of persons engaged in the learned professions, or in literature, art, and science; and classed as "scientific persons," officers of literary and scientific societies, curators of museums, analytical chemists, and a certain number who styled themselves naturalists, botanists, geologists, mineralogists, chronologists, and civil engineers. These, however, with a number of professors and teachers, pursue science as a vocation. We prefer drawing our materials from the membership of our learned societies. Many of their members are, it is true, professors and teachers, yet they appear in a more congenial character as members of our academies, or fellows of our learned societies; and though it can scarcely be said that their members are in all cases absolutely men of science, or that all the men of science in the country are to be found in their lists, in them we have, at least, a goodly band of men associated together for the advance of science. Judging from the facilities such societies afford for the association of persons of kindred mind and taste, the common use of technical libraries and instruments, and the publication of their transactions, we are safe in assuming that they attract, at least, the greater number of men anxious to labour for the promotion of science.

The article then goes on to list the membership of the learned societies — 306 in the Meteorological Society, 3,595 in the Horticultural Society and so on — and draws its conclusion:

Were all who in London and the provinces are associated for the promotion of science carefully calculated, we should find that there are now about 120 learned societies, with an aggregate of 60,000 members; and deducting from the number at least one-fourth for members who belong to more than one society, we arrive at the interesting fact that there are, in the United Kingdom, 45,000 men representing the scientific world, or in the proportion of fifteen in every ten thousand of the entire population; the "upper ten thousand" of the aristocracy of learning being thus three times as many as the " upper ten thousand " of the aristocracy of wealth.

And there is a triumphal peroration:

A brilliant future opens itself for the cultivation of science. Happy will it be when " many shall run to and fro, and knowledge shall be increased ;" happy when men will realise that " pleasure is a shadow, wealth is vanity, and power a pageant ; but knowledge is ecstatic in enjoyment, perennial in fame, unlimited in space, and infinite in duration." Truly, in the performance of this sacred office the man of science "fears no danger, spares no expense, looks in the volcano, dives into the ocean, perforates the earth, wings his flight into the skies, enriches the globe, explores sea and land, contemplates the distant, examines the minute, comprehends the great, ascends to the sublime, no place too remote for his grasp, no heaven too exalted for his reach."   LEONE LEVI

VOL I, 99 1869

Here is NATURE'S *regally dismissive comment on a plan by one M. Boutet to construct a Channel bridge without any underground work: two mighty piers were to be floated out on the tide on large sheet-iron buoys and then sunk.*

Certainly, we have here Baron Munchausen over again. These cast-iron piers, with a base of 390 ft. by 260 ft., over 200 ft. high, we are informed, are to weigh about 2,500 tons. What is the thickness of their metal to be? Information is wanting on this point; but an iron structure of these dimensions, to bed itself on the bottom of the Channel, could not be designed of less than ten times the weight named.

Assuming, for the sake of argument, that the rise of the tide would float that structure away by the means above described—and our business is to analyse the project as it is, not to suggest or attempt to improve on it—may we not ask with surprise, where would the centre of gravity of this floating structure be? Its centre of gravity would be about ninety feet above the level of the water, and at least one hundred feet *above* the centre of displacement. Why do our ships not upset, what insures their stability, and why do they right themselves? Mainly, because their centre of gravity is in its lowest position, *below* their centre of displacement. Here, however, we have a floating structure in which the centre of gravity would be enormously *above* the centre of displacement, and in its highest position. A slight oscillation, a breath of wind would overturn it, and suppose it could be floated away from shore, it would topple over—right itself upside down; the "sheet-iron buoys" would be uppermost, and the structure below them, forming a gigantic wreck somewhere in the Channel.

So much, then, about the piers. It may give the ordinary reader an idea of the character of this scheme. Shall we say anything about the 9,000 and odd feet clear span? At first sight it appears to be a typographical error; surely 900 and odd feet were meant; but then we meet with the fact of the Channel being divided into *ten* spans, so there is no getting out of it.

The whole proposition is the offspring of a highly imaginative mind. Of all the schemes or suggestions to cross the Channel by rail, this is the most incoherent. *There is nothing in it*—not one point of merit. It is not bold, because it lacks the spirit of boldness, viz. Sense. Not a trace of an engineer's mind is to be found in it. Our asylums produce innumerable schemes of this kind,

but they are not permitted to disturb the public mind. It is a relief to have done with it. We are glad to say there are several projects which do not lack either sense or ability on the part of the originators. Some of them appear practicable, and one or two highly promising of success, and these will form the subject of our next communication.

VOL I, 161 1869

J. J. Sylvester was an eminent mathematician with wide interests and an idiosyncratic wit. In addition to his many and varied contributions to his subject (some of the best in partnership with his friend Arthur Cayley), he is remembered for his treatise on The Laws of Verse, *and indeed he took his poetical efforts very seriously. He also translated verse from several languages under the pen-name "Syzygeticus". (He was, it seems, responsible for introducing the term "syzygy" into mathematics.) He spent two long periods in the United States and achieved recognition late in Britain, partly, it appears, for reasons of anti-Semitism. Here he is in full flow in a polemic (which originated in a presidential address to the British Association for the Advancement of Science) against T. H. Huxley, who had displeased him with some ill-judged comments (as Sylvester felt) about the nature of mathematics.*

IT is said of a great party leader and orator in the House of Lords that, when lately requested to make a speech at some religious or charitable meeting, he declined to do so on the ground that he could not speak unless he saw an adversary before him—somebody to attack or reply to. In obedience to a somewhat similar combative instinct, I set to myself the task of considering certain recent utterances of a most distinguished member of this Association, one whom I no less respect for his honesty and public spirit than I admire him for his genius and eloquence, but from whose opinions on a subject which he has not studied I feel constrained to differ. Göthe has said—

"Verständige Leute kannst du irren sehn
In Sachen nämlich, die sie nicht verstehn."

Understanding people you may see erring—in those things, to wit, which they do not understand.

I have no doubt that had my distinguished friend, the probable President-elect of the next Meeting of the Association, applied his uncommon powers of reasoning, induction, comparison, observation, and invention to the study of mathematical science, he would have become as great a mathematician as he is now a biologist; indeed he has given public evidence of his ability to grapple with the practical side of certain mathematical questions; but he has not made a study of mathematical science as such: and the eminence of his position and the weight justly attaching to his name, render it only the more imperative that any assertions proceeding from such a quarter, which may appear to be erroneous, or so expressed as to be conducive to error, should not remain unchallenged or be passed over in silence.

He says "mathematical training is almost purely deductive. The mathematician starts with a few simple propositions, the proof of which is so obvious that they are called self-evident, and the rest of his work consists of subtle deductions from them. The teaching of languages, at any rate as ordinarily practised, is of the same general nature—authority and tradition furnish the data, and the mental operations are deductive." It would seem from this that, according to Prof. Huxley, the business of the mathematical student is from a limited number of propositions (bottled up and labelled ready for future use) to declare any required result by a process of the same general nature as a student of language employs in declining and conjugating his nouns and verbs : that to make out a mathematical proposition and to construe or parse a sentence are equivalent or identical mental operations. Such an opinion scarcely seems to need serious refutation. The passage is taken from an article in *Macmillan's Magazine* for June last, entitled "Scientific Education—Notes of an After-dinner Speech," and I cannot but think would have been couched in more guarded terms by my distinguished friend had his speech been made before dinner instead of after.

VOL 1, 237 1869

And here, in the second part of his long lecture, which bears reading in toto, *is his tribute to his subject.*

When followed out in this spirit, there is no study in the world which brings into more harmonious action all the faculties of the mind than the one of which I stand here as the humble representative and advocate. There is none other which prepares so many agreeable surprises for its followers, more wonderful than the transformation scene of a pantomime, or, like this, seems to raise them, by successive steps of initiation, to higher and higher states of conscious intellectual being. This accounts, I believe, in part for the extraordinary longevity of all the greatest masters of the Analytical art, the Dii Majores of the mathematical Pantheon. Leibnitz lived to the age of 70 ; Euler to 76 ; Lagrange to 77 ; Laplace to 78 ; Gauss to 78 ; Plato, the supposed inventor of the conic sections, who made mathematics his study and delight, who called them the handles or aids to philosophy, the medicine of the soul, and is said never to have let a day go by without inventing some new theorems, lived to 82 ; Newton, the crown and glory of his race, to 85 ; Archimedes, the nearest akin, probably, to Newton in genius, to 75, and might have lived on to be 100, for aught we can guess to the contrary, when he was slain by the impatient and ill-mannered sergeant sent to bring him before the Roman General, in the full vigour of his faculties, and in the very act of working out a problem ; Pythagoras, in whose school, I believe, the word mathematician (used, however, in a somewhat wider than its present sense) originated, the second founder of geometry, the inventor of the matchless theorem which goes by his name, the precognizer of undoubtedly the miscalled Copernican theory, the discoverer of the regular solids and the musical canon (who stands at the very apex of this pyramid of fame), if we may accept the tradition, after spending 22 years studying in Egypt and 12 in Babylon, opened school when 56 or 57 years old in Magna Græcia, married a young wife when past 60, and died, carrying on his work with energy unspent to the last, at the age of 99. The mathematician lives long and lives young ; "the wings of his soul do not early drop off, nor do its pores become clogged with the earthy particles blown from the dusty highways of vulgar life."
Some people have been found to regard all mathematics, after the 47th proposition of Euclid, as a sort of morbid secretion, to be compared only with the pearl said to be generated in the diseased oyster or, as I have heard it described, "une excroissance maladive de l'esprit humain."

VOL 1, 261 1870

NATURE *adds as a footnote an anecdote that Sylvester's remarks drew from the Russian physicist M. H. Jacobi.*

"En causant un jour avec mon frère défunt sur la nécessité de contrôler par des expériences réitérées toute observation, même si elle confirme l'hypothèse, il me raconta avoir découvert un jour une loi très-remarquable de la théorie des nombres, dont il ne douta guère qu'elle fût générale. Cependant par un excès de précaution ou plutôt pour faire le superflu, il voulut substituer un chiffre quelconque réel aux termes généraux, chiffre qu'il choisit au hasard, ou, peut-être, par une espèce de divination, car en effet ce chiffre mit sa formule en défaut ; tout autre chiffre qu'il essaya en confirma la généralité. Plus tard il réussit à prouver que le chiffre choisi par lui par hasard, appartenait à un système de chiffres qui faisait la seule exception à la règle.
"Ce fait curieux m'est resté dans la mémoire, mais comme il s'est passé il y a plus d'une trentaine d'années, je ne rappelle plus les détails.　　　"M. H. JACOBI
"Exeter, 24 Août, 1869."

VOL 1, 263 1870

The year 1869 saw the opening of the Suez Canal. NATURE *was on hand with its commendation.*

THE ISTHMIAN WAY TO INDIA

THE Canal has been opened. The flotilla, with its noble, royal, imperial, and scientific freight, has progressed along the new-made way from sea to sea. From Port Saïd, that new town between the sea and the wilderness, with its ten thousand inhabitants, and acres of workshops and building-yards, and busy steam-engines, the naval train floated through sandy wastes, across lakes of sludge and lakes of water filled from the Salt Sea ; past levels where a few palm-trees adorn the scorched landscape ; past hill-slopes on which the tamarisk waves its thready arms ; past swamps where flocks of flamingoes, pelicans, and spoonbills, disturbed by the unwonted spectacle, sent up discordant cries ; through deep excavations of hard sand or rock ; across the low flat of the Suez lagoons, where Biblical topographers have searched for the track of the children of Israel ; and so to the "red" waters of the great Gulf of Arabia. The flotilla has done its work : the Canal has been opened ; and the distance by water to India is now 8,000 miles, instead of the 15,000 miles by the old route round the Cape of Good Hope.
It is a great achievement. So great, that we need not wonder that the capital of 8,000,000*l.* sterling with which it was commenced in 1859 was all expended, and as much more required, before the work was half accomplished. And perhaps we ought not to be too much overcome with pity for the 20,000 unlucky Egyptians—natives of the house of bondage—pressed every month up to the year 1863 by their paternal Government to labour, wherever required, along the line of excavations. How persistent are Oriental customs ! Here we have in modern days— the days of power-looms, of steam printing-presses, and under-sea telegraphs—a touch of the old tyranny, the taskmasters and the groanings, associated in our memories with the very earliest of Egyptian history.

VOL 1, 110 1869

Here is a reactionary squib from Charles Darwin's great rival and friend, Alfred Russel Wallace.

Government Aid to Science

I VENTURE to hope that you will allow me space in your columns to express opinions on this subject which are not popular with scientific men, and which are evidently opposed to your own views as indicated in your recent article on Science Reform.
The public mind seems now to be going mad on the subject of education ; the Government is obliged to give way to the clamour, and men of science seem inclined to seize the opportunity to get, if possible, some share in the public money. Art education is already to a considerable extent supplied by the State,—technical education (which I presume means education in "the arts") is vigorously pressed upon the Government,—and Science also is now urging her claims to a modicum of State patronage and support.
Now, sir, I protest most earnestly against the application of public money to any of the above specified purposes, as radically vicious in principle, and as being in the present state of society a positive wrong. In order to clear the ground let me state that, for the purpose of the present argument, I admit the right and duty of the State to educate its citizens. I uphold national

education, but I object absolutely to all sectional or class education; and all the above-named schemes are simply forms of class
education. The broad principle I go upon is this,—that the
State has no moral right to apply funds raised by the taxation
of all its members to any purpose which is not directly available
for the benefit of all. As it has no right to give class preferences
in legislation, so it has no right to give class preferences in the
expenditure of public money. If we follow this principle,
national education is not forbidden, whether given in schools
supported by the State, or in museums, or galleries, or gardens,
fairly distributed over the whole kingdom, and so regulated as to
be equally available for the instruction and amusement of all
classes of the community. But here a line must be drawn. The
schools, the museums, the galleries, the gardens, must all alike be
popular (that is, adapted for and capable of being fully used and
enjoyed by the people at large), and must be developed by means
of public money to such an extent only as is needful for the
highest attainable *popular* instruction and benefit. All beyond
this should be left to private munificence, to societies, or to the
classes benefited, to supply.

Vol 1, 288 1870

*The Mint has always engaged the attention of scientists, most
notably Isaac Newton, who held office as its Master. Is the
trial of the Pyx still enacted?*

THE TRIAL OF THE PYX

THE trial of the Pyx is the formal testing of the coin
of the realm, to ensure its being of the requisite
weight and fineness. The name is derived from the Pyx,
or chest, in which the coins selected for the purpose are
contained. The first trial of the Pyx took place in the
ninth and tenth years of Edward I. And as the last
observance of this ancient ceremony was held during the
past week, a few brief notes may not be without interest.

The authority under which the trials were made varied
considerably. First, the members of the King's Council,
then the Barons of the Exchequer constituted the court,
King James I. presiding at one trial. The court now consists of several members of the Privy Council, under the
presidency of the Lord High Chancellor and a jury selected
from the Hon. Company of Goldsmiths.

Last week the high officers of the Mint assembled at
the Treasury, and in their presence the Lord Chancellor
charged the jury to examine the coin of the late
Master of the Mint, Thomas Graham, F.R.S., and to
ascertain whether it was within the latitude of "remedy"
allowed by law.

This remedy amounts to 12 grains on each troy pound
of gold coin, or to 0·257 grain on each sovereign; and 24
grains on each pound troy of silver coin. Portions cut
from standard test-plates were handed to the jury who
adjourned to Goldsmiths' Hall. They then opened the
Pyx-chest and tested the coin by weight; having done
this, a certain number of gold coins were melted into an
ingot, which was then assayed; the same process being
adopted with the silver coin. In the present instance the
Pyx represented a coinage of 14 millions gold and 1 million
of silver coin; the verdict of the jury being, that the coin
both as to weight and fineness was within the remedy
allowed by law. The details, however, were most favourable to the late illustrious Master who has so lately
passed away.

An adverse verdict would probably have been followed
by no more serious penalty than the forfeiture of the
Master's sureties, but it is interesting to note that in
the reign of Henry I. the money was so debased as to
call for the exemplary punishment of the "Moneyers,"
while in Anglo-Saxon times the chief officer or Reeve
would have been punished by the loss of his hand
should he fail to clear himself of the charge of producing
false coinage.

Vol 1, 429 1870

*The obituary of Captain Brome expresses the generous fire of
indignation, often seen in the pages of* NATURE, *at high-
handed actions of officialdom.*

WITH great regret we have to record the death of
Captain Fred. Brome, formerly Governor of the
Military Prison on Windmill Hill, Gibraltar, and well
known to many of our geological and archæological
readers as the able and indefatigable explorer of the
ossiferous caves and fissures of the rock.

His explorations, an account of which, so far as they
related to the human remains and relics, was published in
the Transactions of the Congress of Prehistoric Archæology for 1868, were commenced in April, 1863, and unre-
mittingly continued, often under considerable difficulties,
to December, 1868, when he was most unaccountably
removed from the post he had so long and so well
occupied.

The amount of labour and responsibility thus voluntarily undertaken by Captain Brome, solely in the interest
of science, and without any personal motive whatever, can
scarcely be imagined, nor can the value of the results obtained by him be easily over-estimated.

A more striking instance of self-devotion to a purely
scientific object can nowhere be found.

The results of Captain Brome's work may be said to
have afforded all, or nearly all, the knowledge we possess
of the priscan population of the Rock of Gibraltar, and
have added enormously to our materials for determining
the nature of its quaternary fauna, as disclosed in the
ossiferous breccia and other contents of the rock fissures,
from the examination of which Cuvier truly anticipated
that the most important information would be derived.

Captain Brome's death occurred, we are sorry to say,
under very melancholy circumstances. Having been removed from the post which he had so long and so usefully
filled, and for which, from his great experience, extraordinary energy, and high sense of duty, he was so admirably qualified, he was appointed, on coming to England, Governor of the Military Prison at Weedon. Here
he hoped to find an asylum for his family, and some
compensation for the sacrifices he had been compelled
to make in leaving Gibraltar.

But this was not to be. Amongst the numerous reduc-
tions of late effected in our military establishments, the
disestablishment of the prison at Weedon was one. The
notice that his services would be no longer required was
received by Captain Brome a short time since, and it
seems to have so affected him, from the apprehension that
his family would thus be deprived of all support—and this
after a public service of thirty years—that, although a
strong and vigorous man, he gradually sank, from mental
depression, as it would seem, and he may truly be said to
have died of a broken heart on the 4th March, leaving
a widow and eight children, we fear wholly unprovided
for.

A more melancholy case, and one more deserving of the
sympathy of the scientific world, and, as we should venture to hope, of the consideration of the authorities at the
War Office, it is impossible to conceive. G. BUSK

Vol 1, 509 1870

Each issue of NATURE *ran reports of the proceedings of
learned societies in Britain and around the world. Here
James Prescott Joule makes his appearance.*

Literary and Philosophical Society, February 22.—Dr.
J. P. Joule, president, in the chair, referred to the observations
he had made in former years on the progressive rise of the
freezing point of one of his thermometers. He had made a
further observation, and found that a rise—unmistakeable, though
very small—was still taking place after a lapse of twenty-six years
since the bulb was blown.

Vol 1, 520 1870

Expressions of solicitude about the health of men of science were frequent.

WE regret to learn that Mr. Archibald Geikie, who recently left England to investigate the Geology of the Lipari Islands, was prostrated by fever as soon as he arrived there, and is in such a weak state of health, that he has been ordered back to England.

VOL 2, 51 1870

Peevish comparisons of how French and German scientists were treated, relative to the British, recurred with great regularity.

AN Imperial decree has been published in Paris, ordering that the Minister of Fine Arts shall henceforth bear the title of Minister of Literature, Science, and Art, and also that his department shall include the superintendence of the Institut de France, Academie des Sciences, the libraries, learned societies, and the like. When shall we get *our* Ministry of Literature, Science, and Art?

VOL 2, 51 1870

A waspish letter:

The Royal Society

I CANNOT but think that the list of candidates recommended by the Council for election into the Royal Society published in your last number will be read by the outside world with considerable surprise. I look in vain in it for the names of two men, at least, of world-wide reputation, and well known as no mere *dilettanti* in their respective sciences, who were among the candidates, while the names of others are found there, which are on everybody's lips with the thought, What have they done to merit the scientific distinction which is looked on by every lover of science as almost an opening of the gates of paradise? Is it possible for us outsiders to learn anything of the considerations which govern the election? NOT AN F.R.S.

VOL 2, 48 1870

The fall of a meteorite (or aërolithe*) in Libya is reported to have caused consternation among the tribesmen, who discharged their rifles at the monster. Note that no correspondence in foreign languages was ever translated; this would presumably have been seen as an aspersion on the education of* NATURE *readers.*

THE Director of the Meteorological Office has forwarded the following extract from a letter from M. Coumbary, Director of the Imperial Meteorological Observatory at Constantinople for publication:—

Constantinople, 9 mars, 1870

Mon cher Monsieur,—Je saisis l'occasion qui m est offerte pour vous transmettre la communication que vient de nous faire M. Carabella, Directeur des Affaires Etrangères du Vilayet de Tripoli de Barbaru.

"Tripoli, 2 février, 1870

"Le Mutasserif de Mourzouk (Fezzan), latitude 26° N., longitude 12° E. de Paris, nous fait savoir que vers le 25 décembre, 1869, il est tombé à l'est de la ville, vers le soir, un immense globe de feu, mésurant un mètre à peu près de diamètre, et qu'au moment où il a touché terre il s'en est détaché de fortes étincelles qui, en se produisant, claquaient comme des coups de pistolet, et exhalaient une odeur que l'on n'a pas specifiée. Cet aérolithe est tombé à peu de distance d'un groupe de plusieurs arabes, parmi lesquels se trouvait le Chiok-el-Veled de Mourzouk. Ceux-ci en ont été tellement éffrayés qu'ils ont immédiatement déchargé leurs fusils sur ce monstre incompréhensible. Son Excellence Ali Riza Pacha a écrit à Mourzouk pour faire transporter ici l'aérolithe ; au cas probable où il soit trop pesant on le mettra en pièces ; nous vous enverrons tout cela. Il y a un mois de voyage d'ici à Mourzouk. Ce n'est donc que dans deux mois à peu près que nous pourrons vous faire cette expédition. S'il peut vous être de quelque intérêt de le savoir, je vous dirai que quelques voyageurs du Waddad que j'interrogeais m'ont dit que le Sultan du Waddad et tous les grands personnages de sa cour ont des poignards, des sabres et des lances faits avec du fer tombé du ciel, et qu'il en tombe de grandes quantités dans ce pays-là. (Sd.) "L. CARABELLA"

Je crois devoir vous informer qu'au reçu de cette lettre et à la suite des démarches nécessaires, S.A. le Grand Vizir a bien voulu faire donner ordre immédiatement par télégraphe à Tripoli, pour que l'on prenne les mesures nécessaires afin que ce météorite nous parvienne intact.—Recevez, cher monsieur, &c., (Sd.) ARISTIDE COUMBARY

VOL 1, 538 1870

John Tyndall, a hot-tempered Irish Orangeman and alpinist, was not only one of the great experimental physicists of the nineteenth century, but a scientific polymath. While professor at the Royal Institution in London, he pursued his interests in light-scattering (the Tyndall effect). Scattering of sunlight by the particles that polluted the air of Albemarle Street led him to a scheme for destroying atmospheric organic matter by heat. This in turn caused him to think about the dissemination of airborne bacteria, and he became a fervent supporter of Louis Pasteur and an opponent of spontaneous generation of life. Bacteriological warfare, however, as prefigured here, he clearly saw as a jest.

PROF. TYNDALL will have much to answer for in the results that may be expected from the spread of his "dust and disease" theory. It is stated by the *Athenæum* that a new idea has been broached in a recent lecture by Mr. Bloxam, the lecturer on chemistry to the department of artillery studies. He suggests that the committee on explosives, abandoning gun cotton, should collect the germs of small-pox and similar malignant diseases, in cotton or other dust-collecting substances, and load shells with them. We should then hear of an enemy dislodged from his position by a volley of typhus, or a few rounds of Asiatic cholera. We shall expect to receive the particulars of a new "Sale of Poisons" Act, imposing the strictest regulations on the sale by chemists of packets of "cholera germs" or "small-pox seed." Probably none will be allowed to be sold without bearing the stamp of the Royal Institution, certifying that they have been examined by the microscope and are warranted to be the genuine article.

VOL 1, 562 1870

Here is the origin of one of the great British institutions, the Natural History Museum in South Kensington.

Copy of a Memorial presented to the Right Hon. the Chancellor of the Exchequer

SIR,—It having been stated that the scientific men of the Metropolis are, as a body, entirely opposed to the removal of the Natural History Collections from their present situation in the British Museum, we, the undersigned Fellows of the Royal, Linnean, Geological, and Zoological Societies of London, beg leave to offer to you the following expression of our opinion upon the subject.

We are of opinion that it is of fundamental importance to the progress of the Natural Sciences in this country, that the administration of the National Natural History Collections should be separated from that of the Library and Art Collections, and placed under one officer, who should be immediately responsible to one of the Queen's Ministers.

We regard the exact locality of the National Museum of Natural History as a question of comparatively minor importance, provided that it be conveniently accessible and within the Metropolitan district.

GEORGE BENTHAM, F.R.S., F.L.S., F.Z.S.
WM. B. CARPENTER, M.D., F.R.S., F.L.S., F.G.S.
W. S. DALLAS, F.L.S.
CHARLES DARWIN, F.R.S., F.L.S., F.Z.S.
F. DUCANE GODMAN, F.L.S., F.Z.S.
J. H. GURNEY, F.Z.S.
EDWARD HAMILTON, M.D., F.L.S., F.Z.S.
JOSEPH D. HOOKER, M.D., F.R.S., F.L.S., F.G.S.
THOS. H. HUXLEY, F.R.S., V.P.Z.S., F.L.S., F.G.S.
JOHN KIRK, F.L.S., C.M.Z.S.
LILFORD, F.L.S., F.Z.S.
ALFRED NEWTON, M.A., F.L.S., F.Z.S.
W. KITCHEN PARKER, F.R.S., F.Z.S.
ANDREW RAMSAY, F.R.S., V.P.G.S.
ARTHUR RUSSELL, M.P., F.R.G.S., F.Z.S.
OSBERT SALVIN, M.A., F.L.S., F.Z.S.
P. L. SCLATER, F.R.S., F.L.S., F.Z.S.
G. SCLATER-BOOTH, M.P., F.Z.S.
S. JAMES A. SALTER, M.B., F.R.S., F.L.S., F.Z.S.
W. H. SIMPSON, M.A., F.Z.S.
J. EMERSON TENNENT, F.R.S., F.Z.S.
THOMAS THOMSON, M.D., F.R.S., F.L.S.
H. B. TRISTRAM, M.A., F.L.S.
WALDEN, F.Z.S., F.L.S.
ALFRED R. WALLACE, F.R.G.S., F.Z.S.

Eating habits in Belgium:

La Petite Culture en Belgique

I ENCLOSE drawings of earthern pots, which I observed nailed against the south side of a farm-house near this. These pots are for sparrows' nests, and the young, when fledged, are taken

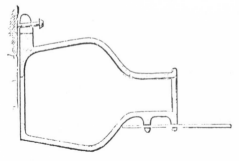

and eaten. I think this form of "La petite culture" cannot be commended in a country so swarming with insects as Belgium, and I infer from the careful make of the pots that the custom is not a new one, though it may be new to some of the readers of NATURE.   N. A. STAPLES

Louvain, June 4

Putting the universities in their place:

NOTES

WELL-WISHERS of the University of Oxford will rejoice to hear that the honorary degree of D.C.L. has been offered to Mr. Darwin. The state of Mr. Darwin's health unfortunately precludes him from accepting the proffered honour, but the scientific naturalists of this and other countries will none the less appreciate the compliment which has been paid to their great leader. It is all the more graceful as Mr. Darwin is not an Oxford, but a Cambridge man, a circumstance which the University of Cambridge seems to have forgotten ; though by-and-by it will be one of her claims not to be herself forgotten.

Henry Charlton Bastian was a tireless proponent of spontaneous generation or panspermia. He was willing to take on Louis Pasteur and conducted an interminable polemic, denouncing the experiments with the famous sealed, swan-necked flasks (their contents devoid of life to this day). Bastian was enraged at the verdict of the commission appointed to decide between Pasteur and his adversary, Felix Pouchet (who, feeling that the cards were stacked against him, thought the better about facing up to an interrogation and simply took to his heels). Here is the start of Bastian's first broadside in NATURE. *John Tyndall, who in the next issue wrote an admiring report of Pasteur's work on the diseases of silkworms (pébrine and flâcherie), remarked with obvious irritation about the spontaneous generation debate that "it is much to be desired that some really competent person in England should rescue the public mind from the confusion now prevalent regarding this question."*

FACTS AND REASONINGS CONCERNING THE HETEROGENOUS EVOLUTION OF LIVING THINGS

IN all ages it has been believed by many that Living things of various kinds could come into being *de novo*, and without ordinary parentage. Much difference of opinion has, however, always prevailed as to the kinds of organisms which might so arise. And, although received as an article of faith by many biologists—perhaps by most—in the earlier ages, this doctrine or belief has, in more recent times, been rejected by a very large section of them. Definitely to prove or disprove the doctrine in some of its aspects is a matter of the utmost difficulty, and there are reasons enough to account for the wave of scepticism on this subject, which has been so powerful in its influence during the last century. The notions of the ancients were altogether crude, and founded upon insufficient proofs. It was not in their power to settle such a question ; and when the inadequacy of the evidence on which they had relied became known, then much doubt was thrown also on the truth of the conclusion at which they had arrived. All this was natural enough. When, therefore, about a century ago, the rude microscopes of the time began to reveal a multitude of minute organisms whose existence had been hitherto unsuspected ; when more facts became known concerning the various modes of reproduction amongst living things ; and, above all, when the philosophical creeds of the day were supposed to be irreconcilable with such a doctrine, then a growing scepticism in the minds of many gradually developed into an utter disbelief in the possibility of the occurrence of what was called "spontaneous generation."

This was the state of things anterior to and during the time of the celebrated controversy between the Abbé Spallanzani and John Needham. Then it was that the former of these two champions, with the view of accounting for phenomena which would otherwise have necessitated his admission of the doctrine which he rejected, recklessly launched upon the world the *hypothesis* that multitudinous, minute, and almost metaphysical "germs" existed everywhere- ready to burst into active Life and development whenever they came under the influence of suitable conditions. Armed with this all-powerful *Panspermic* hypothesis, Spallanzani argued against the conclusions of Needham. His views on this subject were supported by the still more extravagant theories of Bonnet. The doctrine of "*L'Emboîtement des germes*" was the production of an unbridled fancy, and might, perhaps, never have been elaborated, had not the Leibnitzian doctrine concerning "Monads," as centres of force and activity, been already in existence, and at the time all-powerful in the philosophical world.

The controversy which was initiated by these two pioneers in microscopical research they were unable to terminate—the enigma which they sought to solve has, since their time, still pressed for solution, and still the tendency has been to solve it after one or other of the modes by which they attempted to account for the occurrence of the phenomena in question. It is and has been contended, on the one hand, that Living things can originate *de novo*, and without ordinary parentage ; it is contended, on the other, that this is impossible—that every Living thing is the product or off-cast of a pre-existing Living thing, and that those which appear to arise *de novo* have, in reality, been produced by the development of some of the myriads of visible or invisible "germs" which pervade the atmosphere.

And here, further into Bastian's long article, is an example of his tone in debate.

Unfortunately for the cause of Truth, people have been so blinded by his skill and precision as a mere experimenter, that only too many have failed to discover his shortcomings as a reasoner.

But it will already have been perceived by the attentive reader, that it was not necessary for me—in my endeavour to establish as a Truth the great doctrine which M. Pasteur has striven to repudiate—to show the inconclusiveness of his reasonings on that branch of the subject to which I have just been alluding. I have striven rather to show in their true light the real nature of such modes of reasoning, which are I fear only too likely to be repeated by others. So long as people are unable readily to appreciate the worthlessness of arguments like these, they will never be likely to penetrate through the clouds of controversy which envelope this subject. Their mental vision will be blinded, and the truth will remain hidden from them.

Vol 2, 170 1870

St George Mivart was another great polemicist of the time. Initially a Darwinist, he came increasingly to doubt whether natural selection accounted satisfactorily for evolution and was troubled by conflicts with the Roman Catholic religion to which he was an early convert. He evidently had a gift for making enemies and when he denounced the Church for its obscurantist pronouncements on evolution, his books were placed on the Index and he was excommunicated. In a Nature *leader in which he adverts to the physical characteristics of primates, he encapsulates his philosophy like this: "To prevent misconceptions, I may add that fully recognising the truth of Mr Darwin's appreciation of man's zoological position, which I have ever maintained and indeed laboured to support, I never the less completely differ from him when I include the totality of man's being. So considered, Science convinces me that a monkey and a mushroom differ less from each other than do a monkey and a man." Darwin nevertheless held Mivart in greater respect than any other of his adversaries. Here is the beginning of the article, just sufficient to show Mivart's style of debate.*

THE VERTEBRATE SKELETON

SKELETAL archetypes, and "theories of the skull," have of late years gone much out of fashion. The view which made each man a potential Briareus as to limbs, seems itself to be considered as no longer having a leg to stand upon. The fortress of the "Petrosal" has long been carried by assault, and is peaceably and securely occupied; and although we have had lately a brilliant passage of arms apropos of the "auditory ossicles" from which the unlucky Sauropsida retired with broken "hammers" and diminished "anvils;" yet the once widespread interest in skeletal controversies seems to have long subsided. The old war-cries are no longer heard, the question "Is the post-frontal a parapophysis?" falls on indifferent or averted ears, and we fear that even not a few of our anatomists call into daily functional activity a mandible, to the true nature and homologies of which they are comparatively indifferent.

What was the surprise of some, then, who last year witnessed, in the theatre of the Royal College of Surgeons, an unlooked-for resurrection. Some rubbed their eyes—could they have had a long sleep, and was it still the year 1849 instead of 1869? A quasi-vertebrate theory of the skull once more ! Again an exposition of cranial hœmal arches !

" Jam redit et Virgo, redeunt Saturnia regna."

Vol 2, 291 1870

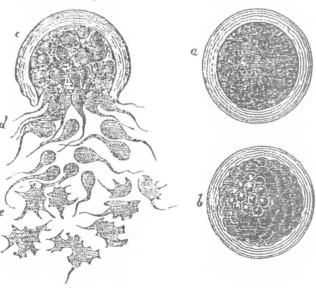

Martial concerns were well represented in the early volumes of NATURE, *for this was the high noon of Empire. This is an extract from one of a series of long articles entitled "The Science of War".*

The possibility of discovering what takes place in the bore of a cannon at the time of its discharge, and of ascertaining how fast the shot travels, is a subject which has long attracted the attention of artillerymen, and among others that of a talented officer, Captain Andrew Noble, of the Elswick Ordnance Works. This gentleman's labours have recently been crowned with success ; and an apparatus has been devised of which one hardly knows whether to admire more its exceeding delicacy or its wonderful results. With its aid the examination of gunpowder is now being conducted with comparative ease, and what is still more important, with unerring certainty.

A detailed description of the instrument, which has received the name of Chronoscope, would necessitate more space than we have here at our disposal, but its main features, and the principle upon which it is based, are easily explained. The tube or bore of a gun is fitted inside at certain intervals with metal rings (to the number of six or eight) the outside margins of which are sharpened into so many knife-edges. On a shot passing along the bore and through these rings, the edges of the latter are jammed down upon and made to cut through the ends of various insulated wires, one of which is placed under each ring. If we now suppose each of these wires to be in connection with an electric battery, it follows as a matter of course that as one wire after another is cut through, and the insulation consequently removed, an electric current passes ; so that if there are six rings and wires fitted at intervals in the tube of the gun, the passage of a shot along it would be instrumental in producing six electric sparks following rapidly one upon the other.

We now understand how the shot is made to tell the tale of its flight ; but there remains yet to be explained how the story is written down. This recording of the signals is accomplished by a very simple arrangement. A series of metal discs, one in connection with each wire, is made, by means of a clock-work arrangement, to revolve at a certain rapid velocity, say at the rate of 1,000 inches in a second ; the surface of the discs is of polished silver, coated with lamp-black, and as soon as the desired speed has been attained, the gun, which is in

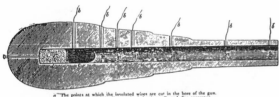

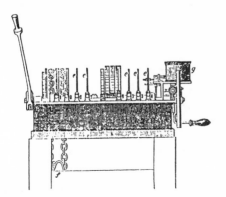

a The points at which the insulated wires are cut in the bore of the gun.
b The insulated wires leading to the induction coils and disc apparatus.
c The shot.
d The cartridge.
e Revolving discs.
f Weight for setting machinery in motion.
g Stop clock to record the number of revolutions.

electrical communication with the instrument, is fired. As the shot traverses the first ring, No. 1 wire is cut through, and a spark thereupon hops over to the recording disc, removing a little of the lamp-black covering, and thus marking the place by laying bare a minute spot of bright metal. No. 2 wire, when cut by the second ring, leaves a similar record upon another disc precisely in the same manner ; and so on with Nos. 3, 4, 5, and 6, the relative position of the six spots on the six discs indicating exactly the velocity with which the shot has passed the six different rings or stations.

* * *

The great importance of this beautiful invention need not be dilated upon by us, as the value of its aid in experiment is at once apparent to the veriest tyro in gunnery. As a measurer of time and speed of the most perfect character, its delicacy is certainly unsurpassed ; for, by merely dividing every inch of the discs into a thousand divisions or degrees, we are at once enabled to calculate with precision to the millionth part of a second.

VOL 2, 379 1870

Anecdotes about inheritance of acquired characteristics were prominent in the correspondence columns.

Hereditary Deformities

IN the lessons in Ethnology in " Cassell's Popular Educator," it is stated, on the authority of Dr. Theodor Waitz, and the Secretary of the Anthropological Society, that "an officer, whose little finger had accidentally been cut across, and had, in consequence, become crooked, transmitted the same defect to his offspring. Another officer wounded at the battle of Eylau, had his scar reproduced on the foreheads of his children." And again, " In Carolina, a dog which had accidentally lost its tail transmitted the defect to its descendants for three or four generations." Do these stories rest on a good foundation ? We know that congenital peculiarities of form and disposition are transmitted from parent to offspring, but that an accidental *deformity* should be so transmitted is a very different affair, and if substantiated would introduce *Accidental Distortion* as a co-worker with natural selection in the modification of species.

Faversham, Kent, Aug. 27   WM. FIELD

VOL 2, 376 1870

But the next comes from a highly respectable French physiologist and a discoverer of hormones. His reputation suffered in later years from his extreme Lamarckian views. Charles Edouard Brown-Séquard endorsed testis transplants for prolonging life and virility; he had injected himself with extracts of dog testes and made this material available to physiologists and doctors, having found that it reversed the ravages of age, improved his eyesight and memory and, moreover, lengthened the trajectory of his urine.

As the subject of hereditary deformities is attracting some attention in our columns, it may be worth while to call attention to Brown-Séquard's experiments on epileptic guinea-pigs detailed at the recent meeting of the British Association. Dr. Brown-Séquard produced epileptic fits in the guinea-pigs either by the section of one-half of the spinal cord, or by the division of the sciatic nerve on one or both sides. During the fits it sometimes happens that the hind foot gets between the teeth, and is bitten. The animal, on recovery from the fit, tastes the blood, and if it be one in which the sciatic nerve has been divided, proceeds to nibbl off the two outer toes, which have entirely lost their sensibility from the operation on the nerve. In breeding from pairs of this kind, the offspring is without the two toes of which the

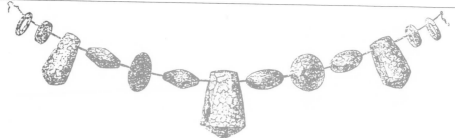

parents have deprived themselves; and in these cases all the off-spring become, as they grow up, perfectly epileptic; while in ordinary cases epilepsy is only rarely transmitted hereditarily. Other peculiarities existing in these epileptic guinea-pigs were also found to be transmitted to their offspring; and in dissection of the hereditarily malformed animals, a node was found on the sciatic nerve corresponding to that formed after section of the nerve in the parent.

VOL 3, 14 1870

What follows savours of racialist superstition. Charles Leland was a writer of comic verses, still to be found in anthologies (especially the ballad "Hans Breitmann's Barty").

Pangenesis

ON the introduction of Mr. Charles G. Leland, the author of the famous "Breitman Ballads," who was present at the reading of Mr. Galton's paper on Pangenesis before the Royal Society on the 30th ult., I have seen Mr. Lewis Ware, a young American gentleman who has been studying science in Paris since 1868.

By him I am informed that M. Leconte (I presume the physiological chemist of that name) is accustomed to mention in his lectures that he had frequently transfused the blood of one kind of animal into the veins of another; but it does not appear, in reference to those experiments, that any subsequent effects were noticed, as regards the offspring of such animals.

M. Leconte, however, further relates that once, not by way of experiment, but in order to save life, endangered, it must be supposed, from the occurrence of previous hemorrhage, he transfused into the veins of a white man blood drawn from a negro, and that the subsequent offspring of this white man by a white mother were *swarthy* in complexion.

Now, I cannot find any *published* record of M. Leconte's operation and its singular consequences, and it is impossible at the present moment to reach him by letter. I desire therefore to give publicity to what *seems* to be a conclusive proof of the theory of "pangenesis," with the view of eliciting a confirmation or refutation of the statement from some one who may chance to read this note, and who may have the necessary opportunities and leisure for further inquiry into the particulars of so very remarkable an incident. It is obvious that the number of children so affected, and the coincidence or absence of other changes in the hair, the form of skull, &c., require to be investigated, and the credibility of the parents fully authenticated.

10, Savile Row, W., April 8   JOHN MARSHALL

VOL 3, 467 1870

A long book review by the redoubtable Edwin Ray Lankester contains the following expression of disapproval, couched in terms that editors today would see it their duty to expunge.

It required, however, a man of considerable knowledge of the subject to write such a book worthily, and we doubt whether Dr. Nicholson, though he deserves credit for enterprise, was quite the man to undertake it. He excuses himself for shortcomings in plan and execution in his preface, on the score of leading a busy life. Now is it, we would ask, for men who lead lives devoted to other objects than the pursuit of zoology, to bring out educational works on that branch of science?

VOL 2, 491 1870

The author of the book strikes back —

Dr. Nicholson's "Zoology"

I NOTICE in NATURE for Oct. 20, a review by Mr. E. Ray Lankester, of a Manual of Zoology recently published by me, and I crave a small portion of your space to say a few words thereon. Upon Mr. Lankester's zoological strictures on my work I will not enter, partly because the public verdict on the merits of my work has already been very emphatically and decisively expressed; partly because the sins laid to my charge are chiefly of *omission* and not of *commission*, and are, therefore, more or less inevitable in a work of such limited compass; and partly because it must be patent to everyone how much more admirably the work, unfortunately left to me, would have been discharged by Mr. Lankester himself.

In the matter of *Greek*, however, Mr. Lankester really must excuse me if I decline to bow to his superior knowledge. I am well aware that he probably entertains a fresher recollection of his school days than I can boast of, and I might, therefore, without shame, have pleaded guilty to some obliviousness of Greek roots. Mr. Lankester, however, has been singularly unlucky in the point of attack chosen by him. He takes upon himself to condemn the whole of the glossary to my work, because he finds the *twelfth* word of the same ("actinomeres") derived from the Greek word *aktin*, and he is good enough to add the information that "there is no such Greek word as *aktin*." Now, any decent lexicon would have informed Mr. Lankester that *aktin* is not only good Greek, but that it is the original form of the word, and that *aktis* was employed for the first time by Pindar, not, therefore, till about 450 B.C.

In conclusion, if I may be permitted to make a suggestion, I would recommend Mr. Lankester, in his capacity as critic and appraiser of the work of other men, not to judge in future of the value of a haystack by the first straw that he may happen to pull out of it; or, if he must do this, to be very sure before giving his opinion to the public, that it *is* a straw that he has succeeded in laying hold of.

Newhaven, Edinburgh   H. ALLEYNE NICHOLSON

— but, it seems, ineffectually:

DR. NICHOLSON'S extraordinary assertions as to the supposed word "aktin" really demand no serious discussion, which, indeed, would be out of place in NATURE. A reference to Liddell and Scott's Lexicon will conclusively demonstrate to any person interested in the matter that he is entirely wrong. The following additional blunders in Dr. Nicholson's glossary will enable your readers more fully to judge of his accuracy, and it will require considerable boldness to attempt to justify them by reference to imaginary archaic forms :—1. In several places we find Dr. Nicholson giving "poda" as the Greek for "feet," a gross grammatical fault. 2. "Pseud*os*" is given as the adjective corresponding to the English word "false." 3. "Enchuma" is said to be a Greek word meaning "tissue." It has not this meaning. Dr. Nicholson's mistake arises from ignorance of the origin of the signification of the word "parenchyma." 4. "Laima" is given in several places in the glossary for "throat," in place of "laimos."

It is improbable that these are anything but a fraction of Dr. Nicholson's etymological misrepresentations. Mistakes in the glossary of a zoological work are not of very great importance, and would not in this case have demanded notice had they not been fair samples of the general character of the book in which they occur.

I much regret that the fact of the writer's name being appended to the notice of Dr. Nicholson's work should have led him into the region of personalities, whither I do not intend to follow him.

E. RAY LANKESTER

VOL 3, 86 1870

The fateful year 1870 saw the beginning of the siege of Paris.
NATURE *wanted it to be known that the savants were acquitting themselves nobly.*

THE bastion in front of Fort Bicêtre, known as Bastion No. 87, is manned by the members of the Ecole Polytechnique. The professors of the college have consented to serve under their former pupils, wherever these have been selected as lieutenants. In this bastion may be seen MM. Bertrand, Bonnet, Langier, Frémy, Tissot, Laguerre—all members of the Institute, professors at either the College de France or at the Sorbonne—daily at their posts in the bastion, which has already acquired the reputation of being one of the best mounted among the fortifications of Paris.

VOL 2, 521 1870

Two letters about cockroaches follow, the first gruesome, the second charged with the classical scholarship that readers relished.

The Cockroach

I HAVE only to-day noticed the Rev. C. J. Robinson's letter on this subject in your issue of the 29th Sept. A friend of mine, whom I have known all my life, who occupied an important trust as Bank Manager in India last year, and who is at present home on sick leave, assures me that Dr. Norman Macleod is wrong when he denies the nail-nibbling propensities of the cockroach. My friend had been in Kurachee for some time, and on his journey from that town to Bombay by sea he was annoyed one night in his berth by some insect crawling over his face; half asleep and half awake he put up his hand to his face and sent the insect to the foot of his berth. Shortly after he was awoke by a pain at his great toe, and on looking at it he discovered that a cockroach had nibbled off all the nail down to the quick. JAMES DURIE
Aberystwith, Oct. 8

Were Cockroaches known to the Ancient Greeks and Romans?

YOUR correspondent, Rev. C. J. Robinson, drew attention in your columns (NATURE, Sept. 29) to the question whether these troublesome insects were known to the Ancient Greeks and Romans; he says, "there is a good deal to lead one to suppose that the μυλακρίς mentioned by Aristotle, and the *Blatta pistrinorum* of Latin writers was the same as our loathsome pest." I think Mr. Robinson is mistaken in supposing that the μυλακρίς is mentioned by Aristotle, at least I can find no mention made of this insect in the writings of the Stagirite. The word μυλακρίς, meaning some kind of insect, occurs in the fragments of Aristophanes preserved by Pollux, who amongst other meanings of the term gives the following one:— ζῷόν η ἐν τῷ μύλωνι γινόμενον, and then quotes this couplet from Aristophanes,

"Ινα ξυνῶσιν ᾧπερ ἥδεσθον βίῳ,
Σκώληκας ἐσθίοντες, καὶ μυλακρίδας.

"where they may partake of the food of which they are fond, eating worms and *mylocrides*." It would not be possible to say what the μυλακρίς here denotes, but from the creature being often produced in mills, it may possibly mean a "Cockroach," though a "meal-worm" (*i.e.*, the larva of the beetle, *Tenebrio mulitor*) would suit equally well. The Greeks, however, had a word which may well represent the Cockroach, though it is even here impossible to speak with certainty. The word, σίλφη, it is probable denotes this insect. Aristotle (Hist. Anim. viii. 19. § 4) uses the word once; he enumerates the *silphe* amongst insects which cast their skins. The Scholiast on the "Peace" of Aristophanes says the *silphe* is an ill-smelling insect (δυσώδμος). Aetius (8. 33.) speaks of "the fat of the stinking *silphe* which inhabits houses." The epigrammatist Evenus (Analect. i. p. 167) speaks of the *silphe* of the booksellers' shops, and applies to it the epithets, page-eating (σελιδηφάγος), destructive (λωβήτειρα), black-bodied (μελαινόχρως)." Lucian speaks of the mere book collector as providing pastime for mice and habitations for *silphe*, and cuffs his slaves for not keeping the mice and *silphe* away. (Advers.

Indoct, iii. 114, Ed. Hemsterhus). The Scholiast here gives a description of the *silphe* which Schneider with some reason refers to some kind of *Lepisma*. Ælian (H. A. i. 37) says that the *silphe* infest swallows' nests; these cannot be cockroaches. Galen and Paulus Aegineta apply the epithet, βδέουσαι, to the *silphai*. Dioscorides (ii. 38) says that the inside of the *silphe* found in bake-houses when pounded with oil is good for pains in the ear. This leads me to the *Blatta* of the Romans. "On pulling off," says Pliny, "the head of a *blatta* it gives forth a greasy substance, which, beaten up with oil of roses, is said to be wonderfully good for affections of the ears." He speaks of the disgusting nature of this insect, one kind of which is known by the name of *Myloecon*, and found in mills (Nat. Hist. xxix. 39). In another place (xi. 34) Pliny says, "It is the nature of the *blatta* to seek dark corners and to avoid the light; they are very often found in baths." According to Virgil, "the light-avoiding *blattæ*" find their way into bee-hives (Geor. iv. 243). Horace (Sat. ii. 3, 119) ridicules an old miser for sleeping on straw and leaving his bed clothes in his chest, the food of *blattæ* and *tineæ*, "Blattarum ac tinearum Epulæ." Martial (Lib. iv. Ep. 37.) says unless his books are well put together they become the prey of *tineæ* and *blattæ*.

Constrictos nisi das mihi libellos
Admittam tineas trucesque blattas.

From the above passages it will be seen that the *blatta* was a destructive insect to clothes, books, &c., that it avoided the light, and was fond of warm places, that it frequented mills and exuded a greasy substance from its head, that it was a disgusting creature (probably in allusion to the smell) all of which particulars are true of cockroaches, and as there are many species of the family, and are widely distributed over all parts of the globe and must have been known to the ancients, I think there is good reason for concluding that the cockroach was known to the Greeks by the name of σίλφη, and to the Romans by that of *blatat*. W. HOUGHTON

VOL 3, 27 1870

Balloons acquired practical importance during the siege of Paris, for they became the sole means of communicating with the provinces. Frequently they fell within the Prussian lines, and ideas about how they might be steered were debated in the columns of NATURE *by physicists, including Hermann von Helmholtz. This account by W. de Fonville, an intrepid balloonist (or aeronaut, as they were then called) and frequent correspondent to* NATURE *for some decades, tells of the celebrated escapade of the great French republican politician Léon Gambetta, minister of the interior in the Government of National Defence, who got away from Paris by air to organise resistance in the country.*

BALLOON ASCENTS FOR MILITARY PURPOSES

AS soon as the war broke out, balloons, which had been so long forgotten by statesmen, were recalled to their memory by hundreds of projectors. Some of the schemes suggested were of the wildest description; and scientific men took advantage of this circumstance to reject everything connected with aëronautics. But surprises and reverses became so frequent in the French army, that it became evident that any apparatus able to carry observers would be considered as a preserver from such disgraces. As soon as it was clear that the Prussians were intending to besiege Paris, the Minister of War issued orders for the construction of a captive balloon, intended to watch the movements of any besieging army moving round the capital; but instead of having recourse to Mr. Giffard, the constructor of so many magnificent balloons, it was resolved to employ MM. Godard and Nadar. Paris was divided into aërial districts, the first being given to Nadar and the other to Godard. Nadar then received orders to establish his balloon on the foot of Buttes Montmartre, and Godard close to the Montsouris Meteorological Observatory on the banks of the small streamlet Bièvre, where it crosses the fortifications. The balloons intended to be attached were not made on purpose, they merely used old ones which were worn out; the gas-pipes were also not sufficiently large, and the gas-pressure was very low, so that when the first attempts at inflating were made, the Godard balloon took more than three days to be filled; and, when filled, was tossed so heavily by the wind, that it was necessary to let the gas escape. Nadar was still more unfortunate, and could not arrive even at the inflating of his balloon, except after immense labour, by laying a pipe along the ground for a space of more than 300 yards. Moreover, when the first balloon was floated, it was as late as the 4th of September. I then ordered Godard to continue his inflating process. Many scientific bodies met, and deliberated upon the modes of improving captive balloon ascents; but none of the members had ever ascended, and hence their practical knowledge was so small as to amount practically to nothing.

* * *

When the investment of Paris was completed, the question naturally arose of using balloons for carrying messages, the resolution having been taken by the minister, M. Rampont, Post-office Director, to summon to his office several aëronauts, Nadar, Artoise, myself, and a few gentlemen supposed to be acquainted with aërostatics; and the ascent was decided upon in a long discussion.

The first who ascended, was Durioff with his own balloon, famous from several ascents. Durioff started up into the air early in the morning, and employed an immense lifting power, the wind blowing strongly besides, and Durioff disappeared like a dream. He was alone in his car, carrying a bag of letters, with plenty of ballast; I protested in the most urgent manner against sending into the air a single man unassisted, but without any success. The advice was neglected in consequence of the success of the first operations. Reverses were necessary to call postal authority to a better sense of the real state of things. M. Garnier Pagès, a member of the new Government, invented the carrying by balloons of aërial pigeons, and the second balloon ascent was the occasion of the first pigeon expedition.

One of the aëronauts known to our readers is Mangin, the proprietor of the unfortunate "Union," of which the wreck was fully described, who tried an ascent a few days after Durioff. He made a foolish agreement with the Post Office, to carry with his poor worn-out balloon a weight of 1,000lbs., but the balloon was unable to retain a single puff of gas, and the attempt was doomed to failure.

Two or three days afterwards, Mangin tried another ascent with the "City of Florence," a large balloon of 1,200lbs. capacity and belonging to Eugène Godard.

The "City of Florence" was inflated and fitted up by its proprietor, and left the ground on the morning of a clear day, with a light north-easterly wind. It carried with Mangin a medical man practising at Lyons, with a special mission from the Government for the eastern departments. The ascent succeeded very well, and Dr. Lutz was landed safely. But the landing of Lutz gave rise to a singular circumstance. A Prussian spy, having read in the papers that Lutz had come down from the heavens, presented himself at Dijon as the real Lutz, and acted in accordance with that suggestion. The fraud was not discovered without some delay and some trouble, but owing to some peculiar circumstances it was at last exposed, the false Lutz was seized, tried by a court-martial, condemned to death, and shot on the spot.

* * *

Next to Godard's singular ascent, we must mention the one executed by Trignot for carrying Gambetta to his post at the head of the Government. An accident took place in the air while it was open, and the balloon emptied itself at an extraordinary rate, landing, against the will of the aëronauts, in Prussian territory. If sharpshooters had not come to the rescue, Gambetta would have been made a prisoner. Kératry was in the same manner sent in a balloon, and succeeded in escaping after some adventurous feats.

Vol 3, 115 1870

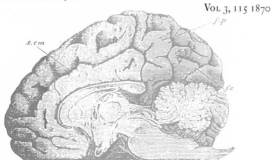

The savants are in action during the siege. The eminent chemist Marcellin Berthelot was in charge of the supply of explosives, as president of a scientific defence committee and senator, and disposed of wide political powers. He appears as a character — Professor Bertheroy — in Emile Zola's novel Paris.

M. Berthelot, although not a member of the French Institute, is the president of a standing committee for using scientific discoveries in the defence of Paris. That committee holds its sittings at the Ministry for Public Instruction, of which M. Jules Simon is the head. Many valuable suggestions have been adopted by that committee, which was closely connected with the Committee of Barricades, presided over by Rochefort.

* * *

M. Jamin, the celebrated Professor of Natural Philosophy at the Sorbonne and member of the French Institute, has enlisted as a private in the artillery of the National Guard, and is doing his duty regularly, although begging for a weekly authorisation from the lieutenant in command to enable him to attend the sittings.

* * *

The two Becquerels, father and son, have left Paris, and their place is filled by supernumeraries. Some papers have remarked very sharply upon it, and asked for the discharge of the younger.

Vol 3, 132 1870

The Victorian taste for exploration and adventure was reflected in periodic travel pieces in Nature.

LETTERS FROM CENTRAL AFRICA
Seriba Ghattas in Djur, *July* 29, 1870

AFTER an absence of nearly eight months I have arrived here once more, considerably reduced in bulk in consequence of the privations and fatigues which I have had to undergo, but otherwise thoroughly well and active. A poultry yard and a milch cow, which I intend to provide myself with, will, in addition to a few weeks' rest, restore my lost strength completely. The journey to the Niam-Niam country, which I undertook as the guest of my friend, Mohammed Abu Tsammat, with his ivory caravan of 300 men, and whose acquaintance I made during the river journey, was successfully completed, as we had no losses to deplore, except a few female slaves who were taken away whilst fetching water; and besides the wounding of the leader, Mohammed, only one of my people was injured by an arrow, which struck him in the arm, but fortunately the wound was speedily healed.

* * *

After passing through another desert for several days, we reached the territory of the Mombuttu King, Munsa, whose residence was the most southern point reached by me, situated a little beyond the third degree north latitude. The southern part of it lies on the great Uelle river, which appears to me to be the upper Chari, flowing into the Tschad lake, and which resembles the Blue Nile, near Chartum.

* * *

I could fill volumes were I to relate all my experiences at the court of this wild brown Cæsar, covered all over with red copper spangles, and looking like a well-furnished kitchen; of his numerous wives, painted in all the colours of the rainbow; of his immense palace, resembling a railway station, one of the rooms of which, and where I was first received, being 100 feet long by 50 feet broad, and 40 feet high. It would be impossible for me, however, to pass over in silence the horrible cannibalism which is here, as well as among the real Niam-Niams, everywhere in vogue. Munsa dines off human flesh every day of his life; the Mombuttu people make regular battues upon the wilder negro races in the south, where those that are killed are at once cut up, the fat is melted down, and the flesh dried. Those that are captured are driven off to be slaughtered at convenience.

The Niam-Niams are thrown more upon their own resources. If, however, there should happen to be a cessation of internecine feuds, they attack the Nubian caravans, although it should be to their interest to keep the peace, as they are well paid for their ivory and provisions with copper and glass beads, and their chiefs receive rich presents. It is true the Nubians are not so philanthropic in their scribes, but in respect to the Niam-Niams, nothing can be said against them, as hostilities would destroy the object they have in view. The buried ivory cannot be discovered by any divining rod; there are no cattle to be stolen; and the women and children always hide themselves at once, and in time, in the impenetrable thicket of the woods, so that no booty is to be obtained in slaves. It is, therefore, however improbable it may sound, the Niam-Niams who, entirely through their horrible lust for human flesh, commence the war. "Flesh, flesh!" is their war-cry, and a few female slaves, at least, who have lost their way in fetching water, are sacrificed to their cannibalism.

VOL 3, 215 1871

The celebrated economist William Stanley Jevons was also admired in his time for his work in "equational logic". His "logical piano", a device for carrying out logical operations, each defined by a key on a keyboard, was displayed at the Royal Society to great acclaim. Jevons published often in NATURE. *Here is a sample of his wide interests. Those who have read Oliver Sacks's collection of essays* The Man Who Mistook His Wife for a Hat *will recall that the twin brothers, both idiot-savants, studied by Sacks, had a preternatural ability to discriminate numbers of objects: when Sacks dropped a matchbox on the floor, they both instantly called out the number of spilled matches (more than a hundred).*

THE POWER OF NUMERICAL DISCRIMINATION

IT is well known that the mind is unable through the eye to estimate any large number of objects without counting them successively. A small number, for instance three or four, it can certainly comprehend and count by an instantaneous and apparently single act of mental attention. The limits of this power have been the subject of speculation or experiment among psychologists, and Sir William Hamilton thus sums up almost the whole of what is known about it :—

"Supposing that the mind is not limited to the simultaneous consideration of a single object, a question arises, How many objects can it embrace at once? . . . . I find this problem stated and differently answered by different philosophers, and apparently without a knowledge of each other. By Charles Bonnet, the mind is allowed to have a distinct notion of six objects at once; by Abraham Tucker the number is limited to four; while Destutt Tracy again amplifies it to six. The opinion of the first and last of these philosophers appears to me correct. You can easily make the experiment for yourselves, but you must beware of grouping the objects into classes. If you throw a handful of marbles on the floor, you will find it difficult to view at once more than six, or seven at most, without confusion; but if you group them into twos, or threes, or fives, you can comprehend as many groups as you can units, because the mind considers these groups only as units; it views them as wholes, and throws their parts out of consideration. You may perform the experiment also by an act of imagination."

And here, after analysis of the results, is his conclusion.

When we take into account the direction of the errors, the results are as follows :—

5	6	7	8	9	10	11	12	13	14	15
+ ·06	+ ·09	+ ·05	0·0	− ·05	− ·27	− ·46	− ·51	− ·85	− ·93	− 1·27

Thus there is a clear tendency to over-estimate small numbers and to under-estimate large ones. There is an evident inclination towards those medium numbers which most frequently recurred : how far this discredits the experiments I cannot undertake to say, but it is an instance of that inevitable bias in mental experiments against which it is impossible to take complete precautions.

My conclusion that the number five is beyond the limit of perfect discrimination, by some persons at least, is strongly supported by the principles of rhythm. All the kinds of time employed by musicians depend upon a division of the bar into two or three equal parts, or into multiples of these. Music has, indeed, been composed with the bar divided into five equal parts, but no musicians have yet been found capable of performing it (Rees' Cyclopædia, RHYTHM). Short runs, indeed, consisting of five or even seven equal notes, are not unfrequently employed by the best musicians, but it is to be doubted whether the ear can grasp them surely. I presume it is beyond doubt that 6, 8, 9, or more equal notes in a bar are always broken up by the hearer, if not by the performer, into periods of 2, 3, or 4. Quinary music, even if it could be executed, would be ill appreciated by the hearers, and, though all the powers of the human mind may be expected to progress in the course of ages, quinary rhythm belongs to the music of the distant future.

W. STANLEY JEVONS

VOL 3, 281 1871

Etymology was a discipline that fell within NATURE'S *purview. This letter is typical.*

On the Derivation of the name "Britain"

HAVING been from home, I did not see the letter of "A. H." in your publication of March 16 until yesterday. His only objection to my derivation of the name *Britain* is that the word *tin* in his opinion was "not used in this island so early as the argument for its forming part of the word *Britain* requires. The following remarks will show that it *must* have been used in this island quite as early as the word *Britain*.

His assertion that "our word *tin* is of comparatively modern formation," cannot be established. It *must* have been familiar to the Cornish centuries before Diodorus Siculus described St. Michael's Mount, in Cornwall, under its name of *Iktin*, from whence tin was exported by the Phœnicians as far back as the time of Moses (Numb. xxxi. 22); and from none but the Phœnicians could the Cornish have derived the word *tin*—for that metal, as well as its name, was unknown to them before they were visited by the Phœnicians. The name *Iktin* (Tin-port) by which the Mount was called in the time of Diodorus, proves the existence of the word *tin* prior to that period, and the present Cornish word *stean* can only be a corruption of the very anciently adopted word *tin*—a corruption arising probably during the Roman period, so that instead of *tin* being a corruption of the Welsh *ystaen*, or of the Latin *stannum*, as ".A. H." imagines, the reverse is evidently the case.

Assuming, with most authors, the original Phœnician word to be *tin*, that name continues unchanged in the Saxon, English, Dutch, Danish, and Icelandic languages; but the Swedish name is now *tinn*; the German, *zinn*; the French, *étain* and *tain*; the Latin, *stannum*; the Italian, *stagno*; the Spanish *estaño*; the Portuguese, *estanho*; the Irish, *-stan*; the Welsh, *ystaen*; the Cornish, *stean*; the Armoric, *stean* and also *staen*—the initial letter or sound *s* in each of the last nine names being a mere prefix, as in the modern word *sneeze* for *neeze* (Job xli. 18). With this exception, and except the ordinary terminations of the Latin, Italian, Spanish, and Portuguese names, these thirteen different spellings are merely the different ways in which different nations of Europe pronounce the Phœnician word *tin*.

Diodorus speaks of *Iktin* as an island adjoining *Britain ;* and this island (for it *is* an island two-thirds of the day) was no doubt long before his time called sometimes *Iktin* and sometimes *Bretin* ;—*Iktin* when it was regarded as a "port," and *Bretin* when regarded as a "mount"—*ik* being the Cornish for "port," and *bre* the Cornish for "mount." It was however most generally known as a mount, and as the most remarkable object in *Mount*'s Bay, to which it has therefore given its *English* name, having long before the Christian era, in all probability, given its ancient Cornish name of *Bretin* to the island in which we live.

Plymouth, March 29   RICHARD EDMUNDS

A leader condemns the prevarications that followed the proposal to establish a school of physical sciences at Newcastle upon Tyne, despite the offer by the Duke of Northumberland to cap any sum that could be raised publicly. The state of science teaching at the University of Durham, with which the Newcastle enterprise was to be associated, was probably not untypical of that in many British universities.

At this time the scientific instruction of the University amounted to twelve lectures annually from a Reader in Chemistry, some teaching from a Reader in Natural Philosophy, and a very few lectures from a Reader in Medicine. After the sittings of the Commission, the Senate took powers to establish a school of physical science in the faculty of Arts and to endow chairs therein, but as the stipends for these chairs were not forthcoming, their powers were never practically exercised. The readership in natural philosophy for some reason became vacant a few years afterwards, and as no fresh appointment was made, teaching in that department ceased even nominally to exist. We may be spared going into details as to the extent of the classes in the other two subjects ; suffice it to say, that at the present time the physical science teaching supposed to be accomplished in Durham by the University cannot be regarded as anything more than the merest apology.

We have omitted any mention of Astronomy. Durham University has an Observatory and an Observer,—it has also a "Professor of Mathematics and Astronomy." We trust we do the present worthy occupant of the chair no injustice in supposing that his occupation has rested in the former rather than the latter branch of knowledge. With the Readership in Hebrew also on his hands, more than this could scarcely be expected.

Charles Piazzi Smyth was a leading figure in astronomy circles. He was director of the Edinburgh Observatory and made observations on comets, the solar spectrum and the radiant energy of the Moon. He also dabbled extensively and farcically in mathematical studies of the pyramids. His strictures on anonymous peer review have echoes in our time. Here he gets to the nub of his long complaint.

I cannot but wonder more than ever, and even with exceeding admiration, at what *any* Scientific Societies in the present day have got to do with that accursed thing in all national history represented by Secret Committees, secret members, secret judgments, *veiled* prophets, who may, and—as would most clearly be shown if the whole correspondence in this case were to be published—who do, blunder utterly in understanding a plain sentence of simple English, who likewise enact a mistaken rule to tie down some astronomers in their own business, prove themselves totally void of Christian charity and gentlemanly feeling, and all the time require the incense of passive obedience to their partial edicts and strange behests.

Is not this then a matter just as important as any that can occur in the interests of true science and unalloyed, for the Royal Commission now sitting on Scientific Education and the Advancement of Science to take account of ! For, if that Commission fitly and fully represents the general government in these times of this free and enlightened land in which we live, it would seem to be one of their holiest duties to the nation at large, to see that a base political method of a past and exploded era of our history, after being driven with ignominy out of every other branch of government, be not allowed to linger in sequestered nooks and dark corners of State-supported or State-aided societies for scientific pursuits.

     C. PIAZZI SMYTH

Opinion in 1869 was divided about whether science could be made amenable to women's fickle intelligences; and if the attempt was to be made, at which stratum of society should it be directed? The author of this letter shares, it seems, the view of Oscar Wilde that if education did have any effect in England it could only damage the upper classes and would probably lead to acts of violence in Grosvenor Square.

Lectures to Ladies

No one can appreciate more heartily than I do the excellent article on "Lectures to Ladies" which appeared in NATURE No. II. ; but I feel far from sanguine of success attending the efforts there referred to. If we put aside the impulse of dilettantism and the spirit of rivalry as against men, there will, let us hope, be left a very fair residue in the shape of love of learning, for learning's sake, as a reason for attendance ; and it is only this pure love of learning which can make such lectures in the long run successful. It cannot, however, be such a love which brings to the lectures of the University College Professors, Lady Barbara, who sneers aloud when the lecturer wisely lays a sure foundation of elementary facts and ideas ; or which carries to South Kensington the Hon. Miss Henrietta, who tosses her head when she finds the great Mr. Huxley paddling about in that common river the Thames, and treating his audience as if they were little girls at the Finsbury Institution.

I very much fear that the Lady Barbaras of the present generation are beyond redemption, and that many earnest men are wasting their strength in trying to win the minds of intellectual coquettes.

There is an order of women, however, having in their number, as I know full well, some of the brightest and best of the women of England, to whom such lectures would be as manna in the wilderness. To women struggling, as many of us are, to get their daily bread by the hard task of teaching, and in the struggle getting glimpses of the sweetness and the light of real knowledge, the chance of listening to real teachers would be an inestimable boon. These are the women to whom, it must be remembered, second two miles of the road were quite straight ; so that I could easily have seen the dog if he had been merely running a comparatively short distance in front of the horses. Why this animal should never have returned to his former home on his own account, I cannot suggest ; but I think it was merely due to an excessive caution which he also manifested in other things. Be the explanation of this, however, what it may, as a fact he never did venture to come back upon his own account, notwithstanding there never was a subsequent occasion upon which any of his former friends went to the town but the terrier was sure to return with them, having always found some way of escape from his intended imprisonment.

Regent's Park, N.W.   GEORGE J. ROMANES

The nature of colour perception, which so intrigued James Clerk Maxwell, was taken up by the Honourable J. W. Strutt, soon to undergo metamorphosis into Lord Rayleigh (Third Baron). Here he describes his famous experiments with the rotating coloured discs and anticipates many much later discoveries about human colour receptors.

SOME EXPERIMENTS ON COLOUR

THE theory of colour perception, although in England it has not yet made its way into the text-books, still less into the popular works on science, is fully established with regard to many important points. It is known that our perception of colour is threefold, that is, that any colour may be regarded as made up of definite quantities of three primary colours, the exact nature of which is, however, still uncertain. More strictly stated, the fundamental fact in the doctrine of colour is that, between any four colours whatever given, as well in quantity as in quality, there exists what mathematicians call a linear relation, that is, that either a mixture of two of them (in proper proportions) can be found identical, so far as the eye is able to judge, with a mixture of the other two, or else that one of them can be matched by a mixture of the other three. There are various optical contrivances by which the mixture spoken of may be effected. In the year 1857, Mr. Maxwell published an account of some experiments with the colour top undertaken to test the theory. From six coloured papers, black, white, red, green, yellow, and blue, discs of two sizes were prepared, which were then slit along a radius so as to admit of being slipped one over the other. Any five out of the six being taken, a match or colour equation between them is possible. For instance, if yellow be excluded, the other five must be arranged so that a mixture of red, green, and blue is matched with a mixture of black and white. The large discs of the three colours are taken and slipped on to each other, and similarly the small discs of black and white. When the small discs are placed over the others and the whole made to rotate rapidly on any kind of spinning machine, the colours are blended, those of the large discs and those of the small, each into a uniform tint.

By adjustment of the discs an arrangement may be found after repeated trials, such that the colour of the inner circle is exactly the same both in tint and luminosity with that of the outer disc. The quantities of each colour exposed may then be read off on a graduated circle, and the result recorded. For instance (the circle being divided into 192 parts), eighty-two parts red mixed with fifty-six green and fifty-four blue, match thirty-seven parts white mixed with 155 black. In this way Maxwell observed the colour equations between each set of five, in all six sets formed by leaving out in turn each of the six colours. Moreover, for greater accuracy each set was observed six times, and the mean taken. But according to the theory these six final equations are not all independent of each other, but if any two of them are supposed known, the others can be found by a simple calculation. Accordingly, the comparison of the calculated and observed equations furnishes a test of the theory ; but in practice, in order to ensure greater accuracy, instead of founding the calculations on two of the actually observed equations chosen arbitrarily, it is preferable to combine all the observations into two equations, which may then be made the basis of calculation. In this way, a system of equations is found necessarily consistent with itself, and agreeing as nearly as possible with the actually observed equations. A comparison of the two sets gives evidence as to the truth of the theory according to which the calculations are made, or if this be considered beyond doubt, tests the accuracy of the observations. In Maxwell's experiments the average difference between the calculated and observed systems amounted to ·77 divisions of which the circle carried 100. So good an agreement is regarded by him as a confirmation of the whole theory ; but it seems to me, I confess, that only a very limited part of it is concerned. The axioms, in virtue of which it is permitted to combine the colour equations in the manner required for the calculations, are only such as the following :—If colours which match are mixed with colours which match, the results will match. It is difficult to imagine any theory of colour which will not include them. What proves the threefold character of colour—the most important part of the doctrine—is simply the fact that with any five-coloured papers *whatever* a match can be made, while with less than five it cannot (except in certain particular cases). In regard to this point the value of the quantitative experiments is rather that they show of what sort of accuracy the eye is capable in this kind of observation.

Those to whom the subject is new may think at first that if colour be threefold a match ought to be possible between any *four* colours. And so it is possible if there is no other limitation ; but in experiments with the revolving discs, we are subject to a limitation, being obliged to fill up the whole circumference somehow. The difficulty will clear itself up, when it is remembered that one of the five colours may be black, so that with any *four* colours and *black* a match can be made with revolving discs.

He then records his quantitative observations on a fine day and on a cloudy day, discusses the use of liquid colour filters, such as litmus solution, and draws his conclusions:

Impartial observers, unprejudiced by the results of mixing pigments, or, on the other hand, by experiments on the spectrum, see, so far as I can make out, no connection between the four principal colours—red, yellow, green, and blue. It seems to them quite as absurd that yellow should be compounded of red and green, as it most unquestionably is, as that green should be a compound of blue and yellow, though many have accepted the latter alternative on the authority of painters, and some have even worked themselves into the belief that it is only necessary to look at the colours in order to recognise the compound nature of green. My own prejudice would be on the other side, the result of experiments on the compound yellow, which is seen so easily to pass into green on the one side or red on the other. The most impartial opinion that I can form is that there is no real *resemblance* between any of the four, and if this be so it is certainly a most remarkable, if not unaccountable, fact. The difficulty is not so much that we are unable to analyse the compound sensation, as to explain why our inability is limited to yellow (and white). For everyone, I imagine, sees in purple a resemblance to its components red and blue, and can trace the primary colours in a mixture of green and blue. Sir John Herschel even thinks that our inability to resolve yellow leaves it doubtful whether our vision is trichromic or tetrachromic, but this seems to me to be going much too far. Surely the fact that the most saturated yellow can be compounded of red and green, deprives it of any right to stand in the same rank with them as primary colours, however little resemblance it may bear to them and blue. Besides, if yellow is to be considered primary, why not also white, which is quite as distinct a sensation as any of the others? Undoubtedly there is much that is still obscure in the mutual relations of the colours—why, for instance, as mentioned by Sir John Herschel, a dark yellow or orange suggests its character so little as to be called by a new name (brown), while a dark blue is blue still. But difficulties such as these should make us all the more determined to build our theories of colour on the solid ground that normal vision is threefold, and that the three primary elements of colour correspond nearly with red, green, and blue.

J. W. STRUTT

VOL 3, 234 1871

chapter two
1871-1873

James Clerk Maxwell dilates on colour vision and versifies. Henry Roscoe expounds his plans for a people's university. William Crookes expatiates on the "psychic force", Herbert Spencer hectors his readers about the true meaning of Darwinism and the chauvinistic physicist P. G. Tait taunts the author of a German physics treatise. Charles Babbage dies. Charles Darwin offers his thoughts on animal instincts. Henry Bessemer invents a cabin that rocks when the ship rolls and the sunspot cycle, we are told, causes cholera.

1871-1873

Ulysses S. Grant became president of the United States. In Russia, Fyodor Dostoevsky published *The Possessed*, and an industrial and agricultural depression settled on Britain. At Ujiji, Henry Morton Stanley finally caught up with Dr David Livingstone (he presumed).

In this fragment from a long lecture given before the British Association for the Advancement of Science and reprinted in Nature *(after the communication on the same topic (p.18) from J. W. Strutt),* James Clerk Maxwell *displays his accustomed geniality and grace of style.*

ON COLOUR VISION

ALL vision is colour vision, for it is only by observing differences of colour that we distinguish the forms of objects. I include differences of brightness or shade among differences of colour.

It was in the Royal Institution, about the beginning of this century, that Thomas Young made the first distinct announcement of that doctrine of the vision of colours which I propose to illustrate. We may state it thus :—We are capable of feeling three different colour-sensations. Light of different kinds excites these sensations in different proportions, and it is by the different combinations of these three primary sensations that all the varieties of visible colour are produced. In this statement there is one word on which we must fix our attention. That word is, Sensation. It seems almost a truism to say that colour is a sensation .

* * *

In the same way most blue and yellow paints, when mixed, appear green. The light which falls on the mixture is so beaten about between the yellow particles and the blue, that the only light which survives is the green. But yellow and blue light when mixed do not make green, as you will see if we allow them to fall on the same part of the screen together.

It is a striking illustration of our mental processes that many persons have not only gone on believing, on the evidence of the mixture of pigments, that blue and yellow make green, but that they have even persuaded themselves that they could detect the separate sensations of blueness and of yellowness in the sensation of green.

We have availed ourselves hitherto of the analysis of light by coloured substances. We must now return, still under the guidance of Newton, to the prismatic spectrum. Newton not only

Untwisted all the shining robe of day,

but showed how to put it together again. We have here a pure spectrum, but instead of catching it on a screen, we allow it to pass through a lens large enough to receive all the coloured rays. These rays proceed, according to well-known principles in optics, to form an image of the prism on a screen placed at the proper distance. This image is formed by rays of all colours, and you see the result is white. But if I stop any of the coloured rays, the image is no longer white, but coloured ; and if I only let through rays of one colour, the image of the prism appears of that colour.

I have here an arrangement of slits by which I can select one, two, or three portions of the light of the spectrum, and allow them to form an image of the prism while all the rest are stopped. This gives me a perfect command of the colours of the spectrum, and I can produce on the screen every possible shade of colour by adjusting the breadth and the position of the slits through which the light passes. I can also, by interposing a lens in the passage of the light, show you a magnified image of the slits, by which you will see the different kinds of light which compose the mixture.

The colours are at present red, green, and blue, and the mixture of the three colours is, as you see, nearly white. Let us try the effect of mixing two of these colours. Red and blue form a fine purple or crimson, green and blue form a sea-green or sky-blue, red and green form a yellow.

Here again we have a fact not universally known. No painter, wishing to produce a fine yellow, mixes his red with his green. The result would be a very dirty drab colour. He is furnished by nature with brilliant yellow pigments, and he takes advantage of these. When he mixes red and green paint, the red light scattered by the red paint is robbed of nearly all its brightness by getting among particles of green, and the green light fares no better, for it is sure to fall in with particles of red paint. But when the pencil with which we paint is composed of the rays of light, the effect of two coats of colour is very different. The red and the green form a yellow of great splendour, which may be shown to be as intense as the purest yellow of the spectrum.

I have now arranged the slits to transmit the yellow of the spectrum. You see it is similar in colour to the yellow formed by mixing red and green. It differs from the mixture, however, in being strictly homogeneous in a physical point of view. The prism, as you see, does not divide it into two portions as it did the mixture. Let us now combine this yellow with the blue of the spectrum. The result is certainly not green ; we may make it pink if our yellow is of a warm hue, but if we choose a greenish yellow we can produce a good white.

Vol 4, 13 1871

It is not clear what the upshot was of the proposed monument to Victorian idealism, described in a leader by Henry Roscoe, the great Mancunian physical chemist and man of conscience.

THE PEOPLE'S UNIVERSITY

A GIGANTIC and imposing educational scheme is about to be launched, which, whether it proves feasible or not, must attract the attention and enlist the sympathy of all well-wishers to the intellectual development and material welfare of the country. This is no less an idea than the establishment of a National Working Men's University, which is to be founded with special reference to instruction in those subjects which have a direct bearing on the arts and manufactures. That our workmen are, as a rule, altogether ignorant of the scientific principles upon which the processes they ought to guide and govern are dependent, and that England in this respect stands in a much inferior position to continental nations, is now a well-recognised fact. The result of this lamentable ignorance is stated by certain authorities to be severely felt in those of our trades and manufactures in which we have to compete with other nations; and although this conclusion has been denied by many, yet concerning the necessity for scientific education amongst our artisans there has never been a difference of opinion. The question then arises, How are we to bring to our rising artisans on an extended and national scale the knowledge of scientific principles which they so much need, and for which the best of their class show so much desire and even aptitude? One solution to this problem is being attempted by the scheme of a National University for Industrial and Technical Training. The proposal is to establish a metropolitan institution in which complete and thorough instruction in all those branches of knowledge which are of importance to our manufacturing industry shall be given.

VOL 4, 41 1871

The fragments that follow are from a characteristic piece of naturalist travel writing. The author is Chas. F. Tyrwhitt-Drake; the place is the Desert of Tih.

Bustard (*Otis hubara*) Ar. *Hubara*. I noticed a few of these birds in the Tîh; the Arabs say that the lesser bustard (*Otis tetrax*) which is also occasionally found there, is the young of the larger, but does not attain its full growth for two years. They also say that these birds, when attacked by a falcon, will cover it with their fæces, and so drive it off.

* * *

Fox, Ar. *Taáleb, Abou'l Husein*. In the East, as in Europe, this animal is looked upon as the type of cunning, and numberless stories are current concerning it. The following are examples:—

When a fox is over much troubled with fleas, he plucks out a mouthful of his hair, and then he takes to the water, holding the tuft in his mouth; all the fleas creep up on to this to escape drowning, and the fox then drops it into the stream and retires, freed from his enemies.

The celebrated Arabic author and theologian, Esh Shafiey, relates that when in Yemen, he and his fellow travellers prepared two fowls for dinner one day, but the hour of prayer coming on, they left them on the table and went to perform their devotions; meanwhile a fox came and stole one. After their prayers were finished, they saw the fox prowling about with their chicken in his mouth, so they pursued him and he dropped it; on coming up nearer to it, however, they found it only to be a piece of palm fibre, which the fox had dropped to attract our attention, and had, in the meantime, crept round and carried off the second chicken and left them dinnerless.

The fox is said to feign death, and to inflate his body, and when any animal, prompted by curiosity, comes to look at him, he springs up and seizes it.

The fable of the fox and stork is changed to the fox and raven; the former invites the latter to dinner, and gives him soup in a shallow wooden bowl; the raven returns the compliment, and pours out some wheat over a *silleh* bush. The *silleh* is one of the most thorny of the desert plants.

Another story told of the fox is, that one day he met five slaves, who were travelling with a large supply of food and other goods; he joined them, and after a time they reached a well, but had no rope wherewith to draw up the water. The fox suggested that they should throw down the meal and that one of their number should go down and knead it, which was accordingly done. After a while the fox said to the four who remained above, "Your comrade must have found a treasure, why don't you go down and share it?" This hint was enough, and they all hurried down, while the fox decamped with their goods and chattels.

A fox's gall is said to be a specific for epilepsy, and his fat for the gout.

VOL 4, 52 1871

Yellow rain was discovered, it seems, in 1871 —

Yellow Rain

THE following notice will perhaps be of some interest to the readers of NATURE. In December 1870, after a heavy rain at Rosario de Cucuta (New Granada), a great many small round specks of a yellow clayish substance were found on the leaves of plants that had been exposed to the rain. A sample of this substance was sent to Dr. A. Rójas, of this town, who forwarded it to me in order to examine it under the microscope. It proved to be composed almost entirely of a species of *Triceratium*, and another of *Cosmarium*, which must have been carried away by a violent storm from their lacustrian abodes.

Caracas (Venezuela), April 1871   A. ERNST

VOL 4, 68 1871

— and also two millennia before:

This I think is the most probable meaning, and is, perhaps, the earliest "yellow" shower on record. The colour, as stated by the latter, is nearly correct, as shown by the investigation of MM. De Candolle and Prevost, who discovered, microscopically, that the red snow was due to the presence of small globules of a bright red colour, which were surrounded by a gelatinous membrane, transparent and slightly yellow, and were mixed with fragments of moss and dust. "An examination of the crimson snow found by Captain Ross in the Arctic regions by M. De Candolle proved it to be identical with the Alpine red snow; the globular bodies are of a vegetable nature, and were once thought to belong to the *Uredo*, but M. Bauer disproved this, and named the plant *Protococcus nivalis*. There are cases, however, in which the presence of animalcules gives a reddish tinge to snow." ‡

Honeydew is mentioned by Pliny (Bk. XVIII. c. 28), who states that a great many of the ancients affirmed that dew burnt up by the scorching sun is the cause of honeydew on corn.

In the *Chronicum Scotorum* is the earliest direct record of a "shower of honey" I know of.§ It says: "A.D. 714 . . . . it rained a shower of honey upon Othan Bec. . . . ."

When it is known that any sudden appearance, giving a colour to the ground, or prominent places, or on trees, &c., is generally thought to have descended from above, this passage is quite intelligible. The "shower of honey" was nothing more than a "secretion of *aphides*," whose excrement has the privilege of emulating the sugar and honey in sweetness and purity." ‖ Some contend that it is due solely to the exudation of the saccharine juices of trees; but, feasible as this may seem, it is not sufficient to account for this phenomenon, which often extends over very large tracts of land. If the exudation is promoted by the aphides, and the dew increased by their own excrement, then this explanation is, I believe, the true one. The former view is not to be discarded without some consideration ; for one observer states that, in the course of thirty years he had attended to this subject, he had never met with any honeydew which did not seem to him to be clearly referable to aphides as its origin. ¶ This view does not go counter to what I conceive to be the correct one ; for exudations do take place, and the quantity of "dew" can be increased by the aphides.

There is a very curious account given in a now little known work of what was considered the real cause of "honeydew," but I will not trespass further upon the valuable space of this journal in quoting it. I give the title at foot.*

More can be said upon this interesting subject ; and on another occasion I hope to resume the investigation, by attempting to explain the "yellow rains" of a different kind to those treated of in this letter.   JOHN JEREMIAH

43, Red Lion Street

Vol 4, 161 1871

Sir William Crookes was a peerless experimentalist in physics and physical chemistry, whose Achilles' heel was a penchant for the occult. In old age, he made, as will emerge, a fearful fool of himself, but at the outset some at least of the Victorian physicists were willing to give him the benefit of the doubt. One such was Balfour Stewart, professor at Owens College in Manchester, who made the following comment in a leader.

MR. CROOKES ON THE "PSYCHIC" FORCE

WITH a boldness and honesty which deserve the greatest respect, Mr. Crookes has come forward as an investigator of those mysterious phenomena which have now been so long before the public that it is unnecessary to name them, more especially as their generally received name is very objectionable.

Two things have contributed to retard our knowledge of these strange events. In the first place, until lately few men of name have been associated with their occurrence, so that outsiders have not had the facts put before them in a proper manner. In the next place, we are inclined to endorse the remark of Mr. Crookes, that men of science have shown too great a disinclination to investigate the existence and nature of these alleged facts, even when their occurrence had been asserted by competent and credible witnesses.

Vol 4, 237 1871

J. P. Earwaker, a regular commentator recruited by Norman Lockyer, was less sanguine.

On account of these and many other objections, I am forced to the conclusion previously stated, that these experiments were inaccurately performed—the details were not sufficiently examined, nor obvious errors apparently avoided, so that until they are repeated in the presence of other scientific men, they are not worthy of scientific consideration. We have read of the same phenomenon over and over again described as due to spiritual manifestations—many of them, as is well known, performed through the same agency—a medium—as those in this case. The British Association is about to meet. Let Mr. Crookes but repeat any one of the experiments at one of the evening *soirées*, and, if he can do this, he will make the Edinburgh Meeting for ever memorable, and will have earned for himself the undying reputation of having been the first to discover that in the midst of apparent humbug true science really and truly did exist.   J. P. EARWAKER

Vol 4, 279 1871

Not long after NATURE published an admiring report on the safety of gun-cotton, based on tests in which rifle bullets were fired into boxes containing nitroglycerine on the one hand and gun-cotton on the other, the following catastrophe was recorded. To judge from the report in NATURE, the amount of weapons research in England at the time must have been inordinate.

THE GUN-COTTON EXPLOSION AT STOWMARKET

THE disastrous explosion of gun-cotton, which occurred on Friday last on the premises of the Patent Safety Gun-cotton Company, is a calamity of unusual significance. Besides the large number of killed, amounting, we believe, to five and-twenty persons in all, there were as many as seventy maimed and injured, many of them too, in such a manner as only violent explosions are known to torture and lacerate their victims ; and when it is taken into consideration that in all probability a dozen tons of the material actually exploded, the grave nature of the accident is in truth not surprising. The whole group of factory-buildings and out-houses were levelled to the earth at one fell swoop, and for miles away the effect of the catastrophe was acutely felt.

But it is not only from a social point of view that the affair is to be deplored. As a result seriously affecting the science of explosives, the occurrence is peculiarly unfortunate ; for the belief in the safety of gun-cotton as an industrial and military agent will now be gravely shaken. It is all very well for scientific men to adduce a plausible reason for the occurrence, and to prove conclusively that with due care and precaution a disaster of this nature could not possibly have happened ; but the public unfortunately will not be satisfied with a theoretical assurance of this kind ; and indeed measures should certainly be taken, not only to guard against such wholesale death and destruction, but to render the same absolutely impossible.

Vol 4, 309 1871

This examination paper, set to boys at Clifton College (a school always noted for its strength in science), suggests that the standards of the day were remarkably high.

MAGNETISM

1. Soft iron can never be permanently magnetised, yet a piece of soft iron in contact with a magnet becomes a magnet. Why ?
2. What do you understand by *coercive force*, and magnetic *saturation* ?
3. How is magnetism influenced by heat ?
4. Mention the substances which are attracted by a magnet in addition to iron.
5. State one or more of the methods by which steel bars may be magnetised.
6. What is the *declination* or *variation* of the magnetic needle, and the present extent of it ?

CHEMISTRY

a. For First and Second Sets, Modern Side only.

1. Mention the oxygen compounds of phosphorus, and the action of water upon them.
2. Give an account of arsenicum and its chief characteristics.
3. What are the constituents and characteristics of arseniuretted hydrogen ?
4. What is "white arsenic," and how may it be prepared ?
5. You are given a liquid suspected to contain arsenic ; by what means would you examine it ?

b. For all other Forms.

1. Ammonia gas, and hydrochloric acid gas, are brought into contact : what is the resulting compound, and to what may it be compared ?
2. What is *ammonium* ? Describe the formation and appearance of ammonium amalgam.

3. What do you know of chloride of nitrogen?
4. What is nitric acid, and by what means may it be procured?
5. State the action of nitric acid upon metals,—copper, tin, antimony,—and the general tendency of the acid.

BOTANY

Third A. and B. only. (To be written on separate paper only).

1. Describe the following forms of roots :—*tap, napiform, premorse, tubercular.*
2. How are fluids absorbed by the roots?
3. Show clearly the true nature of the various forms of the bulb.
4. What is a "rhizome" (or root-stock)? Compare it with a "corm."
5. Give an account of the structure of the stem in a common potato.
6. Why is it that plants and animals have a mutual dependence on each other for their life?

Vol 4, 329 1871

In October 1871, Nature *took note of the death of Charles Babbage, a wayward genius and scourge of the scientific and political establishments. The obituary is more revealing than most others published during this period, which were invariably bland hagiographies (or, as used to be said,* de mortuis nil nisi bunkum). *Ada, Countess Lovelace, the daughter of Lord Byron, who assisted Babbage in his work on the difference engine and has become in recent years something of a feminist totem, is not mentioned. Nor is it recorded that Babbage rejected a knighthood as a vanity inappropriate for a scientist.*

CHARLES BABBAGE
DIED THE 20TH OF OCTOBER, 1871

THERE is no fear that the worth of the late Charles Babbage will be over-estimated by this or any generation. To the majority of people he was little known except as an irritable and eccentric person, possessed by a strange idea of a calculating machine, which he failed to carry to completion. Only those who have carefully studied a number of his writings can adequately conceive the nobility of his nature and the depth of his genius. To deny that there were deficiencies in his character, which much diminished the value of his labours, would be useless, for they were readily apparent in every part of his life. The powers of mind possessed by Mr. Babbage, if used with judgment and persistence upon a limited range of subjects, must have placed him among the few greatest men who can create new methods or reform whole branches of knowledge. Unfortunately the works of Babbage are strangely fragmentary. It has been stated in the daily press that he wrote eighty volumes; but most of the eighty publications are short papers, often only a few pages in length, published in the transactions of learned societies. Those to which we can apply the name of books, such as "The Ninth Bridgewater Treatise," "The Reflections on the Decline of Science," or "The Account of the Exposition of 1851," are generally incomplete sketches, on which but little care could have been expended. We have, in fact, mere samples of what he could do. He was essentially one who began and did not complete. He sowed ideas, the fruit of which has been reaped by men less able but of more thrifty mental habits.

It was not time that was wanting to him. Born as long ago as the 26th of December, 1792, he has enjoyed a working life of nearly eighty years, and, though within the last few years his memory for immediate events and persons was rapidly decaying, the other intellectual powers seemed as strong as ever.

About the difference engine and the analytical engine, the obituarist had this to say:

As early as 1812 or 1813 he entertained the notion of calculating mathematical tables by mechanical means, and in 1819 or 1820 began to reduce his ideas to practice. Between 1820 and 1822 he completed a small model, and in 1823 commenced a more perfect engine with the assistance of public money. It would be needless as well as impossible to pursue in detail the history of this undertaking, fully stated as it is in several of Mr. Babbage's volumes. Suffice it to say that, commencing with 1,500*l.*, the cost of the Difference Engine grew and grew until 17,000*l.* of public money had been expended. Mr. Babbage then most unfortunately put forward a new scheme for an Analytical Engine, which should indefinitely surpass in power the previously-designed engine. To trace out the intricacies of negotiation and misunderstanding which followed would be superfluous and painful. The result was that the Government withdrew all further assistance, the practical engineer threw up his work and took away his tools, and Mr. Babbage, relinquishing all notions of completing the Difference machine, bestowed all his energies upon the designs of the wonderful Analytical Engine. This great object of his aspirations was to be little less than the mind of a mathematician embodied in metallic wheels and levers. It was to be capable of any analytical operation, for instance solving equations and tabulating the most complicated formulæ. Nothing but a careful study of the published accounts can give an adequate notion of the vast mechanical ingenuity lavished by Mr. Babbage upon this fascinating design. Although we are often without detailed explanations of the means, there can be little doubt that everything which Mr. Babbage asserted to be possible would have been theoretically possible. The engine was to possess a kind of power of prevision, and was to be so constructed that intentional disturbance of all the loose parts would give no error in the final result.

And he sums up nobly:

Let it be granted that in his life there was much to cause disappointment, and that the results of his labours, however great, are below his powers. Can we withhold our tribute of admiration to one who throughout his long life inflexibly devoted his exertions to the most lofty subjects? Some will cultivate science as an amusement, others as a source of pecuniary profit, or the means of gaining popularity. Mr. Babbage was one of those whose genius urged them against everything conducive to their immediate interests. He nobly upheld the character of a discoverer and inventor, despising any less reward than to carry out the highest conception which his mind brought forth. His very failures arose from no want of industry or ability, but from excess of resolution that his aims should be at the very highest. In these money-making days can we forget that he expended almost a fortune on his task? If, as people think, wealth and luxury are corrupting society, should they omit to honour one of whom it may be truly said, in the words of Merlin, that the single wish of his heart was " to give them greater minds " ?

Vol 5, 28 1871

Below NATURE *records what must be a low point of French gastronomy.*

A proposition to manufacture artificial milk, brought forward by M. Gaudin, seems worthy of some notice. That gentleman estimated that 500,000 litres per day of milk could be prepared in Paris at an exceedingly trifling cost, which should have all the nutritious qualities of good milk, and which should, besides, be neither unpleasant of taste or smell. An emulsion at a very high temperature is made of *bouillon de viande* prepared from bones, fat, and gelatine, and when cold, a product is obtained resembling in taste stale milk of a cheesy flavour; the components of ordinary milk are all present, the gelatine representing the casein; fat, the butter; and sugar, the sugar of milk. For admixture with coffee, chocolate, soup, &c., the milk is said to be by no means disagreeable.

Many propositions were brought forward to economise the blood from the abattoir, the plan suggested by M. Gaultier of mixing it with flour in the manufacture of bread being perhaps the best and simplest, as the fibrine and albumen, so rich in nitrogen—of which the alimentary properties are well known—are in this way utilised to the highest degree. Less inviting is the proposal of M. Fud to consume the carcases of animals that died of typhus, rhinderpest, and other diseases, the flesh in these instances being, so asserts M. Fud, capable of use as food, if only cooked in a suitable manner.

VOL 5, 45 1871

Nature was ever sedulous in promoting women's rights.

Seldom have medical students given better answers. And yet it has been argued that physiology was far too difficult and technical a subject to be studied even by the students in Arts of our University. Hence women in all ranks of society should have physiology taught to them. It should be an essential subject in their primary, secondary, and higher schools. So strong are my convictions on this subject, that I esteem it a special duty to lecture on physiology to women, and whenever I have done so, have found them most attentive and interested in the subject, possessing indeed a peculiar aptitude for the study, and an instinctive feeling, whether as servants or mistresses, wives or mothers, that that science contains for them, more than any other, the elements of real and useful knowledge.

VOL 5, 74 1871

The voluble philosopher Herbert Spencer was one of the last universal men. He was highly contentious and sensitive, and engaged in much hair-splitting in his disputations with Charles Darwin and T. H. Huxley. The debate, as in the pronouncement by Spencer that follows, seems to have taken on an almost theological aspect. It was Spencer who coined the phrase "survival of the fittest", and Huxley believed that Spencer used him as a devil's advocate: "He thinks that if I can pick no holes in what he says he is safe. But I pick a great many holes, and we agree to differ." And Darwin opined in a letter to Joseph Hooker: "If he had trained himself to observe more, even at the expense, by the law of balancement, of some loss of thinking power, he would have been a wonderful man."

THE SURVIVAL OF THE FITTEST

LAST summer a discussion took place in your pages on the expression, "Survival of the Fittest," and on the principle it formulates. Though, as being responsible for this expression, there seemed occasion for me to say something to dissipate the errors respecting it, I refrained from doing so, for the reason that the rectification of misstatements and misinterpretations is an endless work, which it is almost useless to commence.

In your last number, however, the question has cropped up afresh in a manner which demands from me some notice. A Professor is tacitly assumed to be an authority in his own department; and a statement made by him respecting the views of a writer on a matter coming within this department, will naturally be accepted as trustworthy. Hence it becomes needful to correct serious mistakes thus originating.

In your abstract of Prof. E. D. Cope's paper, read before the American Association for the Advancement of Science, I find the following sentences :—

"This law has been epitomised by Spencer as the 'Preservation of the Fittest.' This neat expression, no doubt, covers the case, but it leaves the origin of the fittest entirely untouched."

There are here two misstatements, the one direct and the other indirect, which I must deal with separately.

So far as I can remember, I have nowhere used the phrase, "Preservation of the Fittest." It is one which I have studiously avoided; and it belongs to a class of phrases for the avoidance of which I have deliberately given reasons in "First Principles," sec. 58. It is there pointed out that such expressions as "Conservation of Force," or "Conservation of Energy," are objectionable, because "conservation" implies a conserver, and an act of conserving—implies, therefore, that Energy would disappear unless it was taken care of; and this is an implication wholly at variance with the doctrine enunciated. Here I have similarly to point out that the expression "Preservation of the Fittest" is objectionable, because in like manner it supposes an act of preserving—a process beyond, and external to, the physical processes we commonly distinguish as natural; and this is a supposition quite alien to the idea to be conveyed. One of the chief reasons I had for venturing to substitute another formula for the formula of Mr. Darwin, was that "Natural Selection" carries a decidedly teleological suggestion, which the hypothesis to be formulated does not in reality contain; and a good deal of the adverse criticism which the hypothesis has met with, especially in France, has, I think, arisen from the misapprehension thus caused. The expression, "Survival of the Fittest," seemed to me to have the advantage of suggesting no thought beyond the bare fact to be expressed; and this was in great part, though not wholly, the reason for using it.

Prof. Cope's indirect statement, that I have said nothing to explain "the origin" of the fittest, is equally erroneous with his direct statement which I have just corrected. In the "Principles of Biology," sec. 147, I have contended that no "interpretation of biologic evolution which rests simply on the basis of biologic induction, is an ultimate interpretation. The biologic induction must be itself interpreted. Only when the process of evolution of organisms is affiliated on the process of evolution in general, can it be truly said to be explained. . . . We have to reconcile the facts with the universal laws of re-distribution of matter and motion." After two chapters treating of the "External Factors" and "Internal Factors," which are dealt with as so many acting and reacting forces, there come two chapters on "Direct Equilibration" and "Indirect Equilibration"—titles which of themselves imply an endeavour to interpret the facts in terms of Matter, Motion, and Force. It is in the second of these chapters that the phrase "Survival of the Fittest" is first used; and it is there used as the most convenient physiological equivalent for the purely physical statement which precedes it.

Respecting the adequacy of the explanation, I, of course, say nothing. But when Prof. Cope implies that no explanation is given, he makes still more manifest that which is already made manifest by his mis-quotation—either that he is speaking at second hand, or that he has read with extreme inattention.   HERBERT SPENCER

Athenæum Club, Jan. 29

VOL 5, 263 1871

Next a medical curiosity:

THE *British Medical Journal* says that the people of Rome are very much interested just now in the fate of a poor fellow, Cipriani, who has swallowed a fork in public, prongs downwards, and who is now suffering, in consequence, agonies which are the subject of daily bulletin. Some comfort may be derived by his friends from the record lately published of Mr. Lund's patient at Manchester, who survived swallowing a dessert knife six inches long ; and from the perusal of a recent article in the *Journal de Médecine et de Chirurgie*, in which instances are cited where the alimentary canal has safely supported the most unexpected foreign bodies—among others, lizards, a file, a tea-spoon, a bat ; and, finally, from the whimsical but melancholy instance of a man who, to amuse himself, swallowed successfully and safely a five-franc piece, a closed pocket-knife, and a coffee-spoon, but killed himself at last in the vain effort to digest a pipe.

VOL 5, 411 1872

And the antics of the proliferating natural history societies:

THE Perthshire Society of Natural Science held its Annual Meeting on March 7, when Colonel H. M. Drummond Hay was elected president in the room of Dr. Buchanan White, who has held the office for five years. This enterprising society must be congratulated on the work it has done in the exploration of the natural history of the county, and in the commencement of the publication of so valuable a work as the *Fauna Perthensis*, and the promotion of so useful a periodical as the *Scottish Naturalist*. Botany seems, however, up to the present time, to have been neglected by the Society, which is to be regretted in a county with so rich and interesting a flora. The Society has also held "a meeting for investigation into the qualities, as articles of food, of certain Perthshire animals," commonly known as a "Frog-supper." Among the articles of the bill of fare were—Pâté d'Ecureuil, Matelot de Grenouille, Alouette à la Crapaudine, Ecureuil au naturel.

VOL 5, 450 1872

The newspaper article (perhaps a third leader in The Times?*) that so excites the editorial ire is actually quite funny, but shows all the same how little has changed in 125 years.*

NEWSPAPER SCIENCE

WHETHER some knowledge of Science or some love for scientific truth will ever penetrate the masses, may well be questioned when we read such an article as the following, which appeared in the daily paper boasting the largest circulation in the world, and which we reprint almost entire as a curiosity of newspaper literature :—

"What is a Joule ?—or who is he, if a Joule is a human being, and not a vegetable—a weapon of offence, or something to drink, or a Phantom ? And if Joule be human, why did he not consider that human reason is fallible, and human patience exhaustible, when he penned, or got somebody else to pen, a maddening article which has appeared in the *Nautical Magazine*, from which we gather that the transformations of energy are in their nature similar to the operations of commerce ; but with this difference, that in thermodynamics the relative values never vary. This, it seems, is the universal theorem of a Joule ; and a red-hot poker must always bear the same relation to sixpence as the contents of a tea-kettle at boiling point bear to a five-pound note. . . . Under the new dispensation the sovereign, 'to which all other forms of energy can be referred,' is to be an unit of heat. On the obverse is stamped 'Joule's equivalent,' and on the other side is inscribed 772 foot-pounds. One unit of heat is the amount required to raise the temperature of one pound of water one degree, and the equivalent for this coin is 772 foot-pounds of work—that is, the work required to be expended to raise one pound weight 772 feet. . . . But what is the new 'Joule's equivalent' to be made of ?—cobwebs, leather, or fresh butter ?—and who wants to raise a pound weight 772 feet ? As a problem of proportion, the theory is, of course, philosophical enough ; but it would be just as easy to fix a unit of cold as well as a unit of heat ; and, under any circumstances, until Joule comes into the open and tells us who he is, what he means, and when his equivalents are to be put into circulation, society, we fear, will decline to recognise a sovereign as a Joule, or thirty shillings as a Joule and a half."

Now, with the mental condition of the man who could pen such an article as this we have nothing to do ; he may go on writing according to his lights every day of the week, and no one but his own friends need interfere to stop him. But there are one or two considerations which arise from the perusal of it not without their importance.

In the first place, bearing in mind the contempt for Science so often apparent in the public utterances of men of high calibre—instances occur to us as we write, and probably will to our readers, of men of the highest culture in literature or art, who never allude to scientific work or to scientific teachers without a scarcely disguised sneer at the inferior part which they play in the national economy—we may, after all, be content that Science is alluded to at all in a paper possessing so large a circulation. The next consideration is one to which we attach the highest importance.

Surely it is now time that scientific men themselves should take a little more trouble than they do—we know it is asking a good deal from them—in the matter of bringing their own work, and the importance of it to the community, before such audiences as the daily papers afford. Were they to do this, the labours of our great scientific teachers—our Huxleys, Tyndalls, and Carpenters—would be enormously lightened. If we hear of an attendance of several thousands at a penny lecture by Huxley at Manchester, or a Sunday afternoon lecture in St. George's Hall by Carpenter, we fancy a love of science is spreading with rapid strides ; but the fact is that the strides are not so rapid as they might be, because the labourers on whom progress depends are so few and the area of their lecture work is restricted, whereas many newspapers, on the other hand, number their readers by hundreds of thousands. Until scientific men do this, we must be content with the present state of things. It is in no spirit of invidious comparison that we may remind our readers of the frequent extracts which appear in our columns from *Harper's Weekly*, a political and general paper of very large circulation in the United States, the scientific department of which, containing information of the highest value, is edited by one of the most eminent scientific men of America.

And so to a majestic peroration:

On the whole we prefer the author of " What is a Joule ?" to such a man as this, because we believe he does less harm, and is less likely to mislead " able editors."

There is one grain of comfort even in the imbecilities and inanities of would-be humorous writers in news-papers, that at least they have woke up to the idea that a scientific discovery is worth laughing at. This is a step gained. Twenty years ago, even ten years ago, the name of even so distinguished a scientist as Dr. Joule would have been utterly unknown to the herd of newspaper writers. We must be thankful for even this much ; and look hopefully forward to the good day coming when Science will take her place by the side of her sisters, Art and Literature, as equally deserving of popular culture.

Vol 5, 457 1872

Samuel Morse died a disappointed man, for his love was painting and his work on the electric telegraph was undertaken in the hope that it would give him the freedom to pursue his vocation. "Painting", he wrote, "has been a smiling mistress to many, but she has been a cruel jilt to me. I did not abandon her; she abandoned me." His efforts led him into conflict with a greater man, Joseph Henry.

PROFESSOR S. F. B. MORSE

INTELLIGENCE has already been received in this country of the death of Samuel Finley Breese Morse, the eminent electrician, who died at New York on the 2nd inst. at the age of eighty-one. Prof. Morse was the son of the Rev. Jedediah Morse, well known as a geographer, and was born at Charlestown, Massachusetts, on the 27th of April, 1791. He was educated at Yale College, but, having determined to become a painter, he came to England in 1811, formed a friendship with Leslie, and in 1813 exhibited at the Royal Academy a colossal picture of " The Dying Hercules." He returned to America, and for a few years followed the profession of a portrait painter. In 1829 he again visited England, and on his return voyage was accompanied by Prof. Jackson, the eminent American chemist and geologist, through whose influence he turned his attention to the conduction of electricity through metallic wire, a subject in which the chemical tastes displayed by him while at College gave him additional interest, and to which he now devoted the whole powers of his mind.

Between 1835 and 1837 Prof. Morse invented several machines which more or less foreshadowed the electric telegraph ; and obtained from Congress a vote of 30,000 dollars, with which to make an experimental essay be-tween Washington and Baltimore. The first electric telegraph completed in the United States was the line between these cities, which was finished in 1844. Since that time the Recording Electric Telegraph of Morse has been adopted over the whole country, and at the time of his death there were not less than twenty thousand miles of electric wires, stretching over the States between the Atlantic and the Pacific Ocean.

Prof. Morse received during his life recognition of his services to science from a large number of foreign Govern-ments and scientific societies, not the least remarkable being the one inspired by the late Emperor of the French. At his suggestion delegates from France, Russia, Sweden, Belgium, Holland, Austria, Sardinia, Tuscany, the Holy See, and Turkey, met at Paris, and voted an award of 400,000 frs. to Prof. Morse as a testimonial of appreciation of his services.

A record of Prof. Morse's scientific career would not, however, be complete, without referring to a controversy which some years ago occupied the attention of the scientific world in the United States, in which he was engaged with Prof. Henry, now President of the Smithsonian Institution at Washington. So much personal matter was introduced into the dispute that a special committee of the Board of Regents of the Smithsonian Institution was appointed to investigate the matter, the report of which now lies before us. The result of this investigation is summed up as follows :—

" We have shown that Mr. Morse himself has acknow-ledged the value of the discoveries of Prof. Henry to his electric telegraph ; that his associate and scientific assis-tant, Dr. Gale, has distinctly affirmed that these dis-coveries were applied to his telegraph, and that previous to such application it was impossible for Mr. Morse to operate his instrument at a distance ; that Prof. Henry's experiments were witnessed by Prof. Hall and others in 1832, and that these experiments showed the possibility of transmitting to a distance a force capable of producing mechanical effects adequate to making telegraphic signals ; that Mr. Henry's deposition of 1849 . . . . is strictly correct in all the historical details, and that, so far as it relates to Mr. Henry's own claim as a discoverer, is within what he might have claimed with entire justice ; that he gave the deposition reluctantly, and in no spirit of hostility to Mr. Morse ; that on that and other occasions he fully admitted the merit of Mr. Morse as an inventor ; and that Mr. Morse's patent was extended through the influence of the favourable opinion expressed by Prof. Henry."

The conclusion therefore which must be arrived at, and it is one of no small importance in the history of electrical and telegraphic science, is that to Prof. Henry, and not to Prof. Morse, is unquestionably due the honour of the discovery of a principle which proves the practicability of exciting magnetism through a long coil, or at a distance, either to deflect a needle or to magnetise soft iron.

Prof. Morse's services to science as a successful applier of this principle in its practical details are so unquestion-able, that we feel we are but doing a duty in setting this question right on this side the Atlantic.

Vol 5, 509 1872

A treatise on the cause of cholera emanated from the Russian Imperial Academy of Sciences. The author's name is not revealed, but Mr B. G. Jenkins, "of the Inner Temple", who read the paper to the Historical Society, also communicated it to Nature. *Here is the denouement of the story.*

You are all probably aware that the great astronomer Schwabe discovered that the sun-spots have what is called a ten-year period ; that is, there is a minimum of spots every ten years. It was also discovered that the diurnal variation in the amount of declination of the magnetic needle has a ten-year period. The same was proved in regard to earth currents, and also auroræ. The maxima and minima of the four were found to be contemporaneous. This was a great result ; but Professor Wolf, on tabulating all the sun-spots from the year 1611, discovered that the period was not ten years, but 11·11 years. This period is now the accepted one for the sun-spots, and it has been established for the magnetic declination, and by Wolf for the auroræ. Now, it is a curious fact that the last year of every century, as 1800, has a minimum of sun-spots, so that the minima are 1800, 1811·11, 1822·22, 1833·33, &c. The maxima do not lie midway between the minima, but anticipate it by falling on the year 4·77 after a minimum ; for example, 1800 was a minimum year, then 1804·77 was a maximum year. Now, cholera epidemics have, I believe, a period equal to a period and a half of sun-spots. Reckoning then from 1800, we get a period and a half the date 1816·66, which was shortly before the great Indian outbreak ; another period and a half gives 1833·33, a year in which there was a maximum of cholera ; another, 1849·99, that is, 1850, a year having a maximum of cholera ; another, 1866·66, a year having a maximum of cholera ; another, 1883·33, as the year in which there will be a cholera maximum. It follows from what has been already said that 1783·33 would be a year in which cholera was at a maximum. Now it is a fact that in April 1783 there was a great outbreak of the disease at Hurdwar.
" I would call attention to the parallelism of increase and decrease of these curves. I am not, however, prepared to say that sun-spots originate cholera ; for they may both be the effects of some other cause, which may indeed be

the action of the other planets upon the earth and upon the sun. If that be the case—and I see no reason why it should not—we may then have an explanation of the minor periods and of the large period of 56 years, which Wolf believes he has detected, and also of the minor periods observed in cholera-epidemics.
" My own opinion, derived from an investigation of the subject, is that each planet, in coming to and in going from perihelion—more especially about the time of the equinoxes—produces a violent action upon the sun, and has a violent sympathetic action produced within itself—internally manifested by earthquakes, and externally by auroral displays and volcanic eruptions, such as that of Vesuvius at the present moment ; in fact, just such an action as develops the tail of a comet when it is coming to and going from perihelion ; and when two or more planets happen to be coming to or going from perihelion at the same time, and are in, or nearly in, the same line with the sun—being of course nearly in the same plane—the combined violent action produces a maximum of sun-spots, and in connection with it a maximum of cholera on the earth. The number of deaths from cholera in any year—for example, the deaths in Calcutta during the six years 1865–70—increased as the earth passed from perihelion, especially after March 21, came to a minimum when it was in aphelion, and increased again when it passed to perihelion, and notably after equinoctial day ; thus affording a fair test of my theory."

Vol 6, 27 1872

James Clerk Maxwell, one of the most attractive figures in nineteenth-century science, was a man of abundant wit and a versifier who found his way into many collections of comic verse (most widely with "Rigid Body Sings"). From time to time he contributed anonymously to Nature, *as here.*

ELECTRIC VALENTINE

TELEGRAPH CLERK ♂ TO TELEGRAPH CLERK ♀

" THE tendrils of my soul are twined
    With thine, though many a mile apart ;
And thine in close-coiled circuits wind
    Around the magnet of my heart.

" Constant as Daniell, strong as Grove ;
    Seething through all its depths, like Smee ;
My heart pours forth its tide of love,
    And all its circuits close in thee.

" O tell me, when along the line
    From my full heart the message flows,
What currents are induced in thine ?
    One click from thee will end my woes."

Through many an Ohm the Weber flew,
    And clicked this answer back to me—
" *I am thy Farad, staunch and true,*
    Charged to a Volt with love for thee."

$$\frac{dp}{dt}$$

[NOTE BY THE EDITOR—

Ohm   = Standard of resistance.
Weber = Electric current.
Volt = Electromotive force.
Farad = Capacity (of a condenser).

Velocity of Puck, $\dfrac{\text{Once round the Earth}}{\text{40 minutes.}}$

  „   of Ohm, $\dfrac{\text{Quadrant of meridian of Paris}}{\text{1 second.}}$

∴ 1 Ohm = 600 Pucks.]

Vol 6, 83 1872

Here is shown a device for measuring musical intervals. The vibrations are recorded by a tuning fork on a smoked drum.

A wire five, six, eight, ten, &c., metres long, suspended by narrow strips of caoutchouc, is soldered at one end to a small plate of brass, L, placed between the sounding-board of a stringed instrument and the foot of the bridge, the other end being slightly clasped to a heavy stand S. Near the fixed point a small piece of tinsel (*c*) is soldered on, and to this is attached a feather (*b*), by means of a little soft wax (by this arrangement a greater amplitude of vibration is attained than if the feather were directly attached to the wire). The musician stands in such a position that the wire may not impede the movements of his bow, and plays fragments of simple melodies in slow time (each note lasting at least a second). The vibrations of the strings are transmitted to the bridge, the metal plate, the wire, and, lastly, to the feather, which vibrates synchronously. It only remains to trace these vibrations.

VOL 6, 86 1872

The pocket spectroscope has been a travelling companion to many physicists, up to our own time, when R. S. Mulliken described how he lay in bed in a hotel room in Chicago and observed through the window with his pocket spectroscope the spark spectrum generated by the trams which were keeping him awake. James Prescott Joule in his native Manchester made the observations below.

Spectrum of Lightning

I HAD a good view of the spectra of lightning during the storm of yesterday. Frequently there was only one bright line visible, this being coincident with the nitrogen line. At other times there were several bright lines, sometimes with, and at other times without, the nitrogen line. Several flashes showed a continuous spectrum without visible lines. My instrument was a small direct-vision spectroscope, but sufficiently powerful to divide the sodium line.   J. P. JOULE

Broughton, Manchester, June 19

VOL 6, 161 1872

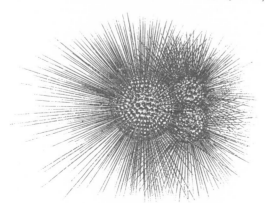

The Scottish physicist and bigot P. G. Tait was a master of vituperation. Here he amuses himself at the expense of the hapless Professor Zöllner. Tait also, it should be remarked, had drawn ridicule on himself by publishing a violently chauvinistic history of thermodynamics, which exalted Joule and Lord Kelvin while derogating the Germans, namely Rudolf Clausius and Julius Robert von Mayer.

SENSATION AND SCIENCE

THERE lies before us, as we write, a work of exceptionally high merit as a mere literary composition, entitled, *Ueber die Natur der Cometen, Beiträge zur Geschichte und Theorie der Erkenntniss.* Von J. C. F. Zöllner, Professor an der Universität, Leipzig. The title does all it can to indicate the sensational character of the work, which deals not alone with the nature of comets, the inferiority of British to German physicists, and the grave offence of which a German is guilty when he sees anything to admire except at home ; but also with the errors of Thomas Buckle, the relations of Science to Labour and Manufacture, and the analogies of development in Languages and in Religions !

It is impossible for us in a brief article to give the reader even a general glance at the numberless sources of amusement which the work affords. We will, therefore, confine our detailed remarks to a few of the parts which have most interested us, merely premising that we cannot pretend to give anything resembling a complete analysis of the contents of this astounding volume.

There follows some extensive quotations in the original German and then this:

We now come to the richest part of the volume, the preface and introduction, written (as we are told) later than the rest, and therefore when the author had managed thoroughly to divest himself of all the usual amenities, as well as of regard for at least the scientific character of certain living philosophers.

From the introduction we paraphrase as follows (the passage follows some fierce remarks about Dr. Tyndall) :—

" I can assert that, when I read the addresses of Sir W. Thomson and Prof. Tait to the British Association, and when on my return to Leipzig I found on my table among the scientific novelties the German edition of their ' Natural Philosophy,' edited by Helmholtz and Wertheim (including particularly section 385), then, indeed, the appearance of my work seemed to be a *Naturprocess*; something necessary in the chain of scientific development, of which even I myself scarce knew how it had arisen, and what was my share in it. In fact, the desire to bring to light in this book what is more or less struggling to appear in German science, what is bringing out a hollow sound now from one string, anon from another ; this desire, I say, has been with me to the latest scratch of my pen. I therefore doubt not that, simultaneously with mine, other heads have been working at the same problem, and perhaps in unconscious coincidence have arrived at the same solution. May then such facts ever more forcibly impress us with the conviction that the claims of personal services belong much more to the Age and to the Race than to the individual, and that no ever so clear conscious selection of means can be compared with that wonderful harmony with which Nature seeks to farther, and at the same time more surely to reach, her to us unknown ends."

To understand the bearing of the above passage, which is simply a literal assertion of

Deutschland, Deutschland über Alles,
Ueber Alles in der Welt,

the reader must refer to the preface, where he will find (along with much metaphysics) a war-dance over the mangled scientific reputation of Sir W. Thomson. The celebrated "moss-grown fragments from the ruins of another world" was a joke taken in earnest by many even in this country; so we can hardly blame Prof. Zöllner for falling into the trap; but why "bewachsene" instead of "bemooste" in translating the passage for thy country-men, O Zöllner?

And to conclude:

It would next be our task to show how heartily Helmholtz is pitched into for having sanctioned by his name the German translation of the work in question, and for his worthy recognition of Sir W. Thomson's scientific discoveries; but enough—Deutschland über Alles, and down with every Deutscher who sees aught to admire or to respect beyond the limits of Germany!

VOL 6, 177 1872

Henry Morton Stanley's journey across Africa in search of David Livingstone was one of the great newspaper scoops of the century. It was closely followed by NATURE.

DR. LIVINGSTONE'S DISCOVERIES

FROM Mr. Stanley's despatches to the *New York Herald*, which, by the courtesy of the English representative of that paper, have appeared in the *Times*, we gather some important and definite information as to the exact nature of Livingstone's discoveries; and more than this, we have a full explanation of the circumstances which kept our great traveller so long out of the reach of civilisation, and of the work he still hopes to accomplish

Mr. Stanley's account of his meeting with Livingstone is a touching one. After many delays, on the 3rd of November, 1871, he came in sight of the outlying houses of Ujiji, and, anxious to enter the African town with as much *éclat* as possible, he disposed his little band in such a manner as to form a somewhat imposing procession. At the head was borne the American flag; next came the armed escort, who were directed to discharge their fire-arms with as much rapidity as possible; following these were the baggage men, the horses, and asses; and in the rear of all came Mr. Stanley himself. The din of the firing aroused the inhabitants of Ujiji to the fact that strangers were approaching, and they flocked out in great crowds, filling the air with deafening shouts, and beating violently on their rude musical instruments.

As the procession entered the town Mr. Stanley observed a group of Arabs on the right, in the centre of whom was a pale-looking, grey-bearded, white man, whose fair skin contrasted with the sunburnt visages of those by whom he was surrounded. Passing from the rear of the procession to the front, the American traveller noticed the white man was clad in a red woollen jacket, and wore upon his head a naval cap with a faded gilt band round it. In an instant he recognised the European as none other than Dr. Livingstone himself; and he was about to rush forward and embrace him, when the thought occurred he

was in the presence of Arabs, who, being accustomed to conceal their feelings, were very likely to found their estimate of a man upon the manner in which he conceals his own. A dignified Arab chieftain, moreover, stood by, and this confirmed Mr. Stanley in his resolution to show no symptoms of rejoicing or excitement. Slowly advancing towards the great traveller, he bowed and said, " Dr. Livingstone, I presume?" to which address the latter, who was fully equal to the occasion, simply smiled and replied "Yes." It was not till some hours afterwards, when alone together, seated on a goat skin, that the two white men exchanged those congratulations which both were eager to express, and recounted their respective difficulties and adventures.

VOL 6, 184 1872

Towards the countries of the East, a rather patronising tone was commonly adopted.

SCIENCE IN JAPAN

PROF. W. E. GRIFFIS writes us a very encouraging letter from Fukuwi, Japan, where he is giving practical instruction in a chemical laboratory established a year ago. Sixty students attend his daily lectures on chemistry and physics, properly illustrated by experiments, and twelve students do actually practise in the chemical laboratory. What he says of Japan is equally true here in the United States, only that the rubbish of astrology and Chinese philosophy, which prevent rapid progress there, are here represented by notions not less common nor less obstinate. He says :—"In teaching physical science in Japan, one has need to begin at the lowest foundation, to demonstrate everything, and to clear away much rubbish of astrology, Chinese notions of philosophy, falsely so called, &c.; yet the students are fairly intelligent, and promise hopefully to fill, in some measure, the greatest educational need of the country—good teachers."

The following will also merit attention :—"It may please you to know that Japan, just entering upon her course of modern civilisation, has begun by not only assigning a foremost place to physical science in her schools, but has already established several laboratories, in which students receive practical instruction from German and American professors. The chief laboratory in Osaka is presided over by a German professor, having nearly one hundred students. Another laboratory, it is expected, will be established in Yeddo. There is one in the province of Kaga, in charge of a German professor; another, also under a German, is at Shidzoka, in the province of Suruga. The laboratory in Fukuwi, province of Echeyen, has been established nearly a year." This is the laboratory of Prof. Griffis, above spoken of.

It gives us, indeed, great pleasure to record these significant evidences of progress in the far-off Japan. These facts, as well as many others, show that at length commerce, the arts, and physical science, have commenced their missionary career in Japan, and will soon introduce the blessings of civilisation in that great country leaving the Japanese and Chinese gods to take care of themselves, if they can.

VOL 6, 352 1872

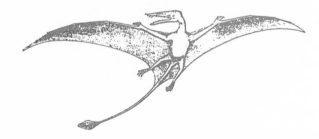

Michael Faraday was not long dead when NATURE *was launched. He cast a long shadow and there were many tributes to and reminiscences of the great man. This anecdote is taken from one such encomium. Mr Abbott, an old friend of Faraday's, was refused access to the laboratory when he called at the Royal Institution and could get no message taken in.*

It was near one o'clock, Anderson (Faraday's faithful assistant) would soon be going to his dinner, when probably he might catch sight of Mr. Faraday coming upstairs. Mr. Abbott waited; punctually at one Anderson emerged from the laboratory, Mr. Faraday followed, and, recognising his old friend at once, begged him to join them at dinner. "For," added Faraday, "I am a Goth you know, and always dine in the middle of the day." At dinner Faraday told Mr. Abbott a characteristic story about Anderson; how one morning during his glass experiments he found his assistant had been stoking the furnace all night long. Faraday had told him to keep the fire up, and omitting to release him in the evening, Anderson, with his soldier's excellent experience, stuck to his post till he received the next orders from his master. The fact that this simple obedience was all the assistance Faraday ever had increases the astonishment with which one regards the extent of his labours. The secret of the massiveness of Faraday's work was no doubt that he felt he had *one* aim before him, and therefore he rigidly kept from himself everything that would fritter away his time; political and commercial matters were passed by; he had his warfare to accomplish, and "no man that warreth entangleth himself with the affairs of this life."

* * *

With all this stern reality there was also a fine poetic fancy in Faraday. To him the Universe was no machine. His was "a face-to-face, heart-to-heart, inspection of things," and this, as Carlyle says, is "the first characteristic of all good thought in all times." Scientific phraseology never hid from him the grandeur and mystery that at bottom lies in everything. A thunder-storm was to him no mere affair of positive and negative electricity, no mere discharge of electric potential, but something infinitely beyond all this—" a window through which he looked into Infinitude itself." Dr. Gladstone tells us "he

would stand gazing at the lightning, a stranger to fear, with his mind full of lofty thoughts, or perhaps of high communings. Sometimes, too, if the storm were at a little distance, he would summon a cab, and in spite of the pelting rain, drive to the scene of awful beauty." A friend thus met him once at Eastbourne, "in the thick of a tremendous storm, rubbing his hands with delight because he had been fortunate enough to see the lightning strike the church tower."

We are told that a new fact "seemed to charge him with an energy that gleamed through his eyes and quivered through his limbs." The writer remembers an illustration of this, when Dr. Tyndall brought Mr. Faraday into the laboratory to look at his new discovery of calorescence. As Faraday saw for the first time a piece of cold, black platinum raised to a dazzling brightness when held in the focus of dark rays, a point undistinguishable from the air around, he looked on attentively, putting on his spectacles to observe more carefully, then ascertained the conditions of the experiment, and repeated it for himself; and now quite satisfied he turned with emotion to Dr. Tyndall and almost hugged him with pleasure. And so on another occasion, when Prof. Plücker was showing in the laboratory some lovely experiments with vacuum tubes, Faraday literally danced with delight round the electric discharge, exclaiming, as he gazed at the moving arches of light, "Oh! to live always in it!"

VOL 6, 412 1872

It is surprising that modern arithmetical notation had not yet come into use in 1872.

Millions of Millions

WHY do not Messrs. Ranyard and Co. adopt the late Benjamin Gompertz's most convenient notation of prefixing a circle to the first significant figure, or suffixing a circle to the last significant figure having therein a digit for the number of zeros employed?

Thus: ·⑤718 is ·00000718

And   718⑥ is 718000000

S. M. DRACH

74, Offord Road, N., Sept. 17

VOL 6, 434 1872

This remarkable invention by Henry Bessemer (founder of the Iron and Steel Institute and chiefly now remembered for the Bessemer converter, which turns pig iron into steel) seems to have vanished into the mists of history.

MR. BESSEMER'S SALOON STEAMER FOR THE CHANNEL PASSAGE

THE prevention of sea-sickness by means of a swinging cabin has nothing novel about it, but the originality and inventive merit in the suspended saloon devised by Mr. Bessemer, and now about to be actually constructed in a ship specially designed for it by Mr. Reed, the late Chief Constructor of the Navy, are of the highest order. The association of those names is in itself a sufficient guarantee that the idea will be carried into execution with complete security as respects the safety of the passengers and the seaworthiness of the ship, and a full knowledge of the scientific principles involved.

Persons suffering from sea-sickness complain not only of giddiness arising from themselves and everything about them being continually in motion, but also in particular of a qualm which comes over them every time the ship, or the part of it on which they are standing, is descending, sinking, as it were, from under their feet. An approach to this qualm is commonly felt in a garden swing during the descent, and also in jumping from considerable heights. There can be very little doubt that this is due to the fact that the intestines are then wholly or partially relieved from their own weight, and therefore exercise an unusual pressure against the stomach, liver, and diaphragm. This pressure produces the qualm, and its rapid and frequent alternations cause sufficient irritation to produce in most people sea-sickness, and in some persons more serious effects. Physiologists are by no means agreed as to how much of sea-sickness is due to this cause, and how much to the reaction upon the stomach of the brain-disturbance, due in part, perhaps, to the actual motion of the head, but largely to the optical effect of the motion. It is pretty certain that all these causes contribute to produce the effect of sea-sickness. It is beyond doubt that they all aggravate it.

Mere swinging cots or small cabins go but a very little way to remedy any of these evils. Even if suspended in two directions, like a compass or barometer upon jimbals, the translatory motion, whether up or down, or to and fro, remains wholly unaltered, and even the oscillatory motion is not got rid of, but only altered in character, being reduced to a minimum at a point near the middle of the ship. The distressing effect upon the eye of the relative motion of surrounding objects also remains. These effects will not be wholly eliminated by Mr. Bessemer's invention ; but some of them will be very much reduced, and it remains to be seen whether the reduction is sufficient to get rid of the sickness.

The design, as settled by Mr. Bessemer and Mr. Reed, includes the construction of large steam vessels of light draught, 350 feet long, 40 feet beam, drawing 7 feet of water, and worked by two pairs of paddle-wheels. In the middle of each of these is provided a well, or hole, for the reception of a saloon 70 feet long, 20 feet wide, and 20 feet high, constructed so as to form a box girder in itself, and suspended at its extremities upon a pair of trunnions, on which it can turn, so that it may be kept steady as the vessel rolls from side to side. The saloon is not allowed to swing quite freely, but its motion is controlled by hy-draulic machinery, acting either upon a rocking arm or a tangent bar (it does not appear as yet which has been selected), which enables a man to regulate its position at his discretion. This man sits opposite a spirit level, and, by merely turning a handle which opens certain valves, can keep the bubble of the spirit level at zero, so as to keep the saloon virtually upright at all times. The chief novelty of the invention consists in two points—the great size of the swinging cabin or saloon, and the controlling of its motion by hand, instead of trusting to self-adjustment. Both these are very important improvements on the simple swinging cabin.

VOL 7, 41 1872

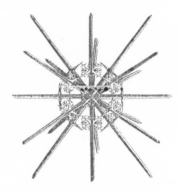

Sergeant-Major Reid seems to have possessed greater scientific curiosity and knowledge than one commonly associates with the profession of arms. This correspondence ran to several exchanges.

Flight of Projectiles— A Query

I SHALL feel thankful to any of your numerous mathematical correspondents who will kindly favour me with a simple formula for determining the deflection in the flight of a leaden cylindro-conoidal projectile—the time of flight of which is known—caused by wind of known force acting at different angles to the vertical plane of the trajectory, with an application of the formula to the following cases. Any other cause of deviation, such as that due to rotation, &c., may be neglected :—

Suppose the bullet to be 1·27″ long, and its diameter ·447″, weight 480 grs. and the wind to be of force 4, approximate pressure 4 lbs. per square inch, what is the deviation?

1. When the wind acts at right angles to the trajectory?
2. When it acts at any angle less than a right angle, say 45°?

ROBERT REID

School of Musketry, Hythe, Feb. 10

VOL 7, 341 1872

Here is the beginning of a leader by Charles Darwin, in which he airs his interest in animal instincts.

ORIGIN OF CERTAIN INSTINCTS

THE writer of the interesting article in NATURE of March 20 doubts whether my belief " that many of the most wonderful instincts have been acquired, independently of habit, through the preservation of useful variations of pre existing instincts," means more than " that in a great many instances we cannot conceive how the instincts originated." This in one sense is perfectly true, but what I wished to bring prominently forward was simply that in certain cases instincts had not been acquired through the experience of their utility, with continued practice during successive generations. I had in my mind the case of neuter insects, which never leave off-spring to inherit the teachings of experience, and which are themselves the offspring of parents which possess quite different instincts. The Hive-bee is the best known instance, as neither the queen nor the drones construct cells, secrete wax, collect honey, &c. If this had been the sole case, it might have been maintained that the queens, like the fertile females of humble-bees, had in former ages worked like the present neuters, and had thus gradually acquired these instincts ; and that they had ever afterwards transmitted them to their sterile off-spring, though they themselves no longer practised such instincts. But there are several species of Hive-bees (*Apis*) of which the sterile workers have somewhat different habits and instincts, as shown by their combs. There are also many species of ants, the fertile females of which are believed not themselves to work, but to be served by the neuters, which capture and drag them to their nests ; and the instincts of the neuters in the different species of the same genus are often different. All who believe in the principle of evolution will admit that with social insects the closely allied species of the same genus are descended from a single parent-form ; and yet the sterile workers of the several species have somehow acquired different instincts. This case appeared to me so remarkable that I discussed it at some length in my " Origin of Species ;" but I do not expect that anyone who has less faith in natural selection than I have, will admit the explanation there given. Although he may explain in some other way, or leave unexplained, the development of the wondrous instincts possessed by the various sterile workers, he will, I think, be compelled to admit that they cannot have been acquired by the experience of one generation having been transmitted to a succeeding one. I should indeed be glad if anyone could show that there was some fallacy in this reasoning. It may be added that the possession of highly complex instincts, though not derived through conscious experience, does not at all preclude insects bringing into play their individual sagacity in modifying their work under new or peculiar circumstances ; but such sagacity, as far as inheritance is concerned, as well as their instincts, can be modified or injured only by advantage being taken of variation in the minute brain of their parents, probably of their mothers.

And he concludes with an anecdote:

I am tempted to add one other case, but here again I am forced to quote from memory, as I have not my books at hand. Audubon kept a pinioned wild goose in confinement, and when the period of migration arrived, it became extremely restless, like all other migratory birds under similar circumstances ; and at last it escaped. The poor creature then immediately began its long journey on foot, but its sense of direction seemed to have been perverted, for instead of travelling due southward, it proceeded in exactly the wrong direction, due northward.

CHARLES DARWIN

VOL 7, 417 1873

There were many occasions on which a public injustice to one of their brethren roused the scientists of Victorian Britain to action. Joseph Hooker, director of the Royal Botanic Gardens at Kew, seems to have been shabbily treated by the bureaucrat at the Ministry of Works, to

whom he was answerable. The circumstances, with extracts from the correspondence pertaining to the affair, were meticulously chronicled in NATURE *by the signatories — a galaxy of scientific luminaries. The article concludes like this:*

It but rarely falls in either with our duties or our desires to meddle in public questions ; and not until we found Dr. Hooker maimed as regards his scientific usefulness—not until we saw the noble establishment of which he has hitherto been the living head in peril of losing services which it would be absolutely impossible to replace ; not, indeed, until we had observed a hesitation upon your part which we believe could only arise from lack of information—did the thought of interference in this controversy occur to us. Knowing how difficult it must be for one engrossed in the duties of your high position to learn the real merits of a conflict like that originated by the First Commissioner of Works, we venture to hope that you will not look with disfavour on an attempt to place a clear and succinct statement of the case before you.

That statement invites you respectfully to decide whether Kew Gardens are or are not to lose the supervision of a man of whose scientific labours any nation might be proud ; in whom natural capacity for the post he occupies has been developed by a culture unexampled in variety and extent ; a man honoured for his integrity, beloved for his courtesy and kindliness of heart ; and who has spent in the public service not only a stainless but an illustrious life. The resignation of Dr. Hooker under the circumstances here set forth would, we declare, be a calamity to English science and a scandal to the English Government. With the power to avert this in your hands, we appeal to your justice to do so. The difficulty of removing the directorship of Kew from the Department of Works cannot surely be insuperable ; or if it be, it must be possible to give such a position to the Director, and such definition to his duties, as shall in future shield him from the exercise of authority which has been so wantonly abused.

CHARLES LYELL
CHARLES DARWIN
GEORGE BENTHAM, Pres. Linn. Society
HENRY HOLLAND, Pres. Royal Institution
GEORGE BURROWS, Pres. Roy. Coll. of Physicians
GEORGE BUSK, Pres. Roy. Coll. of Surgeons
H. C. RAWLINSON, Pres. Roy. Geogr. Society
JAMES PAGET
WILLIAM SPOTTISWOODE
T. H. HUXLEY
JOHN TYNDALL

VOL 6, 215 1872

Paul Broca, the French surgeon and anatomist of the brain (and another savant who bore himself heroically during the siege), became increasingly engrossed in later life with anthropology. This is a sample of his style, taken from a long article.

The alveolar process of the old man is oblique, but the upper part of the face is vertical, and the facial angle is very open. The forehead is wide, by no means receding, but describing a fine curve ; the amplitude of the frontal tuberosities denotes a large development of the anterior cerebral lobes, which are the seat of the most noble intellectual faculties. If the Cromagnon Troglodytes are still savages, it is because their surrounding conditions have not permitted them to emerge from barbarism ; but they are not doomed to a perpetual savage state. The development and conformation of their brain testify to their capability for improvement. When the favourable opportunity arrives, they will be able to progress towards civilisation. These rough hunters of the mammoth, the lion, and the bear, are just what ought to be the ancestors of the artists of the Madeleine.

VOL 7, 426 1873

There is no mistaking where the political sympathies of
NATURE and its correspondents lay in regard to the
Franco-Prussian War. Accounts such as this of science
flourishing in adversity repeat themselves in later
conflicts.

SCIENCE IN PARIS DURING THE SIEGE

IN a somewhat striking passage in the *De Augmentis
Scientiarum* Francis Bacon contrasts the endurance
of monuments of learning with that of those raised by
the hand of man. The verses of Homer, he reminds us,
have endured for more than twenty centuries, during which
time numberless palaces, temples, and cities have disappeared from the face of the earth. Some such reflection
as this may have induced the members of the *Académie
des Sciences* of Paris to continue their weekly meetings
with perfect regularity during the bombardment. While
everything else was turned upside down, while a dynasty
was passing away, while sons and brothers were perishing around them, an enemy at their gates, want within
their walls, and missiles of war threatening themselves
and their household gods, these men continued their usual
studies. We are reminded of Archimedes at the siege of
Syracuse, save that there we have but one man, while here
it is a large body of the intellectual flower of the country.
Some of the more active of the members are men who
have attained that philosophic calm, which not even the
terrors of war can dispel, nothing diverts them from the
even tenor of their way—

> Si fractus illabatur orbis,
> Impavidum ferient ruinæ.

We frequently meet with the names of Dumas, Elie de
Beaumont, and Chevreul in the *Comptes Rendus* published
during the siege. The youngest of these men was born in
the last year of the last century; they have seen every
phase of Parisian life, are men of infinite experience and
learning, the very soul of the Academy; they have held
office under various Governments, have seen more than
one revolution; barricades and street fighting, and the
Parthian cap are no novelty to them; but with all
their experience they had never beheld a bombardment
of Paris; yet, all honour to them, they did not abate one
jot of their Academic work. Perhaps the members may
have felt it a relief to have to deal with immutable and
indestructible facts, while everything around them was so
mutable and perishing. Perhaps they remembered a
saying of one of their countrymen :—" L'homme n'a pas de
self-critérium. L'indestructibilité du fait est le critérium
unique, infaillible, absolu, multiple, un, présent dans tous
les ordres des connaissances."*

The papers read before the Academy during the period of
the siege relate for the most part to matters connected
with war and to the food resources of a besieged city.
M. Payen writes on hippophagy, M. Frémy on the use of
osseine as food, M. Riche on the use of black puddings of
ox-blood. In the *Revue des Cours Scientifiques* we also find
some important papers by M. Bouchardat on the food supply
of Paris, and a paper by the same author on the sanitary
condition of the city during the siege, and during the same
months of the preceding year. M. Berthelot has contributed some important papers on the force of various
explosive substances, both solid, liquid, and gaseous.
These papers are well worthy the attention of our war
authorities. There is also a paper on dynamite, by M.
Champion, and on the ignition of gunpowder at a distance by means of electricity. The subject of balloons
and ballooning, of course, engages a good deal of discussion. M. Marey contributes several important papers
on the motions of birds during flight, accompanied by
graphic representations of them, registered somewhat
after the manner of the vibrations of a tuning fork, and
shown by sinuous lines. The diagram representing the
vertical oscillations of a wild duck during flight is very

striking. Beyond these papers there is nothing of much
importance.

Here, for example, are the principal papers of one number (December 5th) taken at random :—

M. Milne-Edwards discusses the nutritive value of organic substances contained in bones, and the proper
rations for sustaining the human body in a perfect state
of health; M. Chevreul makes observations upon M.
Frémy's paper on the use of osseine as food; M. Gazeau
details various experiments on the nutritive properties of
cocoa leaves; M. Montier treats of the specific heat of
gases under constant volume; M. Riche of the preparation of osseine and gelatine; and M. Castelhaz of the
refining of crude tallow.

The future historian of science will wonder when he
reads in the *Comptes Rendus* for January 9, 1871 :—

M. Chevreul donne lecture à l'Académie de la déclaration
suivante :

"Le jardin des plantes médicinales, fondé à Paris par édit du
Roi Louis XIII. la date du mois de janvier, 1626.

"Devenu le Muséum d'Histoire naturelle par décret de la
Convention du 10 de juin 1793.

"Fut bombardé,

"Sous le règne de Guillaume Ier roi de Prusse, Comte de
Bismark chancelier,

"Par l'armée Prussienne, dans la nuit du 8 au 9 de janvier,
1871.

"Jusque-là, il avait été respecté de tous les partis et de tous
les pouvoirs nationaux et étrangers.

"E. CHEVREUL, *Directeur.*"

He will grieve when he reads " M. Le Président annonce
à l'Académie la douloureuse nouvelle, malheureusement
très-probable, de la mort du . . . ." occurring too
often in what should be only a record of the living
and of their work. So we grieve : and yet more when we
see the intellectual resources and energies of a great
country paralysed, and the whole current of its active
thought diverted no man knows whither ; while its schools
and colleges are empty, and many of those who should
be in them have been killed untimely to satisfy the
necessities of war.   G. F. RODWELL

VOL 3, 490 1871

*Lord Kelvin notwithstanding, inventors of perpetual
motion machines could still get a hearing (of a kind)
in NATURE.*

Perpetual Motion

PROBABLY your sense of justice will induce you to insert some
very brief remarks on your review of my article in the *Quarterly
Journal of Science*. The tone of the review is a penalty which
all who venture to impugn commonly accepted theories must be
prepared to submit to. Heresy in science meets with as little
mercy as heresy in theology. I confess that in one sense of the
word I am consciously a perpetual-motionist, but not in the sense
of believing that any merely mechanical contrivance can produce
perpetual motion. That there are forces in nature which can and
do produce it, is a matter of daily, yearly, and secular experience.
If I am a perpetual-motionist in this sense, I am in good company. You will find that Sir W. Thomson, in the *Philosophical
Magazine* for February 1854, described a machine by which a
steam-engine or water-wheel could produce thirty-five times the
heat commonly considered as equivalent to the force used ; or the
corresponding amount of cold. At that time, then, two years
after his paper read to the British Association (to which you refer
me), he certainly did not hold such an opinion with regard to the
mechanical equivalent of heat as to exclude the possibility of such
an engine.

The final judgment of the question I confidently leave to time
and facts. When any of the "grand founders of a rapidly progressive science" can spare time from their investigations to refute
my fallacies, I shall gladly retract them.   H. HIGHTON

VOL 3, 368 1871

chapter three
1873-1876

PLANS ARE AFOOT FOR A BALLOON CROSSING OF THE ATLANTIC. CHARLES DARWIN'S FOLLOWER GEORGE
J. ROMANES WEIGHS THE EVIDENCE THAT ANIMALS HAVE A SENSE OF HUMOUR. RAPHAEL MELDOLA
DISCUSSES THE CHEMISTRY OF CREMATION. ARE SCORPIONS SUICIDE-PRONE? *NATURE* GAZES WITH
WONDER UPON THE REMINGTON TYPEWRITING MACHINE.

1873-1876

Benjamin Disraeli became prime minister of Britain and saw to the passage of the Factories Act, which forbade the employment of children less than nine years old. The drainage system for London was completed. In France, the Third Republic was proclaimed.

Hyperbole on the part of scientists defending their calling was not unknown in the nineteenth century, as the prolegomenon of this leader suggests.

A VOICE FROM CAMBRIDGE

IT is known to all the world that science is all but dead in England. By science, of course, we mean that searching after new knowledge which is its own reward, a thing about as different as a thing can be from that other kind of science, which is now not only fashionable, but splendidly lucrative—that "science" which Mr. Gladstone and Mr. Lowe always appeal to with so much pride at the annual dinner of the Civil Engineers—and that other "science" prepared for Jury consumption and the like.

It is also known that science is perhaps deadest of all at our Universities. Let any one compare Cambridge, for instance, with any German university; nay, with even some provincial offshoots of the University in France. In the one case he will find a wealth of things that are not scientific, and not a laboratory to work in; in the other he will find science taking its proper place in the university teaching, and, in three cases out of four, men working in various properly appointed laboratories, which men are known by their works all over the world.

VOL 8, 21 1873

Here now is Henry Roscoe in his reminiscence of Justus von Liebig with an obiter dictum *to the same effect.*

It was not in Liebig's nature to spare either private persons or Governments when he thought that science would be advanced by plain speaking. In his two papers on "Der Zustand der Chemie in Oestreich" *(1838)*, and in "Preussen" *(1840)*, whilst he points out the shortcomings of both countries, bravely asserts, in the strongest terms, the dependence of national prosperity upon original research, a subject concerning which in England, *most people, thirty years later (to our shame be it said) are altogether in the dark !*

VOL 8, 27 1873

The plan for a balloon crossing of the Atlantic, more than a century before Richard Branson, had as its justification a test of the hypothesis of an aerial gulf stream. The article describes the hazards of the enterprise — mitigated by the inclusion in the passenger list of "six very powerful and experienced carrier pigeons", each bearing on the breast in indelible ink the outline of a balloon and on the wings the legend "Send news attached to the nearest newspaper", which would convey the message that the aeronauts had landed in the drink — and continues:

There can be no doubt that this daring expedition, whether it descends without mishap on the shores of Europe, or comes to grief in the middle of the Atlantic, will add something to our knowledge of the atmosphere; but many will no doubt think that all the knowledge that will be acquired by this sensational and hazardous method might be acquired by safer and more ordinary methods. We certainly, with all our heart, wish the enterprise complete success; but we think it · very pertinent to refer to some remarks on the project in *La Nature* by the experienced balloonist, M. G. Tissandier. After referring to the theory of the easterly current in the atmosphere, M. Tissandier says, "We leave to the

aëronaut all the responsibility of this hypothesis, which appears to us to be based upon vague conjectures; we should have a little more confidence in the resources which he expects to find above the Gulf Stream. This warm river, which traverses the extent of the Atlantic, should draw along with it a current of air, which the aërial navigator might take advantage of.

VOL 8, 365 1873

In the end, perhaps fortunately, nothing came of the plan.

IT seems that the projected balloon voyage from New York to Europe is not now likely to take place. An attempt was made to inflate the balloon on the 10th, but it failed, owing to a high wind. The attempt was renewed on the 12th, but a rent appeared and the operation was abandoned. Mr. Wise, the aëronaut, had foreseen this result, owing to the imperfect manner in which the balloon was constructed; and indeed from what has been stated, it would seem Science may be congratulated that an enterprise in which newspaper advertising had so much to do, has been thus liberated from the responsibility of having to answer for a much more serious disaster, which, we repeat, need not be risked at all so far as Science is concerned.

VOL 8, 436 1873

The intelligence of animals exercised amateur and indeed professional naturalists and was the subject of many letters to the editor. Here is a typical example.

Ingenuity in a Pigeon

THE following facts (having been witnessed by myself) may, perhaps, be considered worthy of insertion in your journal, as bearing on the subject of "Perception and Instinct in the Lower Animals," which has lately been brought into such prominent notice.

On the Richmond road (Surrey), at about a mile from the town, stands an old roadside inn, yclept "The Black Horse," owned by one R. Ketley. Attached to the house are a number of domestic pigeons of various breeds, chiefly "Pouters."

Having occasion to wait for my pony to be harnessed at this inn a few years since, my attention was directed by a gentleman (a resident of the neighbourhood) with whom I was acquainted, to the strange conduct of one of these birds.

A number of them were feeding on a few oats that had been accidentally let fall while fixing the nose-bag on a horse standing at bait. Having finished all the grain at hand, a large "Pouter" rose, and flapping its wings furiously, flew directly at the horse's eyes, causing that animal to toss his head, and in doing so, of course shake out more corn. I saw this several times repeated; in fact, whenever the supply on hand had been exhausted.

I leave it to your readers to consider the train of thought that must have passed through the pigeon's brain before it adopted the clever method above narrated, of stealing the horse's provender.

Was not this, indeed, something more than mere instinct?

RICHARD H. NAPIER

Upton Cottage, Bursledon, Southampton, Aug. 13

VOL 8, 324 1873

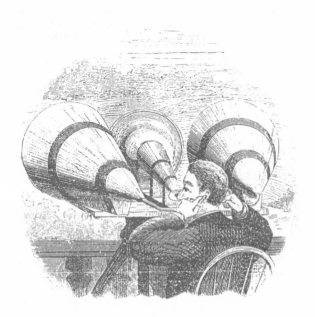

The history of science was undoubtedly regarded as a part of the study of science, and it featured regularly in the columns of Nature. *Sedley Taylor was a leading astronomer of the period.*

GALILEO'S WORK IN ACOUSTICS

IN looking through the "Dialoghi delle Nuove Scienze" of Galileo, I came unexpectedly on a passage * containing two remarkable discoveries in acoustics, which I should have confidently referrred to a much later age. For the sake of such of your readers as may share the same erroneous impression, I hope you will allow me to give, in NATURE, a short account of these results.

The first is a perfectly accurate explanation of the phenomenon called "resonance." Every pendulum has a fixed period of oscillation peculiar to itself. Even when the "bob" is of considerable weight it is possible to set it swinging through a large arc by merely blowing against it with the mouth, provided the successive puffs are properly timed with reference to the pendulum's period of vibration. In the same way a single ringer can, by regular pulling, throw the heaviest bell into oscillations of such extent as to be capable of lifting half-a-dozen men who should hang on to its rope, off the ground all together. When a string of a musical instrument is struck, its vibrations set the air in its vicinity trembling, and the tremors thus set up spread themselves out through space. If they fall on a second wire in unison with the first, and therefore prepared to execute its vibrations in the same period, the effects of the successive impulses are accumulated, and the wire's oscillations can be distinctly seen to go on dilating until they have attained an extent equal to those of the wire originally struck.

Anyone who looks into the chapter on resonance in the "Tonempfindungen" will see that the account of the phenomenon given by the greatest living acoustician is, in principle, identical with that of Galileo.

The second point to which I wish to draw attention is an experiment involving the earliest direct determination of a vibration-ratio for a known musical interval. Galileo relates that he was one day engaged in scraping a brass plate with an iron chisel, in order to remove some spots from it, and noticed that the passage of the chisel across the plate was sometimes accompanied by a shrill whistling sound. On looking closely at the plate, he found that the chisel had left on its surface a long row of indentations

parallel to each other and separated by exactly equal intervals. This occurred only when the sound was heard : if the chisel traversed the surface silently, not a trace of the markings remained. It was found that a rapid passage of the chisel gave rise to a more acute, a slower to a less acute, sound, and that, in the former case, the resulting indentations were closer together than they were in the latter. After repeated trials two sets of markings were obtained which corresponded to a pair of notes making an exact fifth with each other ; and, on counting the number of indentations contained in a given length of each series, it appeared that for 30 of the lower sound there were 45 of the higher, which numbers are in the exact proportion (2 : 3), which connects the lengths of two equally tense wires, giving that interval. Galileo, who had felt a tremor pass from the chisel to his hand at each experiment, inferred that what really determined a musical interval was the ratio of the numbers of vibrations performed in equal times by its constituent notes, and that that ratio was inversely as that of the lengths of the wires producing them. In order to bring out the crucial nature of his experiment, he goes on to remark, with extreme acuteness, that there was, prior to it, no reason for regarding the relations known to connect musical intervals with the *lengths* of wires as in any exclusive sense *representing* such intervals. With equal propriety might the ratio of the *tensions* under which two wires of equal lengths emitted sounds forming an interval be taken as its representative. In this case we should obtain the inverse square root of the ratio resulting from the former mode of comparison. Thus Galileo's experiment alone supplied decisive ground for concluding that the relations of *length* between similarly circumstanced wires, likewise governed those of *period* between corresponding aërial vibrations.

Prof. Tyndall, in referring to the above experiment, has described it as performed "by passing a knife over the edge of a piastre" (" Sound," 2nd ed., p. 51). This is an obvious mistake caused by incorrect translation. Galileo was scraping " una piastra d'ottone," *i.e.*, not " a piastre," but " a plate of brass." An excellent numismatist assures me that the *material* mentioned is alone decisive of the point, the piastre in Galileo's time being invariably made of *silver*.   SEDLEY TAYLOR

Sir Bartle Frere was a builder of Empire and General Sir Garnet Wolseley was W. S. Gilbert's "model of a modern Major-General".

IN the discussion which followed, Sir Bartle Frere's address, at the opening of the African Section of the Society of Arts, Mr. Hyde Clarke read a letter from Lieut. Maurice, Private Secretary to Sir Garnet Wolseley, dated " Head Quarters, Yancoomassie," from which we take the following extract ; it may prove of some interest to students of the Science of language :— " A somewhat curious piece of word-coining, which has fallen under our notice here, may interest you in connection with the broader aspects of the subject of which you write. The Ashantees having experience of our rockets only as they come to them in destructive form at the end of their journey, call them by the sound they make, 'Schou-schou,' or something of the kind. The Fantees, on the other hand, adopt bodily into their Language our own names for those things which they have not seen before. Thus to the Houssa or the Fantee, in speaking to one another, our rockets are named rockets, while their enemies call them schou-schou. It is possible that as war has not been in savage times an uncommon condition of mankind, analogous causes for different names having been adopted by different nations may have been not unfrequent in the past."

Mary Somerville was a venerated mathematician, author of
Celestial Mechanics, *who gave her name to Somerville College, the first women's college at Oxford. The Marquis de Laplace himself averred that she was the only woman ever to have read his works, which, moreover, were not understood by twenty men in France as well as she understood them. She died three years after the birth of* NATURE. *Her biography by her daughter received this long and admiring notice in* NATURE.

There was nothing exceptional in her bringing up, or her opportunities. In fact, no woman of her time and station could have had a more typical experience of life than she had. She was born nearly a century ago, in 1780, and spent her childhood and youth in Scotland, within an ordinary circle of the upper middle-class society of her age and country, and therefore very closely circumscribed by lines of defence against innovations and social changes of any kind. Her father, Captain Fairfax (a brave officer who commanded the *Repulse* during the war), received the news of her having taught herself the first six books of Euclid with the remark— " We must put a stop to this, or we shall have Mary in a strait-jacket one of these days. There was ' X,' who went raving mad about the longitude !" This gallant captain was, moreover, a genuine good Tory, who took decided views in regard to all questions involving a departure from established precedents, and when his young daughter ventured to express her admiration for the short-cut hair, which was then the badge of a Liberal in politics, he exclaimed, " By G—, when a man cuts off his queue the head should go with it." Her mother, who found all her intellectual cravings amply satisfied with the reading of her Bible, a volume of sermons, and a stray copy of a newspaper, fully concurred in her husband's views of the education suited to young women, and was at great pains to thwart her daughter's unladylike taste for pursuits regarded at the time as the exclusive privileges of men, and to keep her mind and hands closely fettered by the bonds of a household possessed of very limited pecuniary means.

* * *

... at the age of 17 the possession of a copy of " Bonnycastle's Algebra," procured for her by her uncle and future father-in-law, Dr. Sutherland—the only one of her relations who did not absolutely oppose her efforts to acquire knowledge—enabled her to solve the mystery of the X's and Y's ; and from that hour till the day of her death, mathematics, in one shape or other, may be said to have formed part of her daily existence. For more than half a century they were the staple occupation of her morning hours when the duties of her house and family had been disposed of ; at a very advanced age she began and mastered the study of Quaternions, and other forms of modern mathematics, and at 89 she " still retained facility in the calculus."

The restless activity of her intellect had indeed never slumbered. When she received her first lessons in painting and music, she had begun at once to try and trace out the scientific principles on which these arts are based, and never rested till she had gained some knowledge of the laws of perspective and of the theory of colour, and had learnt to tune her own instruments. In later years

she may be said to have been always in the van of discovery—not indeed as an originator but as the readiest and aptest of students—and from the time when Young showed her how he conducted the experiments by which he claimed to have discovered the undulatory theory of light, and Wollaston made her one of the very first witnesses of the seven dark lines crossing the solar spectrum, whose detection laid the basis of some of the most wonderful cosmical discoveries of this or any age, Mary Somerville, to the last day of her long life of nearly 92 years, followed with quick and appreciative understanding every step in the advance of modern research. Age could not quench the fire of her intellect, and even in her 92nd year, when the Blue Peter, as she quaintly remarks, had long been flying at her foremast, and she had soon to expect the signal for sailing, she could interest herself in the phenomena of volcanic eruption, speculate on their effects, and follow with lively sympathy the progress of scientific inquiry, and the issues of passing events.

In reading the personal recollections of this wonderful woman nothing strikes one more than the ordinary and even commonplace conditions under which her great intellect advanced to maturity. In her case the only exceptional features were her natural gifts and her perseverance in cultivating them ; and this is precisely the point that should not be lost sight of. Mary Somerville will always present a noble instance of what a woman has been capable of achieving, but it would be straining the argument too far to say that we are justified from her special case to draw general conclusions in regard to women's aptitude for the study of the higher forms of physical science.

VOL 9, 418 1874

Here is Charles Darwin again, with another observation on animal behaviour.

Flowers of the Primrose destroyed by Birds

FOR above twenty years I have observed every spring in my shrubberies and in the neighbouring woods, that a large number of the flowers of the primrose are cut off, and lie strewn on the ground close round the plants. So it is sometimes with the flowers of the cowslip and polyanthus, when they are borne on short stalks. This year the devastation has been greater than ever; and in a little wood not far from my house many hundred flowers have been destroyed, and some clumps have been completely denuded. For reasons presently to be given, I have no doubt that this is done by birds; and as I once saw some greenfinches flying away from some primroses, I suspect that this is the enemy. The object of the birds in thus cutting off the flowers long perplexed me. As we have little water hereabouts, I at one time thought that it was done in order to squeeze the juice out of the stalks; but I have since observed that they are as frequently cut during very rainy, as during dry weather. One of my sons then suggested that the object was to get the nectar of the flowers; and I have no doubt that this is the right explanation. On a hasty glance it appears as if the foot-stalk had been cut through; but on close inspection, it will invariably be found that the extreme base of the calyx and the young ovary are left attached to the foot-stalk. And if the cut-off ends of the flowers be examined, it will be seen that they do not fit the narrow cut-off ends of the calyx, which remains attached to the stalk. A piece of the calyx between one and two-tenths of an inch in length, has generally been cut clean away; and these little bits of the calyx can often be found on the ground; but sometimes they remain hanging by a few fibres to the upper part of the calyx of the detached flowers. Now no animal that I can think of, except a bird, could make two almost parallel clean cuts, transversely across the calyx of a flower. The part which is cut off contains within the narrow tube of the corolla the nectar; and the pressure of the bird's beak would force this out at both the cut-off ends. I have never heard of any bird in Europe feeding on nectar; though there are many that do so in the tropical parts of the New and Old Worlds, and which are believed to aid in the cross-fertilisation of the species. In such cases both the bird and the plant would profit. But with the primrose it is an unmitigated evil, and might well lead to its extermination; for in the wood above alluded to many hundred flowers have been destroyed this season, and cannot produce a single seed. My object in this communication to NATURE is to ask your correspondents in England and abroad to observe whether the primroses there suffer, and to state the result, whether negative or affirmative, adding whether primroses are abundant in each district. I cannot remember having formerly seen anything of the kind in the midland counties of England. If the habit of cutting off the flowers should prove, as seems probable, to be general, we must look at it as inherited or instinctive; for it is unlikely

that each bird should have discovered during its individual life-time the exact spot where the nectar lies concealed within the tube of the corolla, and should have learnt to bite off the flowers so skilfully that a minute portion of the calyx is always left attached to the foot-stalk. If, on the other hand, the evil is confined to this part of Kent, it will be a curious case of a new habit or instinct arising in this primrose-decked land.

Down, Beckenham, Kent, April 18   CH. DARWIN

VOL 9, 482 1874

A savage controversy between Herbert Spencer and P. G. Tait, Edward Frankland, the chemist, and several other prominent figures concluded with the following anonymous contribution.

I THINK it is positively due, not only to the writer of the now famous article in the *British Quarterly Review*, but to Newton's memory and to Science itself, that the correspondence which has been going on should not seem to terminate as a drawn game, at any rate in the opinion of some bystanders, who may from their antecedents be presumed competent to judge.

That Mr. Spencer will ever be convinced is, I suppose, hopeless; I at any rate am not going to try to convince him. But I can assure the *British Quarterly Reviewer* that he has my very deepest sympathy in his argument with an antagonist who is at once so able a master of fence as Mr. Spencer, and yet is so intensely unmathematical, it would seem, as to pass from "exact quantitative relation" to "proportionality;" or as to talk of the effect of a force, without defining how the effect is to be measured, without feeling the slightest difficulty.

Nor does it seem that Mr. Frankland, in NATURE, vol. ix., p. 484, is quite justified in his conclusion that the truth lies *between* the two opposite views. And his own view is in fact entirely coincident with the Reviewer's, except, perhaps, on a point which is not relevant to the controversy, viz. how far the experimental proof of the so-called physical axioms is *complete*.

Will it comfort the Reviewer if I tell him some of my own experience? I, too, read Spencer after my degree; and on the first reading of the "First Principles" came to the sad conclusion that I had not understood any mathematics properly; so much fresh light seemed to be thrown on them. I read it again, and more critically, and doubted whether Spencer was quite correct. I read it again, and concluded that he was wrong in his physics and mathematics. I ought to add that I too was, like the Reviewer,   A SENIOR WRANGLER

VOL 9, 500 1874

Charles George Gordon, or Chinese Gordon as he was known to his numberless adulators, was the archetypal Victorian intellectual eccentric soldier-hero, who was to die theatrically at the fall of Khartoum — an affront to British arms soon fearsomely avenged.

COL. GORDON'S JOURNEY TO GONDOKORO

WE have been favoured with the following remarks concerning Colonel Gordon's journey to Gondokoro. Colonel Gordon, "His Excellency, the Governor-general of the equator!" arrived at Khartoun on March 13, and had with him a *Pall Mall Gazette* of Feb. 13; he writes on the 17th from Khartoum as follows:—

"At this season of the year the air is so dry that animal matter does not decay or smell, it simply dries up hard; for instance a dead camel becomes in a short time a drum.

"The Nile, flowing from the Albert Nyanza below Gondokoro, spreads out into two lakes; on the edge of these lakes aquatic plants, with roots extending 5 ft. into the water, flourish; the natives burn the tops when dry, and thus form soil for grass to grow on; this is again burnt, and it becomes a compact mass. The Nile rises and floats out portions, which, being checked in a curve of the channel, are joined by other masses, and eventually

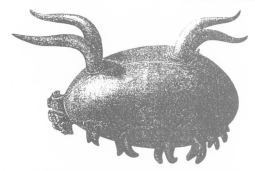

the river is completely bridged over for several miles, and all navigation is stopped.

" Last year the governor of Khartoum went up with three companies and two steamers, and cut away large blocks of the vegetation ; at last one night the water burst the remaining part, and swept down on the vessels, dragging them down some four miles, amidst (according to the Governor's account) hippopotami, crocodiles, and large fish, some alive and confounded, others dead or dying, the fish being crushed by the floating masses. One hippo was carried against the bows of the steamer and killed, and crocodiles 35 ft. long were killed : the Governor, who was on the marsh, had to go five miles on a raft to get to the steamer.

"The effect of these efforts of the Governor of Khartoum is that a steamer can now go to Gondokoro in twenty-one days, whereas it took months formerly to perform the same journey."

Colonel Gordon left Khartoum on March 21, and in his last letter from Fashoda, 10° N., he touches on some of the scenes on the banks of the river—the storks, which he was in the habit of seeing arrive on the Danube in April, laying back their heads between their wings and clapping their backs in joy at their return to their old nests on the houses, now wild and amongst the crocodiles 2,000 miles away from Turkey ; the monkeys coming down to drink at the edge of the river, with their long tails, like swords, standing stiff up over their backs ; the hippos and the crocodiles. Such scenes to a lover of nature, as Col. Gordon is, doubtless would serve to make up in some measure for the loss of civilised society and comforts.

VOL 10, 72 1874

The suicidal propensities of the scorpion engaged the attention of NATURE *readers intermittently for decades and many of the creatures were sacrificed in the attempt to resolve the question. This letter seems to be the first of many.*

Suicide of a Scorpion

I SHALL feel obliged if you will record in NATURE a fact with reference to the common Black Scorpion of Southern India, which was observed by me some years ago in Madras.

One morning a servant brought to me a very large specimen of this scorpion, which, having stayed out too long in its nocturnal rambles, had apparently got bewildered at daybreak, and been unable to find its way home. To keep it safe, the creature was at once put into a glazed entomological case. Having a few leisure minutes in the course of the forenoon, I thought I would see how my prisoner was getting on, and to have a better view of it the case was placed in a window, in the rays of a hot sun. The light and heat seemed to irritate it very much, and this recalled to my mind a story which I had read somewhere, that a scorpion, on being surrounded with fire, had committed suicide. I hesitated about subjecting my *pet* to such a terrible ordeal, but taking a common botanical lens, I focused the rays of the sun on its back. The moment this was done it began to run hurriedly about the case, hissing and *spitting* in a very fierce way. This experiment was repeated some four or five times with like results, but on trying it once again, the scorpion turned up its tail and plunged the sting, quick as lightning, into its own back. The infliction of the wound was followed by a sudden escape of fluid,

and a friend standing by me called out, " See, it has stung itself ; it is dead ; " and sure enough in less than half a minute life was quite extinct. I have written this brief notice to show (1) That animals may commit suicide ; (2) That the poison of certain animals may be destructive to themselves.

Bridge of Allan, N.B., Oct. 23   G. BIDIE

VOL 11, 29 1874

This is evidently a challenge to NATURE *readers.*

An Anagram

THE practice of enclosing discoveries in sealed packets and sending them to Academies, seems so inferior to the old one of Huyghens, that the following is sent you for publication in the old conserved form :—

$A^8C^3DE^{12}F^1GH^6I^6L^3M^3N^5O^6P$
$R^4S^5T^{14}U^6V^2WXY^2$ WEST

VOL 10, 480 1874

Animal locomotion became a preoccupation of physiologists and physicists. The following is an example of the ingenuity that was brought to its study. First the horse:

Fig. 4 will give an idea of the instrument employed in studying the complicated problem of quadrupedal action, in which it will be seen that the movements of each foot communicate, through elastic tubes, movements to the which is the most common ; *A* indicating the time, and *B* indicating the number of feet which support the body at each instant of the step. From it the left hind-foot is

FIG 4.

seen to reach the ground before any of the others, and to produce the first sound : the second is caused by the simultaneous impact of the right hind and left fore feet ; and the third by the right fore-foot, which the animal always his hand, has it tied, as a knapsack, on his back, and he sets the recording watchwork in motion with his teeth. Notwithstanding the difficulties of the experiment, very successful tracings have been obtained, which show that the full gallop is really a gallop in four-time, in which, although the fore-feet hit the ground with a fair interval, the hind feet hit it nearly simultaneously. The time of complete suspension is extremely short.

VOL 10, 499 1874

And then a bird:

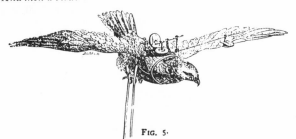

FIG. 5.

Fig. 5 shows a buzzard saddled with the machinery which, by means of the two tubes running downwards from it, transmits the vertical and horizontal movements of its wing to the recording apparatus, which is not represented. In the study of the more intricate points the necessary instruments are so heavy that the whole bird has to be partially supported. This is done by attaching it to the extremity of a long lever which revolves, with scarcely any friction, on a pivot. This is found not seriously to interfere with the normal flight of the bird.

Vol 10, 518 1874

Next an early contribution from George J. Romanes. Darwin wrote a letter of encouragement to Romanes after reading an article of his in Nature, *and so confirmed him in his choice of career as a biologist. Romanes later aligned himself more or less with Herbert Spencer in his reservations about natural selection. He wrote three successful books on comparative psychology, the best known of which is* Animal Intelligence. *His account of his terrier's behaviour betrays that lack of self-consciousness which marks the true scholar.*

Sense of Humour and Reason in Animals

In the recently published edition of the "Descent of Man" there is some additional matter concerning the above subjects, and as the following illustrative cases fell under my own observation, I think it is worth while to publish them as supplementary to those adduced by Mr. Darwin.

Several years ago I used to watch carefully the young Orang Outang at the Zoological Gardens, and I am quite sure that she manifested a sense of the ludicrous. One example will suffice. Her feeding-tin was of a somewhat peculiar shape, and when it was empty she used sometimes to invert it upon her head. The tin then presented a comical resemblance to a bonnet, and as its wearer would generally favour the spectators with a broad grin at the time of putting it on, she never failed to raise a laugh from them. Her success in this respect was evidently attended with no small gratification on her part.

I once had a Skye terrier which, like all of his kind, was very intelligent. When in good humour he had several tricks, which I know to have been self-taught, and the sole object of which was evidently to excite laughter. For instance, while lying upon one side and violently grinning,* he would hold one leg in his mouth. Under such circumstances nothing pleased him so much as having his joke duly appreciated, while if no notice was taken of him he would become sulky. On the other hand, nothing that could happen displeased him so much as being laughed at when he did not intend to be ridiculous. Mr. Darwin says :— "Several observers have stated that monkeys certainly dislike being laughed at" (p. 71). There can be little or no doubt that this is true of monkeys ; but I never knew of a really good case among dogs save this one, and here the signs of dislike were unequivocal. To give one instance. He used to be very fond of catching flies upon the window-panes, and if ridiculed when unsuccessful, was evidently much annoyed. On one occasion, in order to see what he would do, I purposely laughed immoderately every time he failed. It so happened that he did so several times in succession—partly, I believe, in consequence of my laughing—

and eventually he became so distressed that he positively *pretended* to catch the fly, going through all the appropriate actions with his lips and tongue, and afterwards rubbing the ground with his neck as if to kill the victim : he then looked up at me with a triumphant air of success. So well was the whole process simulated, that I should have been quite deceived, had I not seen that the fly was still upon the window. Accordingly I drew his attention to this fact, as well as to the absence of anything upon the floor ; and when he saw that his hypocrisy had been detected, he slunk away under some furniture, evidently very much ashamed of himself.

Vol 12, 66 1875

A gentle admonition from a physicist:

Insects and the Colours of Flowers

There is one point connected with Mr. Darwin's explanation of the bright colours of flowers which I have never seen referred to. The assumed attractiveness of bright colours to insects would appear to involve the supposition that the colour-vision of insects is approximately the same as our own. Surely this is a good deal to take for granted, when it is known that even among ourselves colour-vision varies greatly, and that no inconsiderable number of persons exist to whom, for example, the red of the scarlet geranium is no bright colour at all, but almost a match with the leaves.   RAYLEIGH
Whittinghame, Preston Kirk

Vol 11, 6 1874

In deprecating the ecologically improvident process of cremation, the distinguished organic chemist Raphael Meldola reveals himself as a prototypic green. A hundred and twenty years later, a letter to Nature *drew attention to another hazard — that of mercury poisoning in the vicinity of crematoria from the vaporisation of dental fillings.*

THE CHEMISTRY OF CREMATION

IN a paper recently published in a German periodical,* on the chemical bearings of cremation, Prof. Mohr calls attention to a point which, so far as we know, has not yet been considered.

He remarks that, in the first place, it is necessary that the combustion of the body should be *complete*. Anything of the nature of distillation gives rise to the production of fetid oils, such as were produced when in early times dead horses were distilled for the manufacture of sal-ammoniac. Such a revolting process is surely not compensated by the small commercial value of the products obtained. To effect complete combustion we must have a temperature such that the destruction is final, nothing remaining but carbonic acid, water, nitrogen, and ash ; for which purpose a complicated apparatus consuming large quantities of fuel will be necessary. The gases produced can only be destroyed by being passed through red-hot tubes to which excess of atmospheric air can gain access.

On comparing the substances produced by such a total decomposition of the body with those produced in the ordinary course of subterranean decay, it will be seen that one compound is totally lost by burning —the ammonia which results from the decomposition of the nitrogenous tissues. This ammonia, escaping into the air or being washed into the soil, is ultimately assimilated by plants—goes to the formation of nitrogenous materials, and thus again becomes available for animals. In the ordinary course of nature a continuous circulation of ammonia between the animal and vegetable kingdoms is thus kept up : if we stop one source of supply of this substance, we destroy the equilibrium—we draw upon the ammoniacal capital of the globe, and in the course of time this loss cannot but react upon animal life, a smaller amount of which will then be possible. There is no compensating process going on in nature as is the case with the removal of atmospheric oxygen by breathing animals—we deduct from a finite quantity, and the descendants of present races will, in time to come, have to bear the sin of our shortsightedness, just as we have had to suffer through the shortsightedness of our ancestors, who destroyed ruthlessly vast tracts of forests, thereby incurring drought in some regions and causing destructive inundations in others.

Another loss of ammonia is entailed by civilisation in the use of gunpowder. Nitre results from the oxidation of ammonia, and is a source of nitrogenous compounds to plants, which again reduce the nitrogen to a form available for ammonia. The nitrogen liberated by the explosion of gunpowder adds to the immense capital of the atmosphere, but is no more available for the formation of plants. Every waste charge of powder fired represents a certain loss of life-sustaining material against which the economy of nature protests. The same is to be said of nitro-glycerine, gun-cotton, &c., which contain nitrogen introduced by the action of nitric acid.

Wood and coal are other illustrations of finite capital. Every pound of these substances burnt in waste—consumed, that is, without being made to do its equivalent of work—is a dead loss of force-producing material, for which our descendants will in the far-distant future have to suffer. The changes brought about by the cessation of one large supply of ammonia may be compared with geological changes which, though of extreme slowness, produce vast changes in the lapse of ages.   R. M.

A tragicomic attempt at unpowered flight, analysed by the balloonist W. de Fonvielle:

THE FLYING MAN

THE fatal experiment made by M. de Groof at Cremorne Gardens could not possibly have led to success. The possibility of directing an apparatus in the air by any mechanical contrivance, without actually using the lifting power of gas, is out of the question, and we do not wish to enter into a discussion on that point. But several interesting problems may be examined *à propos* of the inquest held by the coroner on the death of the unfortunate man.

De Groof's wings, irrespective of their motive power, may be regarded as two imperfect parachutes intended to diminish his rate of falling, and, if kept horizontal, prevent it increasing above a certain rate. It remains to see if their surface was large enough to keep that velocity within reasonable limits. The wings of De Groof were 30 ft. by 4 ft. ; but being irregularly shaped, we may suppose the surface of each was 100 sq. ft., or in round numbers 200 sq. ft. for the two. The weight of the machine not being far from 4 cwt. if we include the man, we may say in gross numbers that each square foot had a kilogramme to support, which is more than ordinary ; the parachute maker taking 1 kilogramme for each square metre, which is about ten times smaller.

But to ascertain if the velocity, although being larger than under ordinary circumstances, was really dangerous we must go to the formulæ established by General Didion and quoted by Poucelet—

$$R = 1.936 \, (A \, 0.036 + 0.084 \, v^2)$$

Under the above circumstances, R the rate of falling is always inferior to the value of x given by the equation

$$10 = 1.936 \, (0.036 + 0.084 \, x^2)$$

a being obviously enough the velocity for which $R =$ to the weight pressing on the unit of surface. When the motion is such the velocity cannot be increased. If we make the calculation it is easy to see that the velocity is about 7 metres per second, almost $=$ the fall from 3 metres to the ground. It is large, but not too large for a practised jumper, if he were clever enough to keep his balance, which is not very easy, it must be confessed.

This cogitation on ancient numerals shows another facet of Victorian scholarship, by no means neglected in NATURE.

The Origin of our Numerals

MR. DONISTHORPE'S ingenious construction of our numerals by corresponding numbers of lines (NATURE, vol. xii. p. 476) induces me to offer a few remarks on this subject, which has a literature of its own. There can be no doubt, I believe, that our forms were derived directly from the Arab series called Gobar; that the Arabs had them from the Indians, and the Indians from the Chinese. My esteemed friend Dr. Wilson, of Bombay, published a "Note on the Origin of the Units of the Indian and European numerals," in 1858,* in which he showed the derivation of some of our numerals from ancient Indian forms found on cave inscriptions of Western India, on the Bhilsa Topes, and on coins. My remarks are founded wholly on the forms given in this note, which is little known, I believe, in England.

Dr. Wilson obtains our first four numeral forms from the Chinese, traced through different Indian script characters nearly as supposed by Mr. Donisthorpe. One, two, three horizontal bars and a square for 4. He also finds the eight in the forms ⊓⊔, ⊔⊔, and ∞ on the cave inscriptions.

Before proceeding to the other numerals I wish to notice a rule which may be deduced from the consideration of the changes in the forms of numerals in passing from one people to another, that the same form may be turned through angles of 90° or 180°, and may be inverted or reversed without altering its value. Even the same people have used a form turned in different ways for the same numeral. The Arabs used their 2, 3, and 4 in two ways, making angles of 90° with each other; the 2, 4, and 5 of Sacro Bosco and Roger Bacon were the Indian script Modi (and ours) turned through 180°, or upside down; other examples will be noticed.

The most important derivation by Dr. Wilson is that from the Chinese ✚ ten; this is found on the Bhilsa Topes with a circle round it (Dr. Wilson thinks to distinguish it from the oldest form of K found on the cave inscriptions). The nine is found on the Bhilsa Topes as ⊕, or one under ten, and on old coins thus: ℭℬ. The Indian caves give half of ten ⊣, ⊢, for five (as V is the half of the Roman ten, X). It is from this form that Dr. Wilson derives the Indian Modi and Nagari fives ५, ᛦ, ᛉ. It is here that I venture to differ slightly from Dr. Wilson. One of the cave forms of four is ५, which Dr. Wilson interprets (as in the case of nine) one under five, or five less one; now this form without the under bar, as well as the other forms of five, are, it seems to me, the halves not of the cross (✚) merely, but of the cross and circle thus: ⊕, ⊙, which are as nearly as possible two half diameters and half circumference. The form ⋊ is, I believe, the origin of our four, and not the Chinese or Indian square, as supposed. This I think will be evident when we compare the Arab four (५) with the Indian four above. The Arab four is also employed thus: ⋌, which inverted gives ⋋, a sufficiently near approximation to our four.

Dr. Wilson has not been able to find the origin of our seven, but this is obtained from his Arab seven ∧, by turning it round (⅂) and making one leg shorter than the other, nearly resembling the Gobar seven ⌐. We may also find an earlier source in the Chinese seven turned round 180°, ⋣, which is almost exactly the German written seven. Neither six nor seven is to be found on the cave inscriptions. In Dr. Wilson's Arab series the Indian five ५ is used for six, and the Gobar six, as well as ours, may be taken from the Nagari seven ৬. We may also find an origin in the Chinese six ⋇, by omitting the horizontal bar, as in the case of the seven. That such liberties

were taken is evident on a consideration of the five of Sacro Bosco and Roger Bacon (५), the Indian five *without the bar*, and turned round 180°. If there is any merit in these suggestions it belongs to Dr. Wilson.

JOHN ALLAN BROUN

VOL 13, 47 1875

John Tyndall had entered the spontaneous generation debate exasperated, it seems, by the caperings of the infuriating Henry Charlton Bastian (p. 10). His report of the outcome of a series of experiments that concurred fully with Louis Pasteur's conclusions elicited a letter of gratitude from the great man, which NATURE published, as was its custom, in the original. It begins—

[The following letter has been sent us for publication by Prof. Tyndall.—ED.]

PERMETTEZ-MOI de vous dire combien je suis charmé que vous apportiez dans la question de la génération spontanée la grande autorité de votre esprit philosophique et de votre rigeur expérimentale. C'est tout à la fois un honneur pour mes recherches et une vive satisfaction personelle que les conclusions auxquelles vous êtes arrivé s'accordent si bien avec celles de mes propres travaux, malgré la différence des méthodes que nous avons suivies. Le tour piquant que vous avez su donner à vos expériences les fera pénétrer plus avant que les miennes dans l'esprit de tout lecteur que n'égarent pas les idées *à priori*.

— and concludes optimistically:

Elle est inattaquable, cette conclusion que j'ai déja formulée: dans l'état actuel de la science, l'hypothèse de la génération spontanée est une chimère. Votre bien dévoué,

Paris le 8 Fevrier, 1876   L. PASTEUR

VOL 13, 305 1876

Here is some doggerel by a forgotten poetaster, not, it must be said, of James Clerk Maxwell's quality. There are 13 verses, of which these are the first three and the last.

SONG OF THE SCREW

A MOVING form or rigid mass,
  Under whate'er conditions,
Along successive screws must pass
  Between each two positions.
It *turns around* and *slides along*—
This is the burden of my song.

The *pitch of screw*, if multiplied
  By angle of rotation,
Will give the distance it must glide
  In motion of translation.
Infinite pitch means pure translation,
And zero pitch means pure rotation.

Two motions on two given screws,
  With amplitudes at pleasure,
Into a third screw-motion fuse;
  Whose amplitude we measure
By parallelogram construction
(A very obvious deduction).

* * *

But time would fail me to discourse
  Of Order and Degree.
Of Impulse, Energy, and Force,
  And Reciprocity.
All these and more, for motions small,
Have been discussed by Dr. Ball.

VOL 14, 30 1876

Paul Broca hoped that a sufficiently minute anatomical inspection of the brain might reveal something about the seat of human intelligence and talents. Whenever possible the brains of famous people were anatomised and the results were often reported in NATURE.

AT a public dinner given by the Anthropological Society of Paris, a proposal of a singular nature, signed by MM. Hovelacque, Dally, Mortillet, Broca, Topinard, and others, was circulated for additional signatures. Each of these gentlemen promises to write a will directing that his brain be sent to the Anthropological Society for inspection and dissection. It is thought that by procuring the thinking organ of persons whose habits and works are perfectly known, some light might be thrown on the laws of physico-mental organisation. The scheme having been published in several Parisian papers, has provoked a furious attack from the *Univers*.

VOL 14, 581 1876

The Remington typewriter represented the kind of technical advance that appealed irresistibly to the Victorians, and it was received with rapture. One of the first to become proficient in its use was Sir George Stokes, Lucasian Professor of Mathematics at Cambridge. NATURE *gave a long description of the machine, concluding like this:*

THE REMINGTON TYPE-WRITING MACHINE

The whole instrument is not larger than a sewing-machine. Its cost is twenty guineas. It only writes in capitals, the total number of keys being forty-four, arranged in four rows of eleven in each. Its simplicity is the best guarantee of its durability.

As to the "typoscript" (in contradistinction to the manuscript of ordinary handwriting), there is no comparison between its clearness and that of average penmanship. It has, in fact, all the appearances of print, with its many advantages as regards legibility, compactness, and neatness. Errors, if detected soon enough, can be corrected by the repetition of the word or sentence, and the subsequent obliteration, upon reperusal, of the faulty lines. The ink employed can be transferred like transfer ink.

The principal question which this beautiful and ingenious little instrument suggests to our minds is, whether it would not be better for every one of us to learn the Morse telegraph language, and employ it for writing upon all occasions instead of the cumbrous letters now in vogue. Thought is more quick than formerly. Germany is rapidly rejecting its archaic type ; why should we not go further and write in Morse, where spots and horizontal lines do duty for all necessary signs, and type-writers of the simplest form would be required ?

VOL 14, 44 1876

This early heuristic version of an artificial heart seems to have been well ahead of its time.

The artificial heart is constructed with caoutchouc cavities, supplied with valves to represent those in the human circulation. The *auricle* is covered with netting, to which four parallel cords, running through holes in the big board, are attached. The cords are fixed on a square piece of wood, which is kept in position by a spiral spring, and in connection with the moving board by the thread S.O. The *ventricle* has over it a case (white in the figure) to the edges of which cords are fixed, which are attached at their other ends to a board, which is put into communication with the moving board by means of the hooks and elastic rings (F), and the cord S.V. It is evident that any strain on the cords S.V. or S.O. will compress the auricle (o) and the ventricle (v) against the main board to which they are attached, and so produce a systole of these viscera. A magnified view of this artificial heart, into the cavities of which recording "ampoules" have been introduced, is given in Fig. 2.

VOL 13, 182 1875

This undemanding way of arriving at an upper limit for the weight of an atom still falls short by some 12 orders of magnitude.

THE extraordinary divisibility of matter is well illustrated by a lecture experiment recently described to the Berlin Chemical Society by M. Annaheim. He employs the strong colouring power of fuchsin and cyanin. To form an idea what quantities of colouring matter were still perceptible by the eye, he dissolved 0·0007 gramme of fuchsin (a particle about 0·5 mm. diameter) in spirit of wine, and diluted the solution to the extent of 1,000 cubic centimetres. Thus in each centimetre there was still 0·0000007 gramme colouring matter. If this liquid be put in a burette of about 1 cm. diameter, it appears strongly coloured on a white ground, and the colour can be distinctly seen from a distance. If a drop from the burette (there are thirty-five of them in a cub. ctm.) be now let fall into a small dry test-tube of about 0·8 cm. diameter, the red colour is still evident if the tube be held obliquely on white paper, and looked at parallel to the paper, while a second tube with pure spirit of wine is held near for comparison. It follows from this, that with the naked eye one can still perceive 0·00000002 gramme fuchsin. Assuming that one drop of the solution only contains one molecule of colouring matter (and so much must in all circumstances be present), the absolute weight of an atom of hydrogen is inferred to have the astonishingly small value of 0·000000000059 gramme (viz. 0·00000002 : 337·5 ; molecular weight = 337·5). M. Annahein makes a similar experiment with cyanin, and infers the absolute weight of an atom of hydrogen to be 0·000000000054 gramme, which closely agrees with the former estimate. From these experiments, then, it is mathematically certain, that the absolute weight of an atom of hydrogen cannot be greater than 0·0000000005 gramme.

VOL 14, 537 1876

A biography of Alexander von Humboldt in two volumes, compiled by a committee of his compatriots and rendered into English, received an approving anonymous notice, spread over three pages of NATURE. *Here is a description of Humboldt's triumphal progress through Asiatic Russia during his famous expedition of 12,000 miles.*

Wherever he went crowds of local dignitaries, soldiers and police officers surrounded him. Governors of provinces, commandants of fortresses, superintendents of mines welcomed him with speeches and reports whenever he appeared within the limits of their jurisdiction. Generals supplied him with minutes of the strength of the various brigades under their command, while officers and men in dress uniforms saluted him in military fashion as he passed their posts. At Miask these military marks of respect culminated in the presentation, by the directors of the mines, of a grand cavalry sabre, in honour of his sixtieth birthday. The learned bodies were equally on the alert to show him respect. At Kasan, after incessant feasting and speechifying, the Professors escorted him to his lodgings at 1 A.M. in gala costume, and reappeared in the same attire at 4.30 A.M. to speed his departure to the next station. After enduring a host of similarly oppressive social distinctions, which included at Jekatharinenburg the obligation of leading off a ball in a stately quadrille, and on the Steppes at Orenburg the necessity of presiding over a Kirghis festival at which the men ran races and the Tartar Sultanas warbled sweet songs in his praise, Humboldt had to encounter at Moscow one of the most absurd ordeals to which the fame of his greatness exposed him. On his arrival he was invited to attend a special meeting of the Physical Society, and duly made his appearance at the University, holding in his hand the paper he had prepared to read to the learned members "On the deviation of the Magnet in the Ural." The court, passages, stairs, and halls were crowded with great people, gorgeous with stars and orders, amongst whom stood conspicuous the Professors, wearing long swords girded to their sides, and three-cornered hats tucked under their arms. Speeches of welcome in German, French, and Latin from the Governor-General, the chief clergy, and the deans of the various faculties had to be heard and replied to, and instead of engaging in scientific discussion on magnetic aberration, Humboldt had to listen to a Russian poem in which he was hailed as Prometheus, and to examine a plait of Peter the Great's hair, which was solemnly presented for inspection by the Rector of the University. The "Asie Centrale" and a few very fragmentary works were the immediate results of this most oppressively-honoured expedition, from which, satiated with ceremonials and respect, Humboldt had, in the winter of the same year, 1829, returned to Berlin, which thenceforth to the end of his long life in 1859 became his home.

Dr Johnson's Dictionary defined the lion as "the fiercest and most magnanimous of fourfooted beasts" and the tiger as merely "a fierce beast of the leonine kind", but the learned Mr Haughton knows better. Note the three-figure accuracy with which he has determined the relative strengths of fore- and hind-limbs of the creatures. Had he but had access to the tigon — the tiger-lion cross, on view some years later in the Paris zoo — he might have found that 7.0 men were needed to subdue it, and so clinched the issue.

The Strength of the Lion and the Tiger

IN NATURE, vol. xii., p. 474, in a review of Dr. Fayrer's book on the tiger, doubts are thrown by the reviewer on the statement that the tiger is stronger than the lion. Dr. Fayrer's statement cannot be contradicted by any person well acquainted with both animals. In my book on "Animal Mechanics," published in 1873, I have proved, p. 392, that the strength of the lion in the fore limbs is only 69.9 per cent. of that of the tiger, and that the strength of his hind limbs is only 65.9 per cent. of that of the tiger.

I may add that five men can easily hold down a lion, while it requires nine men to control a tiger. Martial also states that the tigers always killed the lions in the amphitheatre. The lion is, in truth, a pretentious humbug, and owes his reputation to his imposing mane, and he will run away like a whipped cur, under circumstances in which the tiger will boldly attack and kill.

At p. 482 you state that Dr. Bolau, of Hamburg, is about to publish an account of the anatomy of a gorilla which nearly reached Hamburg alive, and was preserved in spirits. Your readers will be glad to learn that he has been anticipated by Prof. Macalister, of Trinity College, Dublin, who has already published a full account of a similar animal, which nearly reached Liverpool alive some years ago, and was dissected by myself and Dr. Macalister. A comparison of his muscles with those of man, chimpanzee, and hamadryas, will be found in my "Animal Mechanics," p. 404 *et seq.*

SAMUEL HAUGHTON

Trinity College, Dublin, Oct. 1

A long and beautifully illustrated five-parter on the history of the telegraph details both its technical and its social aspects. This extract recounts how the invention was first used to collar a fleeing criminal more than 60 years before wireless telegraphy achieved a similar success with the arrest on board ship of the murderer Dr Crippen.

In the late trial of John Tawell, at Aylesbury, for the murder at Salt Hill, near Slough, the Electric Telegraph is frequently mentioned in the evidence, and referred to by Mr. Baron Parke in his summing up. The *Times* newspaper very justly observes 'that had it not been for the efficient aid of the Electric Telegraph, both at the Paddington and Slough stations, the greatest difficulty, as well as delay, would have been occasioned in the apprehension of the prisoner.' Although the train in which Tawell came to town was within a very short distance of the Paddington Station before any intelligence was given at the Slough Telegraph Office, nevertheless, before the train had actually arrived, not only had a full description of his person and dress been received, but the particular carriage and compartment in which he rode were accurately described, and an officer was in readiness to watch his movements. His subsequent apprehension is so well known, that any further reference to the subject is unnecessary.

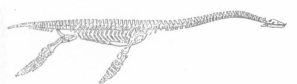

chapter four
1877-1880

Francis Galton considers the genetics of criminality. Jules Verne is commended by the editor of *Nature*. William Ewart Gladstone's thesis that Homer was colour-blind engenders heated debate. *Nature* laments the death of James Clerk Maxwell at the age of 48. Readers are informed of the capacity of René Descartes's cranium and the exploits of Bully, the bull terrier, a very genius among canines.

1877-1880

Work began on the Channel tunnel. Queen Victoria was proclaimed Empress of India. Britain was victorious in the Zulu War. Heinrich Schliemann excavated Mycenae, and General Custer made his last stand. Thomas Alva Edison patented his phonograph and the filament lamp. Cleopatra's Needle, pilfered from Egypt and recovered from the sea-bed of the Bay of Biscay, where it had been carelessly deposited, was erected on the Victoria Embankment in London. The British parliament promulgated the Red Flag Act, which required all motor vehicles to be preceded by a man on foot, carrying a red flag.

Samuel Smiles, author of uplifting tracts entitled
Self-Help, Thrift *and* Duty *and something of a Victorian
icon, was a surgeon and industrious biographer. Here he has
written about a now forgotten and wholly self-taught
naturalist, who in a more meritocratic age would doubtless
have ended up a professor.*

A WORKING NATURALIST

*Life of a Scottish Naturalist : Thomas Edward, Asso-
ciate of the Linnean Society.* By Samuel Smiles.
Portrait and Illustrations by George Reid, A.R.S.A
(London : John Murray, 1876.)

IT is rather a delicate thing and seldom advisable to
publish a full and formal biography of a man with
an account and attempted estimate of his work while he
himself is still alive and in comparative vigour. There
are many sound reasons, however, to justify Mr. Smiles
in telling the wonderful story of the still living Thomas
Edward, the Banff shoemaker and naturalist ; not the
least weighty of these is that it will bring to Edward
some of that *kudos* and cash which he has earned long
ago, and which it would have been well for himself and
for science had they, or at least the latter, reached him
years since.

Thomas Edward, born 1814, is the son of very humble
but thoroughly respectable Scotch parents, who were able
to bestow upon him the scantiest schooling. He was
brought up in Aberdeen, where he was in rapid succession
expelled from three schools on account of his intense inborn passion for " everything that hath life." He used to
take all sorts of birds and beasts and creeping things to
school with him, in his pockets, in boxes, or in bottles.
Tom's specimens would often escape, and the scene may
be imagined when some unconscious urchin realised that
a snail, or a horse-leech, or a " Maggie-mony-feet " (centipede), was crawling up his bare legs. Poor Edward meant
no harm, but it was too much to expect that an ordinary
dame or dominie would, in that remote and unscientific
age, at least, take the trouble to understand the boy's
nature and tendencies and aspirations. The consequence
was severe thrashings and expulsion.

When Edward left school finally he was only six years
old, could read with difficulty, and could hardly be said
to be able either to write or count. After serving for
some time in a mill he was apprenticed to a shoemaker,
Begg, " a low-class Cockney," Mr. Smiles calls him, and
certainly a regular brute, who treated the poor boy and
his birds and beasts in a most cruel fashion. However
Edward managed with this man and a subsequent master
to pick up a fair knowledge of the shoemaking trade, and
after vainly trying to emigrate as a stowaway, removed to
Banff when about twenty years old, where he has lived
ever since working as a journeyman for wages, and
devoting every moment of his leisure to the gratification
of his passion for natural history. It is common enough
for working men, and especially for shoemakers, to take
an interest in certain animals, especially in birds ; but
Edward's fondness for animals was no fancy of this sort.
From almost his infancy he was devoured with a passion
for the observation and possession of animals of all kinds;
to him no living creature was unclean or loathsome, and he
feared to face or handle nothing from a centipede or an
adder to a polecat. While yet a baby in his mother's
arms he nearly put a premature end to his career by

springing to clutch at a passing insect ; and while being
drilled as a militiaman in Aberdeen he made himself liable
to severe punishment by rushing like a madman from the
ranks in chase of a passing butterfly.

As a journeyman shoemaker Edward had to work from
six in the morning till nine at night. Shortly after settling
in Banff he married, luckily for his love of nature, a
prudent and considerate woman, who, instead of thwarting
his eccentricities, did what she could to help him and
enable him to indulge them. " Weel," she said once,
when asked what she thought of his habits, " he took
such an interest in beasts that I didna compleen. Shoemakers were then a very drucken set, but his beasts
keepit him frae them. My man's been a sober man all
his life ; and he never negleckit his wark. Sae I let him
be." " Wise woman ! " justly adds Mr. Smiles. Shoemakers' wages in Banff were very low—only a few shillings
a week. For many years Edward earned only about 10s.
a week, and yet on this he managed to rear, without incurring debt, a thoroughly respectable and honest family
of about a dozen children, who have repaid their parents'
care by doing what they could to comfort their old age.
Edward being a man who had a proper sense of his duty
to his family, seldom thought of allowing his favourite
pursuit to encroach on his long working hours. As his
consuming passion must be satisfied, he took the only
course legitimately open to him ; he gave up his nights
to it. As soon as he got home from work, unless indeed
the weather was unusually bad, he shouldered his gun,
equipped himself in his eight-pocket coat, four-pocket
vest, double-storied hat, and other traps of a rude but
efficient enough kind, and putting his supper of oatmeal
cakes in his pocket, set out to watch and catch the denizens
of the woods, heaths, air, and sea-shore of the region around
Banff. He would prowl about as long as the light permitted,
lie down for an hour or two in a hole, under the lee of a
bush, inside some old ruin, or underneath a flat tombstone in some eerie churchyard, snatch an hour or two's
sleep, and be up again with the first streak of dawn.

The long leader-review, of which this is the beginning, concludes rather poignantly:

But Edward never complained of his lot, and had Mr. Smiles not written the present work, he would have had to stick to his stool to the end. All Edward ever wanted was some way of earning a living that would have enabled him to give more time and attention to his scientific pursuits, and no one will deny that it would have been immensely to the gain of science could his services have been devoted entirely to it, for he was too passionately fond of nature ever to have been spoiled by prosperity. But regrets are now useless ; happily Edward is not beyond the reach of consolation and well-merited reward, and happily he is receiving them. He will be mentioned in the annals of science as an observer of the highest accuracy and originality, who gave up to a parish a genius fitted for an immensely wider sphere. The obvious moral of the work to those who have to spend most of their time in earning their daily bread, as well as to others, we need not point here. Mr. Smiles's work is one of the most interesting biographies ever written, and the illustrations gratuitously contributed by Mr. Reid are a great pleasure. Our readers by buying the book will not only become possessed of a rare treat, but will at the same time help to confer a substantial benefit upon Thomas Edward, the Scottish Naturalist.

Vol 15, 349 1877

This project for an early "light railway" evidently did not reach fruition.

A company is now being formed, we learn from the *Engineer*, to construct a pneumatic railway between the South Kensington Station of the District Railway and the Albert Hall. The line will rise the whole way to the Albert Hall, the ruling gradient being 1 in 48. The train will be blown through the tube by an ejector, in other words, a great centrifugal pump, two feet in diameter, fixed close to the District station, and worked by a pair of condensing engines exerting about 170 indicated horse-power. The tunnel will be of brick, and the floor will be paved. Its cross-sectional area will be 105·5 square feet ; at the end of the train is fixed a screen or piston, with an area of 104 square feet, the difference being allowed for windage. The train will consist of six carriages, of very light build, the rail gauge being four feet. This train will hold 200 passengers, and the total load will be thirty-two tons, or ten tons less than the weight of a single engine on the Metropolitan Railway. The maximum resistance at twenty miles an hour will be about 2,420 lbs., requiring to overcome it a pneumatic pressure of 2·6 ounces per square inch, and 162-horse-power, assuming the useful effect to be sixty per cent.

Vol 16, 217 1877

Nature reported events from distant lands. Was there a stringer in Budapest?

A Hungarian prelate, the Archbishop Louis of Haynald, has constructed an astronomical observatory at his own expense at Kalocsa, lat. 46° 31′, long. 16° 32′. Among the instruments are a Browning telescope, a small (4-inch) Merz refractor, and a Cooke transit instrument. The arrangement of the new observatory is superintended by M. de Konkoly, already known as having built on his own property, O-Gyalla, a well-furnished observatory. We may add that the Archbishop of Haynald has already devoted considerable sums to botanical researches.

Vol 16, 217 1877

Francis Galton, kinsman of Charles Darwin, was one of the great originals of his time. As a young man he had undertaken perilous journeys of exploration and wrote a book full of advice for the like-minded called The Art of Travel: *to keep your clothes dry if caught in a storm, take them off and sit on them; to soften your boots, break an egg into each before putting it on; as a restorative when you are below par, toss off a slurry of gunpowder in warm soapy water, and so on. Galton was an inventor of great ingenuity. He was convinced that intense thought caused the brain to overheat, and he accordingly constructed his Universal Patent Ventilating Hat, from which he was never parted: by squeezing a rubber bulb at the end of a tube, the wearer activated a valve that caused the crown to lift.*

Galton was a pioneer in the application of quantitative measurement to problems of biology and of human affairs. Even religion was not excluded, for Galton suggested that the principle of the efficacy of prayer, for instance, could be tested: since the populace regularly urged God to save the Queen (or King), Galton compiled data on the longevity of British monarchs. Finding no positive outcome, he decided that clergymen and their families would serve as a test case, inasmuch as their prayers for the survival of their loved ones would be informed by fervour, not merely frequency. Prayer, Galton was able to establish, was wholly inefficacious.

Galton's treatise on Hereditary Genius *led to the conclusion that exceptional qualities were inherited and that eugenic breeding (a term Galton coined) could lead to the improvement of the human stock. He believed conversely that criminality too was genetically determined — a doctrine that was to have grievous consequences. Here he is at the end of a long lecture to the British Association for the Advancement of Science, reprinted in* Nature.

I will proceed with what appears to be the history of the criminal class. Its perpetuation by heredity is a question that deserves more careful investigation than it has received, but it is on many accounts more difficult to grapple with than it may at first sight appear to be. The vagrant habits of the criminal classes, their illegitimate unions and extreme untruthfulness, are among the difficulties. It is, however, easy to show that the criminal nature tends to be inherited while, on the other hand, it is impossible that women who spend a large portion of the best years of their lives in prison can contribute many children to the population. The true state of the case appears to be that the criminal population receives steady accessions from classes who, without having strongly marked criminal natures, do nevertheless belong to a type of humanity that is exceedingly ill-suited to play a respectable part in our modern civilisation, though they are well-suited to flourish under half-savage conditions, being naturally both healthy and prolific. These persons are apt to go to the bad ; their daughters consort with criminals and become the parents of criminals. An extraordinary example of this is given by the history of the infamous Jukes family in America, whose pedigree has been made out with extraordinary care, during no less than seven generations, and is the subject of an elaborate memoir printed in the thirty-first annual report of the Prison Association of New York, 1876. It includes no less than 540 individuals of Jukes blood, among whom the number of persons who degraded into criminality, pauperism, or disease, is frightful to contemplate.

It is difficult to summarise the results in a few plain figures, but I will state those respecting the fifth generation, through the eldest of the five prolific daughters of the man who is the common ancestor of the race. The total number of these was 103, of whom thirty-eight came through an illegitimate grand-daughter, and eighty-five through legitimate grandchildren. Out of the thirty-eight, sixteen have been in gaol, six of them for heinous offences, one of these having been committed no less than nine times ; eleven others were paupers or led openly disreputable lives ; four were notoriously intemperate ; the history of three had not been traced, and only four were known to have done well. The great majority of the women consorted with criminals. As to the 85 legitimate descendants, they were less flagrantly bad, for only five of them had been in gaol and only thirteen others had been paupers. Now the ancestor of all this mischief, who was born about the year 1730, is described as having been a hunter and a fisher, a jolly companionable man, averse to steady labour, working hard and idling by turns, and who had numerous illegitimate children, whose issue has not been traced. He was, in fact, a somewhat good specimen of a half-savage, without any seriously criminal instincts. The girls were apparently attractive, marrying early and sometimes not badly; but the gipsy-like character of the race was unsuited to success in a civilised country. So the descendants went to the bad, and the hereditary moral weaknesses they may have had rose to the surface and worked their mischief without a check. Cohabiting with criminals and being extremely prolific, the result was the production of a stock exceeding 500 in number, of a prevalent criminal type. Through disease and intemperance the breed is now rapidly diminishing; the infant mortality has of late been horrible among them, but fortunately the women of the present generation bear usually but few children, and many of them are altogether childless.

This is not the place to go further into details. I have alluded to the Jukes family in order to show what extremely important topics lie open to inquiry in a single branch of anthropological research and to stimulate others to follow it out. There can be no more interesting subject to us than the quality of the stock of our countrymen and of the human race generally, and there can be no more worthy inquiry than that which leads to an explanation of the conditions under which it deteriorates or improves.

VOL 16, 346 1877

This provoked the following reflection in a later issue. Hall porters have always of course been renowned for improbable feats of memory. In one of his letters, however, Evelyn Waugh related how he had gone behind the cloakroom desk at White's Club to retrieve his bowler, while the porter was away from his post, and out of the hat had fluttered a small strip of paper on which was scrawled the single word "florid". As to chess, the record now stands at around 40 simultaneous blindfold games.

A Feat of Memory

THE following feat of memory seems to be worthy of record in your pages. It is new to the writer, though by no means uncommon over here.

Like the country itself, many institutions in the United States run to size in a way apt to astonish the dwellers in our "tight little island." So it is with hotels. Thus at some of them many hundreds of persons are simultaneously dining in one room. At the entrance, the hats, &c., of the guests are deposited with a person in attendance to receive them. He does not check or arrange them in any particular order, and he invariably restores them, each to the right owner, as they emerge from the dining-room. The difficulty of the feat naturally depends on the number of hats in charge at the same time. The most remarkable case which has come under the notice of the writer is at the Fifth Avenue Hotel, New York. There the attendant, who is on duty several hours a day, has sometimes as many as five hundred hats in his possession at one time. A majority of them belong to people whom he has never seen before, and there is a constant flux of persons in and out. Yet even a momentary hesitation in selecting the right hat rarely occurs. The performer at the above hotel says that he forms a mental picture of the owner's face inside his hat, and that on looking at any hat the wearer's face is instantly brought before his mind's eye. It would be interesting to test how far this power is possessed by an average unpractised person when put in the right way of doing it. While many of our ordinary recollections are not visual, at least not consciously so, it appears probable that most cases of extraordinary memory consist in an unusual power of making and retaining visualised impressions. Mr. Galton's interesting paper in NATURE (vol. xxi. p. 252) on "Visualised Numerals" goes a long way to show this to be so in mental arithmetic. Sys-

tems of artificial memory tend towards the same point; for they may be roughly described as mainly resting on the systematic manufacture of artificial visualisations; and the hat feat just narrated falls within the same category.

In working the rich mine which Mr. Galton's genius has discovered, I hope he will explore the vein of chess without the chess-board. As efforts of memory, such performances are as surprising as the numerical feats of Colburn and Bidder. And they notably differ from them in that the highest development is reached, not by young boys, but by men of mature years, who, as players over the board, have reached the front rank. The writer (in last year's *Chess Player's Chronicle*) attempted to give a rough estimate of the number of moves and positions possible at chess. They are of course practically illimitable; and with this fact in mind it is easy to form an idea of the difficulty of playing *twelve* games blindfold against very strong antagonists. This task, however, is often performed by Messrs. Zukertort and Blackburne, beyond question in England, and probably in the world, the greatest adepts in this branch of chess-play. It would be highly instructive to learn by what process, in so far as it is a conscious and describable one, these feats are achieved. If Mr. Galton takes the matter up, no doubt he will, with his usual skill, throw a flood of light upon the subject.

EDWYN ANTHONY

Riggs's Hotel, Washington, March 29

VOL 21, 562 1880

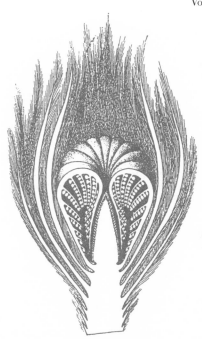

Here is an extract from another of the running debates that so characterised Victorian intellectual life. William Crookes turns on his critic:

It was at the annual dinner of the Fellows of the Royal Society on November 30, 1875, when the royal medal was awarded to me. Dr. Carpenter accosted me with great apparent cordiality, and said, "Let us bury the hatchet! Why should scientific men quarrel?" I signified my full acceptance of the offered peace, and great was my surprise soon after to find that, unmindful of the understood compact, he had exhumed his hatchet and was dealing me unexpected and wanton strokes, tempered by a certain amount of half praise which reminds me of the sort of caressing remonstrance of Majendie in the pre-anæsthetic days, to the dog which he had on his operating table—"*Taisez vous, pauvre bête!*"

In all seriousness, however, I must again ask, what is the meaning of the "personal antagonism," and the persistent attacks which Dr. Carpenter, for the last six years, has directed against me? In his recently published book, in the *Nineteenth Century*, and in his last letter to you, the key-note struck in the *Quarterly Review* six years ago is sustained. We have the same personalities, the same somewhat stale remark about my double nature, and the same exuberance of that most dangerous and misleading class of averments, half truths. Dr. Carpenter, indeed, condescends to admit that I have pursued "with rare ability and acuteness a delicate physical investigation in which nothing is taken for granted without proof satisfactory to others as well as to himself," and that I have "carried out a beautiful inquiry in a manner and spirit worthy of all admiration;" but, after granting so much, he dissembles his love and proceeds to "kick me down stairs." I am damned with faint praise, and put to rights in such a school-masterly style, that I could almost fancy Dr. Carpenter carries a birch rod concealed in his coat-sleeve. He admits that in an humble and subordinate sphere I have done useful work, only I must not give myself airs on that account. Dr. Carpenter reminds me of Dr. Johnson defending Sir John Hawkins, when he was accused of meanness. "I really believe him," said Johnson, "to be an honest man at the bottom; but to be sure he is penurious, and he is mean, and it must be owned he has a degree of brutality, and a tendency to savageness, that cannot easily be defended." In the same magnanimous spirit Dr. Carpenter allows that I have contributed a trifle to science, but he does not forget to add that I am the victim of cerebral duplicity, and I am again held up to illustrate the sad result of neglecting to train and discipline "the *whole* mind during the period of its development," &c.

I have, it appears, two allotropic personalities, which I may designate, in chemical language, Ortho-Crookes and Pseudo-Crookes. The Ortho-Crookes, according to Dr. Carpenter, has acquired "deserved distinction as a chemist." He carries out a "beautiful inquiry in a manner and spirit worthy of all admiration." He has shown "ability, skill, perseverance, and freedom from prepossession." He pursues "with rare ability and astuteness a delicate physical investigation." He has the "spirit of the true philosopher," and he has "deservedly" received "from the Royal Society the award of one of its chief distinctions."

But Pseudo-Crookes, whose career Dr. Carpenter has evidently watched almost from his cradle—as he professes to know the details of his early education—unfortunately took a "thoroughly unscientific course," and developed into a "specialist of specialists." He had "very limited opportunities" and "never had the privilege of associating" with scientific men, although he displayed "*malus animus*" "towards those with whom he claims to be in fraternity." He is "totally destitute of any knowledge of chemical philosophy, and utterly untrustworthy as to any inquiry" not technical. His "assertions" are "well known in the scientific world to be inconsistent with fact." He enters on inquiries "with an avowed foregone conclusion of his own." He has "lent himself to the support of wicked frauds." He has "prepossessions upon which clever cheats play." His "scientific tests" are not "worthy of trust." He is a believer in "day dreams," and the supporter of a "seething mass of folly and imposture;" whilst, to crown all, he actually thinks that the radiometer is driven "by the direct impetus of light." In short, this Pseudo-Crookes is a compound of folly and knavery such as has rarely, if ever, previously been encountered.

WILLIAM CROOKES (The Ortho-Crookes?)
London, October 29

Norman Lockyer, the editor, reviews with approval the works of Jules Verne:

Hector Servadac, or the Career of a Comet.—From the Earth to the Moon.—Around the Moon.—Twenty Thousand Leagues under the Sea.—Around the World in Eighty Days.—The Fur Country.—A Winter amid the Ice, &c., &c. (London : Sampson Low and Co.)

THESE remarkable works, which we owe to the genius of Jules Verne, the first-named being that which has last appeared, are well deserving of notice at our hands, for in the author we have a science teacher of a new kind. He has forsaken the beaten track, *bien entendu;* but acknowledging in him a travelled Frenchman with a keen eye and vivid imagination—and no slight knowledge of the elements of science—we do not see how he could have more usefully employed his talents.

In Hector Servadac, *Verne causes a comet to rip from the Earth the whole of Algeria, which is borne away into solar orbit. His heroes recover from their shock and set off in search of fellow men. But first lunch. Lockyer quotes:*

"'By jingo !' he exclaimed, 'this is a precious hot fire.' Servadac reflected. In a few minutes he said :—
"'It cannot be that the fire is hotter, the peculiarity must be in the water.'
"And, taking down a centigrade thermometer which he had hung upon the wall, he plunged it into the skillet. Instead of 100°, he found that the instrument registered only 66°.
"'Take my advice, Ben Zoof,' he said; 'leave your eggs in the saucepan a good quarter of an hour.'"

Verne introduces a puff for the metric system of measures:

A new point in favour of the metric system is here introduced; for our astronomer, anxious to determine the density and mass of *Gallia*, as the fragment had now been named (this is more pardonable than Gallium), finds that not only the metre of the archives, but all other measures whatever had disappeared. He shows that—

10 5-franc pieces 37 mm. in diameter
10 2-franc   ,,   27 mm.   ,,
20 50-cent.   ,,   18 mm.   ,,

exactly make a metre.

Then comes the dissertation on gravity:

"'Now, gentlemen,' said the Prof. Rosette, 'we are in a position to complete our calculations; we can now arrive at Gallia's attraction, density, and mass.'
"Every one gave him their complete attention.
"'Before I proceed,' he resumed, 'I must recall to your minds Newton's general law, "that the attraction of two bodies is directly proportional to the product of their masses, and inversely proportional to the square of their distances."'

* * *

"As he spoke the professor designedly kept his eyes fixed upon Ben Zoof. He was avowedly following the example of Arago, who was accustomed always in lecturing to watch the countenance of the least intelligent of his audience, and when he felt he had made his meaning clear to him, he concluded that he must have succeeded with all the rest. In this case, however, it was technical ignorance, rather than any lack of intelligence, that justified the selection of the object of this special attention.

And so on. Lockyer sums up:

We are glad to have such books to recommend for boys' and girls' reading. Many young people, we are sure, will be set thirsting for more solid information.

Lockyer was nothing if not opinionated. He was not alone in holding that artists must take note of scientifically established fact and was not above lecturing them on their shortcomings. Here is part of his analysis of the paintings displayed at the Royal Academy summer exhibition.

I next come to those pictures which I think are inaccurate in colour.

86. "Christiana with her Family, accompanied by Mercy, arrive at the Slough of Despond : Mercy finds a way across"—R. Thorburn, A. Impossible cloud colours. Clouds bluer than sky and atmosphere nowhere.

146. "Solitude"—P. F. Poole, R.A. Impossible green sky and cloud.

201. B. Riviere, A. Unnatural moonlight and impossible pea-soup shadows. The softness and colour of the latter suggest that Mr. Riviere has never studied moonlight.

231. "David, the Future King of Israel, while a Shepherd at Bethlehem"—J. R. Herbert, R.A. Colour impossible both in quantity and quality.

240. "A Dream of Ancient Egypt : the Morning of the Exodus"—Andrew MacCallum. I should like to hear the painter lecture on the connection of the colours of bodies with the light which falls upon them.

298. "Jarl Hacon in the Pentland Firth"—J. Hope M'Lachlan. High blotches of red over green and yellow impossible, and brick-dust beams of light proceeding from nothing still more impossible.

309. "The Sunrise Gun, Castle Cornet, Guernsey"—Tristram Ellis. Sky colour good; impossible colour of water under sky conditions given.

353. "After the Rain"—W. H. W. Foster. Unnatural sunset, colour and distribution of light wrong.

424. "The Last Journey"—Clara Montalba. Impossible green sky; the sun is neither setting nor set.

483. "An Autumn Sunrise" — Cecil G. Lawson. Interesting as a foretaste of the future when the sun shall have cooled.

525. A. Dixon. Green hopelessly wrong.

542. "The Dee Sands"—J. W. Oakes, A. Sky colours impossible with so high a sun.

555. "The Last of the Wreck"—E. Ellis. Green clouds !

630. "An Incident by the Wayside"—Mark Anthony. Impossible blue sky.

These, then, are the pictures I shall use as texts in my future notes.   J. NORMAN LOCKYER

The proposition about Homer's colour blindness was advanced by the one-time and future prime minister, William Ewart Gladstone, in his book about the poet (published by Macmillan). This started a debate that allowed the physicists and others to display in remorseless detail their grasp of classical Greek.

COLOUR BLINDNESS IN RELATION TO THE HOMERIC EXPRESSIONS FOR COLOUR

IN an article on "The Colour Sense" in the number of the *Nineteenth Century* for October last, Mr. Gladstone points out certain peculiarities, very remarkable and very difficult to account for, in the expressions for colour used by Homer. "Although," he says, "this writer has used light in its various forms for his purposes with perhaps greater splendour and effect than any other poet, yet the colour adjectives and colour descriptions of the poems are not only imperfect but highly ambiguous and confused," And again—"We find that his sense of colour was not only narrow, but also vague, and wanting in discrimination."

The article is an expansion of a chapter in the same author's "Studies on Homer and the Homeric Age," published in 1858 (vol. iii. page 457), from which the proposition is quoted ; " That Homer's perception of the prismatic colours, or colours of the rainbow, and *à fortiori* of their compounds, were, as a general rule, vague and indeterminate." Mr. Gladstone gives many examples illustrating these opinions, and by powerful and ingenious reasoning, he endeavours to establish from them the general conclusion that "the organ of colour was but partially developed among the Greeks of the heroic age."

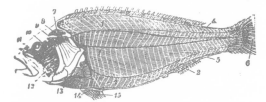

Here is another small sample to give the tone of the argument, from an article that continues for several pages of small print.

SO far as I can follow Mr. Gladstone's investigations, it appears to me that Homer has exactly fulfilled all the conditions mentioned in the previous article. As many references are made to natural objects which have the same colours now as they had in his time, I am able, with my colour-blind experience, to judge what sensations they would present to his eyes, supposing him colour-blind, and I can thus form a judgment of the appropriateness and consistency of his descriptions on that hypothesis. I can clearly trace the existence of two groups of epithets, which, so far as I can see, are kept fairly distinct, and the words in which are never mixed up with the ideas belonging to the contrary group. The epithets are—

For the group of the yellow sensation : ξανθός, ἐρυθρός, φοῖνιξ, ῥοδόεις, χλωρός, κυάνεος, and perhaps οἶνοψ.

For the group of the blue sensation : πορφύρεος and ἰοειδής.

For neutral sensations, irrespective of the words λευκός and μέλας (which may be left out of consideration altogether, the use of them being normal, and the vision of the colour-blind in regard to them being normal also) there is the epithet πολιός, on which an important element of the argument hangs.

Claude Bernard, the father of modern physiology, was hounded through much of his working life by the anti-vivisectionists, among whom were numbered his wife and daughter. The wheeze described below probably met with less protest: the siege of Paris was not kind to animals, all of which — from Castor and Pollux, the two elephants in the zoo, to sewer rats (according to Charles Joseph Bonaparte, a far greater delicacy than young chickens) — got eaten. A contemporary diary described how a large cat was being fattened to be served up at Christmas, surrounded by mice-like sausages; and Henry Labouchère, having dined off a piece of spaniel, recorded that dog epicures held poodle to be the best while bulldog was coarse and flavourless.

In his lecture on Claude Bernard, M. Paul Bert narrated a singular stratagem which was invented by Bernard during the last Franco-German war, and might be utilised without difficulty, under similar circumstances. It was proposed to re-victual Paris, which was strictly blockaded by German forces. A large number of cattle had been collected, waiting for an opportunity to cross the German lines. But a difficulty was to silence these animals, as their cries would attract the attention of the enemy. Claude Bernard proposed to practise upon them the section of the nerve which enables them to emit their usual cries. The operation is so easy that it could be executed in a few seconds by an ordinary butcher. None of the animals appeared to suffer in any way by the mutilation which had made them mute. But the military movement proved a failure, and for other causes the re-victualling could not take place.

VOL 19, 325 1879

Elie Metchnikoff, whom we shall meet again, was an erratic genius, clearly admired by a brother-naturalist.

RESEARCH UNDER DIFFICULTIES

THE following short preface to a very valuable account of the stages of development from the egg of one of the centipedes (Geophilus), no member of which group had been studied previously to this account, gives so convincing a picture of the enthusiasm for investigation which may animate the modern naturalist, that it is worthy of a place in NATURE for the encouragement of the "craft." Elias Metschnikoff has during the past fifteen years worked more assiduously with the microscope at the observation of the minute details of embryology than any other student. To him we are indebted for our first accurate knowledge of this subject in the case of many important animal forms, *e.g.*, sponges, various jelly-fishes, marine worms, the scorpion, and the book-scorpions, various insects, crustaceans, starfishes, and ascidians. One result has been the injury of his eyesight. In reading to-day his memoir on Geophilus, published in 1875 (*Zeitschr. für wiss. Zoologie*), it occurred to me that the following passage has more than technical interest :—

"After having for many years sought in vain for material suited for the investigation of the embryology of the centipedes, I chanced to obtain a quantity of the eggs of Geophilus. My find, however, took place under such circumstances, and these interfered so much with my investigation, that I feel justified in describing them more minutely. For some considerable time I had been afflicted with a chronic affection of the eyes, and consequently commenced in the spring of the present year a journey to our south-eastern steppes in order to turn my attention to anthropological studies. Instead of taking with me as in previous years all the apparatus necessary for microscopical research, I took this time on my journey only anthropological measuring instruments. When, then, I was in the neighbourhood of Manytsch, nearly in the heart of the Kalmuk steppes, and was visiting a small forest plantation, I discovered quite unexpectedly a number of eggs of Geophilus which had been deposited under the bark of a rotten tree-stem where the females were watching over them. I gathered up the precious material, and having packed it carefully in two bottles, set off with all speed to Astrachan, in order there to set about the microscopic investigation of the eggs. But when, after four days' travelling I arrived in a Russian village, Jandiki, near the shore of the Caspian Sea, and inspected my two bottles, I found in them only a couple of dead, opaque eggs, all the others having entirely disappeared. Fortunately I succeeded in Jandiki, where there is also a small plantation, in obtaining fresh material of the same kind, and this I brought in good condition to Astrachan, making the journey by steamboat. In the town of Astrachan I was able to borrow a Hartnack's microscope from a medical man practising there, and on a second journey took it with me to Jandiki. In this way I was enabled to make out the chief features of the developmental history of Geophilus by the use of my less seriously affected left eye. At the same time, in spite of the very favourable character of the Geophilus-eggs for microscopic research, I could not bring my work to the desired degree of completeness."

Determination and pluck have their scope in embryology !

E. RAY LANKESTER

VOL 19, 342 1879

Here is James Clerk Maxwell, atoning for a disobliging review.

Guthrie's " Physics "

SOME weeks ago (p. 311) you published in NATURE a review by Prof. Maxwell of a little book of mine on Practical Physics. It is not my intention to complain in any way of the review, partly because it would be a profitless trespass on your space, but mainly because, while the tone is unfavourable, the instances adduced by the reviewer go a long way to confute his own statements in all cases where there is any connection between the two.

Some well-meaning friend has composed and sent me a copy of the inclosed. There appear to be various opinions as to the authorship. It has even been suggested that Prof. Maxwell, with that sense of humour for which he is so esteemed, and with a pardonable love of mystification, is himself the author.

February 24   FREDK. GUTHRIE

REMONSTRANCE TO A RESPECTED DADDIE ANENT HIS LOSS OF TEMPER

Suggested by Prof. CLERK MAXWELL'S *review of* GUTHRIE'S " PHYSICS "

Worry, through duties Academic,
    It might ha'e been
That made ye write your last polemic
    Sae unco keen :

Or intellectual indigestion
    O' mental meat,
Striving in vain to solve some question
    Fro' "Maxwell's Heat."

Mayhap that mighty brain, in gliding
    Fro' space tae space,
Met wi' anither, an' collidin',
    Not face tae face.

But rather crookedly, in fallin'
    Wi' gentle list,
Gat what there is nae help fro' callin'
    An ugly twist.

If 'twas your "demon" led ye blindly,
    Ye should na thank him,
But gripe him by the lug and kindly
    But soundly spank him.

Sae, stern but patronising daddie !
    Don't ta'e 't amiss,
If a puir castigated laddie
    Observes just this :—

Ye 've gat a braw new Lab'ratory
    Wi' a' the gears,
Fro' which, the warld is unco sorry,
    'Maist naught appears.

A weel-bred dog, yoursel' must feel,
    Should seldom bark.
Just put your fore paws tae the wheel,
    An' do some Wark.

$$d \sqrt{\frac{m}{n}}.$$

VOL 19, 384 1879

Another example of one of NATURE *authors' favourite themes:*

Intellect in Brutes

DR. RAE has so fairly disposed of Mr. Henslow's examples of so-called "practical" and "abstract reasoning" that further comment is unnecessary. As, however, the subject of intellect in brutes is on the *tapis*, I will give an instance of sagacity in a dog that finally set at rest any doubts I ever entertained that the difference between human and animal intelligence is one of degree only.

If you have space for it, the accompanying plan will be of great value in describing the circumstances.

Mr. J. W. Cherry, of the Madras Forest Service, was owner of the dog in question, a bull terrier, called "Bully." We

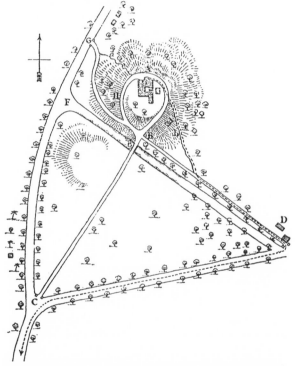

lived in the bungalow (A), the compound of which was bounded south and west by public roads (D C) and (G F C) both leading to the cantonment of Mangalore in the direction C. There were three gates into the compound at (C) (D) and (G), the main approach to the Bungalow leading over a bridge (B), that spanned a branch public road (F D). The compound was filled with trees and shrubs, and bordered by dense lantana hedges, so that with the exception of a portion of the western road at F, neither of the cantonment roads were visible from the bridge, nor could the foot-paths (a) and (b) be seen thence.

Now Bully had a lady friend (canine) living in the cantonment, and at times she was so attractive that absences without leave on the part of the dog were frequent. After one of these excursions Bully had been brought back, and chained up for the night. Next morning, while his master and I were sitting at early breakfast, it was decided that he should be released, and to effectually stop further delinquency, a peon was sent down to the bridge with orders to intercept him if he started for the cantonment.

Bully was brought in and unchained; he had that unmistakable air of detected guilt deservedly punished, and spent some time in begging for scraps from the table in a most deprecating manner. Shortly, however, he strolled into the verandah, and then down the front steps on to the gravel walk. After wandering about aimlessly for a few minutes, he quietly started off down the approach (A H B). We followed, keeping out of his sight. At the turn of the road Bully met the unexpected apparition of the peon standing on the bridge. In a moment, though not a word was spoken by the man, the dog turned and came straight back to the room, whither we had in the meantime slipped back unobserved, and re-entered it wagging his tail violently and looking exceedingly sheepish. He now lay down and closed his eyes. The cocked ears showed that sleep was mere pretence, and he soon rose again, went out into the front garden, and hunted for buried bones, purely imaginary ones, I believe. His search gradually led him down the hill by the foot-path (a),—we keeping him in sight, as before—and he finally reached the road at the bottom. There all disguise was dropped, and he started off for the cantonment. As he neared the spot (F) the peon espied him, and shouted out his name. He turned at once, climbed the hill, and came into the bungalow, where the same farce of repentance was gone through.

Bully now seemed to have made up his mind that escape was impossible; he lay down on a mat in the verandah, and remained there for a long time. But for the persistent cock of the ears we should have imagined the animal really asleep. Mr. Cherry eventually went to his office-room, and I remained in the verandah reading the morning paper, and occasionally glancing at Bully. He lay very still, but once or twice I detected him opening his eyes and raising his head to look round him. Each time he caught my eye he wagged his tail vehemently for a moment or two, and then resorted to his sham sleep.

It may have been for half-an-hour, or thereabouts, that this state of things continued. I then became interested in an article in the paper, and when I next looked up Bully was gone. I called Mr. Cherry, and the house was searched. No Bully. The peon was sent for and interrogated; he had not seen the dog. As a last resource inquiry was made of the horsekeepers down at the stables (D). The reply was—"Yes, the dog had passed through the gate (D) some time before." Taking advantage of my occupation and the absence of his master, Bully had left the house, and taken his way to the cantonment by the only path by which he could have escaped unnoticed by the peon—that shown by the dotted line.

In this necessarily short account I have hardly done justice to Bully's diplomatic powers, but most of your readers will appreciate the intelligence that led the dog to successfully elude the watch set over him.

E. H. PRINGLE

VOL 19, 458 1879

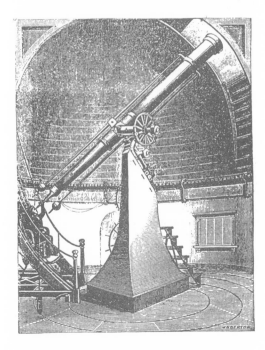

The Duke of Argyll was a frequent contributor to Nature. *He was a considerable naturalist, but an implacable opponent of Charles Darwin and appears to have made a comprehensive fool of himself on regular occasions.*

A Carnivorous Goose

I INCLOSE to you an account of a Golden Eagle, which I have reason to know to be authentic. The possibility of a bird so purely graminivorous as a goose being taught to eat flesh, and acquiring the power of digesting it, is extremely curious. It is well known, however, that cows are largely fed on fish offal in Scandinavia, and I have heard of a Highland cow devouring a salmon which an unwary angler had hid among fern on the banks of a river in Sutherland.   ARGYLL

Isola Bella, Cannes, April 7

"March, 1879.—There is in the possession of W. Pike, Esq., at Glendarary, in the Island of Achil, Co. Mayo, a Golden Eagle, now about twenty-five years old, which was taken from the nest and brought up in confinement. This eagle, in the spring of 1877 laid three eggs, which Mr. Pike took away, replacing them with two goose-eggs, upon which the eagle sat, and in due time hatched two goslings. One of these died, and was torn up by the eagle to feed the survivor, who, to the great tribulation of its foster-parent, refused to touch it, together with the other flesh with which the eagle tried to feed it, Mr. Pike providing it with proper food. The eagle, however, in course of time, taught the goose to eat flesh, and (the goose having free exit and ingress to the eagle's cage) always calls it by a sharp bark whenever flesh is given to it, when the goose hastens to the cage and greedily swallows all the flesh, &c., which the eagle, tearing its prey to pieces, gives it.

"I saw them in May, 1878, when, the goose being a year old, had made a nest in the eagle's cage, and laid eleven eggs, and the two birds were sitting side by side on the nest. I hear from Mr. Pike that he did not allow them to hatch out, fearing that it might interfere with their attachment to one another.

"The eagle is very tame and fond of Mr. Pike ; he goes into the cage, and it allows him to handle it as he likes, but will not allow any one else near it. It never attempts to get out of the hole made for the goose to go in and out."

Gustave Le Bon was an all-purpose savant. What conclusions were drawn from his measurements of cranial capacities is not recorded.

THE cranium of Descartes is often adduced as an exception to the general rule that a great mind requires a large brain. This statement seems to have rested on no exact measurement, and Dr. Le Bon resolved recently to test its accuracy. The result is that he finds the cubic capacity of Descartes' skull to be 1,700 centimetres, or 150 centimetres above the mean of Parisian crania of the present time. At the same time Dr. Bordier has recently found the average capacity of the skulls of thirty-six guillotined murderers to be 1,547·91 c.c., the largest reaching the high figure of 2,076 c.c.

The Reverend Charles Lutwidge Dodgson, better known as Lewis Carroll, was a mathematician long before he produced the Alice books. His writing was as entertaining as his lectures were dismal. He observed ruefully that he had an audience of 95 people for his first lecture but that for his last, 25 years later, there were two. Here is the genial start of an anonymous review of his book on geometry.

A few words by way of introduction. Mr. Dodgson has been a teacher of geometry at Oxford, we believe, for nearly five-and-twenty years, and during that time has had frequent occasion to examine candidates in that subject. For a great part of the above-stated period things went pretty smoothly, and King Euclid held undisputed sway in the "Schools ;" but eleven years ago a troubler of the geometrical Israel came upon the scene, and read a paper before the Mathematical Society, entitled "Euclid as a Text-Book of Elementary Geometry." The agitation thus commenced acquired strength, and at length, in consequence of a correspondence carried on in these columns, the Geometrical Association was formed. A prime mover in this matter was that Mr. Wilson who wrote the paper, and subsequently brought out the geometry cited. Mr. Dodgson is one of the gentlemen opposed to this change, and the moving cause of the present Iliad is the "vindication of Euclid's masterpiece." Another consequence of the agitation is that many have tried their prentice hands on the production of new geometries—"rivals," our author calls them—"forty-five were left in my rooms to-day." Can we wonder then, that, his soul being stirred within him, he should overhaul a selection of them to see what blots he could "spot" in them? He might well have taken for his motto one once familiar to us—

> " If there's a hole in a' your coats,
> I rede ye tent it ;
> A chiel's amang ye takin' notes,
> An' faith he'll print it ! "

The review then gives a taste of the Dodgson style:

"Minos"—who argues for Mr. Dodgson—himself seems to think that his remarks will now and again be considered hypercritical. Take the following :—

Niemand (the general representative of the "rivals," quoting from the "Elementary Geometry"). Two straight lines that meet one another form an angle at the point where they meet (p. 5). *Min.* Do you mean that they form it "*at* the point," and nowhere else ? *Nie.* I suppose so. *Min.* I fear you allow your angle no magnitude, if you limit its existence to so small a locality ! *Nie.* Well, we *don't* mean "nowhere else." *Min.* (*meditatively*). You mean *at* the point—and *some*where else ? *Where* else, if you please ? *Nie.* We mean—we don't quite know why we put in the words at all. Let us say "Two straight lines that meet one another form an angle." *Min.* Very well. It hardly tells us what an angle *is*, and, so far, it is inferior to Euclid's definition ; but it may pass. Again (p. 73), *Nie.* reads, P. 5, Ax. 5, "Angles are equal when they could be placed on one another so that their vertices would coincide in position, and their arms in direction." *Min.* "Placed on one another !" Did you ever see the child's game, where a pile of four hands is made on the table, and each player tries to have a hand at the top of the pile ? *Nie.* I know the game. *Min.*

Well, did you ever see both players succeed at once? *Nie.* No. *Min.* Whenever that feat is achieved you may *then* expect to be able to place two angles " on one another !" You have hardly, I think, grasped the physical fact that, when one of two things is *on* the other, the second is *underneath* the first. But perhaps I am hypercritical.

VOL 21, 240 1879

James Clerk Maxwell, the pre-eminent physicist of his time, died in 1879 at the age of 48. He seems to have been an exceptionally sweet-natured man and altogether devoid of vanity. This is the conclusion of a long obituary notice.

About five or six days before his death, when he was suffering from such extreme weakness that he could say very little, after lying motionless with his eyes closed for some time, he presently looked up and remarked, "'Every good gift and every perfect gift is from above, and cometh down from the Father of lights, with whom is no variableness, neither shadow of turning.' Do you know that is a hexameter ?

' πᾶσα δόσις ἀγαθὴ καὶ πᾶν δώρημα τέλειον,'

I wonder who composed it."

His knowledge of hymns and hymn-writers was very extensive, and he took great pleasure during his illness in reciting from memory some of his favourites among the writings of Richard Baxter, George Herbert, and others.

To attempt to give any adequate idea of his contributions to science in a sketch like the present would be but to mislead the reader. His great work on "Electricity and Magnetism," the second edition of which is now in the press, is the admiration of mathematical physicists. More generally known are his treatise on the Theory of Heat, and his little text-book entitled "Matter and Motion" which was published by the S.P.C K. One of his earliest papers on the "Theory of Rolling Curves," was communicated to the Royal Society of Edinburgh by Professor Kelland, and read on February 19, 1849, when Clerk Maxwell was an Edinburgh student barely eighteen years of age. His paper on the "Equilibrium of Elastic Solids," above alluded to, was read before the same society on February 18, 1850. His paper on the "Transformation of Surfaces by Bending" was read before the Cambridge Philosophical Society on March 13, 1854, about two months after taking his degree. This was followed in December, 1855, and February, 1856, by papers on "Faraday's Lines of Force." In 1857 he obtained the Adams Prize, in the University of Cambridge, for his paper on the "Motions of Saturnian Rings." His paper on the "Theory of Compound Colours, and the Relations of the Colours of the Spectrum," which was chiefly instrumental in gaining the Rumford Medal, was read before the Royal Society on March 22, 1860. His "Dynamical Theory of the Electromagnetic Field," including a brief sketch of the Electromagnetic Theory of Light, was read before the Royal Society on December 8, 1864. The results of Clerk Maxwell's experiments on "The Viscosity and Internal Friction of Air and other Gases," were made known to the Royal Society in the Bakerian Lecture read, February 8, 1866. Then follow his Royal Society papers "On the Dynamical Theory of Gases," in May, 1866, and "On a Method of Making a direct Comparison of Electrostatic with Electromagnetic Force, with a Note on the Electromagnetic Theory of Light," in June, 1868. Lately he took great interest in Graphical Statics, and contributed a long paper "On Reciprocal Figures, Frames and Diagrams of Forces," to the Royal Society of Edinburgh, in December, 1869. Among his most recent papers are a paper on "Stresses in Rarefied Gases arising from Inequalities of Temperature," read

before the Royal Society on April 11, 1878, and a paper on "Boltzmann's Theorem," read before the Cambridge Philosophical Society. It would take too long to enumerate his articles and reviews published in the *Philosophical Magazine* and in NATURE. His contributions to the ninth edition of the "Encyclopædia Britannica" include the articles "Atom," "Attraction," "Capillary Action," "Constitution of Bodies," "Diagrams," "Diffusion," "Ether," "Faraday," and "Harmonic Analysis." "Harmonic Analysis" was the last article he wrote.

One of the most remarkable of his works is the recently-published volume of the Electrical Researches of the Hon. Henry Cavendish, of which Prof. Maxwell is the editor. The MSS. are in the possession of the Duke of Devonshire, and are now at Chatsworth. They were entrusted by him to Prof. Maxwell shortly after the completion of the Cavendish Laboratory. Some of Cavendish's experiments were repeated by Prof. Maxwell with all the appliances of modern apparatus, and others were carried out by his pupils.

Most of the apparatus which he employed in his researches has been presented by Prof. Clerk Maxwell to the Cavendish Laboratory, together with many of his books. He always regarded the laboratory with great affection, and the University owes much to his liberality. One of the most interesting pieces of his handy-work now preserved in the laboratory is a plaster model of Prof. Willard Gibbs's thermodynamic surface, described in the fourth edition of "Maxwell's Theory of Heat." All the lines on the surface are drawn by his own hand, many of them being mapped out by placing the surface obliquely in the sunshine and marking the boundary between light and shade. Another valuable model constructed while Prof. Maxwell was at Cambridge is his dynamical illustration of the action of an induction coil in which two wheels represent by their rotation the primary and secondary currents respectively, the wheels being connected through a differential gearing to which a body of great moment of inertia is attached, the rotation of which represents the magnetism of the coil. A friction break represents resistance, and a spring may be attached to the secondary wheel to represent the capacity of a condenser placed in the secondary circuit. Among other valuable pieces of apparatus presented by Prof. Maxwell to the laboratory are the receiver, plates, and inertia bar employed in his researches on the viscosity of air and other gases, his colour-top, portions of the "colour-box," including the variable slits, with the wedge for measuring their width, a polariser and analyser made of thin films of stretched gutta percha, the mechanism for illustrating the motion of Saturnian rings, a real image stereoscope, and the dynamical top, whose moments of inertia about three axes, which are at right angles to each other, can be so varied by means of screws that the axis of rotation can be made that of greatest or of least moment of inertia. When the axis of rotation is the mean axis, the motion of the top is, of course, unstable. When Prof. Maxwell came to Cambridge in 1857 to take his M.A. degree, he brought this top with him from Aberdeen. In the evening he showed it to a party of friends in college, who left the top spinning in his room. Next morning he espied one of these friends coming across the court, so jumping out of bed, he started the top anew, and retired between the sheets. The reader can well supply the rest of the story for himself. It is only necessary to add that the plot was completely successful.

Prof. Clerk Maxwell's papers will be placed in the hands of Prof. Stokes, who is one of his executors, in order that they may be published or catalogued and preserved in such a way as to be readily available to those wishing to consult them.

The death of James Clerk Maxwell is a loss to his University and to the world too great for words. He rests from his labours, but his works will follow him.

WM. GARNETT

But not even obituaries are out of bounds to the pedant.

The "Hexameter," Πᾶσα δόσις ἀγαθή . . .

THERE is an obstacle in the way of regarding this passage (James i. 17) as a hexameter quoted by the Apostle from some poet, as the late lamented Prof. Clerk Maxwell is reported in Mr. Garnett's interesting notice of his life, work, and, not least, his character, to have suggested. The final syllable of δόσις is short, as the accentuation of πρᾶξις and similar verbal nouns proves. *Arsis*, as in "Βέλος ἐχεπευκές," Il. α, 51, can hardly be pleaded.   J. J. WALKER
University Hall, W.C., November 17

VOL 21, 57 1879

The German expert on felines seems to doubt whether the tailless cat really exists, and he places the Isle of Man in the Pacific.

GEOGRAPHICAL NOTES

THE Germans have so deservedly earned a distinguished reputation as scientific geographers, that it is quite pleasing to catch one very seriously tripping in geographical matters. In Philip Leopold Martius's "Das Leben der Hauskatze und ihrer Verwandten" (Weimar : B. F. Voigt, 1877), in the part of the work treating of the varieties of the domestic cat, appears (s. 61) the following extraordinary statement : "Die schwanzlose Katze von der Insel Man *im stillen Ocean* wenn nicht das *Kap Man auf Borneo* darunter zu verstehen, ist wohl noch nie zu uns nach Deutschland gekommen, obgleich sie auf der Katzenausstellung in London einst vertreten war." The author goes on to express his earnest wish that a pair of these great rarities, Manx cats, may be procured and exhibited at some zoological garden. Manxmen will hardly thank him for placing their native isle in the Pacific Ocean and confounding them with Polynesians, but the suggestion as the result of ponderous research that after all perhaps such a place as the Isle of Man does not exist, but that its mythical development has arisen from a mistake as to a cape of the same name in Borneo is too delicious altogether, and so ingenious and thoroughly German that it must needs be recorded for the benefit of the readers of NATURE.

VOL 21, 49 1879

Arachnids too have intelligence.

Intellect in Brutes

SOME time since I observed the following conduct of two spiders, which will show how they sometimes overcome difficulties in the way of capturing their prey. A rather large house-spider had its web in the corner of a room, and during the summer it feasted upon the insects that were unlucky enough to be caught. One evening I noticed a large dipterous insect strike the web ; the spider darted out and succeeded in fastening one foot of the fly. The spider then kept running back and forth, attaching a thread to a wing, then to a leg, which was soon broken by the violent efforts of the fly to release itself. The spider worked without ceasing for over half an hour to secure its victim ; it then quitted operations, and retired to a distant corner of its web. After seeming to reflect for a while what was best to do, it left the web, went up the wall eight or ten inches distant, and entered a crack in the ceiling. I supposed at the time that the spider had been injured in the scuffle, but what was my surprise after a few moments to see the spider coming back, and close behind another followed ; the two went on the web near the centre, and stopped side by side, apparently consulting as to the best mode of attack. Then at the same instant both spiders darted upon the insect, one towards the head, the other towards the tail. So rapid were their movements I could hardly follow them. In a short time the insect was securely fastened. Both spiders then returned to the centre of the web. Soon after the friendly spider went to the crack in the ceiling, while the other enjoyed the feast alone.   A. M.
North Manchester, Indiana, U.S., February 25

VOL 21, 494 1880

Now a scientific feat of detection. The effect of surface layers was already known. Benjamin Franklin would astonish his friends by stilling turbulent water when he passed his stick over the surface: the stick was hollow and contained oil, which he would release at the right moment.

Remarkable Discovery of a Murder in Bermuda

THE following account of a murder which was committed in Bermuda in the autumn of 1878 is taken from a letter written to Gen. Sir J. H. Lefroy, C.B., F.R.S., lately Governor of these islands, and author of the "Annals of Bermuda," by the Attorney-General of the islands, Mr. S. Brownlow Gray. The mode of discovery of the crime is so remarkable that I think it ought to be put on record, and Sir J. H. Lefroy has kindly permitted me to make extracts from the letter for that purpose. I believe no account of the circumstances of the case has as yet been published in Europe. There seems to be no likelihood as to mistake regarding the facts. The special occurrence could probably only happen in the tropics in warm water.
H. N. MOSELEY

"In the autumn of 1878 a man committed a terrible crime in Somerset, which was for some time involved in deep mystery. His wife, a handsome and decent mulatto woman, disappeared suddenly and entirely from sight, after going home from church on Sunday, October 20. Suspicion immediately fell upon the husband, a clever young fellow of about thirty, but no trace of the missing woman was left behind, and there seemed a strong probability that the crime would remain undetected. On Sunday, however, October 27, a week after the woman had disappeared, some Somerville boatmen looking out towards the sea, as is their custom, were struck by observing in the Long Bay Channel, the surface of which was ruffled by a slight breeze, a long streak of calm such as, to use their own illustration, a cask of oil usually diffuses around it when in the water. The feverish anxiety about the missing woman suggested some strange connection between this singular calm and the mode of her disappearance. Two or three days after—why not sooner I cannot tell you—her brother and three other men went out to the spot where it was observed, and from which it had not disappeared since Sunday, and with a series of fish-hooks ranged along a long line dragged the bottom of the channel, but at first without success. Shifting the position of the boat, they dragged a little further to windward, and presently the line was caught. With water glasses the men discovered that it had caught in a skeleton which was held down by some heavy weight. They pulled on the line ; something suddenly gave way, and up came the skeleton of the trunk, pelvis, and legs of a human body, from which almost every vestige of flesh had disappeared, but which, from the minute fragments remaining, and the terrible stench, had evidently not lain long in the w..!er. The husband was a fisherman, and Long Bay Channel was a favourite fishing-ground, and he calculated, truly enough, that the fish would very soon destroy all means of identification ; but it never entered into his head that as they did so their ravages, combined with the process of decomposition, would set free the matter which was to write the traces of his crime on the surface of the water. The case seems to be an exceedingly interesting one ; the calm is not mentioned in any book on medical jurisprudence that I have, and the doctors seem not to have had experience of such an occurrence. A diver went down and found a stone with a rope attached, by which the body had been held down, and also portions of the scalp and of the skin of the sole of the foot, and of clothing, by means of which the body was identified. The husband was found guilty and executed."

VOL 22, 170 1880

A prototypic form of broadcasting:

A VERY curious telephonic experiment has been made in Switzerland on the occasion of the Federal *fête* of singers. A telephone had been placed in the Zürich Festhalle and two conductors connected with the Bâle telegraphic office, where a large audience had congregated. The distance from Bâle to Zürich is about 80 kilometres. The Bâle audience enjoyed the singing about as well as if they had been placed in the upper circle of an ordinary Opera House. At the end of the performance they proved their satisfaction by clapping hands, which the telegraphic wires transmitted with perfect fidelity to the Zürich performers.

VOL 22, 329 1880

The taste for animal stories was then, as now, unquenchable.

THE subject of a depraved taste in animals is an interesting one, which has not been studied as much perhaps as it might. In human beings it would seem to depend on ill-health of either body or mind, but in animals it would seem as if it might be present and the animal enjoy good health. One remarkable instance in an herbivorous animal we can vouch for. It occurred in a sheep that had been shipped on board one of the P. and O. steamers to help to supply the kitchen on board, but while fattening it developed an inordinate taste for tobacco, which it would eat in any quantity that was given to it. It did not much care for cigars, and altogether objected to burnt ends ; but it would greedily devour the half-chewed quid of a sailor or a handful of roll tobacco. While chewing there was apparently no undue flow of saliva, and its taste was so peculiar that most of the passengers on board amused themselves by feeding it, to see for themselves if it were really so. As a consequence, though in fair condition, the cook was afraid to kill the sheep, believing that the mutton would have a flavour of tobacco. Another very remarkable case has just been communicated to us by Mr. Francis Goodlake : this time a flesh-eating animal in the shape of a kitten, about five months old, who shows a passionate fondness for salads. It eats no end of sliced cucumber dressed with vinegar, even when hot with cayenne pepper. After a little fencing it has eaten a piece of boiled beef with mustard. Its mother was at least once seen to eat a slice of cucumber which had salt, pepper, and vinegar on it. The kitten is apparently in good health, and its extraordinary taste is not easily accounted for. Even supposing it once got a feed of salmon mayonaise, why should it now select to prefer the dressing to the fish ?

VOL 22, 329 1880

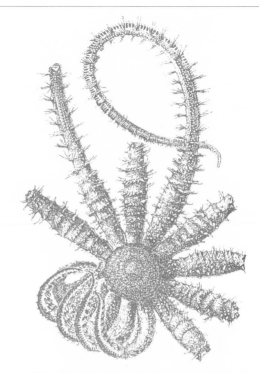

It was H. L. Mencken's view that the names of Thomas Alva Edison and Alexander Graham Bell would remain ever accursed, for they had done more than anyone else to add to life's damned nuisances. Here Bell is commended by a well known and very vocal experimental physicist, Silvanus P. Thompson.

THE PHOTOPHONE

EARLY in the summer of the current year it was announced in the columns of NATURE that Prof. Graham Bell had made a discovery which, for scientific interest, would rival the telephone and phonograph, and that he had deposited in a sealed packet, in the custody of the Smithsonian Institute, the first results of his new researches ; that announcement has now received its due fulfilment in the lecture by Prof. Bell to the American Association for the Advancement of Science, on Selenium and the Photophone, which will be found on another page. In spite of those who ingenuously attempted at the time of our announcement to forestall Prof. Bell and to discredit the idea that he had done anything new, the discovery, which he has now published, is a startling novelty. The problem which he has attacked is that of the transmission of speech, not by wires, electricity, or any mechanical medium, but by the agency of light. The instrument which embodies the solution of this principle he has named the Photophone. It bears the same relation to the telephone as the heliograph bears to the telegraph. You speak to a transmitting instrument, which flashes the vibrations along a beam of light to a distant station, where a receiving instrument reconverts the light into audible speech. As in the case of that exquisite instrument, the telephone, so in the case of the photophone, the means to accomplish this end are of the most ridiculous simplicity. The transmitter consists of a plane silvered mirror of thin glass or mica. Against the back of this flexible mirror the speaker's voice is

The Articulating Photophone. The Transmitter.

directed; a powerful beam of light is caught by a lens from the sun and directed upon the mirror, so as to be reflected straight to the distant station. This beam of light is thrown by the speaker's voice into corresponding vibrations. At the distant station the beam is received by another mirror, and concentrated upon a simple disk of hard rubber fixed as a diaphragm across the end of a hearing-tube. The intermittent rays throw the disk into vibration in a way not yet explained, yet with sufficient power to produce an audible result, thus reproducing the very tones of the speaker. Other receivers may be used, in which the variation in electrical resistance of selenium under varying illumination is the essential principle. The experimental details have been worked out by Prof. Bell in conjunction with Mr. Sumner Tainter. They have discovered that other substances beside hard rubber, gold, selenium, silver, iron, paper, and notably antimony, are similarly sensitive to light. This singular production of mechanical vibrations by rays of light is even more mysterious than the production of vibrations in iron and steel by changes of

magnetisation. It was indeed this latter fact which led the discoverers to suspect the analogous phenomenon of photophonic sensibility in selenium and in other substances. Hitherto, in consequence of the mere optical difficulties of managing the beam of light, the distance to which sounds have been actually transmitted by the Photophone is less than a quarter of a mile, but there is no reason to doubt that the method can be applied to much greater distances, and that sounds can be transmitted from one station to another wherever a beam of light can be flashed; hence we may expect the slow spelling out of words in the flashing signals of the heliograph to be superseded by the more expeditious whispers of the Photophone.

We congratulate Prof. Bell most sincerely on this addition to his well-won laurels, and venture to predict for his photophone a great, if not a widely-extended, future of usefulness.   Silvanus P. Thompson

Vol 22, 481 1880

Here now an accomplished, if unappealing, exercise in bacteriology:

The feet of certain individuals are characterised by a peculiar powerful and fœtid odour, which is really connected with the moisture that soaks the soles of the stockings and the inside of the boots. The moisture, which comes from the skin of the soles, especially from that of the heels, has no offensive smell whilst it is exuding, but it rapidly acquires the characteristic odour when taken up by the stocking.

The fluid is an admixture of sweat with serous exudation from the blood, occurring in persons whose feet sweat profusely, and who, from much standing or walking, acquire an erythematous or eczematous condition of the skin of the soles, the local eczema or erythema being favoured by the softening and macerating effect of the sweat on the epidermis.

When a small portion of the sole of the wet stocking was teased out in water, the drop of water was found to be swarming with micrococci.

A second generation of the organism, which the author calls *Bacterium fœtidum*, was obtained by placing a small piece of the wet stocking in a test-glass, charged with pure vitreous humour. This and succeeding generations were cultivated at a temperature which varied between 94° and 98° F. The successive generations were obtained by inoculating pure vitreous humour, with requisite precautions.

In twenty-four hours the surface of the vitreous humour was always found covered with a delicate scum, which in forty-eight hours was compact and tolerably resistant.

In the scum of one day's growth and in the fluid below it organisms were found as cocci, single and in pairs, in transition stages towards rod formation, as single and jointed rods, and as elongated single rods. Many of the rods were actively motile.

* * *

The bacterium grows in turnip infusion less actively than in vitreous humour. The observations were not sufficiently extended to determine whether the bacterium forms spores when cultivated in turnip infusion, but they sufficed to show that if such a formation takes place, it occurs much less actively than when the cultivation is in vitreous humour.

The fœtid odour of the stocking was reproduced in the cultivation glasses, although the strength of the odour diminished in successive generations.

Dr. Thin stated at the meeting that an antiseptic treatment by which the bacteria were killed in the stockings and inner surface of the soles of the boots completely destroyed the fœtor.

Vol 22, 209 1880

The Articulating Photophone. The Selenium Receiver.

chapter five
1880–1885

THE PHYSICIST CHARLES VERNON BOYS MAKES HIS FIRST CONTRIBUTION TO THE COLUMNS OF *NATURE* (ON HOW
TUNING-FORKS AFFECT SPIDERS) AND CHARLES DARWIN PENS HIS LAST: THE NEXT ISSUE CARRIES DARWIN'S OBITU-
ARY. KARL PEARSON, MATHEMATICIAN, GENETICIST AND EUGENICIST, DISPLAYS HIS CLASSICAL SCHOLARSHIP. AN
EARLY FAX MACHINE MAKES ITS APPEARANCE. EADWEARD MUYBRIDGE INVENTS THE PHOTO FINISH AND FRANCIS
GALTON DEVISES THE IDENTISCOPE AND PLOTS THE WEIGHT OF ENGLISH NOBLEMEN THROUGH THE YEARS.

1880-1885

W. E. Gladstone was back as prime minister of Britain, supplanting for a second time Disraeli. President James Abram Garfield was shot, and Mount Krakatoa erupted, causing a cataclysmic tidal wave. France presented the Statue of Liberty to the American people, and the Metropolitan Opera opened in New York and the Natural History Museum in London. Work stopped on the Channel tunnel.

Punch *finds diversion in science:*

SONGS OF THE SCIENCES—I. ZOOLOGY

WE must regard it as a noteworthy sign that science has begun to percolate so through society generally that it has reached the pages of *Punch*. Almost every week we find a bit of more or less telling waggery, and last week the first of a series of "Songs of the Sciences" appeared, which we reproduce :—

Oh ! merry is the Madrepore that sits beside the sea,
The cheery little Coralline hath many charms for me ;
I love the fine Echinoderms of azure, green, and grey,
That handled roughly fling their arms impulsively away :
Then bring me here the microscope and let me see the cells,
Wherein the little Zoophite like garden floweret dwells.

* * *

Then study well zoology, and add unto your store,
The tales of Biogenesis and Protoplasmic lore :
As Paley neatly has observed, when into life they burst,
The frog and the philosopher are just the same at first.
But what's the origin of life remains a puzzle still,
Let Tyndall, Haeckel, Bastian go wrangle as they will.

VOL 23, 148 1880

Charles Vernon Boys was an experimental physicist of unbounded ingenuity and one of NATURE's *favourite contributors over six decades. His curiosity was not limited to physics.*

THE INFLUENCE OF A TUNING-FORK ON THE GARDEN SPIDER

HAVING made some observations on the garden spider which are I believe new, I send a short account of them in the hope that they may be of interest to the readers of NATURE.

Last autumn, while watching some spiders spinning their beautiful geometrical webs, it occurred to me to try what effect a tuning-fork would have upon them. On sounding an A fork and lightly touching with it any leaf or other support, of the web or any portion of the web itself, I found that the spider, if at the centre of the web, rapidly slews round so as to face the direction of the fork, feeling with its fore feet along which radial thread the vibration travels. Having become satisfied on this point, it next darts along that thread till it reaches either the fork itself or a junction of two or more threads, the right one of which it instantly determines as before. If the fork is not removed when the spider has arrived it seems to have the same charm as any fly : for the spider seizes it, embraces it, and runs about on the legs of the fork as often as it is made to sound, never seeming to learn by experience that other things may buzz besides its natural food.

If the spider is not at the centre of the web at the time that the fork is applied, it cannot tell which way to go until it has been to the centre to ascertain which radial thread is vibrating, unless of course it should happen to be on that particular thread or on a stretched supporting thread in contact with the fork.

If when a spider has been enticed to the edge of the web the fork is withdrawn and then gradually brought near, the spider is aware of its presence and of its direction, and reaches out as far as possible in the direction of the fork ; but if a sounding fork is gradually brought near a spider that has not been disturbed, but which is waiting as usual in the middle of the web, then instead of reaching out towards the fork the spider instantly drops—at the end of a thread of course. If under these conditions the fork is made to touch any part of the web, the spider is aware of the fact and climbs the thread and reaches the fork with marvellous rapidity. The spider never leaves the centre of the web without a thread along which to travel back. If after enticing a spider out we cut this thread with a pair of scissors, the spider seems to be unable to get back without doing considerable damage to the web, generally gumming together the sticky parallel threads in groups of three and four.

By means of a tuning-fork a spider may be made to eat what it would otherwise avoid. I took a fly that had been drowned in paraffin and put it into a spider's web and then attracted the spider by touching the fly with a fork. When the spider had come to the conclusion that it was not suitable food and was leaving it, I touched the fly again. This had the same effect as before, and as often as the spider began to leave the fly I again touched it, and by this means compelled the spider to eat a large portion of the fly.

The few house-spiders that I have found do not seem to appreciate the tuning-fork, but retreat into their hiding-places as when frightened ; yet the supposed fondness of spiders for music must surely have some connection with these observations, and when they come out to listen is it not that they cannot tell which way to proceed ?

The few observations that I have made are necessarily imperfect, but I send them, as they afford a method which might lead a naturalist to notice habits otherwise difficult to observe, and so to arrive at conclusions which I in my ignorance of natural history must leave to others.

C. V. BOYS

Physical Laboratory, South Kensington

VOL 23, 149 1880

Frank Buckland was the engagingly odd son of a dangerously eccentric father, William Buckland, the geologist. Father and son took an interest in all living creatures, which they kept as pets or ate. William opined that of all foods, mole was the most repulsive, until he tried stewed bluebottles. He was said also to have eaten the embalmed heart of Louis XIV. When the Bucklands were shown a blood stain on a cathedral floor in Italy on the spot where a saint had been martyred and were told that it renewed itself every night, one of them, dipping his finger into the fluid and licking it, declared it to be no more than bats' urine. Frank continued the tradition, serving to his guests dinners of boiled or sautéed slugs, earwigs, mice en croûte *and cuts of animals that had died at the zoo. He was an early proponent of fish farming, became inspector of fisheries and wrote a four-volume work called* Curiosities of Natural History.

THE LIFE OF FRANK BUCKLAND

Life of Frank Buckland. By his Brother-in-Law, George Bompas. (London : Smith, Elder, and Co., 1885.)

FEW Englishmen were unacquainted with the central figure of this admirably written memoir. His ubiquity as a lecturer and inspector, the happy self-forgetfulness and adaptability of manner which associated him with royal princes as readily as with seaside fishermen, and the strong personality by which he permanently impressed all who came in contact with him, made him beyond all other men of his time the representative and the preacher of the subject to which he devoted all the energies of his life. That subject was natural history, a term not without meaning even in the present day of minute and subdivided scientific work, but conterminous with science half a century ago, when comparative anatomy was hardly known, when the microscope was costly and imperfect, when the provinces of nature had not been mapped nor its workers differentiated.

Frank Buckland was born a naturalist, into a home crammed with animals, living, preserved, fossil; his mother a woman of rare intellectual accomplishment and scientific taste, his father the first geologist of the age. At three years old he could " go through all the natural history books in the Radcliffe Library "; at four we find him lispingly explaining to a Devonshire parson who had brought with pride to Dr. Buckland " some very curious fossils," that they were the vertebræ of an Ichthyosaurus; at five he is rapturous over the teleology of the " tongue-bone " in the skeleton of a whale; and in the archæology of Worcester Cathedral can find only one object of interest—the figure of a lady who had been starved by a disease in the throat.

At twelve he went to Winchester, not the least barbarous school of that barbarous scholastic time. He was " launched," and " tin-gloved," and " toe-fit-tied," and " tunded," and " clowed," and " watched out " at cricket, and " kicked in " at foot-ball, living for two or three years the wretched life of a college junior amid a mob of boys not overlooked by any master and influenced by the bad traditions of a savage past. He used to say that it had done him good, had cured him of "bumptiousness" and arrogance, but he cherished painful memories of individual tyrants and of special acts of tyranny, and was wont when a senior boy to criticise with a bitterness alien from the ordinary conservatism of schoolboys the coarseness of a system which turned a gentleman's son, bred in the refinement of a cultured home, into an abject domestic serf.

Buckland's fagging days over, he was happy, for he could follow his bent undisturbed, and the pages which describe his later Winchester life are amongst the most amusing in the biography. Fond of school work he was not; he was, in fact, looked upon as a " thick," and his compulsory fagging experiences had given him a dislike for games. But he wired trout and eels in the clear Itchen streams, dug out mice on " Hills," chased badgers on Twyford Down, skinned and dissected cats, moles, and bats, articulated skeletons, baked squirrel-pies, and cooked mice in batter. A buzzard, an owl, and a racoon tenanted his lockers in " Moab," a viper lived in his " scob " amongst his books, his hedgehogs kept open a perpetual fosse at the base of the college wall, and a regiment of tame jackdaws looked up to him as their patron. On " Saints' days " he attended the Winchester Hospital, bringing back gruesome fragments of humanity in his pocket-handkerchief, talked médical language, treated confiding boys professionally. Applying for admission to the sick house on behalf of a patient who had partaken too generously of " husked gooseberry fool," he informed the surprised second master that the invalid had a

" stricture of the colon; " he was wont to offer sixpence to any junior who would allow himself to be bled; and he treated surgically a football-wounded shin with such results that the leg when shown eventually to a doctor was pronounced to be in imminent danger of amputation.

The Winchester life found fuller development at Oxford. No one who knew Frank Buckland there will forget those merry breakfasts in the corner of Fell's Buildings; Frank in the blue pea-jacket and the German student's cap, blowing blasts out of a tremendous wooden cow-horn; the various pets who made it difficult to speak or move: the marmots, and the dove, and the monkey, and the chamæleon, and the snakes, and the guinea-pigs, and the after-breakfast visits to the eagle or the jackal or the bear or the pariah dog in the little yard outside. His Long Vacations were spent in Germany, whence he brought back little besides collections of red slugs and green frogs; in 1848 he entered at St. George's Hospital, and in 1854 was gazetted Assistant-Surgeon to the second Life Guards.

The next eight years were very pleasant ones. His father's position as Dean of Westminster threw open to him all the best society in London: we read of parties at Miss Burdett-Coutts's, at the Duke of Wellington's, at Chief Baron Pollock's; microscopic evenings at Dr. Carpenter's; walks around the Abbey with Prince Albert; conversations with Sir B. Brodie, Mr. Gladstone, Whewell, Whately, Prof. Owen, Sedgwick, Bunsen, Ruskin. He was beginning to feel his strength and strike out his line in life; in these years he wrote his first magazine article, delivered his first lecture, published his first book. In 1865 he resigned his commission, married, took the house in Albany Street which he has made historic, started *Land and Water*, devoted himself to fish culture, became Inspector of Fisheries, and worked in his vocation till 1880, when he died at the age of fifty-four, worn out by excessive overwork and by the exposure to wet and cold in all seasons which his professional duties, as he interpreted them, involved.

His power as a lecturer was unrivalled. He could keep an audience in ecstasies of laughing enjoyment for two hours at a stretch. He had inherited his father's remarkable felicity of illustration; his own keen delight in his subject was contagious, his comedy incessant and irresistible. Never was a memory more stored with interesting facts. He was all eyes; noted everything, remembered everything, used everything. Through London streets, as he surveyed them from his favourite seat on the knife-board of an omnibus, on the walls of exhibitions, on sea-coast, river-shore, and hill-side, in the belfry at Ross, by Dean Gainsford's grave—phenomena which others overlooked or passed as trivial were by him pounced upon and analysed and made to bear fruit in discovery and correlation and historical association and practical scientific use. Of human prodigies in every department he was the recognised Proxenus and patron. Miss Swann the giantess and her husband Captain Bates the giant, and the Two-headed Nightingale, and the Siamese Twins, and the New Zealand Chiefs, and Fatima, and Zariffa, and Julia Pastrana the hairy woman, and Benedetti the sword swallower, and the Wild Man of the Woods, and the man who could sing two notes at once, and the man who could drink a bottle of milk under water,—all looked up to him as a father, or sat as guests at his table. He came by degrees to be accepted as an *Arbiter monstrorum;* as the necessary referee whenever any strange revelation or any novel puzzle presented itself in the world of nature. If a whale ran on shore at Gravesend, or a dolphin at Herne Bay; if an unusual sturgeon or tunny was consigned to a London fishmonger; if the lawyers at Nisi Prius were at issue whether a hole in a ship's bottom could have been made by the beak of a swordfish, or the Gloucester Magistrates hesitated over the identity of elvers with young eels; if a sick porpoise arrived at the Zoological Gardens in a condition requiring brandy and water to be exhibited internally and caustic applied without; if the Chief Rabbi felt searchings of heart as to whether oysters might for edible purposes be inserted in the Mosaic catalogue of things that creep; if a sea-lioness were ill in the Aquarium, or a plague of frogs occurred at Windsor; if search were required for John Hunter's coffin in St. Martin's Church, or the skeleton of William Rufus had to be exhumed in Winchester Cathedral,—it was inevitable that Frank Buckland should be telegraphed for first of all. And the influence he exerted was often highly beneficial. To his interference we owe the close time for seals and the Bill for the preservation of marine birds. A description in *Land and Water* of a neglected Museum at Canterbury shamed the Curator into setting it to rights; his good-humoured criticism, from a naturalist's point of view, of the pictures in the Royal Academy, taught the artists beneficially that an eye as keen as Ruskin's was noting their performances in a region beyond Ruskin's reach.

His home in Albany Street was one of the sights of London; but to enter it presupposed iron nerves and a stomach like those of Horace's reapers. Iron nerves—for, introduced at once to some five-and-twenty poor relations, exempt from shyness and deeply interested in your dress and person, to Jacko, and the Hag, and the Nigger, and Jenny, and Tiny, and the parrot and the jaguar, and the laughing jackass, and Jemmy the suricate, and Dick the bear, and Arslan the Turkish wolf-dog, you felt, like Jaques in the play, as if another flood were toward, and the animals were parading for admission. Dura ilia—for the genius of experiment, supreme in all departments of the house, was nowhere so active as at the dinner-table. We read of panther chops, rhinoceros pie, bison steaks, kangaroo ham, horse's tongue, elephant's trunk; of whale boiled with charcoal to refine the flavour; of tripang and lump-fish; of stewed whelks and land-snails, roasted hedgehog, potted ostrich. We notice in the diary such entries as "seedy from lump-fish;" "very poorly indeed, effects of horse;" and we sympathise with a departing guest who notes—"tripe for dinner—don't like crocodile for breakfast."

He was the Samson of science; the "Sunny One" amongst *savants*, as was Manoah's son amongst judges; roars of genial laughter accompany the heroism and the feats of both. But the comic recollections which surround him ought not to mask the serious admiration which is his due;—first, as a public teacher, circulating popular science, generating field clubs and microscopical societies, preparing a public to appreciate and to support the more purely scientific labourer; secondly, as a material benefactor, raising in fifteen years the commercial value of English and Scottish salmon to the extent of 100,000*l.* per annum; thirdly, as having in a manner rare, if not unique, passed behind the veil which

hangs between us and the animal creation. He understood their gestures and expressions as we interpret those of one another, and they understood him in their turn; the creatures at the Gardens, the beasts at Jamrach's, the pets at home, seemed to know him in a human fashion; his dying words—"God is so good to the little fishes that I do not think He will let their inspector suffer shipwreck at the last"—show his identity of feeling with them; no one could talk to him long without a strangely new and reverential sense of brotherhood with these existences who were to him so entirely fraternal as people of his Father's pasture and sheep of his Father's hand. Science has had very many greater sons; none more simple, modest, blameless; none more genial, more humane, or more beloved.

W. TUCKWELL

Vol 32, 385 1885

Shelford Bidwell was a lawyer whose love was physics, a discipline in which he was largely self-taught. While practising as a barrister, he would attend the meetings of the Physical Society, which in due time he contrived to join (not in those days merely a matter of forking out the subscription). He became eventually its president. One of his most striking inventions was this precursor of today's fax machine.

TELE-PHOTOGRAPHY

WHILE experimenting with the photophone it occurred to me that the fact that the resistance of crystalline selenium varies with the intensity of the light falling upon it might be applied in the construction of an instrument for the electrical transmission of pictures of natural objects in the manner to be described in this paper.

In order to ascertain whether my ideas could be carried out in practice, I undertook a series of experiments, and these were attended with so much success that although the pictures hitherto actually transmitted are of a very rudimentary character, I think there can be little doubt that if it were worth while to go to further expense and trouble in elaborating the apparatus excellent results might be obtained.

The nature of the process may be gathered from the following account of my first experiment. To the negative (zinc) pole of a battery was connected a flat sheet of brass, and to the positive pole a piece of stout platinum wire ; a galvanometer was interposed between the battery and the brass, and a set of resistance-coils between the battery and the platinum-wire (see Fig. 1, where B is the battery, R the resistance, P the wire, M the brass plate, and G the galvanometer). A sheet of paper which had been soaked in a solution of potassium iodide was laid upon the brass, and one end of the platinum wire previously ground to a blunt point was drawn over its surface. The path of the point across the paper was marked by a brown line, due, of course, to the liberation of iodine. When the resistance was made small this line was dark and heavy ; when the resistance was great the line was faint and fine ; and when the circuit was broken the point made no mark at all. If we drew a series of these brown lines parallel to one another, and very close together, it is evident that by regulating their intensity and introducing gaps in the proper places any design or picture might be

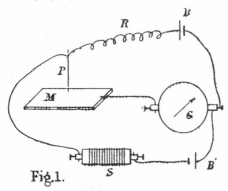

Fig. 1.

represented. This is the system adopted in Bakewell's well-known copying telegraph. To ascertain if the intensity of the lines could be varied by the action of light, I used a second battery and one of my selenium cells, made as described in NATURE, vol. xxiii. p. 58. These were arranged as shown in Fig. 1, the *negative* pole of the second battery, B', being connected through the selenium cell S with the platinum wire P, and the positive pole with the galvanometer G. The platinum point being pressed firmly upon

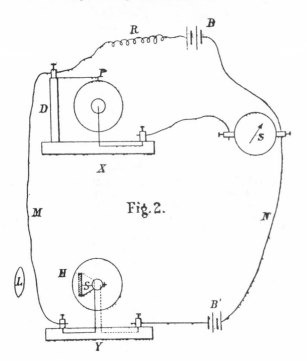

Fig. 2.

the sensitised paper and the selenium exposed to a strong light, the resistance R was varied until the galvanometer needle came to rest at zero. If the two batteries were similar this would occur when the resistance of R was made about equal to that of the selenium cell in the light. The point now made no mark when drawn over the paper. The selenium cell was then darkened, and the point immediately traced a strong brown line ; a feeble light was next thrown upon the selenium, and the intensity of the line became at once diminished. Lastly, a screen of black paper, having a large pin-hole in the middle, was placed at a short distance before the selenium, and the image of a gas-flame was focussed upon the outer surface of the screen, a small portion of the light passing through the pin-hole and forming a luminous disk upon the selenium. The galvanometer was again brought to zero, and, as before, the platinum point made no mark. When however the gas-flame was shaded a firm and steady line could be drawn ; and when the light was interrupted by moving the fingers before the pin-hole a broken line was produced. For this last operation a very sensitive paper was required, and it was found necessary to move the platinum point slowly.

In consequence of the very satisfactory results of these preliminary experiments I made a pair of "tele-photographic" instruments, of which the receiver was slightly modified from Bakewell's form. They are of rude construction, and I shall say nothing more about them except that on January 5 they produced a "tele-photograph" of a gas-flame, which was good enough to induce me to make the more perfect apparatus now to be described.

The transmitting instrument consists of a cylindrical brass box four inches in diameter and two inches deep, mounted axially upon a brass spindle seven inches long, and insulated from it by boxwood rings. The spindle is divided in the middle, its two halves being rigidly connected together by an insulating joint of boxwood. One of the projecting ends of the spindle has a screw cut upon it of sixty-four threads to the inch ; the other end is left plain. The spindle revolves, like that of a phonograph, in two brass bearings, the distance between which is equal to twice the length of the cylinder ; and one of the bearings has an inside screw corresponding to that upon the spindle. At a point midway between the two ends of the cylinder a hole a quarter of an inch in diameter is drilled, and behind this hole is fixed a selenium cell, the two terminals of which are connected respectively with the two halves of the spindle. The bearings in which the spindle turns are joined by copper wires to two binding screws on the stand of the instrument. The transmitter thus described is represented in diagrammatic section at Y (Fig. 2), where H is the hole in the cylinder and S the selenium cell.

The receiving instrument, shown at X (Fig. 2) contains another cylinder similar to that of the transmitter, and mounted upon a similar spindle, which however is not divided, nor insulated from the cylinder. An upright pillar D, fixed midway between the two bearings, and slightly higher than the cylinder, carries an elastic brass arm fitted with a platinum point P, which presses normally upon the surface of the cylinder. To the brass arm a binding screw is attached, and a second binding screw in the stand is joined by a wire to one of the brass bearings.

To prepare the instruments for work they are joined up as shown in Fig. 2, two batteries, a set of resistance coils, and a galvanometer being used, in exactly the same manner as in the preliminary experiments. The cylinder of the transmitting instrument Y is brought to its middle position, and a picture not more than two inches square is focussed upon its surface by the lens L. The pictures upon which I have operated have been mostly simple geometrical designs cut out of tinfoil and projected by a magic lantern. It is convenient to cover a portion of the cylinder with white paper to receive the image. The comparatively large opening H is covered with a piece of tin-foil, in which is pricked a hole which should be only just large enough to allow the instrument to work. [I have not been able to reduce it below one-twentieth of an inch, but with a more sensitive selenium cell it might with advantage be smaller.] The hole is then brought, by turning round the cylinder, to the brightest point of the picture, and a scrap of sensitised paper, in the same condition as that to be used, being placed under the point P of the receiver, the resistance R is adjusted so as to bring the galvanometer to zero. When this is accomplished the two cylinders are screwed back as far as they will go, the cylinder of the receiver is covered with sensitised paper, and all is ready to commence operations.

The two cylinders are caused to rotate slowly and synchronously. The pin-hole at H in the course of its spiral path will cover successively every point of the picture focussed upon the cylinder, and the amount of light falling at any moment upon the selenium cell will be proportional to the illumination of that particular spot of the projected picture which for the time being is occu-

FIG. 3.—Image focussed upon Transmitter.

pied by the pin-hole. During the greater part of each revolution the point P will trace a uniform brown line; but when H happens to be passing over a bright part of the picture this line is enfeebled or broken. The spiral traced by the point is so close as to produce at a little distance the appearance of a uniformly-coloured surface, and the breaks in the continuity of the line constitute a picture which, if the instrument were perfect, would be a monochromatic counterpart of that projected upon the transmitter.

An example of the performance of my instrument is shown in Fig. 4, which is a very accurate representation of the manner in which a stencil of the form of Fig. 3 is reproduced when projected by a lantern upon the transmitter. I have not been able to send one of its actual productions to the engraver, for the reason that they are exceedingly evanescent. In order to render the paper sufficiently sensitive, it must be prepared with a very strong solution (equal parts of iodide and water), and when this is used the brown marks disappear completely

FIG. 4.—Image as reproduced by Receiver.

in less than two hours after their formation. There is little doubt that a solution might be discovered which would give permanent results with equal or even greater sensitiveness, and it seems reasonable to suppose that some of the unstable compounds used in photography might be found suitable; but my efforts in this direction have not yet been successful.

VOL 23, 344 1881

The Raj took the education of his subjects seriously.

CHANDA SINGH, a blind student of St. Stephen's College, Delhi, is, according to the account given in *Allen's Indian Mail*, a prodigy. He cannot read or write, but possesses such a strong memory as to be able to repeat all his text-books, English, Persian, or Urdu, by rote, and to work out sums in arithmetic with remarkable rapidity. The unusual intensity of his mental powers is shown by his ability to multiply any number of figures by another equally large. At the last University examination he was examined *vivâ voce* by order of the Director of Public Instruction of the Punjaub, and he stood twenty-seventh in the list of successful candidates. On the recommendation of the same official, the judges of the local court have allowed him to appear at its law examination. Memory, as is well known, is wonderfully developed in Orientals, owing to the system of education which has obtained amongst them; but cases like Chanda Singh must be very rare even in the East.

VOL 30, 592 1884

Samuel Butler was a formidable Victorian intellectual and his most famous novels, Erewhon *and* The Way of All Flesh, *were greatly admired by his contemporaries. George J. Romanes seems to show in this review the hostility of the professional towards the interloper.*

UNCONSCIOUS MEMORY

Unconscious Memory, &c. By Samuel Butler. Op. 5. (London : David Bogue, 1880.)

MR. BUTLER is already known to the public as the author of two or three books which display a certain amount of literary ability. So long therefore as he aimed only at entertaining his readers by such works as " Erewhon," or " Life and Habit," he was acting in a suitable sphere. But of late his ambition seems to have prompted him to other labours; for in his " Evolution, Old and New," as well as in the work we are about to consider, he formally enters the arena of philosophical discussion. To this arena, however, he is in no way adapted, either by mental stature or mental equipment; and therefore makes so sorry an exhibition that Mr. Darwin may well be glad that his enemy has written a book. But while we may smile at the vanity which has induced so incapable and ill-informed a man gravely to pose before the world as a philosopher, we should not on this account have deemed " Unconscious Memory" worth reviewing. On the contrary, as a hasty glance would have been sufficient to show that the book is bad in philosophy, bad in judgment, bad in taste, and, in fact, that the only good thing in it is the writer's own opinion of himself—with all that was bad we should not have troubled ourselves, and that which was good we should not have inflicted on our readers. The case, however, is changed when we meet, as we do, with a vile and abusive attack upon the personal character of a man in the position of Mr. Darwin; for however preposterous, and indeed ridiculous, the charges may be, the petty malice which appears to underlie them deserves to be duly repudiated. We shall therefore do our duty in this respect, and at the same time take the opportunity of pointing out the nonsense that Mr. Butler has been writing, both about the philosophy of evolution and the history of biological thought.

* * *

... he has also a great deal to say on the philosophy of evolution. "Op. 4" was called "Evolution, Old and New," and now "Op. 5" continues the strain that was struck in the earlier composition. This consists for the most part in a strangely silly notion that "the public generally"—including, of course, the world of science—was as ignorant of the writings of Buffon, Dr. Erasmus Darwin, and Lamarck as was Mr. Butler when he first read the "Origin of Species." That is to say, "Buffon we knew by name, but he sounded too like 'buffoon' for any good to come from him. We had heard also of Lamarck, and held him to be a kind of French Lord Monboddo; but we knew nothing of his doctrine. . . . Dr. Erasmus Darwin we believed to be a forgotten minor poet," &c. No wonder, therefore, when such was our manner of regarding these men, that we required a Mr. Samuel Butler to show us our error. And no wonder that Mr. Charles Darwin, who doubtless may have peeped into the literature which Mr. Butler has discovered, should so well have succeeded in his life-long purpose of concealing from the eyes of all men how much he owes to his predecessors. No wonder, also, that Mr. Darwin, when he chanced to see an advertisement of a forthcoming work by Mr. Butler with the title "Evolution, Old and New," should have inferred, as Mr. Butler observes, "what I was about," and forthwith began to tremble in dismay that at last the Buffoon, the French Lord Monboddo, and the forgotten minor poet had found a champion to vindicate their claims.

And he concludes savagely:

Mighty champion of the mighty dead! When our children's children shall read in Westminster Abbey the inscription on the tomb of Mr. Samuel Butler, how will it be with a sigh that in their day and generation the world knows nothing of its greatest men! But as it is our misfortune to live before the battle over Mr. Samuel Butler's memory has been fought, we respond to his abounding presumption by recommending him, whatever degree of failure he may have experienced in art, once more to "consider" himself "by profession a painter" —or, if the painters will not have him, to make some third attempt, say among the homœopathists, whose journal alone, so far as we are aware, has received with favour his latest work.

GEORGE J. ROMANES

VOL 23, 285 1881

Edifying home experiments, so beloved of the Victorians, were often recorded in NATURE. *Joseph Antoine Ferdinand Plateau was a physicist of standing, the best that Belgium produced. He contributed to many areas of physics and is commemorated in Plateau's spherule — the subsidiary droplet that detaches from a falling liquid drop — and the Talbot-Plateau law of colour perception. He was blinded, apparently as a result of a physiological experiment that involved staring at the Sun for a full half-minute.*

M. PLATEAU describes as "*un petit amusement*" the following experiment:—A flower like a lily, with six petals each about an inch long, was constructed in outline in thin iron wire; the wire being first slightly peroxidised by dipping for an instant into nitric acid. This wire frame was then dipped into a glyceric-soap-solution, which, when it was withdrawn, left soap-films over the petals. The stalk was then set upright in a support, and it was covered by a bell-glass to protect it from air-currents. In a few moments the most beautiful colours made their appearance. If the solution is in good condition the films will last for hours, giving a perpetual play of colour over the flower.

VOL 25, 112 1881

Alexander Graham Bell, the Edinburgh-bred inventor of the telephone, received from his father, Alexander Melville Bell, an interest in the training of the deaf-mute.

The Pronunciation of Deaf-mutes who have been Taught to Articulate

MY attention has just been drawn to the remarkable statement of M. Hément (*C. R.*, xciii. p. 754), that deaf-mutes who have been taught to articulate speak with the accent of their native district; and to the equally remarkable letter of Mr. WM. E. A. Axon, published in NATURE (vol. xxv. p. 101), in support of the same proposition.

I may say in this connection that I have during the past few years examined the pronunciation of at least 400 deaf mutes who have been taught to speak, without remarking any such tendency as that referred to above. It is true that in a few cases dialectic pronunciations are heard, but it always turns out upon investigation that such children could talk *before they became deaf.* The peculiarity is undoubtedly due to unconscious recollection of former speech, and cannot correctly be attributed to heredity.

M. Emile Blanchard (*C. R.*, xciii. p. 755) has directed attention to the harsh and disagreeable character of the utterance of many deaf-mutes who have been taught to articulate, but it has been found in America that this can be overcome by suitable instruction. I am happy to be able to say that I have heard from congenitally deaf children perfectly distinct and agreeable articulation.

The mouths of deaf children are in no way different from our own.[1] Deaf mutes do not naturally speak the language of their country for the same reason that we do not talk Chinese—*they have never heard the language.* They are dumb simply because they are deaf; and I see no reason to doubt that all deaf-mutes may be taught to use their vocal organs so as to speak at least *intelligibly,* if not as perfectly as those who hear.

In most, if not in all, of our American Institutions for the deaf and dumb, articulation is now taught as a special branch of education; and in many of our schools all instruction is given by word of mouth, as it has been found that large numbers of deaf children can be taught to understand spoken words by watching the movements of the speaker's mouth.

So successful has articulation-teaching proved in America and in Europe, that dumbness will soon be universally recognised as a mark of neglected education.

ALEXANDER GRAHAM BELL, Ph.D.
(Nat. Col. for Deaf Mutes, Washington)

London, December 5

VOL 25, 124 1881

Karl Pearson was a man of omnivorous intellectual appetite, a mathematician with a special interest in applying statistical techniques to biological problems. He was a fervent evolutionist who believed that the process was smooth and continuous, and he became in later life a passionate advocate of eugenic measures to improve the human gene pool. He was the first Galton Professor of Eugenics at University College London. Here he displays his classical virtuosity.

Colour and Sound

SOME weeks ago there appeared an account of a series of experiments connecting colour and sound; the following passage from Prof. Max Müller's Chips, ii. 104, may interest some of your readers:—"That Purûravas is an appropriate name of a solar hero requires hardly any proof. Purûravas meant the same as πολυδευκής, endowed with much light; for though rava is generally used of sound, yet the root ru, which originally to cry, is also applied to colour, in the sense of a loud or crying colour, *i.e.* red (cf. ruber, rufus, Lith, rauda, O.H.G. rôt, rudhira, ἐρυθρός; also Sanskrit ravi, sun)." The following footnote occurs :—"Thus it is said, Rv. vi. 3, 6, the fire cries with light, *sokishâ rârapîti*; the two Spartan Charites are called Κλητά (κλητά, incluta) and Φαεννά, *i.e.* Clara, clear-shining. In the Veda the rising sun is said to cry like a new child (Rv. ix. 74, 1)—I do not derive ravas from rap, but I only quote rap as illustrating the close connection between loudness of sound and brightness of light."

Both Greeks and Latins seem to have used the same words for colour and sound, cf. λαμπρός, λευκός, μέλας, σομφός, φαιός, &c.; clarus, fuscus, candidus, &c. Probably not only colour and sound, but smell, taste, and touch had in early times the like words to express degree; even as we find aspera lingua and odor asper; and as we say "a harsh taste" and "a harsh sound." Tastes and smells will be found to suggest colours to the mind exactly as sounds do. If this be so, may not this apparently curious connection be explained as a sort of "*unconscious philological memory*?"   KARL PEARSON

Inner Temple, January 28

Vol 25, 339 1882

Here is an example of Charles Darwin's unquenchable curiosity and the clarity and directness of his literary style. It was his last communication to NATURE. *The same month, his obituary appeared.*

ON THE DISPERSAL OF FRESHWATER BIVALVES

THE wide distribution of the same species, and of closely-allied species of freshwater shells must have surprised every one who has attended to this subject. A naturalist, when he collects for the first time freshwater animals in a distant region, is astonished at their general similarity to those of his native European home, in comparison with the surrounding terrestrial animals and plants. Hence I was led to publish in NATURE (vol. xviii. p. 120) a letter to me from Mr. A. H. Gray, of Danversport, Massachusetts, in which he gives a drawing of a living shell of *Unio complanatus*, attached to the tip of the middle toe of a duck (*Querquedula discors*) shot on the wing. The toe had been pinched so hard by the shell that it was indented and abraded. If the bird had not been killed, it would have alighted on some pool, and the Unio would no doubt sooner or later have relaxed its hold and dropped off. It is not likely that such cases should often be observed, for a bird when shot would generally fall on the ground so heavily that an attached shell would be shaken off and overlooked.

I am now able to add, through the kindness of Mr. W. D. Crick, of Northampton, another and different case. On February 18 of the present year, he caught a female *Dytiscus marginalis*, with a shell of *Cyclas cornea* clinging to the tarsus of its middle leg. The shell was ·45 of an inch from end to end, ·3 in depth, and weighed (as Mr. Crick informs me) ·39 grams, or 6 grains. The valves clipped only the extremity of the tarsus for a length of ·1 of an inch. Nevertheless, the shell did not drop off, on the beetle when caught shaking its leg violently. The specimen was brought home in a handkerchief, and placed after about three hours in water; and the shell remained attached from February 18 to 23, when it dropped off, being still alive, and so remained for about a fortnight while in my possession. Shortly after the shell had detached itself, the beetle dived to the bottom of the vessel in which it had been placed, and having inserted its antennæ between the valves, was again caught for a few minutes. The species of Dytiscus often fly at night, and no doubt they generally alight on any pool of water which they may see; and I have several times heard of their having dashed down on glass cucumber frames, no doubt mistaking the glittering surface for water. I do not suppose that the above weight of 6 grains

would prevent so powerful an insect as a Dytiscus from taking flight. Anyhow this beetle could transport smaller individuals; and a single one would stock any isolated pond, as the species is an hermaphrodite form. Mr. Crick tells me that a shell of the same kind, and of about the same size, which he kept in water "extruded two young ones, which seemed very active and able to take care of themselves." How far a Dytiscus could fly is not known; but during the voyage of the *Beagle* a closely-allied form, namely, a Colymbetes, flew on board when the nearest point of land was forty-five miles distant; and it is an improbable chance that it had flown from the nearest point.

Mr. Crick visited the same pond a fortnight afterwards, and found on the bank a frog which appeared to have been lately killed; and to the outer toe of one of its hind legs a living shell of the same species was attached. The shell was rather smaller than in the previous case. The leg was cut off and kept in water for two days, during which time the shell remained attached. The leg was then left in the air, but soon became shrivelled; and now the shell being still alive detached itself.

Mr. F. Norgate, of Sparham, near Norwich, in a letter dated March 8, 1881, informs me that the larger water-beetles and newts in his aquarium "frequently have one foot caught by a small freshwater bivalve (*Cyclas cornea*?), and this makes them swim about in a very restless state, day and night, for several days, until the foot or toe is completely severed." He adds that newts migrate at night from pond to pond, and can cross over obstacles which would be thought to be considerable. Lastly, my son Francis, while fishing in the sea off the shores of North Wales, noticed that mussels were several times brought up by the point of the hook; and though he did not particularly attend to the subject, he and his companion thought that the shells had not been mechanically torn from the bottom, but that they had seized the point of the hook. A friend also of Mr. Crick's tells him that while fishing in rapid streams he has often thus caught small Unios. From the several cases now given, there can, I think, be no doubt that living bivalve shells must often be carried from pond to pond, and by the aid of birds occasionally even to great distances. I have also suggested in the "Origin of Species" means by which freshwater univalve shells might be far transported. We may therefore demur to the belief doubtfully expressed by Mr. Gwyn Jeffreys in his "British Conchology," namely, that the diffusion of freshwater shells "had a different and very remote origin, and that it took place before the present distribution of land and water."

CHARLES DARWIN

Vol 25, 529 1882

*The obituary notice was by T. H. Huxley — Darwin's
bulldog, as he styled himself — and he did not fail his master.*

CHARLES DARWIN

VERY few, even among those who have taken the
keenest interest in the progress of the revolution in
natural knowledge set afoot by the publication of the
"Origin of Species"; and who have watched, not without
astonishment, the rapid and complete change which has
been effected both inside and outside the boundaries of
the scientific world in the attitude of men's minds
towards the doctrines which are expounded in that great
work, can have been prepared for the extraordinary manifestation of affectionate regard for the man, and of profound reverence for the philosopher, which followed the
announcement, on Thursday last, of the death of Mr.
Darwin.

Not only in these islands, where so many have felt the
fascination of personal contact with an intellect which had
no superior, and with a character which was even nobler
than the intellect; but, in all parts of the civilised world,
it would seem that those whose business it is to feel the
pulse of nations and to know what interests the masses of
mankind, were well aware that thousands of their readers
would think the world the poorer for Darwin's death,
and would dwell with eager interest upon every incident of his history. In France, in Germany, in
Austro-Hungary, in Italy, in the United States, writers of
all shades of opinion, for once unanimous, have paid a
willing tribute to the worth of our great countryman,
ignored in life by the official representatives of the kingdom, but laid in death among his peers in Westminster
Abbey by the will of the intelligence of the nation.

It is not for us to allude to the sacred sorrows of the
bereaved home at Down ; but it is no secret that, outside
that domestic group, there are many to whom Mr. Darwin's death is a wholly irreparable loss. And this not
merely because of his wonderfully genial, simple, and generous nature ; his cheerful and animated conversation, and
the infinite variety and accuracy of his information ; but
because the more one knew of him, the more he seemed
the incorporated ideal of a man of science. Acute as

were his reasoning powers, vast as was his knowledge,
marvellous as was his tenacious industry, under physical
difficulties which would have converted nine men out of
ten into aimless invalids ; it was not these qualities, great
as they were, which impressed those who were admitted
to his intimacy with involuntary veneration, but a certain
intense and almost passionate honesty by which all his
thoughts and actions were irradiated, as by a central fire.

It was this rarest and greatest of endowments which
kept his vivid imagination and great speculative powers
within due bounds ; which compelled him to undertake
the prodigious labours of original investigation and of
reading, upon which his published works are based ;
which made him accept criticisms and suggestions from
any body and every body, not only without impatience,
but with expressions of gratitude sometimes almost comically in excess of their value ; which led him to allow
neither himself nor others to be deceived by phrases, and
to spare neither time nor pains in order to obtain clear
and distinct ideas upon every topic with which he
occupied himself.

One could not converse with Darwin without being
reminded of Socrates. There was the same desire to
find some one wiser than himself ; the same belief in the
sovereignty of reason ; the same ready humour ; the
same sympathetic interest in all the ways and works of
men. But instead of turning away from the problems of
nature as hopelessly insoluble, our modern philosopher
devoted his whole life to attacking them in the spirit of
Heraclitus and of Democritus, with results which are
as the substance of which their speculations were an-
ticipatory shadows.

The due appreciation or even enumeration of these
results is neither practicable nor desirable at this moment.
There is a time for all things—a time for glorying in our
ever-extending conquests over the realm of nature, and a
time for mourning over the heroes who have led us to
victory.

None have fought better, and none have been more
fortunate than Charles Darwin. He found a great truth,
trodden under foot, reviled by bigots, and ridiculed by
all the world ; he lived long enough to see it, chiefly by
his own efforts, irrefragably established in science, inseparably incorporated with the common thoughts of
men, and only hated and feared by those who would revile,
but dare not. What shall a man desire more than this ?
Once more the image of Socrates rises unbidden, and
the noble peroration of the "Apology" rings in our ears as
if it were Charles Darwin's farewell :—

"The hour of departure has arrived, and we go our
ways—I to die and you to live. Which is the better,
God only knows."   T. H. HUXLEY

Vol. 25, 597 1882

Eadweard Muybridge pioneered the photography of animal movement. He also, it appears, invented the photo finish.

"A Dead Heat"

TELEGRAMS from Paris on Monday state that the "Prix du Jockey Club" had resulted in what is usually called a "dead heat." It is unnecessary for me to inform you, that there can be no such thing as a "dead heat." It is called so, I suppose, in consequence of a disagreement among the judges as to which horse first thrusts his nose beyond the winning-post. Are living judges any longer necessary to determine the results of a race? Five years ago I proposed to prove by indisputable evidence the winner of a trotting match which, in consequence of a dispute among the judges, had to be trotted over again. By means of a single thread stretched across the track, and invisible to either horses or their riders, twenty photographic cameras have been made to synchronously record positions impossible for the eye to recognise. With the aid of photography, the astronomer, the pathologist, the chemist, and the anatomist are enabled to pursue the most complex investigations with absolute confidence in the truth it reveals ; why should those interested in trials of speed not avail themselves of the same resources of science? I venture to predict, in the near future that no race of any importance will be undertaken without the assistance of photography to determine the winner of what might otherwise be a so-called "dead heat."

449, Strand, W.C., May 23. EADWEARD MUYBRIDGE
VOL 26, 81 1882

At about this time there broke out in Britain a passionate altercation about what scientists should call themselves; certainly not "scientists", in the view of a seeming majority. The first shot was fired by NATURE.

WE observe that a correspondent of a daily paper proposes that men addicted to the pursuits of science should be called *scientiates*, after the Italian *scienziati* ; and in like manner the studies of science, *scential* studies (*studj scienziali*). The substitution of the American *scientist* for our unsatisfactory phrase *men of science* is of course much to be deprecated ; perhaps we shall come to accept Sir William Thomson's proposed use of *naturalist* for the designation in question, if its sense may be extended. *Scientific studies* is a phrase which cannot be commended for accuracy.

VOL 26, 352 1882

Francis Galton developed a consuming interest in the quantitative measurement of human attributes — how to express, for example, the degree of similarity between two different faces (see p.72). His long paper on the weight of scions of the nobility is typical of his eccentric originality. The scales are still to be seen in the premises of Messrs Berry (now Berry Brothers and Rudd) in St James's Street.

THE WEIGHTS OF BRITISH NOBLEMEN DURING THE LAST THREE GENERATIONS

IT is of considerable interest to know in an exact way the amount of change that may have occurred in our race during recent generations. I therefore send the following results concerning the changes in weight, which I have calculated from data obligingly furnished to me by Messrs. Berry, of 3, St. James's Street, London. Messrs. Berry are the heads of an old-established firm of wine and coffee merchants, who keep two huge beam scales in their shop, one for their goods, and the other for the use and amusement of their customers. Upwards of 20,000 persons have been weighed in them since the middle of last century down to the present day, and the results are recorded in well-indexed ledgers. Some of those who had town houses have been weighed year after year during the Parliamentary season for the whole period of their adult lives. I examined two of the ledgers at my own house, and was satisfied of their genuineness and accuracy ; also that they could be accepted as weighings in "ordinary indoor clothing" unless otherwise stated. Much personal interest attaches itself to these unique registers, for they contain a large proportion of the historical names in our upper classes.

I have ventured to discuss only a small and definite part of this mass of material, and I selected the nobility for the purpose, because the dates of their births could be easily learnt, which had to be done in order to connect the years in which they were weighed with their ages at the time. They formed a more homogeneous group than one that included younger brothers and men about town, who marry late and lead less regular lives. I therefore begged Messrs. Berry to find a clerk for me who should make the required extracts under their direction in an anonymous form for statistical purposes. I also asked to be furnished with an alphabetical list of the persons weighed, that I might know generally with whom I was dealing, and that each schedule should bear a reference to the folio whence it was extracted, so that, whenever verification was needed, the original might be referred to. All this was done, and I am in possession of 139 schedules referring to as many different persons, namely, 109 peers, 29 baronets (who were added as makeweights), and 1 eldest son of a peer. They were born at various times between 1740 and 1830, or thereabouts. Each schedule gives the age and year of the several weighings, the highest and lowest weights recorded in that year, and a copy of such remarks as were entered at the time about the dress.

Galton now launches into a statistical analysis with figures and tables, and arrives finally at the following conclusion.

Whatever may be the exact significance of these mean values, which is by no means so clear as may at first sight be imagined, and whatever may be their absolute worth, which I do not rate very highly, there can be no doubt as to their differential importance. They show with great distinctness that the noblemen of the generation which flourished about the beginning of this century attained their meridian and declined much earlier than those of the generation 60 years their juniors. They were nearly a stone heavier at the age of 40.

The weights of these two generations were identical at the age of 62 or 63, but at that period of life the earlier generation was declining in weight with almost the exact speed at which the latter was continually rising. The steadiness of the rise of the latter from early manhood to late years is very striking ; it is almost in a straight line. I have not sufficient data to justify me to say when its curve culminates ; I have closed it at 70 with a dotted line.

It is only necessary to add that the ledgers of Messrs. Berry are a quarry from which, with some labour, much further information of the kind just given might be drawn. Perhaps the publication of this paper will suggest methods of treating them that have not occurred to myself.

FRANCIS GALTON

VOL 29, 266 1884

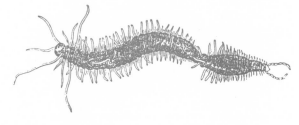

*Another autopsy, another brain, but apparently little
enlightenment about the sources of literary genius:*

It appears from the report of Drs. Brouardel, Segond,
Descout, and Magnin, who conducted the autopsy of Tourguenief,
that the brain of this eminent Russian author weighed 2012
grms. This extraordinary weight, which is only known to have
been exceeded in the case of Rudolphi, is inexplicable, for
Tourguenief, although tall, was not of exceptionally high stature.
The brain is said to have been remarkably symmetrical, and distinguished by the extreme amplitude of the convolutions. According to generally accepted views, however, symmetry of the
circonvolutions is not a favourable cerebral characteristic.

Vol 29, 389 1884

*The French statesman's exiguous brain was fortunately
redeemed by its "very fine" structure.*

As the late M. Gambetta was a member of the Society of
Dissection, an autopsy of his body was made. The weight of
his brain was found to be 1100 grams ; M. Mathias Duval,
Professor in the Faculty of Medicine, found the structure of
the brain to be very fine, and the third convolution, which M.
Broca associates with the speechifying faculty, to be remarkably
developed.

Vol 27, 247 1883

*Here Francis Galton offers his suggestion for resolving a
legal* cause célèbre *that all of England followed with
compulsive interest. Roger Tichborne, heir to a large estate,
was lost at sea. Years later an impostor, the "Tichborne
claimant", afterwards identified as a petty criminal, came
from Australia to claim the inheritance and persuaded
Tichborne's mother that he was indeed her son. The claimant
bore a passable resemblance to the heir, and it was a means of
quantifying the similarity that Galton was advocating. An
etymological footnote to the affair was that a music-hall
performer of diminutive stature, who however resembled the
claimant in his corpulence, became known to his public as
Little Tich, whence the sobriquet Tich for all small men.*

THE "IDENTISCOPE"

It appears from the *Pall Mall Gazette* of October 21
that there is a prospect of "a campaign being run
in the country" on behalf of the "Claimant" by "six of
the best orators whom money can collect, . . . supplied
with a hundred identiscopes." These are optical instruments, containing on the one side a drawing made from
a portrait of the undoubted Roger Tichborne, and on the
other side a drawing made from an equally undoubted
portrait of the Claimant taken nineteen years later, and
the arrangement is such that on looking into the instrument the drawings combine into one. This, it is maintained, leaves no doubt that the two portraits are those of
one and the same individual.
The more important of the questions raised by this
announcement is whether the fact of two genuine portraits
blending harmoniously into a single resultant is stringent
evidence that the portraits refer to the same person. Those
who have examined the optical combinations and photographic composites that I have exhibited at various times
will know that this is not the case. Those who have not
seen them and care to know more about the subject should
look at my "Inquiries into Human Faculty." (Let me take
this opportunity of correcting an error there. The full and
profile composite labelled "two sisters," in the middle of

the upper row of the frontispiece, is really one of three
sisters. I had made many composites of the family, and
by mistake sent the wrong one to the printer.) The reason
why photographic portraits blend so well together is
that they contain no sharp lines, but only shades. The
contour of the face is always blurred, for well-known
reasons dependent on the breadth of the object-glass ;
even the contour of the iris in an ordinary photographic
print looks very coarse and irregular when it is examined
by a low-power microscope. On superimposing a second
portrait, the new shades fall in much the same places as
the former ones ; wherever they overlap they intensify
one another ; where they do not overlap they leave a
faint penumbra which has usually a soft and not un-
pleasing effect. Judging from abundant experience, there
would be no difficulty in selecting photographs of many different persons that should harmonise with the photograph
of the Claimant, and it would be amusing to try strange
combinations. I could suggest one that I think would
succeed excellently : it is of a certain distinguished member
of Her Majesty's——but I must be discreet, though probably if I ever come into possession of suitable photographs
I may make a private experiment.

*There follows a description of the method, which involves
standardising all measurements to the diameter of the iris,
and Galton then sums up:*

I conclude as follows. First, that the fact of two photographic portraits blending harmoniously is no assurance
of the identity of the persons portrayed. Secondly,
when drawings made from portraits are shown to blend it
does not follow that the portraits from which they were
drawn would blend equally well. And lastly, the photographic print of the iris of the eye does not afford a
trustworthy unit of measurement.

Francis Galton

Vol 30, 637 1884

*Why were these experiments on fish not extended to gin and
Scotch?*

So as to further test the efficacy of brandy as a fish restorer,
about which much has lately been said, each fish on showing
signs of exhaustion was removed from the water, dosed with a
small quantity of brandy, and replaced in the tanks from whence
it was taken. The operation proved highly successful, for on
inspection the following day all the objects of the experiment
were found swimming about as usual, and thoroughly restored to
their normal exuberance, with the exception of the dace, which
succumbed to the severe ordeal through which it had passed.

Vol 31, 350 1885

More classical scholarship, revealing an early incendiary episode:

Plutarch on Petroleum

THERE is in "Plutarch's Lives," in the life of Alexander, an interesting notice of the petroleum of Media; I have not found any mention of this passage in "Plutarch" either in encyclopædia or chemical dictionary; I trust, therefore, that you will give me the opportunity of reproducing it in NATURE. I transcribe the passage from the translation of John and William Langhorne (9th edition, London, 1805) :—

"... and in the district of Ecbatana he (Alexander) was particularly struck with a gulph of fire, which streamed continually as from an inexhaustible source. He admired also a flood of naptha, not far from the gulf, which flowed in such abundance that it formed a lake. The naptha in many respects resembles the bitumen, but it is much more inflammable. Before any fire touches it, it catches light from a flame at some distance, and often kindles all the intermediate air. The barbarians, to show the king its force and the subtlety of its nature, scattered some drops of it in the street which led to his lodgings; and, standing at one end, they applied their torches to some of the first drops, for it was night. The flame communicated itself swifter than thought, and the street was instantaneously all on fire."

W. H. DEERING
Chemical Department, Royal Arsenal, Woolwich, May 6

VOL 32, 29 1885

General Charles George Gordon did, of course, die in the Sudan.

ACCORDING to the Swedish papers, on the evening of June 19 a crane was shot at Orkened, in Scania, which had a parchment card tied to its neck with the following lines written in ink :—

> I come from the burning sand
> From Sudan, the murderers' land,
> Where they told the lie,
> That Gordon would die.

The bird had previously been wounded in the wing, and was very exhausted.

VOL 32, 208 1885

Baron Nordenskiöld, the Arctic explorer, was a frequent contributor to NATURE on matters of exploration, geography and archaeology.

AN OLD DRAWING OF A MAMMOTH [1]

AS an addendum to the historical review of the mammoth discoveries in Siberia and the traditions to which they have given rise, which I have rendered in the "Voyage of the *Vega*," I have the pleasure of presenting a curious drawing of the animal, discovered among the Benzelian MSS. in the Linköping library. My attention was directed to the original by the president, Herr Hans Forssell, who, in his memoir of Erik Benzelius the younger, has given an account of the proceedings which it occasioned in the Upsala Scientific Society.[2]

The drawing bears the following inscription :—

"The length of this animal, called Behemot, is 50 Russian ells; the height is not known, but a rib being 5 arsin long, it may be estimated. The greatest diameter of the horn is half of an arsin, the length slightly above four; the tusks like a square brick; the foreleg from the shoulder to the knee 1¾ arsin long, and at the narrowest part a quarter in diameter. The hole in which the marrow lies is so big that a fist may be inserted, otherwise the legs bear no proportion to the body, being rather short. The heathens living by the River Obi state that they have seen them floating in this river as big as a 'struus,' *i.e.* a vessel which the Russians use. This animal lives in the earth, and dies as soon as it comes into the air."

On the reverse of the drawing we read :—

"This drawing and description is given by Baron Kagg, who has just returned from captivity in Russia and Siberia,[1] 1722, in Decembri."

This drawing was exhibited by Benzelius at the meeting of the Upsala Scientific Society, December 14, 1722. The statement referring thereto in the *Journal* of the Society is as follows :—

"Herr Benzelius exhibited a good drawing of an animal, transmitted by Baron Kagg, who has just returned from captivity in Russia and Siberia, which the Siberiaks call Mehemoth or Mammont, which has caused many to believe that it was identical with Behemoth of Job. Herr Prof. Rudbeck and Dr. Martin maintained that it was a sea animal, moreover as Herr Kagg stated that it was found at the River Obi. To this was added that Capt. Lundius had said that its bones were mostly found in the earth by the river. With regard to the animal being drawn with claws, Prof. Rudbeck pointed out that as yet no animal *cornigerum* had been found also to be *unguiculatum*, without being *palmipes* or having skin between the toes like geese, &c. It was decided to write to Herr Kagg, requesting some information about the figure, and asking how he had obtained it, so that it might be ascertained whether it was reliable. There is a

description about this Mehemot in Capt. Müller's account of the Ostiaks."[2]

At a later meeting, January 11, 1723, Dr. Martin stated that he had carefully examined works of zoology, whether there existed any sea animal like that shown at the last conference, but had found nothing like it, although the head—excepting the horns—and probably also the feet and the tail, were like those of the hippopotamus of the River Nile. At the same meeting Benzelius announced that Lieut.-Col. Schönström had promised to forward a whole tusk of this remarkable animal.

On later occasions too the animal was discussed by the Society. Thus on January 18, 1723, a letter was read from the learned linguist, Sparfvenfelt, wherein he explains the derivation of the words Behemoth and Mammont; on February 15 a letter was read from Benzelius, stating that Kagg had received the drawing from a Capt. Tabbert, and that he could give no information as to its correctness. Again, on October 3, Benzelius exhibited a large bone, almost petrified, which was the jaw of a Mammont, or as it was called Behemoth, received from Tobolsk in Siberia, through Capt. Clodt von Jürgensburg, and, on November 22, Benzelius exhibited "part of the tusk of a Behemoth, which was exactly like ivory." Finally, Benzelius communicated with the Russian Chief of Mines, Tatischew, who, in a letter dated May 12, 1725, had given long and important information of the history of the mammoth. This letter is printed in "Acta Literaria Sueciæ" (vol. ii. p. 36, 1725).

A. E. NORDENSKIÖLD

VOL 32, 228 1885

An upsurge of obscurantist Christian fundamentalism in Germany, coupled to anti-Semitism, was led by the Prussian court chaplain Adolf Stoecker. He was a fanatical nationalist, but also a socialist with advanced views on taxation of the rich, which found no favour with Otto von Bismarck. Stoecker was opposed in his bid for election to the Prussian parliament by the representative of the Progressive Party, the eminent pathologist Rudolf Virchow.

OUR readers will doubtless be surprised to learn that the masterly address on Darwin and Copernicus, of which we publish a translation in another column, has called forth much hostile criticism in Germany. It was read before the members of the Berlin Academy of Sciences, of which Prof. Du Bois Reymond is Secretary, at their last annual meeting. Shortly afterwards one of the Clerico-Conservative newspapers of the German capital called attention to what it was pleased to call the public laudation of one of the worst and most dangerous atheists by a member of a public body supported by the State. Many other papers of the same views immediately followed suit; while the notorious Court Chaplain, Stöcker, whose exploits as a Jew-baiter furnished the Berlin correspondents of the daily papers with a good deal of matter about twelve months ago, preached a long sermon against Prof. Du Bois Reymond and his views. His example was followed by other members of the so-called "Orthodox" clergy in Berlin and the provinces. But the Court Chaplain is also a member of the Prussian Parliament; so not content with crushing "atheism" from the pulpit, he put a question in the House on the subject, supported by Herr Windthorst, one of the leaders of the Ultramontane party. They were answered by Prof. Virchow and the Prussian Minister of Public Instruction, thus causing a whole sitting of the Prussian Landtag to be taken up by a debate on the graceful tribute to the memory of Darwin. That such things should take place in Germany, which has always been considered the home of philosophic freedom, really seems to justify the remark of the author of "Darwin and Copernicus," that freedom of thought, which, after taking its rise in England in the middle of the eighteenth century, passed through France to Germany, where it attained a fuller and more systematic development, seems now to be passing away from the latter country again! Let us hope that it is coming to our shores once more, as the Professor says it is.

VOL 27, 565 1883

Here is a countryman's observation on the canny survival strategy of Scottish wildlife.

Intelligence in Animals

I THINK it was about the year 1844 that the Duke of Argyll desired my late father, his factor, to preserve game in the district of Kintyre, Argyllshire. If any steps in this direction had been taken by other proprietors, they were very irregular. My memory goes back to about 1846 and 1848, and at that time the grouse of Kintyre "sat like stones"—they might be shot to dogs from the first to the last day of the season; in fact it was often difficult to get the birds up. With this preservation, grouse increased enormously—and therefore the food supply of the people —to such an extent that the late Sir John Cuningham and my father shot, on one 12th of August, seventy-two brace of grouse. Sir John was a very old man, and insisted on loading his own gun, an old muzzle-loader. My father never shot hard. Now I do not believe any two men with two guns and loaders could do this in the same district with all the improvements in arms and dogs; whilst I have heard my father say that seven brace was a good bag when he was young, before game-preserving.

Grouse yet sit pretty well in Kintyre, and I believe this is the case because it was one of the last districts to preserve and shoot; but the birds are every year becoming wilder, and now in the month of September it is useless to take dogs on the hill, and for two years we, like others, have had to drive them.

I account for this by an alteration in instinct, and I am as sure as any one can be, from observation and the opinion of competent persons, that it is *progressive instinct in successive generations*. Formerly the great enemies of the grouse were ravens, that took their eggs and young birds; foxes, polecats or martin cats, and wild cats, that took them at *night* on the ground; and hawks, that took them on the wing during the day. When man stepped in and altered the balance of Nature, the

Bird that up and flew away,
He lived to breed another day.

No hawk was there to knock him down. He found from experience that flying away before man and his dog came near gave him safety; and his children that inherited the wit or instinct or power of turning heather into nerve-force or intelligent thought—or whatever the straw-splitters like to call it—lived; whilst his brother, that inherited the qualities which kept him hiding in the heather, was shot when forced up.

I had this summer ample corroboration of this theory. About eight years ago I was shooting in the island of Rum; the grouse were not preserved and were extremely tame, so tame in September and October that I had to run after them to make them take the wing, and it was new to dogs. Last year I again shot in the island, and I observed the same tameness in one part of the island, but in another district I observed the grouse were larger, darker, and much wilder. I was puzzled with this until I found out that the late tenant had three years before turned down some English grouse, and in the district where they were so turned down the grouse were very wild.

Knockrioch, April 28   DUNCAN STEWART

VOL 30, 6 1884

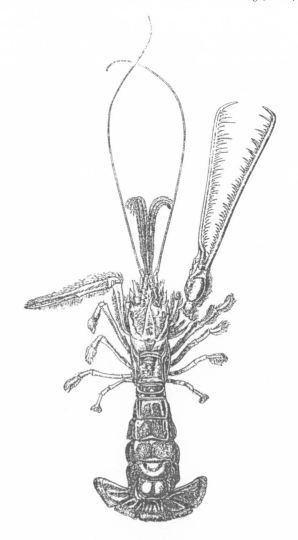

chapter six
1885–1891

John Lubbock finds that animals can count. The Duke of Argyll nettles the biologists. Lord Rayleigh communicates the surface-physics studies of Agnes Pockels. Charles Lutwidge Dodgson (or Lewis Carroll to you and me) constructs an algorithm for finding dates. P. G. Tait enlarges on the physics of golf and Francis Galton on cranial development of Cambridge undergraduates, bright and dull. *Nature* marks the death of the alluring mathematician Sophie Kovalevsky and the hundredth birthday of the great French chemist Michel Eugène Chevreul.

1885-1891

Britain celebrated Queen Victoria's Golden Jubilee and Gladstone his third administration. Louis Pasteur inoculated a young man against rabies. Archduke Rudolph, heir to the throne of Austria, made a melodramatic end of his life in the hunting lodge of Meyerling, and General Gordon met a hero's death at Khartoum, for which an outraged nation exacted a fearful revenge. In the East End of London, Jack the Ripper stalked the fallen women. The Eiffel Tower was completed and the Paris Exhibition was a triumph.

That animals can indeed count was demonstrated by Pavlovian experiments many years later. George J. Romanes saw philosophical problems in the question, but the story of the tramway mules is a good one. Sir John Lubbock, whose publication provoked this letter, was a friend of Charles Darwin, entered politics and became a member of parliament. Darwin deplored the move, which he regarded as a grievous loss to science. But Lubbock reaped his reward when he was elevated to the ermine as Lord Avebury. His descendant in our own day, the one-time Liberal MP Eric Lubbock, inherited his forebear's love of animals, for he attempted to leave his body to the Battersea Dogs' Home as a treat for his canine friends.

Can an Animal Count?

SIR JOHN LUBBOCK, in his interesting paper on animal intelligence (NATURE, vol. xxxiii. pp. 46-7), virtually puts this question with reference to the dog. But the question whether a dog, or any other animal, can count will depend upon what we mean by counting. In the ordinary and correct signification of the term, counting consists in applying conventional signs to objects, events, &c., as when we say "one," "two," "three," to the striking of a clock. Clearly in this sense there is no reason to suppose that any animal can count. But there is another sense in which the term "counting" may be used—*i.e.* as designating the process of distinguishing, with respect to number, between the relative contents of two or more perceptions. While addressing an audience of 100 individuals a lecturer can immediately perceive that it does not contain 1000 ; and even without, in the true sense, counting them may make a tolerably close guess at their number.

* * *

Now, it is clearly only in this way that animals can be supposed to count at all ; and, therefore, the only question is as to how far they are able to take immediate cognisance of the precise numerical content of a perception—or, in the case of a series of events, how far they are able to take similar cognisance of their past perceptions. But, as Sir John Lubbock observes, there is no record of any experiments having been made in this direction. Houzeau (tom. ii. p. 207) says that the mules used in the tramways at New Orleans are able to count five ; for they have to make five journeys from one end of the tramway to the other before they are released, and they make four of these journeys without showing that they expect to be released, but bray at the end of the fifth. If this is really a case of "counting," in the incorrect sense of the term (and not due to observing some signs of their approaching release), it is probably due to their perception of a known amount of fatigue, a known duration of time, or some other such measure.

Several years ago my sister tried to teach an intelligent terrier to fetch a stated number of similar little woollen balls placed in a box at a distance from herself—the number stated, or ordered, being purposely varied from one to six. But although she is good at teaching animals, and here went to work judiciously in ways which I need not wait to describe, the result, as in the case given by Sir John Lubbock, was a total failure.

VOL 33, 80 1885

Women doctors were clearly still not regarded as commonplace in 1887.

M. BÉCLARD has presented some interesting statistics to the Academical Council of Paris on the number of female students in the Faculty of Medicine in the University there. He reports that since Germany closed the doors of its Universities to women, the number in Paris has been constantly increasing. At present the numbers of the various nationalities are : Russians 83, English 11, French 7, Americans 3, Austrians 2, Roumanian 1, and Turk 1. The greater number of these do not pursue their studies as far as the doctor's degree. The large proportion of Russian ladies is due to the closing of the female medical school recently founded at St. Petersburg. M. Béclard thinks that the number of students has now reached the maximum, and will probably decline, since the preliminary studies of the Faculty for both sexes have been ma de alike. VOL 35, 306 1887

The following came only just too late to be an All Fools' Day joke.

A CHINESE native paper published recently a collection of some zoological myths of that country, a few of which are worth noting. In Shan-si there is a bird, which can divest itself of its feathers and become a woman. At Twan-sin-chow dwells the Wan-mu Niao (mother of mosquitoes), a fish-eating bird, from whose mouth issue swarms of mosquitoes when it cries. Yung-chow has its stone-swallow, which flies during wind and rain, and in fine weather turns to stone again. Another bird when killed gives much oil to the hunter, and when the skin is thrown into the water it becomes a living bird again. With regard to animals, few are so useful as the "Jih-kih" ox, found in Kansuh, from which large pieces of flesh are cut for meat and grow again in a single day. The merman of the Southern Seas can weave a kind of silky fabric which keeps a house cool in summer if hung up in one of the rooms. The tears of this merman are pearls. A large hermit-crab is attended by a little shrimp which lives in the stomach of its master; if the shrimp is successful in its depredations the crab flourishes, but the latter dies if the shrimp does not return from his daily excursions. The "Ho-lo" is a fish having one head and ten bodies. The myths about snakes are the strangest of all. Thus the square snake of Kwangsi has the power of throwing an inky fluid when attacked, which kills its assailants at once. Another snake can divide itself up into twelve pieces, and each piece if touched by a man will instantly generate a head and fangs at each end. The calling snake asks a traveller "Where are you from, and whither are you bound?" If he answers, the snake follows him for miles, and entering the hotel where he is sleeping, raises a fearful stench. The hotel-proprietor, however, guards against this by putting a centipede in a box under the pillow, and when the snake gives forth the evil odour, the centipede is let out, and, flying at the snake, instantly kills him with a bite. The fat of this snake, which grows to a great size, makes oil for lamps and produces a flame which cannot be blown out. In Burmah and Cochin-China is a snake which has, in the female sex, a face like a pretty girl, with two feet growing under the neck, each with five fingers, exactly like the fingers of a human hand. The male is green in colour, and has a long beard; it will kill a tiger, but a fox is more than a match for it.

VOL 39, 615 1889

The late nineteenth century was the great age of polar exploration. The journeys were extensively reported in NATURE. *Adolphus Greely led a disastrous expedition to Greenland. It became known that the survivors had sustained themselves by eating the bodies of their dead comrades. In later years Greely ran for political office in the United States, offering himself as an experienced leader of men. His opponent countered: "Adolphus Greely never led more than six men in his life and three of them he ate."*

THE GREELY ARCTIC EXPEDITION
Three Years of Arctic Service. An Account of the Lady Franklin Bay Expedition of 1881-84, *and the Attainment of the Furthest North.* By Adolphus W. Greely, Lieutenant U.S. Army, Commanding the Expedition. Two Vols. (London: Bentley and Son, 1886.)

* * *

In the end they were forced to land at Cape Sabine about the middle of October, and here, with scarcely any shelter, with only about enough food to sustain one man in these regions, and under the most miserable meteorological conditions, on the bleakest spot in all the Arctic, did these men drearily drag themselves through the winter. When at last Commander Schley did reach the spot in June 1884, he found only six out of the twenty-five alive. Yet up to within a few days of the rescue, such observations as were possible were carried on, and the conduct of the men, on the whole, was as noble as could be imagined. This fearful sacrifice of life is deplorable, all the more so when it is remembered that it was due to blundering and half-heartedness on the part of those at home. It is easy to ask whether the gains to science are worth all this sacrifice to human life, but the question is not so easily answered. And whatever the answer is, we may be sure that the Greely disaster will never deter humanity from attempting to find out all about the remotest and most inhospitable corners of its little home.

VOL 33, 481 1886

When the French chemist Michel Eugène Chevreul turned 100, NATURE *marked the event with a small panegyric. In his own country his centenary was greeted by a national celebration, with the unveiling of a statue by the president of the Republic. Chevreul's observations on colour perception — he was director of dyeing at the Gobelins tapestry works in Paris — formed the basis of the technique developed by the post-impressionists, especially Georges Seurat and Paul Signac.*

ON the 28th ult. His Excellency Tcheou Meou-Ki, Director of the Chinese Mission of Public Instruction, paid a visit, with the mandarins attached to his person, to M. Chevreul at his house. He handed to the illustrious chemist a Chinese document expressing in old characters every wish for his happiness and long life. It appears that there is living at this moment in China a Chinese *savant* who at the age of 100 years has just passed his examinations and been admitted a member of the highest academy of the mandarins. The interpreter explained to M. Chevreul that his Chinese visitors considered the fact that two *savants* a hundred years of age were living, one in France and the other in China, was a link connecting the learning of the two countries. When the Chinese Mission had retired, M. Chevreul received a deputation of the inhabitants of the Rue Chevreul, who presented him with a bouquet.

* * *

... M. Chevreul remained in Paris during the siege of 1870-71, working steadily in his laboratory. It was soon after this that an expression in a letter he wrote to a friend led to the honourable title of "Doyen des étudiants de France" being affixed to his name. Although he possesses a large fortune, he still carries on his work at the institutions with which he is connected, and prosecutes his experiments with a juvenile lightness of touch. He is exceedingly temperate, drinking nothing but water or beer, but his longevity is not due to this; he owes it to a robust constitution and to a life wisely ordered, regular, and laborious. "It is," concludes M. Tissandier, "a great and beautiful sight presented by this centenarian, who, like an old oak, shelters under his shadow successive generations. Deaf to the sounds of this world, he has chosen to work alone in his laboratory, where his ever-wakeful intelligence is unceasingly attracted to the rays of eternal truth."

VOL 34, 433 1886

At intervals there appeared articles entitled "Scientific Worthies". These were appreciations, indeed eulogies, of the famous by the equally famous and were in general too fulsome in tone and content for modern tastes. Here is a fragment from one of them, dedicated to the astronomer John Couch Adams.

The discoverer of Neptune has found relaxation from the labours of physical astronomy by little calculations on which we must gaze with astonishment. He has had the curiosity to compute the sums of the reciprocals of the first thousand numbers to 260 places of decimals. We have such confidence in the accuracy of Prof. Adams that we have not thought it necessary to repeat this calculation! He has also taken the trouble to calculate thirty-one of Bernouilli's numbers beyond the point that previous calculators had attained, and he has expressed each of them both as vulgar fractions and as decimals. The sixty-second Bernouilli, the last computed by Adams, runs to 111 places, where fortunately for astronomy the appearance of a recurring figure has terminated this inquiry.

VOL 34, 566 1886

Just as Francis Galton saw most aspects of life and society as amenable to statistical analysis, so others were eager to apply quantitative methods to the most unlikely problems.

A Rule for escaping a Danger

SUPPOSE a weir, AB, across a river, and first let it be at right angles to the direction of the current. Suppose a man in the stream above the weir, nearer to B than to A. Let O be his position, and OX a perpendicular on AB. Then he cannot escape if his velocity, v, is $< \dfrac{BX}{OB} . u$, where u is that of the stream. If his full speed has this critical value, or if there is any uncertainty about his safety, he must swim at right angles to OB.

The rule is obviously correct, for to escape he must clear the nearer end of the weir, and must therefore exert his strength in the direction mentioned. Geometry puts it clearly : Reduce the stream to rest so that the weir is advancing on the man with velocity u. Let P be the point at which the man is overtaken, then, if PN be perpendicular to AB,

$$\frac{OP}{v} = \frac{PN}{u},$$

so that P is on a conic for any given velocity. Varying v, he will escape if the conic reaches the bank. The first to do so touches at the end C of the minor axis, and since CB is a tangent, the angle COB is right. Also now

$$v : u = OC : CB = BX : OB.$$

If the weir slants across the river, the direction of safety is still at right angles to the line joining O to A or B. The swimmer must decide, by looking in both directions, to which bank to direct his efforts. The locus of points for which both directions give the same distance is, to axes through the middle of the weir up and at right angles to the current, of the form

$$(y^2x - 2aby + b^2x)(y^2a - 2bxy + ab^2) = by(x^2 - a^2)^2,$$

a quintic having cusps at A, B.

The rule fails if the change of velocity as one approaches the bank be considerable. One would then strike more across.

If one were being charged by any insensate object, the rule would of course apply.   FRANK MORLEY

Bath College

VOL 35, 345 1887

Charles Vernon Boys remains famous for the bow-and-arrow technique of making quartz fibres. Here is the first report of this inspired method, set in context.

Physical Society, March 26.—Prof. Balfour Stewart, President, in the chair.—The following paper was read :—On the production, preparation, and properties of the finest fibres, by Mr. C. V. Boys. The inquiry into the production and properties of fibres was suggested by the experiments of Messrs. Gibson and Gregory on the tenacity of spun glass, described before the Society on February 12, and the necessity of using such fibres in experiments on which Prof. Rücker and the author are engaged. The various methods of producing organic fibres such as silk, cobweb, &c., and the mineral fibres, volcanic glass, slag wool, and spun glass, were referred to, and experiments shown in which masses of fibres of sealing-wax or Canada balsam were produced by electrifying the melted substance. In producing very fine glass fibres, the author finds it best to use very small quantities at high temperatures, and the velocity of separation should be as great as possible. The oxyhydrogen jet is used to attain the high temperature, and several methods of obtaining a great velocity have been devised. The best results obtained are given by a cross-bow and straw arrow, to the tail of which a thin rod of the substance to be drawn is cemented. Pine is used for the bow, because the ratio of its elasticity to its density (on which the velocity attainable depends) is great. The free end of the rod is held between the fingers, and when the middle part has been heated to the required temperature the string of the cross-bow is suddenly released, thus projecting the arrow with great velocity and drawing out a long fine fibre. By this means fibres of glass less than 1/10,000 of an inch in diameter can be made. The author has also experimented on many minerals, such as quartz, sapphire, ruby, garnet, feldspar, fluor-spar, augite, emerald, &c., with more or less success. Ruby, sapphire, and fluor-spar cannot well be drawn into fibres by this process, but quartz, augite, and feldspar give very satisfactory results. Garnet, when treated at low temperatures, yields fibres exhibiting the most beautiful colours. Some very interesting results have been obtained with quartz, from which fibres less than 1/100,000 of an inch in diameter have been obtained. It cannot be drawn directly from the crystal, but has to be slowly heated, fused, and cast in a thin rod, which rod is attached to the arrow as previously described.

VOL 35, 575 1887

Charles Lutwidge Dodgson published serious business, such as THE HUNTING OF THE SNARK *and this modest algorithm (as well as all his work on logic), under the name of Lewis Carroll; his real name was, it seems, appended only to less important matters, such as Euclid (see p.56).*

TO FIND THE DAY OF THE WEEK FOR ANY GIVEN DATE

HAVING hit upon the following method of mentally computing the day of the week for any given date, I send it you in the hope that it may interest some of your readers. I am not a rapid computer myself, and as I find my average time for doing any such question is about 20 seconds, I have little doubt that a rapid computer would not need 15.

Take the given date in 4 portions, viz. the number of centuries, the number of years over, the month, the day of the month.

Compute the following 4 items, adding each, when found, to the total of the previous items. When an item or total exceeds 7, divide by 7, and keep the remainder only.

The Century-Item.—For Old Style (which ended September 2, 1752) subtract from 18. For New Style (which began September 14) divide by 4, take overplus from 3, multiply remainder by 2.

The Year-Item.—Add together the number of dozens, the overplus, and the number of 4's in the overplus.

The Month-Item.—If it begins or ends with a vowel, subtract the number, denoting its place in the year, from 10. This, plus its number of days, gives the item for the following month. The item for January is "0"; for

February or March (the 3rd month), " 3 "; for December (the 12th month), " 12."

The Day-Item is the day of the month.

The total, thus reached, must be corrected, by deducting " 1 " (first adding 7, if the total be " 0 "), if the date be January or February in a Leap Year: remembering that every year, divisible by 4, is a Leap Year, excepting only the century-years, in New Style, when the number of centuries is *not* so divisible (*e.g.* 1800).

The final result gives the day of the week, " 0 " meaning Sunday, " 1 " Monday, and so on.

EXAMPLES

1783, *September* 18

17, divided by 4, leaves " 1 " over; 1 from 3 gives " 2 "; twice 2 is " 4."

83 is 6 dozen and 11, giving 17; plus 2 gives 19, *i.e.* (dividing by 7) " 5." Total 9, *i.e.* " 2."

The item for August is " 8 from 10," *i.e.* " 2 "; so, for September, it is " 2 plus 3," *i.e.* " 5." Total 7, *i.e.* " 0," which goes out.

18 gives " 4." Answer, " *Thursday.*"

1676, *February* 23

16 from 18 gives " 2."

76 is 6 dozen and 4, giving 10; plus 1 gives 11, *i.e.* " 4." Total " 6."

The item for February is " 3." Total 9, *i.e.* " 2."

23 gives " 2." Total " 4."

Correction for Leap Year gives " 3." Answer, " *Wednesday.*"   LEWIS CARROLL

VOL 35, 517 1887

Alexander Borodin was known to compose only when sickness kept him from the laboratory. His musical friends would greet him with the hope that he was unwell.

WE regret to announce the death of M. Alexander Borodin, Professor of Chemistry at the Medico-Surgical Academy at St. Petersburg, and one of the most eminent Russian musical composers. He died on February 27.

VOL 35, 473 1887

The resourceful M. Arnaudeau envisages a conduit of communication across the Channel, reminiscent of the pneumatic tubes through which bills, payments and change used to be impelled around department stores.

IN a recent number of the *Revue Scientifique*, M. Arnaudeau develops the idea of a double postal tube between Dover and Calais, to be suspended in air. Each tube should be 1 metre in diameter, and of thin metal, allowing the supports to be far apart, say 800 metres. In the tube, a train of ten to fifteen small waggons should run on rails on a floor, the motive power compressed and rarefied air actuating a piston. The lower part of one tube should hold telegraphic, and that of the other telephonic wires. The metallic foundation-piers, some of which should be as much as 70 metres high, should be of truncated-pyramid shape, and capable of floating at first, but gradually filled with masonry and water, and sunk to the bottom. These should support tall pillars having suspension-cables at the top. By the pumping out of the water, these piers could be raised and shifted if necessary.

VOL 36, 349 1887

The physicist P. G. Tait remained a frequent contributor. This, the first of many articles on the physics of golf, reveals an extraordinary passion for what Mark Twain called "a good walk spoiled".

THE UNWRITTEN CHAPTER ON GOLF.

THERE are two ways of dealing with a difficulty—the metaphysical and the scientific way. The first is very simple and expeditious—it consists merely in giving the Unknown a name whereby it may be classified and categorized. Thenceforward the Unknown is regarded as having become part of knowledge. The scientific man goes further, and endeavours to find what lies concealed under the name. If it were possible for a metaphysician to be a golfer, he might perhaps occasionally notice that his ball, instead of moving forward in a vertical plane (like the generality of projectiles, such as brickbats and cricket-balls), skewed away gradually to the right. If he did notice it, his methods would naturally lead him to content himself with his caddie's remark—"Ye heeled that yin," or, "Ye jist slicet it" (we here suppose the metaphysician to be right-handed, as the sequel will show). But a scientific man is not to be put off with such flimsy verbiage as this. He *must* know more. What is " heeling," what is " slicing," and why would either operation (if it could be thoroughly carried out) send a ball as if to cover-point, thence to long slip, and finally behind back-stop? These, as Falstaff said, are " questions to be asked."

As the most excellent set of teeth, if but one incisor be wanting, gives pain rather than pleasure to the beholder; so is it with the works of the magnificent Clark, the sardonic Hutchinson, and the abstruse Simpson. These profess to treat of golf in theory as well as in practice. But in each a chapter is wanting, that which ought to deal with " slicing," " heeling," " toeing," " topping," &c., not as metaphysical abstractions enshrined in homely though unpleasant words, but as orderly (or disorderly) events due to physical causes and capable of receiving a physical explanation. Mayhap, with the aid of scissors and paste, some keen votary of the glorious game will employ this humble newspaper column to stop, however imperfectly and temporarily, the glaring gap which yawns in the work of every one of its exponents! If so, this scrap will not have been written in vain. It may even, in the dim future, lead some athletic pundit to elaborate *The Unwritten Chapter*.

Every one has heard of the uncertain flight of the projectile from Brown Bess, or from the old smooth-bore 32-pounders, and of the introduction of rifling to insure steadiness. Now, all that rifling secures is that the ball shall rotate about an axis nearly in its line of flight, instead of rotating (as the old smooth-bore projectiles did) about an axis whose direction is determined by one or more of a number of trivial circumstances whose effects cannot be calculated, barely even foreseen. Thus it appears that every deviation of a spherical projectile from its line of flight (excluding, of course, that due to gravity) is produced by rotation about an axis perpendicular to the line of flight.

This question was very skilfully treated by Magnus in 1852. He showed by experiment that, when a rotating sphere is exposed to a current of air whose direction is perpendicular to the axis of rotation, the side of the sphere which is advancing to meet the current is subject to greater pressure than is that which is moving in the direction of the current. This difference of pressures tends to make the sphere move in a direction perpendicular at once to the current and to the axis of rotation—the direction, in fact, in which the part of the sphere facing the current is being displaced. But it is a matter of no consequence whether the current of air comes against the sphere, or the sphere moves in the opposite direction (and with the same speed) through still air. Hence Magnus's experimental result amounts to this:—*If a spherical ball be rotating, and at the same time advancing in still air, it will deviate from a straight path in the same direction as that in which its front side is being carried by the rotation.*

* * *

When the toe of the club is turned inwards, the face is pushed tangentially outwards behind the ball, so that the spin and its consequences are exactly the reverse of those just described.

From what has been said above, it is obvious that the flight of a ball, if it be nearly spherical and have its centre of gravity at its centre, depends solely upon the impulse originally given to it. [If the centre of gravity be not in the centre of the ball, it

is only by mere chance (in teeing) that the ball escapes having a rapid rotation given to it, even by the most accurate of drivers. Should it fortunately escape initial rotation, still its flight cannot be regular. A simple and exceedingly expeditious test of this defect consists in placing the ball on mercury in a small vessel. If, in that position, it oscillates rapidly about the vertical, it should be at once rejected as absolutely worthless.] This is a point on which opinions of the wildest extravagance are often expressed. Some balls, it is said, "will not fly," &c. How if they were fired from a blunderbuss? Nobody seems to have made the trial in the only reasonable way—viz. by using a cross-bow or a catapult to give the initial speed. With such an instrument two homogeneous spherical balls of equal size and weight, whatever their other peculiarities, would be despatched under exactly the same conditions, and their behaviour could be *compared*—it would not require to be *contrasted*.

But he is correct (in meaning, though not in his English) who says that some balls "won't drive." It is easy to recognize a good ball by trial, but difficult to define one, at least without periphrasis. A good ball is one which acquires, under given conditions of good driving, as great an initial speed as possible, coupled with the minimum of rotation.

So far as we are aware, all direct scientific experiments on elastic resilience have been made at low speeds, and consequently with but slight distortion of the impinging bodies. But the circumstances of a "drive" in golf are of a totally different character; so that the results of the drive must be themselves regarded as the only data of the requisite kind which we possess. In this matter very valuable data (not for golf alone) might easily be obtained by measuring the height to which a ball rebounds when fired from a powerful catapult against a wooden or stone floor; recording on each occasion the extent to which the springs of the weapon were extended, and the appended weight which would produce the same extension. Some keen golfer may thus find thoroughly useful as well as congenial occupation, when his happy hunting-grounds are inches deep in snow.   P. G. T.

Vol 36, 502 1887

Skis were evidently still a novel concept in Britain.

DR. FRIDTJOF NANSEN, of the Bergen Museum, has announced his intention of attempting to cross the interior of Greenland next summer on *Ski*, viz. the snow-runners found so advantageous during the last Nordenskiöld expedition across that continent. It may be remembered what extraordinary progress the Lapps made at that time on these Scandinavian means of locomotion across snow-fields. Dr. Nansen, who has on a former occasion visited the inland ice in Greenland, has placed his plan before Baron Nordenskiöld, who fully believes in its realization, and is giving Dr. Nansen every assistance. The explorer purposes crossing from the east to the west coast, the reverse of Baron Nordenskiöld's attempt.

Vol 37, 138 1887

That foam-flecked mauler, T. H. Huxley, joins in an attack on the tiresome Duke of Argyll, who had got up the noses of serious scientists for decades. The assault was launched by the geologist T. G. Bonney, an admirer of Charles Darwin (notwithstanding that he was in holy orders). The provocation was a criticism of Darwin's theory of coral reefs. Huxley once owned in a letter to a friend that a reply to one of his philippics had occasioned "such a flow of bile that I have been the better for it ever since". Polemics, he confessed, was to him "as gin is to a reclaimed drunkard."

The Duke of Argyll's Charges against Men of Science.

THE Duke of Argyll's singular passion for besmirching the characters of men of science appears to grow on what it feeds on ; and, as fast as old misrepresentations are refuted, new ones are evolved out of the inexhaustible inaccuracy of his Grace's imagination.

In the last two letters which the Duke of Argyll has addressed to you, he accuses me of having charged the members of the French Institute with having entered into a "conspiracy of silence" in respect of Mr. Darwin's views. I desire to say that the assertion that I have done anything of the kind is untrue and devoid of foundation.

My words, in the passage of which the Duke has cited as much as suited his purpose, stand as follows : "In France, the influence of Elie de Beaumont and of Flourens—the former of whom is said to have 'damned himself to everlasting fame' by inventing the nickname of 'la science moussante' for evolutionism—to say nothing of the ill-will of other powerful members of the Institute, *produced, for a time, the effect of a conspiracy of silence.*"[1] I used the words I have italicized advisedly, for the purpose of indicating that, though the members of the Institute did not enter into a conspiracy of silence, the notorious antagonism of some of them to evolution produced much the same result as if they had done so.

If the Duke of Argyll were properly informed upon the topics about which he ventures to speak so rashly, he would know that M. Flourens wrote a book in vehement denunciation of evolutionism. As I reviewed that book not very long after its appearance, I could not well be ignorant of its existence. And being aware of its existence, I could not possibly have charged M. Flourens with taking any part in a "conspiracy of silence."

The "effect" of the known repugnance to Mr. Darwin's views of some of the most prominent members of the Institute, to which I refer, is the effect upon the younger generation of French naturalists. Considering the influence of the Institute upon scientific appointments, the chances of a candidate known to be an evolutionist would have been small indeed ; and prudence dictated silence.

Mr. Carlyle has celebrated the courage, if not the discretion, of a certain "Rex Sigismundus," who, his Latin being called in question, declared that he was, as a Royal personage, "supra grammaticam." The Duke of Argyll appears to be of King Sigismund's opinion in respect of the obligations which are felt by humbler persons, who have, wittingly or unwittingly, accused their fellows wrongfully ; and I do not suppose that he will descend, on my account, from a position which may be sublime or may be ridiculous, according to one's point of view. The readers of NATURE will choose their own.

T. H. HUXLEY.

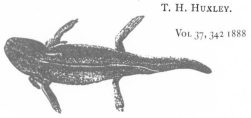

Vol 37, 342 1888

*Here is Francis Galton again, drawing his inferences —
distinctly odd, if valid — from Dr John Venn's measurements
of brain sizes of the bright and the dull at Cambridge. Like the
biceps, the brain, it seems, expands with exercise.*

HEAD GROWTH IN STUDENTS AT THE UNIVERSITY OF CAMBRIDGE.

* * *

Accepting these curves as a true statement of the case—and they are beyond doubt an approximately true statement—we find that a "high honour" man possesses at the age of nineteen a distinctly larger brain than a "poll" man in the proportion of 241 to 230·5, or one that is almost 5 per cent. larger. By the end of his College career, the brain of the "high honour" man has increased from 241 to 249 ; that is by 3 per cent. of its size, while the brain of the "poll" man has increased from 230·5 to 244·5, or 6 per cent.

Four conclusions follow from all this :—

(1) Although it is pretty well ascertained that in the masses of the population the brain ceases to grow after the age of nineteen, or even earlier, it is by no means so with University students.

(2) That men who obtain high honours have had considerably larger brains than others at the age of nineteen.

(3) That they have larger brains than others, but not to the same extent, at the age of twenty-five ; in fact their predominance is by that time diminished to one-half of what it was.

(4) Consequently "high honour" men are presumably, as a class, both more precocious and more gifted throughout than others. We must therefore look upon eminent University success as a fortunate combination of these two helpful conditions.

VOL 38, 15 1888

This commentary, borrowed by NATURE *from its cousin
journal in the United States, shows how little has changed in
politicians' education.*

WE reprint from *Science* of June 1, 1888, the following suggestive paragraph :—"The Committee of the House of Representatives on acoustics and ventilation has actually reported favourably a Bill appropriating seventy-five thousand dollars to subsidize a man who thinks he can construct a steel 'vacuum' balloon of great power. He is to be allowed to use the facilities of one of the navy-yards for the building of his machine, and is to have the money as soon as he has expended seventy-five thousand dollars of private capital upon his air-ship. One of the mathematical physicists of Washington was asked by a member of Congress whether such a balloon could be successfully floated. He set to work upon the problem, and here are some of his results, which are rather curious :—A common balloon is filled with hydrogen gas, which, being lighter than air, causes the balloon to rise and take up a load with it. But, as the pressure of the gas within is equal to the pressure of the atmosphere without, no provision other than a moderately strong silk bag is required to prevent collapse. The inventor of the proposed steel balloon hopes to gain greater lifting-power by using a vacuum instead of gas, the absence of substance of any kind being lighter than even hydrogen gas. But he has to contend with the tendency of the shell to collapse from the enormous pressure of the atmosphere on the outside, which would not be counterbalanced by anything inside of it. The first question which presented itself was, How thick could the metal of the shell be made, so that the buoyancy of the sphere, which would be the most economical and the strongest form in which it could be constructed, would just float it without lifting any load ? The computations showed that the thickness of the metal might be ·000055 of the radius of the shell. For example : if the spherical shell was one hundred feet in diameter, the thickness of the metal composing it could not be more than than one-thirtieth of an inch, provided it had no braces. If it was thicker, it would be too heavy to float. Now, if it had no tendency to buckle, which of course it would, the strength of the steel would have to be equivalent to a resistance of more than 130,000 pounds to a square inch to resist absolute crushing from the pressure of the air on a cross-section of the metal. Steel of such high crushing-strength is not ductile, and cannot be made into such a shell. If the balloon is to be braced inside, as the inventor suggests, just as much metal as would be used in constructing the braces would have to be subtracted from the thickness of that composing the shell. Of course, such a shell would buckle long before the thickness of the metal of which it was composed was reduced to ·000055 of its radius. In other words, it is mathematically demonstrated that no steel vacuum balloon could be constructed which could raise even its own weight. This is an illustration of how intelligently Congress would be likely to legislate on scientific matters unguided by intelligent scientific advice."

VOL 38, 185 1888

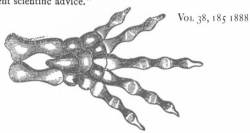

*S. A. Wroblewski is well known for his work on the
liquefaction of gases. He had, it seems, a dramatic life, as
well as death.*

IN the last issue of the Journal of the Russian Chemical and Physical Society there is an interesting article on Prof. S. A. Wroblewski, whose death at Cracow we lately recorded. While a student of the Kieff University, Wroblewski took part in the Polish insurrection of 1863, and was exiled to Siberia, where he had to remain for six years. During his term of exile he elaborated a new cosmical theory, which on his return he hastened to submit to German men of science. Helmholtz received the young man cordially, but advised him to make at the Berlin laboratory certain experiments which would convince him of the erroneousness of his ideas. Wroblewski at once began earnest physical and chemical work, and never afterwards spoke of the theory of his youth. In 1874 he went to Strasburg, and there he published his first serious work, "Ueber die Diffusion der Gase durch absorbirende Substanzen." The flattering opinion expressed about this work by Maxwell in NATURE encouraged Wroblewski to continue physical work on the same lines. He was offered the Chair of Physics at the Cracow University, and the authorities of that institution gave him permission to spend a year at Paris in the laboratory of Sainte-Claire Deville, before beginning his University teaching. There Wroblewski discovered, in the course of his work on the saturation of water with carbonic anhydride under strong pressures, the hydrate of carbonic oxide, and that discovery became the starting-point of a series of works on the condensation of gases. His capital discoveries, made in association with M. Olszanski, which resulted in the condensation of oxygen, azote, and hydrogen, are well known. He was making preparations for an elaborate volume on the condensation of hydrogen, when he perished by accident. While working late in the night in his laboratory, he fell asleep, and in his sleep he overthrew a kerosene lamp. His clothes began to burn, and the wounds thus received resulted four days later in death. The Journal gives a complete list of Wroblewski's works.

VOL. 38, 598 1888

George Francis FitzGerald was born into the Irish Protestant Ascendency and became a professor at Trinity College, Dublin, at the age of 30. He was a many-sided physicist, whose name remains enshrined in the FitzGerald-Lorentz contraction. His relish for a brawl is plain from this spirited review.

PROFESSOR VON " CRANK."

Richtigstellung der in bisheriger Fassung unrichtigen Mechanischen Wärmetheorie und Grundzüge einer allgemeinen Theorie der Aetherbewegungen. Von Albert R. von Miller-Hauenfels, Professor a. D. in Graz. Pp. 256. (Wien : Manz'sche k.k. Hof-Verlags- und Universitäts-Buchhandlung, 1889.)

IT is quite refreshing to come across a real " crank " among the sober Germans. As might be expected, there is a good deal of irregular metaphysics involved in the lucubrations of a German "crank." One would not, however, expect an entire ignorance of the first principles of the mathematics involved. The author of this hardly sufficiently ingenious to be even curious work begins by objecting to the well-known thermodynamic equation for perfect gases—

$$J(C - c) = p\frac{dv}{dt} = R = \frac{pv}{T},$$

because, forsooth, it is not identical with the general differential equation—

$$R = p\frac{dv}{dt} + v\frac{dp}{dt};$$

forgetting that the definition of C, as he himself gives it, assumes that, in the first equation, p is constant. In order to escape this invented difficulty, he loads himself with an equation—

$$JQ = Jcdt + vdp + pdv,$$

which involves the remarkable result that the heat required to warm a gas at constant volume is $JQ = Jcdt + vdp$, while by definition it is $= Jcdt$. It is not necessary to remark that the author carefully neglects to draw this conclusion. His equation is founded on the interesting principle that, when any event produces two different effects on the same organ of sense, each effect must be due to a separate flow of energy.

* * *

And what is the use of spending time looking into such a work as this? It is by studying extraordinary and startling departures from reason, and not the ordinary and familiar ones, that we learn the causes of our aberrations and how to avoid them. It is the same unreasoning prejudice for " I can hardly believe it otherwise," the same neglect to study the meanings of symbols, whether words or letters, the same satisfaction with a theory that leads to some true conclusions, which bristle upon every page of this book, and which are some of the most important factors in the prejudice that ignores the necessity for verification, the muddle-headedness that is content with vague notions, the clinging to an incomplete hypothesis that stands in the way of a true theory, all of which are in all and each of us such bars to progress. If the study of Prof. Miller-Hauenfels' errors leads to even a state of preparedness to look out for similar errors in our own work, the study will have been fruitful.   G. F. F. G.

Vol 40, 244 1889

The Duke of Argyll is once more the Aunt Sally for an assault, this time by Edwin Ray Lankester, who anatomises the Duke's arguments in lengthy detail and concludes:

I venture to point out that the Duke of Argyll has (1) failed to cite facts in support of his assertions of belief in " prophetic germs," and " transmission of acquired characters " when challenged to do so ; (2) that he displays ignorance of two of the most important passages in the works of Lamarck and of Darwin, whom he nevertheless criticizes, and in consequence of his ignorance completely, though unintentionally, misrepresents ; and (3) that he has introduced into these columns a method of treating the opinions of scientific men, viz. by insinuation of motive and by rhetorical abuse, which, though possibly congenial to a politician, are highly objectionable in the arena of scientific discussion.
February 22.   E. RAY LANKESTER.

Vol 41, 416 1890

Now another animal story:

IN the latest of his series of instances—printed in the *American Naturalist*—of the effect of musical sounds upon animals, Mr. R. E. C. Stearns mentions the case of a canary " who is particularly fond of music." This interesting bird belongs to the Rev. Mr. James, who writes as follows :—" Immediately I begin to play upon the flute she chirps about as if enjoying the music. If I open the cage-door and leave her, she will come as near to me as possible, but not attempt to fly to the music ; but if I put her upon my desk, and lay the flute down, she will perch upon the end, and allow me to raise the instrument and play. I often take her into the church and play there upon the organ, and she will perch upon my fingers, notwithstanding the inconvenience of the motion of the hands, and chirp in evident delight at the sweet sounds."

Vol 41, 593 1890

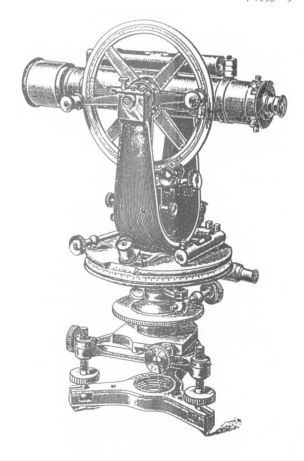

An observation to give the drinker pause:

In *Science* reference is made to a question which may interest 'many of our readers, "Should beer be drunk out of a glass?" Dr. Schultze claims to have established, by a very extended series of experiments, that beer, by as little as five minutes' standing in any glass, even when cold and in the dark, will be materially affected both in taste and odour. By making trial tests on some one hundred persons he sustains his claims. The change is due, as he thinks, to the slight solubility of the glass substance in the beer. Lead is used in the manufacture of glass, making it more easy to manipulate, and from experiments with glass obtained from the leading sources of supply, he determined that one cubic centimetre of beer, by five minutes' standing in glass, dissolved 6–26 ten-millionths of a milligram of the glass substance containg 0–48 thousand-millionths of a milligram of lead oxide. It is this small quantity of glass substance that affects the taste of the beer, and if it contains lead, renders it objectionable for sanitary reasons. By further experiments with vessels of different substances, he comes to the conclusion that gold-lined silver mugs are the best, and he ranks covered salt-glazed stone mugs as good.

Vol 42, 525 1890

The pictures of ears are a selection from a commentary by Edmund Spearman on work by Alphonse Bertillon, the criminologist and pioneer of what he himself called "anthropometry".

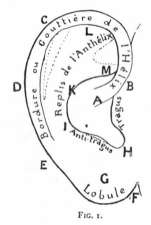

Fig. 1.

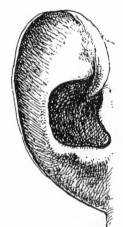
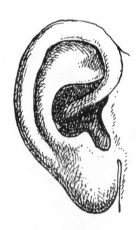

Fig. 2.—Ear showing all the characteristics at a minimum.

Fig. 3.—Ear showing all the characteristics at a maximum.

Vol 42, 644 1890

Altruism among sea urchins?

Mutual Aid among Animals.

Recent discussion of this subject has recalled my attention to an observation made some time ago, while studying the animals of Casco Bay, on the coast of Maine.

Among the specimens brought back from one excursion were four of the common Echini (*E. drobachiensis*). The last one taken had been left exposed to the sun some time before it was noticed and properly cared for.

These four animals were placed alone in a small aquarium, and, as we wished to study the action of the ambulacræ, each was turned mouth up. Soon the action began, with which every naturalist is familiar, and three of the captives slowly rose on edge, and then deliberately lowered themselves into the normal position. The fourth, the injured one, made much less rapid progress : all it could achieve was a slight tipping of its disk. The two nearest Echini, from six to eight inches distant, now moved up and stationed themselves on opposite sides of their disabled comrade.

Fastening their tentacles for a pull, they steadily raised the helpless urchin in the direction in which it had started. As soon as it was possible, one of the helpers moved underneath the edge of the disk on the aboral side, and, when the half-turn was accomplished, the other took station on the oral side, gradually moving back as the object of so much solicitude was very gently lowered to the position nature had made most convenient.

This is the best instance of "giving a lift" I have ever met with among animals of so low a grade. It may not be without interest to others.   Wm. Elder.
Colby University, Waterville, Maine, U.S.A.

Vol 43, 56 1890

Francis Galton reviews a book on criminal characteristics by Havelock Ellis (now remembered principally as an early sexologist). At the end of his review Galton sets out the then widely held views on criminal types.

A fresh indication of frequent misshape in their heads may be derived from the three composite portraits of criminals (who were by no means of a bad order) that are given in this volume. Here the outlines of the heads of the composites are very hazy, testifying to large and *various* differences in the component portraits. These composites show no prevalence of any *special* deformity in head or features.

The hope of the criminal anthropologist is to increase the power of discriminating between the natural and accidental criminal. He aims at being able to say with well-founded confidence of certain men that it is impossible to make them safe members of a free society by any reasonable amount of discipline, instruction, and watchfulness, and that they must be locked up wholly out of the way. Also, to say of some others that it would be both cruel and unwise to treat them as ordinary criminals, because they have been victims of exceptional circumstances : they are not naturally unfit, and therefore still admit of being turned into useful members of society. Extracts are given in this book from the official reports of the prison at Elmira in the United States, where experiments are made in educating prisoners of the latter class. They describe a system of massages and Turkish baths three times a week, courses of literature, æsthetics, and ethics, including a study of Jowett's translation of the " Republic" of Plato, and of the works of Herbert Spencer, together with a gymnasium and a drum corps, suggesting to the unprepared reader a chapter in Gulliver's account of the institutions of Laputa.

Francis Galton.

Vol 42, 76 1890

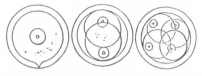

The well known inorganic chemist T. E. (in due course Sir Edward) Thorpe works himself into a chauvinistic passion over what he sees as disrespect to two prominent British figures in the history of chemistry. The final passage of his leader goes like this:

It is altogether beside the question for M. Berthelot now to say in effect :— " Have I not praised your men of science, and thereby drawn down upon myself the wrath of my countrymen ? And yet you are not satisfied ! ' We are sorry for M. Berthelot : he is in the position of the man with many friends, and his friends for the moment are a little angry. He has either not the courage of his convictions, or he has halted between two opinions—with the usual consequences.

With respect to the discovery of the compound nature of water, M. Berthelot now takes up a different position from that which he occupies in " La Révolution Chimique." His contention there was that by every legitimate canon the experiment of June 24, 1783, gives to Lavoisier the priority of discovery. He now admits that Cavendish played " un rôle capital—car il donna le branle aux esprits vers la solution définitive." But how was this possible when Cavendish's memoir was not published until January 1784 ? There is really only one answer—it was given simply by the intervention of Blagden. I repeat that Blagden told Lavoisier of Cavendish's researches and of his conclusions, and that it was in the light of that knowledge that the experiment of June 24, 1783, was made. There can be no question of this. Blagden's testimony, as given in the letter to Crell, is as direct and decisive as it is damning. It was never contradicted by Lavoisier, nor by Laplace, Vandermonde, Fourcroy, Meusnier, or Legendre, who were present on the occasion when Lavoisier himself admits that he received the information. M. Berthelot does not contradict it, but, instead, he asperses the moral character of Blagden. This method of treating a witness whose evidence cannot be rebutted is apt, when unsuccessful, to recoil on him who attempts it. It is perfectly true that Blagden' interpolated the famous passage in Cavendish's memoir :—

" During the last summer, also, a friend of mine gave some account of them [the experiments] to M. Lavoisier, as well as of the conclusion drawn from them. . . . But at that time so far was M. Lavoisier from thinking any such opinion warranted that, till he was prevailed upon to repeat the experiment himself, he found some difficulty in believing that nearly the whole of the two airs could be converted into water."

This passage, however, was inserted with Cavendish's knowledge and consent, and by his assistant and amanuensis, who happened to be the very man who had a personal knowledge of the facts. Assuming the statement to be true, where is the immorality of the proceeding ?

Everything that we can learn authoritatively concerning Blagden goes to show that he was an upright and honourable man. Sir Joseph Banks has testified to his abilities and integrity. Dr. Johnson spoke of his copiousness and precision of communication, with the characteristic addition : " Blagden, sir, is a delightful fellow." Laplace, Cuvier, Berthollet, and Benjamin Delessert, were among his friends.[1] He was rich, and was understood to have speculated to profit in the French funds. For thirteen years he was a Secretary of the Royal Society, and in 1792 he was knighted for his services to science. Every year he spent a considerable time on the Continent, and was frequently in Paris. The gossip of the period states that he aspired to the hand of Madame Lavoisier, who preferred Count Rumford. He died in Berthollet's house at Arcueil, on March 26, 1820. In an obituary notice in the *Moniteur* of September 22, 1820, M. Jomard testifies to his benevolence and generosity, and states that " none of his countrymen have done more justice to the labours and discoveries of the French, or have contributed more than he to the happy relations which have subsisted for six years (1814–20) between the *savans* of the two countries." By his will he provided liberally for his scientific friends : Berthollet, the daughter of Madame Cuvier, and the daughter of Count Rumford, each received £1000 ; and Laplace £100, " to purchase a ring." M. Berthelot asperses the character, not only of Blagden, but also of his countrymen by his insinuations. Would he have us believe that men like Berthollet, Cuvier, and Laplace, would extend their friendship to, and receive pecuniary benefits from, one whom they believed had foully stabbed their compatriot in the back ? It is surely incumbent on M. Berthelot, on every ground, either to substantiate his implications or to withdraw them.

M. Berthelot makes the gratuitous assumption that I am ignorant of the work of Monge. Whether I am or not is altogether beside the mark. There is, indeed, no question of Monge. Monge distinctly disclaims priority to Cavendish, nor did he attempt to establish a right to be considered an independent discoverer of the true nature of water. In his memoir, " Sur le Résultat de l'Inflammation du Gas inflammable et de l'Air dephlogistiqué dans les Vaisseaux Clos," he tells us that the experiments recorded in it were made in June and July 1783, and repeated in October of the same year. " I did not then know," he adds, " that Mr. Cavendish had made them several months before in England, though on a smaller scale ; nor that MM. Lavoisier and Laplace had made them about the same time at Paris in an apparatus which did not admit of as much precision as the one which I employed." I fail to see what M. Berthelot gains by his reference to Monge.

M. Berthelot reproaches Priestley and Cavendish for their adherence to phlogistonism. I say it with all respect, but is it seemly for M. Berthelot, of all men, to cast this stone ? Is not he himself an exemplification of that conservatism which he deplores ? A generation ago the doctrine of Avogadro became the corner-stone of that edifice of which M. Berthelot asserts that Lavoisier laid the foundations. Indeed, the introduction of that doctrine effected a revolution hardly less momentous than that of which Lavoisier was the leader. But what has been M. Berthelot's consistent attitude towards this teaching ? We can illustrate it by a single example. He is the sole teacher in Europe of any position who continues to symbolize the constitution of that very substance of which he claims that Lavoisier discovered the composition by a formula which is as obsolete as any conception of phlogistonism.

T. E. THORPE.

This ingenious device for creating a posthumous effigy of oneself must unaccountably have failed to catch on.

THE curious idea of preserving dead bodies by a galvano-plastic method is not new, but we note that a Frenchman, Dr. Variot, has been lately giving his attention to it (*La Nature*). To facilitate adherence of the metallic deposit, he paints the skin with a concentrated solution of nitrate of silver, and reduces this with vapours of white phosphorus dissolved in sulphide of carbon, the skin being thus rendered dark and shiny. The body is then ready for the electric bath, which is served by a thermo-electric battery, giving a regular adherent deposit of copper if the current is properly regulated. With a layer of ¼ to ¾ mm. the envelope is solid enough to resist pressure or shock. Dr. Variot further incinerates the metallic mummy, leaving holes for the escape of gases. The corpse disappears, and a faithful image or statue remains.

VOL 43, 163 1890

Philip Henry Gosse, who died in 1888, was one of the foremost naturalists of his day and pre-eminent as a populariser. He emerges from his son Edmund's classic memoir, Father and Son, *as a tragic figure, for he ended his days the butt of derision and mirth. Gosse's constant obsession was with his mission, as he saw it, to reconcile the geological record with biblical authority, and he summed up the results of his labours in a book with the provoking title* Omphalos *("the navel"). The navel in question was Adam's and the issue was how — if indeed Adam, as the archetype for humankind, had such an organ — it could have arisen, given that an umbilical cord implied a mother. Gosse's explanation for this*

and for the fossil record was that God had devised two forms of creation, one chronological and set out in the Bible, the other outside time but with the appearance of having evolved over many millennia — a kind of deception on us by God (not elsewhere noted for his sense of humour). Gosse believed that Omphalos *would secure him imperishable fame. Here is how Edmund recorded it in* Father and Son: *"Never was a book cast upon the waters with greater anticipations of success than was this curious, this obstinate, this fanatical volume. My father lived in a fever of suspense, waiting for the tremendous issue. This* Omphalos *of his, he thought, was to bring all the turmoil of scientific speculation to a close, fling geology into the arms of Scripture, and make the lion eat grass with the lamb. He offered it, with a glowing gesture, to atheists and Christians alike. This was to be the universal panacea; this the system of intellectual therapeutics which could not but heal the maladies of the age. But, alas! atheists and Christians alike looked at it and laughed, and threw it away." Edmund wrote a biography of his father 17 years before* Father and Son. *Here then is a passage from the review in* NATURE.

PHILIP HENRY GOSSE.

The Life of Philip Henry Gosse, F.R.S. By his son, Edmund Gosse, Hon. M.A. of Trinity College, Cambridge. (London: Kegan Paul, Trench, Trübner, and Co., Ltd., 1890.)

* * *

His religious views were peculiar, and he gave them a quite peculiar prominence in many of his writings; still he never was subjected to any extreme or very unkindly criticism therefor. As a completely self-taught naturalist he had succeeded in training himself up to a comparatively high standard of knowledge, and at this period all his friends hoped that once time had worn away some of the sorrow from his heart, he would have returned to his studies with renewed zest; and so in time he did, but not before entering into a vague and unsatisfactory series of speculations on the origin and creation of life. Perhaps no work since the "Vestiges of Creation" was received with a greater tempest of adverse criticism than "Omphalos," published by him in 1857. The work of a serious biologist, its at-once-felt unreality, though charming in its way, was clearly not the object aimed at by its author: as a play of metaphysical subtlety, with its postulates true and its laws fairly deduced, it stands complete: Bishop Berkeley would have appreciated this volume, though even he would not have believed in its conclusions. Neither Gosse's friends nor foes seemed to have had any appreciation for it. Here, though we feel bound to make but a passing allusion to this work, we cannot refrain from an expression of regret that Kingsley's letter should have been published without the passage alluding to Newman having been first omitted. The book was a failure from a pecuniary point of view, and yet, though the fact is not alluded to in this memoir, its author contemplated "a sort of supplement, or rather complement, to it, examining the evidences of Scripture; not merely the six-day statements, but the whole tenor of revelation." This never appeared.

VOL 43, 603 1891

The obituary of the mathematician Sophie Kovalevsky, which follows, was by Prince Peter Alekseevich Kropotkin. She was by general consent a mathematician of prodigious gifts, though mathematics seems to have played only a minor part in her turbulent and romantic life. As a child she studied algebra in secret in bed, while her governess slept. The obituary tells the story of the temporary wallpaper, consisting of duplicated notes on differential calculus, pasted on the walls of her nursery, which she believed had some enduring, if subliminal, effect. The German mathematician Karl Weierstrass, who received her reluctantly in Göttingen, quickly recognised her qualities and supported her in her career. She became first assistant to Gösta Mittag-Leffler and then professor in Stockholm. By the time of her death at 41, she seemed to have abandoned mathematics for good and dedicated herself to a second career as a novelist and playwright. According to her biographer her last words were "too much happiness". She and her novelist sister were admirers and friends of Fyodor Dostoevsky and supposedly the models for the two sisters in The Idiot.

PROFESSOR SOPHIE KOVALEVSKY.

THE Swedish papers bring us the sad news of the death of the lady-Professor of Mathematics at the Stockholm University, Mme. Sophie Kovalevsky. She spent her Christmas holidays in the south of France, returned to Stockholm on February 4, and began her course of lectures on the 6th. On the evening of that day she felt ill, and on the 10th she died of an attack of pleurisy. She was born in 1853 at Moscow, and spent her early childhood in a small town of West Russia, where her father, the general of artillery Corvin-Krukowski, was staying at that time ; and afterwards on her father's estate in the same part of Russia. She received her first instruction from her father, but it seems that it was her maternal uncle, an engineer of some renown, Schubert, who awakened in her an interest in natural science. She early lost both her mother and her father, and, having ardent sympathy with the movement which was spreading among Russian youth, she applied for, and at last obtained, permission to study at St. Petersburg. The next year—that is, in 1869, when she was but sixteen years old—she was received as a student at the Heidelberg University, and began the study of higher mathematics. About this time, when extremely young, she married Kovalevsky, the well-known Moscow Professor of Palæontology. From 1871 to 1874, she was again in Germany, this time at Berlin, studying mathematics under Weierstrass ; and at the age of twenty-one, she received the degree of Doctor of Philosophy at Göttingen. Her husband died in 1883, and the next year, in June, she was offered the chair of higher mathematical analysis at the Stockholm Högskola, on condition that she should lecture during the first year in German, and afterwards in Swedish. This she did, and most successfully too—some of her Swedish pupils already being professors themselves. Her chief mathematical papers were : " On the Theory of Partial Differential Equations " (in *Journal für Mathematik*, 1874, vol. lxxx.) : " On the Reduction of a class of Abel's Integrals of the Third Degree into Elliptical Integrals " (in *Acta Mathematica*, 1884, vol. iv.)—both being connected with the researches of Weierstrass ; " On the Transmission of Light in a Crystalline Medium " (first in the Swedish *Förhandlingar*, and next in the *Comptes rendus*, 1884, vol. xcviii.), being part of a larger work in which Mme. Kovalevsky shows the means of integrating some partial differential equations which play an important part in optics ; and " On a Particular Case of the Problem of Rotation of a Heavy Body around a Fixed Point " (in the *Mémoires* of the Paris Academy :

Savants étrangers, vol. xxxi., 1888). The third of these works received from the French Academy the Prix Baudin, which was doubled on account of the " quite extraordinary service " rendered to mathematical physics by this work of Sophie Kovalevsky. She was also elected a Corresponding Member of the St. Petersburg Academy of Sciences.

Besides her mathematical work, Sophie Kovalevsky had lately begun to give literary expression to her ideas. The autobiography of her early childhood (" Reminiscences of Childhood "), published last year in a Russian review, is one of the finest productions of modern Russian literature. In 1887 she published in the Swedish review *Norna* the introduction to her novel, " Væ Victis ! " And in the last issue of the *Nordisk Tidskrift* she brought out, under the pseudonym of Tanya Rerevski, a fragment of a longer novel, " The Family of Vorontsoffs," which she left in manuscript entirely ready for the printer. In her last letter to the writer of these lines in December last, she spoke of bringing out an English version of this novel, which, though written in Russian, could not be published in her mother country.

It need not be said that so highly gifted a woman as Sophie Kovalevsky was modesty itself. She took the liveliest interest in Swedish intellectual life, and had many friends both in Stockholm and in this country, which she visited last year. She leaves a daughter eleven years old. The Swedish papers speak with the greatest sympathy and regret of *their* professor " Sonya " (the little Sophie) Kovalevsky.

In Mme. Kovalevsky's " Reminiscences of Childhood," she records a fact well worthy of note. She was then about ten years old, staying in her father's house in the country. The house was being repaired, and wall-papers were brought from St. Petersburg ; but it so happened that there was no wall-paper for the nursery. So it was papered with the great Ostrogradski's lithographed course upon higher mathematical analysis—a survival of her father's student years ; and the little Sophie, who devoured everything printed within her reach, to the despair of her English governess, was continually reading these mathematical dissertations covered with incomprehensible hieroglyphs. " Strangely enough," she says in her memoirs, " when, at the age of sixteen, I began studying the differential calculus, my teacher was astonished at the rapidity with which I understood him—' just as if it was a reminiscence of something that you knew before,' he said. The continual reading of the wall-papers certainly left some unconscious traces in my childish mind."

P. K.

*Another remarkable woman was Agnes Pockels. At her girls'
school in Braunschweig, she developed an intense interest in
physics, which she also learned from her brother Friedrich, a
physicist who gave his name to the Pockels effect. Agnes was
prevented by the restrictions of the time and place from
attending university. Instead she studied in solitude and in
due course began to use the kitchen as a laboratory. There she
made notable advances in surface physics and foreshadowed
the work by Irving Langmuir and Kathleen Blodgett on
monolayers. At the urging of her brother, she sent an account
of her work (in German) to Lord Rayleigh. Rayleigh did not
fail her and saw to it that a translation was published in
*Nature. Here is the beginning of her paper, with Rayleigh's
letter of introduction.*

Surface Tension.

I SHALL be obliged if you can find space for the accompanying
translation of an interesting letter which I have received from
a German lady, who with very homely appliances has arrived at
valuable results respecting the behaviour of contaminated water
surfaces. The earlier part of Miss Pockels' letter covers nearly
the same ground as some of my own recent work, and in the
main harmonizes with it. The later sections seem to me very
suggestive, raising, if they do not fully answer, many important
questions. I hope soon to find opportunity for repeating some
of Miss Pockels' experiments.   RAYLEIGH.
March 2.

Brunswick, January 10.

MY LORD,—Will you kindly excuse my venturing to trouble
you with a German letter on a scientific subject? Having heard of
the fruitful researches carried on by you last year on the hitherto
little understood properties of water surfaces, I thought it might
interest you to know of my own observations on the subject.
For various reasons I am not in a position to publish them
in scientific periodicals, and I therefore adopt this means of
communicating to you the most important of them.

First, I will describe a simple method, which I have employed
for several years, for increasing or diminishing the surface of a
liquid in any proportion, by which its purity may be altered at
pleasure.

A rectangular tin trough, 70 cm. long, 5 cm. wide, 2 cm. high,
is filled with water to the brim, and a strip of tin about $1\frac{1}{2}$ cm.
wide laid across it perpendicular to its length, so that the
under side of the strip is in contact with the surface of the
water, and divides it into two halves. By shifting this partition
to the right or the left, the surface on either side can be lengthened
or shortened in any proportion, and the amount of the displacement may be read off on a scale held along the front of the
trough.

No doubt this apparatus suffers, as I shall point out presently,
from a certain imperfection, for the partition never completely
shuts off the two separate surfaces from each other. If there is a
great difference of tension between the two sides, a return current often breaks through between the partition and the edge of
the trough (particularly at the time of shifting). The apparatus,
however, answers for attaining any condition of tension which is
at all possible, and in experiments with very clean surfaces
there is little to be feared in the way of currents breaking
through.

I always measured the surface tension in any part of the
trough by the weight necessary to separate from it a small disk
(6 mm. in diameter), for which I used a light balance, with
unequal arms and a sliding weight.

I will now put together the most important results obtained
with this apparatus, most of which, though perhaps not all,
must be known to you.

*Miss Pockels then goes through the characteristics of water-
air and water- solid interfaces. Here is a sample from her
dissertation on the subject.*

IV. *Currents between surfaces of equal tension.*—Between
two normal surfaces, which are unequally contaminated by one
and the same substance, a current sets in from the more to the
less contaminated when the partition is removed; much weaker,
indeed, than that exhibited in the anomalous condition by
differences of tension, but, all the same, distinctly perceptible.
With *equal* relative contamination by the same substance, no
current of course sets in. It is otherwise when the contamination is produced by *different* substances.

I contaminated the surface on one side of the partition by
repeated immersion of a metal plate, on the other by immersion
of a glass plate, which had both been previously carefully cleaned
and repeatedly immersed in fresh water surfaces. I then made
the relative contamination on the two sides equal (*i.e.* = $\frac{1}{2}$) by
pushing in the outer partitions by which the surfaces were inclosed. After the water had been dusted with Lycopodium, the
middle partition was removed. I repeated this experiment eight
times, with different changes devised as checks.

On the removal of the partition a *decided current* set in each
time, *from the surface contaminated by glass to that contaminated
by metal;* and when I replaced the partition after the current had
ceased, and investigated the contamination on both sides, I
always found it greater on the metal than on the glass side.

Thus equal relative contamination by different substances does
not indicate equality of that (osmotic?) pressure which is the
cause of the current between surfaces of equal tension.

For further proof of this result I have made experiments
with other substances; for example, with a floating piece of tinfoil
on one side, and of wax on the other, when, after they had been
acting for a long time, and then the relative contaminations had
been equalized, a current resulted from the wax to the tinfoil;
and again, with camphor on the one, and small pieces of wood
and wax on the other side, which showed a current from the
wax and wood to the camphor.

Since, therefore, the water surface assumes dissimilar qualities
from contact with different substances, the conviction is forced
upon me that it is these bodies themselves (glass, metal, wax,
&c.) which are dissolved, though only feebly in the surface, and
thereby render it capable under sufficient contraction of becoming
anomalous.

*And finally the modest close. This was the first of several
papers on the topic in* NATURE.

What I have further observed regarding solutions in the surface and the like, seems to me less remarkable, and part of it
still very uncertain. I therefore confine myself to these short
indications, but I believe that much might be discovered in this
field, if it were thoroughly investigated. I thought I ought not
to withhold from you these facts which I have observed, although
I am not a professional physicist; and again begging you to excuse
my boldness, I remain, with sincere respect,

Yours faithfully,
(Signed)   AGNES POCKELS.

VOL 43, 437 1891

chapter seven
1891-1899

FRANCIS GALTON QUANTIFIES THE RIGOURS OF THE LAW AND CURES GOUT WITH STRAWBERRIES. HIGH-SPEED PHOTOGRAPHY CAPTURES BULLETS IN FLIGHT AND A TUMBLING CAT. THE BIRD-MAN OTTO LILIENTHAL GRACES THE SKIES. MEN WITH TAILS ARE REPORTED, G. F. FITZGERALD VIEWS SCIENCE AT OXFORD WITH ASPERITY AND HIS FELLOW IRISHMAN JOHN JOLY EXPLAINS THE PHYSICS OF ICE-SKATING. WILHELM KONRAD VON RÖNTGEN DISCOVERS X-RAYS. SOLEMN CEREMONIES MARK THE DEATHS OF TWO GIANTS, LOUIS PASTEUR AND T. H. HUXLEY.

1891-1899

Gladstone formed his fourth and last administration. President Sadi Carnot of France was assassinated and Captain Alfred Dreyfus was found guilty of treason. The Boer War began and so did wireless telegraphy. Louis Lumière presented the first public film show in Paris. Sir William Ramsay and Lord Rayleigh discovered argon, J. J. Thomson the electron, the Curies radium and Ronald Ross the cause of malaria.

Another animal anecdote, worth a letter to NATURE:

A Dog Story.

THE following dog story may interest your readers.

As I went to the train one morning, I saw a brown retriever dog coming full speed with a letter in his mouth. He went straight to the mural letter box. The postman had just cleared the box, and was about 20 or 30 yards off when the dog arrived. Seeing him, the sagacious animal went after him, and had the letter transferred to the bag. He then walked home quietly.

Putney, September 23.   JOHN BELL.

VOL 44, 521 1891

Sir William Crookes reflects on the nature and power of electricity. The story was told at about this time that a Cambridge student, asked in his final viva voce examination what electricity was, hesitated and then stammered that he had known this once but had forgotten. The examiner sighed: "What a pity, what a pity! Only one man in the world who knew what electricity is, and now he has forgotten!" Hilaire Belloc saw the enigma like this:

Here you may put with critical felicity,
The following question: What is electricity?
Molecular Activity, say some;
Others, when asked say nothing and are dumb.

We have happily outgrown the preposterous notion that research in any department of science is mere waste of time. It is now generally admitted that pure science, irrespective of practical applications, benefits both the investigator himself and greatly enriches the community. "It blesseth him that gives, and him that takes." Between the frog's leg quivering on Galvani's work-table and the successful telegraph or telephone there exists a direct filiation. Without the one we could not have the other.

We know little as yet concerning the mighty agency of electricity. "Substantialists" tell us it is a kind of matter. Others view it, not as matter, but as a form of energy. Others, again, reject both these views. Prof. Lodge considers it "a form, or rather a mode of manifestation, of the ether." Prof. Nikola Tesla demurs to the view of Prof. Lodge, but thinks that "nothing stands in the way of our calling electricity ether associated with matter, or bound ether." High authorities cannot even yet agree whether we have one electricity or two opposite electricities. The only way to tackle the difficulty is to persevere in experiment and observation. If we never learn what electricity is, if, like life or like matter, it should remain an unknown quantity, we shall assuredly discover more about its attributes and its functions.

The light which the study of electricity throws upon a variety of chemical phenomena—witnessed alike in our little laboratories and in the vast laboratories of the earth and the sun—cannot be overlooked. The old electro-chemical theory of Berzelius is superseded, and a new and wider theory is opening out. The facts of electro-lysis are by no means either completely detected or co-ordinated. They point to the great probability that electricity is atomic, that an electrical atom is as definite a quantity as a chemical atom. The electrical attraction between two chemical atoms being a trillion times greater than gravitational attraction is probably the force with which chemistry is most deeply concerned.

It has been computed that, in a single cubic foot of the ether which fills all space, there are locked up 10,000 foot-tons of energy which have hitherto escaped notice. To unlock this boundless store and subdue it to the service of man is a task which awaits the electrician of the future. The latest researches give well-founded hopes that this vast storehouse of power is not hopelessly inaccessible.

And to end, an inscrutable allusion to biological electricity:

Electricity seems destined to annex the whole field not merely of optics, but probably also of thermotics.

Rays of light will not pass through a wall, nor, as we know only too well, through a dense fog. But electrical rays of a foot or two wave-length of which we have spoken will easily pierce such mediums, which for them will be transparent.

Another tempting field for research, scarcely yet attacked by pioneers, awaits exploration I allude to the mutual action of electricity and life. No sound man of science endorses the assertion that "electricity is life"; nor can we even venture to speak of life as one of the varieties or manifestations of energy. Nevertheless electricity has an important influence upon vital phenomena, and is in turn set in action by the living being —animal or vegetable. We have electric fishes—one of them the prototype of the torpedo of modern warfare. There is the electric slug which used to be met with in gardens and roads about Hornsey Rise ; there is also an electric centipede. In the study of such facts and such relations the scientific electrician has before him an almost infinite field of inquiry.

The slower vibrations to which I have referred reveal the bewildering possibility of telegraphy without wires, posts, cables, or any of our present costly appliances. It is vain to attempt to picture the marvels of the future. Progress, as Dean Swift observed, may be too fast for endurance. Sufficient for this generation are the wonders thereof.

VOL 45, 63 1891

The hazards of electric telephony:

Engineering of the 29th ult. states that an extraordinary accident had occurred at the London-Paris Telephone Office in the Palais de la Bourse. One of the *employés*, a gentleman named Weller, wished to communicate with the London office on a matter of service. He had already rung up the English officials, and, the bell having sounded in reply, took up the receivers and put them to his ears, when he suddenly sustained a shock of electricity of such severity that it threw him staggering backwards against the door of the telephone cabinet, which, not having been properly fastened, flew open, with the result that he was thrown heavily to the ground. It appears from inquiries that similar accidents, although less serious, have occurred at this telephone office on several previous occasions. The officials attribute them to lightning striking the wire, either at San Gatte, where the submarine cable ends, or at the terminus of the land wire on the Palais de la Bourse. Such accidents, it is declared, might be easily prevented by the simple expedient of erecting lightning conductors at the point where the cable comes ashore, and at the terminus in Paris.

VOL 44, 113 1891

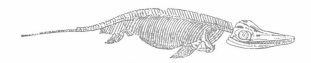

St George Mivart swats a noxious insect:

The Implications of Science.

HITHERTO prevented from again writing, I cannot now remain passive and allow Mr. Dixon to escape from his irrational position under cover of a cloud of verbiage—like a cuttle-fish through water made turbid by its ink.

In my lecture I pointed out that certain truths are implied in all physical science. They *are* so implied. If Mr. Dixon thinks they are not, it is for him to show how experimental science could be carried on, with any real, serious doubt about them. This he has certainly not yet done.

Our knowledge of our own existence "in the present," is knowledge of a particular concrete fact, not of an abstract necessary truth. That "whatever feels, simultaneously exists," *is* such a "necessary truth," but it is an abuse of language to apply that term to anything which may cease to exist the moment after its existence is recognized.

That "nothing can simultaneously be existent and non-existent," does not at all depend upon "terms" or "definitions," but is a law of "things." It would not lose its validity and objective truth, not only if there were no such things as "terms" and "definitions"; it would not lose it if the whole human race came to an end.

I am glad to find my critic does "not doubt" that if he lost an eye his condition would thereby be modified, but if he does not also see that this applies and must apply everywhere and everywhen, I do not envy him his power of mental vision.

Oriental Club, February 2.   ST. GEORGE MIVART.

Vol 45, 343 1892

During the 1890s the views of Jean-Martin Charcot on hysteria attracted much interest — not least from the young Sigmund Freud. Here is a taste of Charcot's therapeutic methods. Gilles de la Tourette gave his name to the syndrome made famous by Oliver Sacks, who saw it all around him on the New York subway.

A VERY original mode of treatment of some nervous complaints has been recently developed by Dr. Charcot, at the Salpêtrière, in Paris (*La Nature*). He was led to it by observing that patients afflicted with *paralysis agitans*, or shaking palsy, often seemed greatly relieved after long journeys by rail or carriage. The greater the train speed and oscillation, the rougher the road and the shaking, the more they liked it and were benefited. Dr. Charcot, taking up the idea, had a chair made, to which a rapid movement from side to side was imparted by electrical agency; like what one sees in processes of sifting by machinery. To a healthy person the experience is execrable; he very soon seeks relief. Not so the patient, however; he enjoys the shaking, and after a quarter of an hour of it, is another man. He stretches his limbs, loses fatigue, and enjoys a good night's sleep afterwards. There are various other nervous diseases to which the method applies. Certain physicians, indeed, have before used such things as tuning-forks and vibrating rods in treatment of neuralgia, &c. A student of Dr. Charcot, Dr. Gilles de la Tourette, has had a vibrating helmet constructed for nervous headaches. It is applied to the head by means of a number of steel strips. Above is a small electric motor making 600 turns a minute; and at each turn a uniform vibration is imparted to the metallic strips, and so to the head. The sensation is not unpleasant; it induces lassitude and sleepiness.

Vol 46, 451 1892

Otto Lilienthal was an insanely brave pioneer of flight. His experiments with gliders eventually cost him his life.

The experimenter in question is Herr Otto Lilienthal, and his success in his so-called "flight" is the result of much thought and considerable practice. The apparatus may be described as a pair of large wings, similar in principle and construction to those of a bird, with two tails at the back, one placed vertically, and the other horizontally. The wings are rigid and fixed, and no motive power at all is used; the whole apparatus weighs twenty kilograms. At the place where the experiments have been carried on, a long sloping hill has been used, with a platform raised about ten metres above the general surface at the top, for the starting point. From this platform the experimenter grips the apparatus between the wings or sails with his hands, and springs off the edge. In the flight he descends at an angle of about 10° to 15°, and the distance covered is sometimes very considerable. In the experiments carried on between Rathenow and Neustadt he covered 80 metres, while from another point he made a flight of 250 metres. The wind of course plays an important part in these flights, but Herr Lilienthal says that with practice one can steer the apparatus well. With the wind blowing stronger on one wing than on the other the equilibrium of the apparatus was found to be greatly disturbed, but this was checked by the movement of the legs, which changed the position of the centre of gravity. In these experiments there is a great opportunity for gaining experience in steering, and it seems very likely that we may learn much thereby.

Here is how he looked in action.

Vol 51, 178 1894

Alfred Russel Wallace shows credulity:

Prenatal Influences on Character.

THE popular belief that prenatal influences on the mother affect the offspring physically, producing moles and other birth-marks, and even malformations of a more or less serious character, is said to be entirely unsupported by any trustworthy facts, and is also rejected by physiologists on theoretical grounds. But I am not aware that the question of purely mental effects arising from prenatal mental influences on the mother has been separately studied. Our ignorance of the causes, or at least of the whole series of causes, that determine individual character is so great, that such transmission of mental influences will hardly be held to be impossible or even very improbable. It is one of those questions on which our minds should remain open, and on which we should be ready to receive and discuss whatever evidence is available ; and should a *primâ facie* case be made out, seek for confirmation by some form of experiment or observation, which is perhaps less difficult than at first sight it may appear to be.

In one of the works of George or Andrew Combe, I remember a reference to a case in which the character of a child appeared to have been modified by the prenatal reading of its mother, and the author, if I mistake not, accepted the result as probable, if not demonstrated. I think, therefore, that it will be advisable to make public some interesting cases of such modification of character which have been sent me by an Australian lady in consequence of reading my recent articles on the question whether acquired characters are inherited. The value of these cases depends on their differential character. Two mothers state that in each of their children (three in one case and four in the other) the character of the child very distinctly indicated the prenatal occupations and mental interests of the mother, though at the time they were manifested in the child they had ceased to occupy the parent, so that the result cannot be explained by imitation. The second mother referred to by my correspondent only gives cases observed in other families which do not go beyond ordinary heredity.

"I can trace in the character of my first child, a girl now twenty-two years of age, a special aptitude for sewing, economical contriving, and cutting out, which came to me as a new experience when living in the country amongst new surroundings, and, strict economy being necessary, I began to try and sew for the coming baby and for myself. I also trace her great love of history to my study of Froude during that period, and to the breathless interest with which my husband and I followed the incidents of the Franco-German war. Yet her other tastes for art and literature are distinctly hereditary. In the case of my second child, also a daughter (I having interested myself prior to her birth in literary pursuits) the result has been a much acuter form of intelligence, which at six years old enabled her to read and enjoy the ballads which Tennyson was then giving to the world, and which at the age of barely twenty years allowed her to take her degree as B.A. of the Sydney University.

"Before the third child, a boy, was born, the current of our life had changed a little. Visits to my own family and a change of residence to a distant colony, which involved a long journey, as well as the work which such changes involve, together with the care of my two older children, absorbed all my time and thoughts, and left little or no leisure for studious pursuits. My occupations were more mechanical than at any other time previous. This boy does not inherit the studious tastes of his sisters at all. He is intelligent and possesses most of the qualifications which will probably conduce to success in life, but he prefers any kind of outdoor work or handicraft to study. Had I been as alive then as I am now to the importance of these theories, I should have endeavoured to guard against this possibility ; as it is, I always feel that it is perhaps my fault that one of the greatest pleasures of life has been debarred to him.

"But I must not weary you by so many personal details, and I trust you will not suspect me of vanity in thus bringing my own children under your notice. Suffice it to say that in every instance I can and do constantly trace what others might term coincidences, but which to me appear nothing but cause and effect in their several developments.

"I will pass on to quote a few passages from letters written to me by two highly intelligent mothers, whom I asked to give me their experiences on this subject, if they had any.

"Mrs. B—— says : 'I can trace, nay, have traced (in secret amusement often), something in every child of mine. Before the birth of my eldest girl I took to ornithology, for work and amusement, and did a great deal in taxidermy too. At the age of three years I find this youngster taking such insects and little animals as she could find, and puzzling me with hard questions as to what was inside them. Later on she used to be seen with a small knife, working and dissecting cleverly and with much care and skill at their *insides*. One day she brought me the tiniest heart of the tiniest lizard you could imagine, so small that I had to examine it through a glass, though she saw it without any artificial aid. By some means she got a young wallaby and made an apron with a pocket inside which she used to call her "pouch." This study of natural history is still of interest to her, though she lacks time and opportunities. Still, she always does a little dissecting when she gets a chance.'

"I never noticed anything about P—— for some years. Three months before he was born a friend, whom I will call Smith, was badly hurt, and was brought to my house to be nursed. I turned out the nursery and he lay there for three months. I nursed him until I could do so no longer, and then took lodgings in town for my confinement. Now after all these years I have discovered how this surgical nursing has left its mark. This boy is in his element when he can be of use in cases of accident, &c. He said to me quite lately, 'How I wish you had made a surgeon of me.' Then all at once the light flashed in upon me, but, alas ! it was too late to remedy the mistake.

"Before the birth of the third child I passed ten of the happiest months of my life. We had a nice house, one side of which was covered with cloth of gold roses and bougainvillea, a garden with plenty of flowers, and a vineyard. Here we led an idyllic life, and did nothing but fish, catch butterflies, and paint them. At least, my husband painted them after I had caught them and mixed his colours. At the end of this time L—— was born. This child excels in artistic talent of many kinds, nothing comes amiss to her, and she draws remarkably well. She is of a bright, gay disposition, finding much happiness in life, even though not always placed in the most fortunate surroundings. Before the birth of my next child, N——, a daughter, I had a bad time. My husband fell ill of fever, and I had to nurse him without help or assistance of any kind. We had also losses by floods. I don't know how I got through that year, but I had no time for reading. N—— is the most prudent, economical girl I know. She is a splendid housekeeper and a good cook, and will work till she drops, but has no taste for reading, but seems to gain knowledge by suction."

If the preceding cases are fully and accurately stated they seem to afford grounds for further investigation. Changes in mode of life and in intellectual occupation are so frequent among all classes, that materials must exist for determining whether such changes during the prenatal period have any influence on the character of the offspring. The present communication may perhaps induce ladies who have undergone such changes, and who have large families, to state whether they can trace any corresponding effect on the character of their children.   ALFRED R. WALLACE.

VOL 48, 389 1893

Advances in photography gave rise to some striking pictures of supersonic missiles in flight. The description is by Charles Vernon Boys, who relished such experimental challenges (see p.79).

Fig. 5 is a photograph of the apparatus set up in one of the passages in the Royal College of Science, in which the experiments were made. It is apparently of the rudest possible construction. The rifle seen on the left of the figure is of course made to rest freely on six points, in order that its position every time it is fired may be the same. The bullet then traverses precisely the same course, so that wires placed in the line between holes in two cards made by one shot will be hit by the next. The two wires which the bullet joins as it passes by are set up in the box seen in the middle of the figure with the lid propped up so as to show the interior. The photographic plate is on the left-hand side and the spark when made is just within the rectangular prolongation on the right-hand side. Paper tubes with paper ends are placed on each side of the box to allow the bullet to enter and leave, and yet not permit any daylight to fall directly on the plate. All is black inside, and so the small amount of light which does enter the box through these holes is not diffused in any harmful manner. The large box at the back is a case 5 ft. long, filled with bran which stops the bullets gently without marking them.

FIG. 5.

The little condenser is just below the rectangular prolongation of the photographic box, the large condenser is the vertical square sheet seen just to the right. The electrical machine used to charge the condensers is seen on the table. It is a very beautiful 12-plate Wimshurst machine made by Mr. Wimshurst and presented to the Physical Laboratory. This machine not only works with certainty but is so regular in its working that no electrometric apparatus is necessary. All that has to be done is to count the number of turns of the handle which are required to produce the sparks at E and E' when the gap at B is not joined, and to count the number which are sufficient to produce a spark at E when the gap at B is suddenly closed. Then if the rifle is fired after any number of turns between these, but by preference nearer the larger than the smaller number, the potential will be right, the spark E, inside the box, and the spark E', which is in sight outside the box, will be let off, and if a plate is exposed a photograph will be taken. If by chance the E' spark is not seen then there is no occasion to waste the plate, another bullet may be fired after resetting the wires and the result will be as good as if one shot had not failed.   VOL 47, 418 1893

The photograph, showing the broadening of the shock-wave angles, and the explanation are from the second part of Boys's article.

In order to illustrate the other fact that the angle of the waves also depends on the velocity of sound in the gas, I filled the box with a mixture of carbonic acid gas, and the vapour of ether, a mixture which is very dense, and through which sound in consequence travels only about half as fast as it does in air, and which will not explode or even catch fire when an electric spark is made within it, or directly act injuriously upon the photographic plate. The increased inclination of the waves is very evident in Fig. 10.

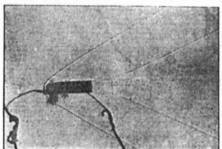

FIG. 10.

VOL 47, 440 1893

This review of a reprinted lecture is by Mrs Grace Frankland, bacteriologist and the wife and collaborator of the chemist Percy Frankland.

WOMEN AND SCIENCE.

THIS little volume is to all intents and purposes a charming and eloquent appeal in support of the claims of women to effectual recognition in the scientific world. In reality it purposes only to give in brief outline the lives of half a dozen women who have rendered important service to mathematical science. But although brief the sketches are so clever that the various characters depicted could scarcely appear more living or real, whilst there is not a single dull sentence to be found in the book.

One of the most interesting of the short studies, because so closely connected with the present, is that of the gifted and fascinating Sophie Kowalevski, who only died three years ago, and who commenced her study of mathematics at the age of fourteen, and at eighteen married Kowalevski, "parce qu'il n'était permis qu'aux dames de suivre les Cours des Universités!" On the presentation of three original theses, the University of Göttingen hastened without further examination to confer the degree of Doctor of Philosophy upon her, and later in life she was appointed to a chair of mathematics in Stockholm. But Sophie Kowalevski was not only a gifted mathematician of whom Kronecker declared "l'histoire des mathématiques parlera comme d'une des plus rares investigatrices," but an accomplished *littérateure*, and the author of numerous books, one of which is entitled "Souvenirs sur George Eliot," whilst "Les Souvenirs d'enfance" is described as a fine bit of psychological study worthy of Tolstoi, or of the new "Immortal" Bourget.

The place of imagination in science, so forcibly insisted upon by Mr. Goschen some years ago in his rectorial address at Edinburgh, is beautifully put in a letter to a novelist friend astonished at her pursuing science and letters simultaneously. "People frequently regard mathematics as a dry and barren science. In reality the pursuit of mathematics demands a great deal of imagination, and one of the greatest mathematicians of our century said, with justice, that it is impossible to be a good mathematician without at the same time having a touch of the poet."

Some sixty or seventy years earlier we read of another highly gifted mathematician, Sophie Germain, who at the same time distinguished herself by her contributions to philosophy. M. Rebière tersely summarises her claims to distinction by thus closing his memoir: "Pour construire la tour Eiffel, les ingénieurs ont utilisé l'élasticité des métaux. On a inscrit sur la tour les noms de 72 savants; on a oublié celui d'une fille de génie, la théoricienne de l'élasticité!"

England is represented by Mrs. Somerville in a very bright and sympathetic little notice, whilst Italy sends her contribution in the shape of "la nobile fanciulla" Marie Agnesi, who Pope Benedict XIV. nominated Professor of Mathematics in the University of Bologna, writing—"It is not you who should thank us; on the contrary, it is we who owe all our thanks to you. From the most remote times Bologna has heard of people of your sex occupying its public chairs. It belongs to you to worthily perpetuate the tradition." In commenting upon this distinction M. Rebière cannot resist telling us of some of the numerous women who have at various times held professorial appointments at Bologna. The list is instructive, and we quote it in full, for we cannot afford to admit women as fellows of any of our learned societies even!—"In languages, philosophy, and theology: Priscopia Cornaro, 'maîtresse des arts libéraux'; Clotilde Tambroni, hellenist, who had Mezoffanti as a pupil. In law: Dotta, daughter of Accurse; Biltizia Gozzadini, in connection with whom a pamphlet was published, *De mulierum doctoratu*; the two sisters, Bettina and Novella Calendrini. It appears that Novella was so beautiful, that it was necessary, in order to avoid distracting the students, to draw a slight curtain between her and the audience. In natural science and medicine: Alexandra Gigliani, Maria Petraccini, Anna Manzolini, and Sybille Mérian. The latter, who was a German, went to study insects at Surinam; she published an important work, and left her collections to the School of Bologna. In physics and mathematics: Laure Bassi, who married Dr. Verati, and who whilst teaching physics during forty years was a model wife and mother; the two astronomers, Thérèse et Madeleine Manfredi, sisters of the Director of the Observatory, who published a volume entitled 'Astronomy for Women.'"

The bust of Marie Agnesi was subsequently placed by Cardinal Dumini in his gallery of distinguished Lombards, and on her tomb these words were inscribed : " Fille remarquable par sa piété, sa science et sa bienfaisance."

We are introduced to a very different woman and mathematician in the person of Madame la Marquise du Châtelet, the friend of Voltaire, and whom the Prince Royal of Prussia familiarly addressed as Vénus Newton !

M. Rebière tells us that she had preserved, in spite of her studies, " une certaine frivolité. Son goût pour la parure et les diamants était très vif. Et puis elle riait de si bon cœur aux marionnettes ! " But whilst indulging in diamonds and puppet-shows, the Marchioness found time to translate Newton's " Principia " from Latin into French, and produced besides numerous learned memoirs, one of which, " Institutions de Physique," was dedicated to her sons in words which, although written more than a century and a half ago, might have been uttered yesterday—" J'ai toujours pensé que le devoir le plus sacré des hommes était de donner à leurs enfants une éducation qui les empêchât dans un âge plus avancé de regretter leur jeunesse, qui est le seul temps où l'on puisse véritablement s'instruire." We find her returning to the same theme in a little essay, " Traité du bonheur," a curious mixture of feelings reflecting very vividly the varying moods of this remarkable woman :—" Nous n'avons rien à faire en ce monde qu'à nous procurer des sensations agréables," she writes; whilst on another page we read, in an eulogistic commentary on the benefits of study more especially to women—" Quand, par hasard, il s'en trouve quelqu une née avec une âme assez élevée, il ne lui reste que l'étude pour la consoler de toutes les exclusions et de toutes les dépendances auxquelles elle se trouve condamnée par état," M. Rebière does not omit to include amongst his memorable women Hypatia, with whose memoir the volume in fact opens.

In conclusion, M. Rebière devotes a couple of pages to suggestions for the making of a book which we fancy would be with difficulty kept within the modest limit of eighty pages, which the little pamphlet before us embraces. " Un livre à faire " remains, says M. Rebière, in which the influence direct and indirect exerted by women on the progress of science might be recorded, a book catholic enough not only to include the *savantes professionnelles*, but the *simples curieuses* or amateurs in science, amongst which George Sand finds a place, the *collaboratrices*, and finally those whose munificence and public spirit have earned for them the well-deserved title of *les protectrices*, instances of which we in this country have fortunately little difficulty in recalling. But possibly the most eloquent tribute which has ever been paid to any woman, and which might appropriately have found mention in M. Rebière's little volume, is that which was so pathetically inscribed by John Stuart Mill on the first page of his essay on " Liberty."

We are glad to learn that meanwhile M. Rebière is compiling a second and more elaborate volume in which women's relation to science will be discussed, upon which subject M. Rebière asks us to mention that he will gratefully receive any notes and suggestions.   G. C. Frankland.

Vol 50, 279 1894

Here are the beginnings in France of rejuvenation therapy with testis extracts, as reported in Nature.

Physiological and therapeutic effects of a liquid extracted from the male sexual gland, by MM. Brown-Séquard and d'Arsonval. Samples of the orchitic liquid for subcutaneous injection were offered to all medical men willing to report upon its effects. Over 1200 physicians availed themselves of this offer, and their results are very encouraging. The malady showing the most striking effect of the remedy was locomotor ataxy, of which 314 out of 342 undoubted cases were cured or considerably improved. Another almost incurable disease which proved very amenable to this treatment was shaking paralysis, of which 25 out of 27 cases were much improved. It appears that the orchitic liquid, though not possessing any direct curative influence upon the various morbid states of the organism, is capable on subcutaneous injection of curing or decidedly ameliorating a great variety of affec-

tions, organic or otherwise. This action is due to two kinds of influence. By the one, the nervous system gains in vigour, and becomes capable of improving the dynamical or organic state of the diseased parts ; by the other, which depends upon the entrance into the blood of new materials, the liquid contributes to the cure of morbid states by the formation of new cellules and other anatomical elements.   Vol 48, 23, 1893

Nature quotes with approval this ode to Hermann von Helmholtz from Punch. *It seems remarkable today that the activities of a scientist (foreign at that) should have received such currency.*

HELMHOLTZ.

What matter titles ? Helmholtz is a name
That challenges, alone, the award of Fame !
When Emperors, Kings, Pretenders, shadows all,
Leave not a dust-trace on our whirling ball,
Thy work, oh grave-eyed searcher, shall endure,
Unmarred by faction, from low passion pure.
To bridge the gulf 'twixt matter-veil and mind
Perchance to mortals, dull-sensed, slow, purblind,
Is not permitted—yet ; but patient, keen,
Thou on the shadowy track beyond the Seen,
Didst dog the elusive truth, and seek in sound
The secret of soul-mysteries profound,
Essential Order, Beauty's hidden law !
Marvels to strike more sluggish souls with awe,
Great seekers, lonely-souled, explore that track,
We welcome the wild wonders they bring back
From ventures stranger than an earthly Pole
Can furnish. Distant still that mental goal
To which great spirits strain ; but when calm Fame
Sums its bold seekers, Helmholtz, thy great name
Among the foremost shall eternal stand,
Science's pride, and glory of thy land.

Vol 50, 530 1894

Yet another animal story:

The following strange incident is described in the *Times* as having occurred in the reptile-house of the Zoological Society's menagerie, the scene being one of the compartments in which the boa-constrictors are confined. Two large boas occupied the chamber, one snake being nine feet and the other eight feet long. When the house was opened in the morning only one boa was found in this cage ; the other had disappeared. Though the survivor was only a foot longer than the other snake, there was no reason to doubt that it had completely swallowed its companion. It was so distended that the scales were almost separated, and it was unable either to coil itself or to move. There is every reason to believe that in accomplishing this almost incredible feat the snake acted by mistake, and that it devoured its companion by what deserves to be called an accident. The larger boa was fed with a pigeon before the house was closed for the night. It swallowed the bird, and the other boa was then given a pigeon, which it had begun to swallow when the snakes were left for the night. It is believed that the larger snake then caught hold of the part of the pigeon which projected from the other's mouth, and gradually enveloped, not only the bird, but the head of the other snake. Once begun, the swallowing process would go on almost mechanically. As the swallowed snake was only one foot less in length than the swallower and of nearly equal bulk, weighing about fifty pounds, the gastric juices must have dissolved the portion which first entered the snake's stomach before the remainder was drawn into the jaws. Though still rather lethargic, the surviving boa is not injured by its meal. It coils itself up without difficulty, and its scales have the beautiful iridescent bloom peculiar to the skin of snakes when in perfect health.

Vol 50, 620 1894

The leisurely, conversational tone of this communication from Lord Kelvin, on board ship, apparently on a cable-laying expedition, catches the spirit of much of the scientific discourse of the period. Mr Maxim was of course the inventor of the Maxim gun (and other technological marvels), later to become Sir Hiram Maxim.

Towards, the Efficiency of Sails, Windmills, Screw-Propellers in Water and Air, and Aeroplanes.

The discussion of this day week, on flying machines, in the British Association was not, for want of time, carried so far as to prove from the numerical results of observation put before the meeting by Mr. Maxim, that the resistance of the air against a thin stiff plane caused to move at sixty miles an hour through it, in a direction inclined to the plane at a slope of about one in eight, was found to be about fifty-three times as great as the estimate given by the old "theoretical" (!) formula, and something like five or ten times that calculated from a formula written on the black-board by Lord Rayleigh, as from a previous communication to the British Association at its Glasgow meeting in 1876.

I had always felt that there was no validity, even for rough or probable estimates, in any of the "theoretical" investigations hitherto published : but how wildly they all fall short of the truth I did not know until I have had opportunity in the last few days, *procul negotiis*, to examine some of the observational results which Maxim gave us in the introduction to his paper. On the other hand, I have never doubted but that the true theory was to be found in what I was taught conversationally by William Froude twenty years ago, and which, though I do not know of its having been anywhere published hitherto, is clearly and tersely expressed in the following sentence which I quote from a type-written copy, kindly given me by Mr. Maxim, of his paper of last week :—

"The advantages arising from driving the aeroplanes on to new air, the inertia of which has not been disturbed, is clearly shown in these experiments."

Founding on this principle, I have at last, I believe, succeeded in calculating, with some approach to accuracy, the force required to keep a long, narrow, rectangular plane moving through the air with a given constant velocity, V, in a direction perpendicular to its length, l, and inclined at any small angle, i, to its breadth, a. In a paper, which I hope to be able to communicate to the *Philosophical Magazine* in time for publication in its next October number, I intend to give the investigation, including consideration of "skin-resistance" and proof that it is of comparatively small importance when i is not much less than 1/10, or 1/20, of a radian, and the "plane" is of some practicably smooth, real, solid material. In the meantime, here is the result, with skin-resistance neglected :—The resultant force (perpendicular, therefore, to the plane) is $2\pi V^2 \sin\theta \cos\theta\, la$; which is $\dfrac{4\pi\cos\theta}{\sin\theta}$ times (or for the case of $\sin\theta = 8$, one hundred times), the old miscalled "theoretical" result.   KELVIN.

Eastern Telegraph Company's Cable Steamer
   Electra ; crossing the mouth of the Adriatic,
    August 17.

Vol 50, 425 1894

The sequence of photographs shown below solves an age-old problem. An accompaying series shows the same event end-on. The Reverend Dr William Archibald Spooner, then in his full pomp as warden of New College, Oxford, was reported to have described a similar event in his best style: "So it popped on its little drawers and away it went."

PHOTOGRAPHS OF A TUMBLING CAT.

M. MAREY'S recent photographs of a falling cat, taken with the view of determining the mechanical conditions which enable the animal to alight on its feet.

* * *

The cat was held by its feet, and was let go in that position. In each of the pairs of figures, the series of images runs from right to left, and the lower is a continuation of the upper. The expression of offended dignity shown by the cat at the end of the first series indicates a want of interest in scientific investigation.

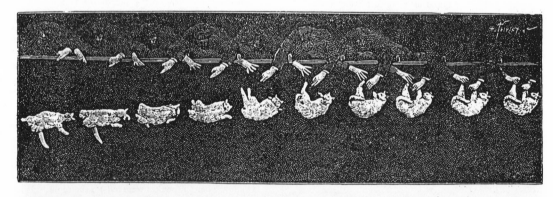

Vol 51, 80 1894

Unease about the complacency of academics at the ancient universities surfaces from time to time, as in this book review by the anonymous Edwin Ray Lankester.

OUR BOOK SHELF.

Harvard College by an Oxonian. By George Birkbeck Hill, D.C.L. (London : Macmillan and Co., 1894.)

DR. BIRKBECK HILL spent two months in 1893 in Cambridge, Massachusetts, and has compiled this little volume giving some account of the history of the celebrated college and university of Harvard. So far as Dr. Hill relies upon previous publications, his account is accurate, but his own observations and impressions are —as is very natural—often quite erroneous. Scant justice is done to the important and costly arrangements for the study of the various branches of the natural sciences which exist either at or in connection with the Massachusetts university. Dr. Hill is not fitted by his own education and experience to report on these matters, nor, indeed, can much value be attached to his somewhat antiquated standpoint as a critic or observer of university institutions. He contrasts Oxford and Harvard at every step, but he fails to give any picture or presentation of the real characteristics of the student's life at Harvard. He does not sufficiently emphasise the fact that the undergraduate at Harvard enjoys the immense benefit of true *university* education, at the hands of distinguished professors, with freedom and independence in regard to his choice and method of study, and as to such personal details of life as board and lodging; whereas the Oxford undergraduate is treated throughout his career as a goose to be nursed, monopolised and plucked by college ushers, who (owing to the system under which they are appointed) are, as a rule, as little capable of good teaching as they are of managing the domestic and disciplinary details of the college-boarding-houses. Dr. Hill notes that the rage for athletics is almost as serious an injury to study at Harvard as it is at Oxford.   L.

Vol 51, 386 1895

Francis Galton found another subject for statistical analysis in the criminal law. The issue does not seem to have lost its relevance a century later.

TERMS OF IMPRISONMENT.

IT would have been expected that the various terms of imprisonment awarded by judges should fall into a continuous series. Such, however, is not the case, as is shown by Table I., which is derived from a Parliamentary Blue-book recently published under the title of " Part I.— Criminal Statistics," p. 215. The original has been considerably reduced in size ; first, by limiting the extracted data to sentences passed on male prisoners without the option of a fine, and, secondly, by entering the number of sentences to the nearest tenth or hundredth, as stated in the headings to the columns. The material dealt with is thereby more homogeneous than in the original, and its significance is more easily seen. The number of cases is amply sufficient to afford a solid base for broad conclusions, there being in round numbers 830 sentences for various terms of years, 10,540 for various terms of months, and 43,300 for various terms of weeks. The diagram drawn from Table I. gives a still clearer view of the distribution of these sentences.

The extreme irregularity of the frequency of the different terms of imprisonment forces itself on the attention. It is impossible to believe that a judicial system acts fairly, which, when it allots only 20 sentences to 6 years imprisonment, allots as many as 240 to 5 years, as few as 60 to 4 years, and as many as 360 to 3 years. Or that, while there are 20 sentences to 19 months, there should be 300 to 18, none to 17, 30 to 16, and 150 to 15. The terms of weeks are distributed just as irregularly. Runs of figures like these testify to some powerful cause of disturbance which interferes with the orderly distribution of punishment in conformity with penal deserts.

On examining the diagram we are struck with the apparent facility of drawing a smooth curve, that shall cut off as much from the hill-tops of the irregular trace as will fill their adjacent valleys. This has been done, by eye, in the diagram, the small circles indicating the smoothed values. Care has been taken that the sums of the ordinates drawn to the smooth curves should be equal to sums of those drawn to the traces, as is shown by the totals in the bottom line of Table I. The smoothed curves may therefore be accepted as an approximate rendering of the general drift of the intentions of the judges as a whole, and show that the sentences passed

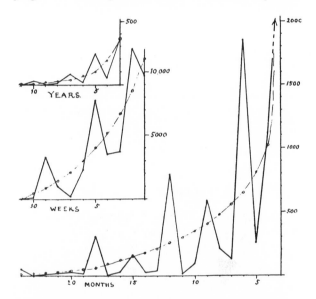

by them severally, ought to be made more appropriate to the penal deserts of the prisoners than they are at present. The steep sweeps of the curves afford a strong testimony to the discriminative capacity of the judges, for if their discrimination had been *nil* and the sentences given at random, those steep curves would be replaced by horizontal lines. We have now to discuss the disturbing cause or causes that stand in the way of appropriate sentences.

The terms of imprisonment that are most frequently awarded, fall into rhythmic series. Beginning with the sentences reckoned in months, we see that their maxima of frequency are at 3, 6, 9, 12, 15, and 18 months, which are separated from one another by the uniform interval of 3 months, or a quarter of a year—a round figure that must commend itself to the judge by its simplicity. And we may in consequence be pretty sure that if the year had happened to be divided into 10 periods instead of 12, the exact equivalent of 3 months, which would then have been 2½ periods, would not have been used in its place. If this supposition be correct, the same penal deserts would have been treated differently to what they are now.

Thus the precise position of the maxima has been apparently determined by numerical fancy, and it seems that the irregularity of the trace is mainly due to the award of sentences being usually in terms of the 3-monthly, but sometimes in that of the 1-monthly, series.

And his conclusion:

I will conclude by moralising on the large effects upon the durance of a prisoner, that flow from such irrelevant influences as the associations connected with decimal or duodecimal habits and the unconscious favour or disfavour felt for particular numbers. These trifles have been now shown on fairly trustworthy evidence to determine the choice of such widely different sentences as imprisonment for 3 or 5 years, of 5 or 7, and of 7 or 10, for crimes whose penal deserts would otherwise be rated at 4, 6, and 8 or 9 years respectively.   FRANCIS GALTON.

VOL 52, 174 1895

NATURE was as much captivated as the rest of society by the thrills of motoring. Here are some of the enthusiastic comments, together with engravings of be-garlanded motorcars, on an early rally in France.

THE RECENT RACE OF AUTO-MOBILE CARRIAGES IN FRANCE.

LAST month a most interesting race of auto-mobile carriages took place in France. The course taken was from Versailles to Bordeaux, and then back to Paris. June 11 was fixed for the day of starting, and forty-six carriages were to have taken part in the race, but only twenty-eight arrived in time, twenty-two of these taking active part, and nine performing the journey within a hundred hours; eight of the latter were worked by petroleum or "gazoline," and one by steam.

FIG. 1.—No. 5. MM. Panhard and Levassor's carriage, worked by gazoline, and to seat two persons (2nd prize, 12,600 francs). Arrived June 13, at 12.57 a.m.

FIG. 2.—No. 16. MM. Peugeot's phaeton, worked by gazoline, and to seat four persons (1st prize, 31,500 francs). Arrived June 14, at 12.2 a.m.

* * *

Taking all the facts into consideration, it appears that the lighter carriages travelled best. This proves the advantage of using petroleum or gazoline, for in order to produce one horse-power it requires per hour $1\frac{11}{144}$ lbs. of gazoline, whereas, if it were worked by steam, at least $6\frac{3}{5}$ lbs. of coal and $39\frac{1}{5}$ lbs. of water would be necessary per hour, and if worked by electricity, there would have to be accumulators of the weight of 220 lbs.

Light carriages have many advantages, for besides having to be less careful about the weight of fuel, they can also have lighter constructed wheels. M. Michelin's carriage, with pneumatic tyres, went the whole distance without an accident, whereas the steam vehicles, one and all, had mishaps, owing almost always to their great weight.

It would take up much time and space to relate the many incidents which occurred; suffice it to say that, apart from ordinary breakdowns, in some towns the travellers were hindered by the inhabitants, in others they were enthusiastically pelted with flowers.

These auto-mobile machines are evidently the carriages of the future. According to the *Times* of July 10, a journey has quite recently been performed in our own country by the Hon. Evelyn Ellis, who was accompanied by Mr. T. R. Simms, managing director of the Daimler Motor Syndicate. The carriage is a four-wheeled dog-cart, and will hold four persons, with room also for two portmanteaus. It was built by Messrs. Panhard and Levassor, of Paris, and is worked by petroleum, the cost

FIG. 3.—No. 15. Worked by gazoline, to seat two persons. Belonging to the sons of Peugeot Brothers (3rd prize, 6300 francs). Arrived June 13, at 6.37 p.m.

being about a halfpenny an hour. The journey undertaken by Mr. Ellis, a distance of fifty-six miles, was performed in five hours and a half.

We understand that the proprietors of the *Engineer* are offering a prize of £1000 to the maker of the fastest going motor.   W.

VOL 52, 300 1895

That year, T. H. Huxley died. NATURE outdid itself in the welter of grief and adulation. Here is how Michael Foster — in his day a name to conjure with, but now chiefly remembered as the founder of the Cambridge School of Physiology, the first modern department of the subject in Britain — recalled his mentor.

TWO scenes in Huxley's life stand out clear and full of meaning, amid my recollections of him, reaching now some forty years back. Both took place at Oxford, both at meetings of the British Association. The first, few witnesses of which now remain, was the memorable discussion on Darwin in 1860. The room was crowded though it was a Saturday, and the meeting was excited. The Bishop had spoken ; cheered loudly from time to time during his speech, he sat down amid tumultuous applause, ladies waving their handkerchiefs with great enthusiasm ; and in almost dead silence, broken merely by greetings which, coming only from the few who knew, seemed as nothing, Huxley, then wellnigh unknown outside the narrow circle of scientific workers, began his reply. A cheer, chiefly from a knot of young men in the audience, hearty but seeming scant through the fewness of those who gave it, and almost angrily resented by some, welcomed the first point made. Then as, slowly and measuredly at first, more quickly and with more vigour later, stroke followed stroke, the circle of cheers grew wider and yet wider, until the speaker's last words were crowned with an applause falling not far short of, indeed equalling, that which had gone before, an applause hearty and genuine in its recognition that a strong man had arisen among the biologists of England.

The second scene, that of 1894, is still fresh in the minds of all. No one who was present is likely to forget how, when Huxley rose to second the vote of thanks for the presidential address, the whole house burst into a cheering such as had never before been witnessed on any like occasion, a cheering which said, as plainly as such things can say : "This is the faithful servant who has laboured for more than half a century on behalf of science with his face set firmly towards truth, and we want him to know that his labours have not been in vain." Nor is any one likely to forget the few carefully chosen, wise, pregnant words which fell from him when the applause died away. Those two speeches, the one long and polemical, the other brief and judicial, show, taken together, many of the qualities which made Huxley great and strong.

This long account of Huxley's career adverts briefly to his passionate rejection of religion — something with which Foster, like many other contemporaries, was obviously uncomfortable.

...while on the objective side his scientific mode of thought thus made him a never-failing opponent of theologic thought of every kind, a common tie on the subjective side bound him to the heart of the Christian religion. Strong as was his conviction that the moral no less than the material good of man was to be secured by the scientific method alone, strong as was his confidence in the ultimate victory of that method in the war against ignorance and wrong, no less clear was his vision of the limits beyond which science was unable to go. He brought into the current use of to-day the term "agnostic," but the word had to him a deep and solemn meaning. To him "I do not know" was not a mere phrase to be thrown with a light heart at a face of an opponent who asks a hard question ; it was reciprocally with the positive teachings of science the guide of his life. Great as he felt science to be, he was well aware that science could never lay its hand, could never touch, even with the tip of its finger, that dream with which our little life is rounded, and that unknown dream was a power as dominant over him as was the might of known science ; he carried about with him every day that which he did not know as his guide of life no less to be minded than that which he did know. Future visitors to the burial-place on the northern heights of London, seeing on his tombstone the lines—

> " And if there be no meeting past the grave,
> If all is darkness, silence, yet 't is rest.
> Be not afraid ye waiting hearts that weep,
> For God still ' giveth his belovèd sleep,'
> And if an endless sleep He wills,—so best"—

will recognise that the agnostic man of science had much in common with the man of faith.

The great evolutionary polemicists of the late nineteenth century sustained into old age the heat of their intellectual passion. Here the pugnacious St George Mivart takes on the fundamentalist Darwinians August Weismann, Ernst Haeckel — the German, and much nastier, opposite number of T. H. Huxley ("Darwin's German bulldog") — and Karl Pearson.

SCIENCE IN THE MAGAZINES.

PROFS. WEISMANN, Haeckel, and Karl Pearson will probably have something to say in reply to a paper which Dr. St. George Mivart contributes to the *Fortnightly*. The paper deals with what is described as " Denominational Science," in which dogma takes the place of facts, and persuasions are given out as if they were demonstrated truths. Dr. Weismann comes under Dr. St. George Mivart's displeasure in this regard ; and a noteworthy characteristic of his is said to be " the confidence with which he propounds hypotheses which are either purely imaginary, or are only supported by an infinitesimal basis of fact, and the readiness with which he comes forward with a fresh gratuitous hypothesis, to replace others which have been refuted by newly-discovered truths." Prof. Haeckel is taken to task for the opinions expressed in his book on " Monism," lately translated into English. The bearing of Dr. St. George Mivart towards the book is indicated by the remark which opens the attack upon some of the points in it. We read : " It is difficult to say whether this small volume is more remarkable for the self-conceit and empty dogmatism, or for the ignorance it displays—ignorance concerning the most fundamental questions of which it treats." To assess these remarks at their proper value, it is necessary to read the article containing them, and the work to which they refer. Prof. Karl Pearson completes the trio upon whose views Dr. St. George Mivart outpours the vials of his wrath. His " Grammar of Science," and his remarks, in the *Fortnightly*, on Lord Salisbury's Oxford address, are given as evidence that " we have in England a denominational writer only second in self-confident dogmatism to Haeckel." All the members of the trio are held up as awful examples of " an unconscious slavery of the intellect to the mere faculty of the imagination, and the consequent presentation of shallow and illogical imaginary phantasms as deep and far-reaching intellectual truths in the form of baseless dogmas of denominational science." Huxley and Karl Vogt are compared by Prof. Haeckel in the *Fortnightly*, the former being given a higher place than the latter, both as regards his philosophical reasonings, and because he showed a much deeper insight into the essence and import of scientific things. Two pages of the six, which form Prof. Haeckel's notice, are taken up with a denunciation of Prof. Virchow's antagonism to Darwinism, and the theory of descent, especially with reference to the most important deduction from the theory—the descent of man from the ape. Virchow's dissent in this matter is used as one of the sticks with which Mr. F. H. Hill belabours agnosticism, and Huxley's support of it, in the *National*, under the title, " Gaps in Agnostic Evolution."

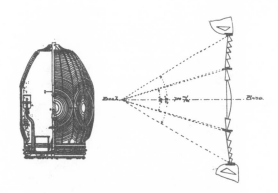

In the same year, Louis Pasteur died, and from the accounts of his obsequies there can be no doubt of the reverence in which he was held by his compatriots. One cannot easily imagine the like today (and especially not in Britain).

General Saussier, surrounded by his staff, and followed by the first division of infantry, preceded the hearse, and behind him came a long line of deputations, many of which had wreaths in their centre. A number of wreaths were borne on litters, and others were carried on six cars, each drawn by a pair of horses. "Along the route from the Rue Dutot to Notre Dame," says the *Times* correspondent, " the compact and silent crowd respectfully uncovered their heads as the hearse passed, and the two thousand soldiers and policemen, drawn up in line to keep the way clear, had absolutely nothing to do. The pall-bearers were M. Poincaré, M. Joseph Bertrand, M. Georges Perrot, Dr. Brouardel, M. Gaston Boissier, and M. Bergeron. After marching for an hour and a half along the left bank of the Seine, the procession reached the square of Notre Dame. The aspect of the Cathedral was most impressive. The presence of President Faure, the Grand Duke Constantine, Prince Nicholas of Greece, Cardinal Richard, the whole of the Diplomatic Corps, the Ministers, the Institute of France, the office-bearers of the Senate and the Chamber of Deputies, the red-robed Judges, the members of the University faculties, in orange, red, and crimson robes, and the other distinguished persons invited—all this display of official mourning was coupled with and yet eclipsed by the profound silence, the manifest grief. The immense crowd was a rare and impressive, if not a unique spectacle."

The Royal Society was represented by Mr. W. T. Thiselton-Dyer, C.M.G., Director of the Royal Gardens, Kew. At the final funeral, which will be held in connection with the Centenary of the Institute, on the 25th inst., several of the Officers and Fellows of the Society will be present, together with many delegates from other of our learned societies.

After the service in Notre Dame, the coffin containing Pasteur's remains was removed to a catafalque outside the Cathedral, and M. Poincaré delivered an oration before it, on behalf of the Government.

Thus does France venerate the memory of her noblest son. But France is not alone in her grief. The human race joins with her in mourning the loss of one who has done so much for humanity and science. The name of him to whom the world owes so much good is imperishable.

Vol 52, 576 1895

From an article on the benefits of spas and of the many "balneo-therapeutic" treatments that they offered, this image of hypochondriacal indulgence:

Vol 60, 418, 1899

Here Francis Galton anticipates the preoccupations of a later generation.

Mr. Francis Galton traces, in the *Fortnightly Review*, a hypothetical discovery of a system of signalling from the planet Mars, and shows how a succession of signals, divided into dots, dashes, and lines of light according to their duration, might be interpreted. Savages can communicate with one another by gestures, deaf mutes by the movements of the lips, and criminals by alphabetical tappings upon the walls of their cells. Mr. Galton shows how, by a kind of Morse code, the Martians could first signal to us the summation of numbers, such as $2 + 3 = 5$, $3 + 3 = 6$, and also the results of multiplication and division. He does not consider the view of the fourth dimensionists, that possibly there are worlds where $2 + 2 = 3$. After the arithmetical rules had been signalled, the supposition is that the relative distances of the planets from the sun were flashed to the earth ; then the relation between the circumference and diameter (π) ; then the area of the circle (πr^2) ; then the names of a number of regular polygons, with the number of sides and area of each. Granting that the Martians were able to make themselves clear so far, they could develop a system of picture-writing. With three varieties of signal, twenty-seven combinations would be possible, and each could represent a particular word or sign. Each side of a polygon with twenty-four sides could, therefore, have a name of its own, and each one would have a definite bearing or direction with reference to the others. All is now plain sailing. The Martians signal the symbols of a number of sides, and, as each is received, a line is drawn in a particular direction. From the formula thus obtained, a picture can be reproduced, as Mr. Galton showed at a Royal Institution lecture in 1893 (see NATURE, vol. xlvii. p. 342). The conclusion is that intelligible messages are possible between planets sufficiently near together for signalling purposes.

Vol 55, 39 1896

The discovery of X-rays, or Röntgen rays, caused an immense stir, and not only among physicists. Here, for instance, is how J. J. Thomson began his Rede lecture at Cambridge, reprinted in NATURE *the year after.*

THE RÖNTGEN RAYS.

Prof. Röntgen, of Wurzburg, at the end of last year published an account of a discovery which has excited an interest unparalleled in the history of physical science. In his paper read before the Wurzburg Physical Society, he announced the existence of an agent which is able to affect a photographic plate placed behind substances, such as wood or aluminium, which are opaque to ordinary light. This agent, though able to pass with considerable freedom through light substances, such as wood or flesh, is stopped to a much greater extent by heavy ones, such as the heavy metals and the bones ; hence, if the hand, or a wooden box containing metal objects, is placed between the source of the Röntgen rays and a photographic plate, photographs such as those now thrown on the screen are obtained. This discovery, as you see, appeals strongly to one of the most powerful passions of human nature, curiosity, and it is not surprising that it attracted an amount of attention quite disproportionate to that usually given to questions of physical science. Though appearing at a time of great political excitement, the accounts of it occupied the most prominent parts of the newspapers, and within a few weeks of its discovery it received a practical application in the pages of *Punch*. The interest this discovery aroused in men of science was equal to that shown by the general public. Reports of experiments on the Röntgen rays have poured in from almost every country in the world, and quite a voluminous literature on the subject has already sprung up.

In view of the general interest taken in this subject, I thought that the Röntgen rays might not be an unsuitable subject for the Rede Lecture.

Vol 54, 302 1896

At about this time, "Röntgen rays" become one of the most numerous entries in NATURE's annual indexes. Whether this is a picture of the hand of Walther Nernst himself is not revealed, but the radiation dose must have been considerable.

A CONTRIBUTION TO THE NEW PHOTOGRAPHY.

NUMEROUS pictures are now being taken by means of the new method. The accompanying illustration, which we owe to the kindness of Prof. Nernst, and the original of which was made by him in the Physical-chemical Laboratory at Göttingen, represents a human hand as

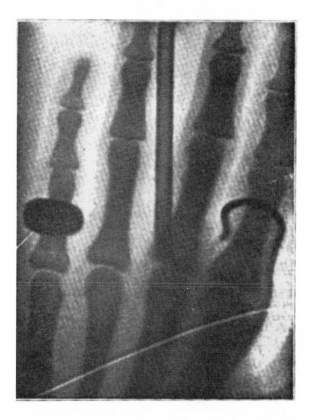

photographed by means of the Röntgen rays. It will be seen that the flesh is very nearly transparent for these rays, while the bones, the gold ring, the piece of wire, and the glass tube are practically opaque. The ring and wire, which were naturally in contact with the flesh of the fingers, appear in the illustration as if suspended in the air   WILLIAM J. S. LOCKYER.

Vol 53, 324 1896

An early indication that X-rays might not be entirely good for you:

Edison reports that his eyes were sore after working for several hours with his fluorescent tubes; but he is not certain that this result is specially attributable to the X-rays. Dr. Wm. J. Morton reports that he sees brilliant flashes of light after he has discontinued work, and, as he has worked with electrical light for many years without injury, he infers that the X-rays are injurious to the eye.

Vol 53, 421 1896

Tailed men seem to have been a recurring theme even before Lord Monboddo disseminated his belief that all human babies were born with tails, which were cut off in a secret conspiracy between midwives.

TAILED men have again turned up. Six years ago, in the course of a visit to the Indo-Chinese region, between 11° and 12° lat. and 104° and 106° long., M. Paul d'Enjoy captured an individual of the Moï race, who had climbed a large tree to gather honey. In descending, he applied the sole of his feet to the bark; in fact, he climbed like a monkey. To the surprise of the author and his Annamite companions, their prisoner had a caudal appendage. He conversed with them, swaggered in his savage pride, and showed that he was more wily than a Mongolian, which, as the author adds, is, however, a very difficult matter. M. d'Enjoy saw the common dwelling of the tribe to which this man belonged, but the other people had fled; it consisted of a long, narrow, tunnel-like hut made of dry leaves. Several polished stones, bamboo pipes, copper bracelets and bead necklaces were found inside; these had doubtless been obtained from the Annamites of the frontier. The Moï used barbed arrows which are anointed with a black sirupy violent poison. The tail is not their only peculiarity: All the Moïs whom M. d'Enjoy has seen in the settlements have very accentuated ankle-bones, looking like the spurs of a cock. All the neighbouring nations treat them as brutes, and destroy these remarkable people, who, the author believes, to have occupied primitively the whole Indo-Chinese Peninsula. The Moï skulls described by MM. Verneau and Zaborowski were certainly by no means those of pure natives: they were taken from graves; but the settled Moï burn their dead, and place the ashes in bamboo pots, or in ratan baskets, considering their spirits as protective divinities. As this somewhat sensational account has been published by our esteemed contemporary *l'Anthropologie* Tome vii., No. 5), we must treat it with respect; and we hope it will not be long before these tailed men are carefully described by a trained scientific observer.

Vol 55, 82 1896

Here is an instance of the perfidy of Her Majesty's ungrateful subjects in the Empire. "Melocheeya" is also the ingredient of a soup with a curious slimy texture.

The specimens referred to in the following letter, which was received from Mr. Kenneth Scott, of Cairo, were carefully examined by Dr. D. H. Scott, of the Jodrell Laboratory, who could only conjecture that they were fragments of the paleæ of some grass. "For some time now malingering Egyptian soldiers have been sent in to the Kasr-el-Aini hospital under my care, suffering from extreme œdema and intense inflammatory injection of the conjunctiva of one or both eyes; the *cornea unaffected.* No discharge from the eye. The condition is entirely unlike that which they also produce by putting in the juice of Euphorbia, slaked lime, seed of 'melocheeya' (? *Corchorus olitorius*) and other things. I obtained the specimens sent you by covering the eye with a thick collodion dressing so as to completely seal it up. The man at the end of five days had evidently feared the inflammation might subside, and therefore raised the dressing and renewed the baneful application, part of which I found on the face of the dressing lying against the eye. I have been entirely unsuccessful in obtaining here any information on the matter, nor have I been able to obtain further quantities of the leaf. The patient either began to fear the consequences of the affair, or his stock of the drug became exhausted, as he in no way interfered with the next collodion dressing which was applied, the eye being quite cured, and the dressing intact after a period of five days."

Vol 58, 255 1898

A dozen didactic quatrains by an astronomer:

BALNIBARBIAN GLUMTRAP RHYME.

(Repeated by the children in the nurseries of Balnibarbi.)[1]

DISTANT scintillating star,
    Shall I tell you what you are?
Nay, for I can merely know
What you were some years ago.

For, the rays that reach me here
May have left your photosphere
Ere the fight of Waterloo—
Ere the pterodactyl flew!

Many stars have passed away
Since your æther-shaking ray
On its lengthy journey sped—
So that you, perhaps, are dead!

Smashed in some tremendous war
With another mighty star—
You and all your planets just
Scattered into cosmic dust!

Strange, if you have vanished quite,
That we still behold your light,
Playing for so long a time
Some celestial pantomime!

But, supposing all is well,
What you're made of, can I tell?
Yes, 'twill be an easy task
If my spectroscope I ask.

There—your spectrum now is spread
Down from ultra-blue to red,
Crossed by dark metallic lines,
Of your cooler layer the signs.

Hence among the starry spheres
You've arrived at middle years—
You are fairly old and ripe,
Of our solid solar type.

Ah, your sodium line is seen
Strongly shifted towards the green.
Hence you are approaching me
With a huge velocity!

But, if some celestial woe
Overtook you long ago,
And to swift destruction hurled
Life on every living world,

Did there in the fiery tide
Perish much of pomp and pride—
Many emperors and kings,
Going to do awful things?

Mighty schemes of mighty czars—
Mighty armies, glorious wars!
From the Nebula they may
Rise to curse a world some day!

                                    G. M. MINCHIN.

---

[1] Balnibarbi is one of the countries visited by Gulliver; the "Glumtrap" is the Balnibarbian equivalent of the English nursery; and the babies of Balnibarbi are brought up on strictly scientific principles—as is evidenced by their knowledge in these verses.

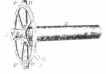

The Irish geologist, physicist and polymath John Joly observes the curious behaviour of a cormorant:

A Shag's Meal.

THE following observation on the habits of the shag (*Phalacrocorax graculus*), which frequents our coasts, is probably of interest in itself and not without bearing on the subject-matter of Mr. Lowe's letter (November 24).

On August 15 last, when at anchor in Wicklow harbour, in the course of a cruise, we noticed a shag alight upon the water at a short distance from our yacht. It was a very calm, bright day. What follows occurred within a distance of a cable's length or thereabouts of our boat; and as I observed the proceedings throughout by the help of a powerful "triëder binocle" (× 9) of Goerz, and my companion, Dr. H. H. Dixon, possesses unusually keen sight and closely followed the events, there is no doubt as to the reality of what is here recorded.

The shag, after swimming about for a little, dived once or twice—apparently fruitlessly—but finally appeared with a large eel in his beak. The eel was big and strong, and twisted into the form of a figure 8, evidently an awkward morsel. The bird kept snapping and shifting it in his beak, till at length with a few violent gulps it swallowed the eel, the latter evidently going down alive. It was to be inferred, in fact, that the shag was not happy with so large a live eel in his crop, for he swam restlessly about, twisting and stretching his neck incessantly. Presently he dived again, was down perhaps twenty seconds, and came up with an eel as large as the first one. This writhed and twisted like its predecessor, and, after much snapping, finally suffered the same fate. The same uneasiness was displayed by the bird, and the bird once more dived.

Dixon and I were expressing some surprise at the rapidity with which our friend had caught the two eels, and also at his very considerable capacity to hold two such large eels—certainly not less than 15 inches long each—when the bird reappeared bearing a third eel, as big as its predecessors and engaged in the same violent resistance. The same snapping, same gulping, same uneasiness and down for a dive once again! This was the third eel.

While we were taking sympathetic breaths with the insatiable shag, the latter reappeared—yet again with a 15-inch eel. Evidently the harbour was so full of 15-inch eels that a shag had only to dive to pick one up. It was also evident that no language could be too strong in which to condemn such unmeasured license. Four 15-inch eels—all swallowed alive—within the space of about four minutes!

But this was only the beginning, as will presently appear. The bird went down again almost immediately after the fourth. We determined to keep careful count and, if possible, get the measure, in eels, of a shag's capacity. Would he bring up another? Yes, there he was again with another 15-inch eel! A very vigorous eel—just like the others in size and appearance, and swallowed in the same manner, after about 30 seconds' resistance. This made five eels.

The question now arose as to what would be the end of this bird. Was he going to die the death of King Henry I. before our eyes? We called him King Henry to distinguish him from other shags.

To make a long story short, we counted *twelve* eels!—all stout 15-inchers. The twelfth seemed, perhaps, rather feebler than the others, but still it nearly got away. For King Henry dropped it in a too vigorous snap, and only recovered it by a prompt plunge forwards. H. R. H. now seemed to reflect that this last misadventure was a warning, swallowed his twelfth, and took flight; disappearing in an easterly direction whence he had come.

There is, of course, only one explanation of all this: the twelve eels were one and the same eel. To suppose the bird caught and devoured twelve eels of this size in as many minutes appears to us incredible. His final appearance as he flew astern of us betrayed no signs of surfeit. He would have had at least two pounds' weight of eel within had he really eaten twelve such eels.

The peculiar procedure of ejecting the prey under water appears very remarkable. Perhaps the head-downward attitude of diving is requisite to effect this.

Has this mode of weakening or playing with his prey been recorded of the shag (or, indeed, of any animal) previously?

Trinity College, Dublin, November 27.   J. JOLY.

The green flash or blue ray at sunrise were topics that nagged at physicists for at least a century, to judge from the regular intervals with which they recurred in the columns of NATURE. *Here Lord Kelvin experiences the manifestation.*

Blue Ray of Sunrise over Mont Blanc.

LOOKING out at 5 o'clock this morning from a balcony of this hotel, 1545 metres above sea-level, and about 68 kilometres W. 18° S. from Mont Blanc, I had a magnificent view of Alpine ranges of Switzerland, Savoy, and Dauphiné; perfectly clear and sharp on the morning twilight sky. This promised me an opportunity for which I had been waiting five or six years; to see the earliest instantaneous light through very clear air, and find whether it was perceptibly blue. I therefore resolved to watch an hour till sunrise, and was amply rewarded by all the splendours I saw. Having only vague knowledge of the orientation of the hotel, I could not at first judge whereabouts the sun would rise; but in the course of half an hour rosy tints on each side of the place of strongest twilight showed me that it would be visible from the balcony; and I was helped to this conclusion by Haidinger's brushes when the illumination of the air at greater altitudes by a brilliant half-moon nearly overhead, was overpowered by sunlight streaming upwards from beyond the mountains. A little later, beams of sunlight and shadows of distant mountains converged clearly to a point deep under the very summit of Mont Blanc. In the course of five or ten minutes I was able to watch the point of convergence travelling obliquely upwards till in an instant I saw a blue light against the sky on the southern profile of Mont Blanc; which, in less than the one-twentieth of a second became dazzlingly white, like a brilliant electric arc-light. I had no dark glass at hand, so I could not any longer watch the rising sun.

KELVIN.

Hotel du Mont-Revard, above Aix-les-Bains,
August 27.

VOL 60, 411 1899

Professor Joly again, this time on the physics of ice-skating:

The Phenomena of Skating and Prof. J. Thomson's Thermodynamic Relation.

IN connection with Prof. Osborne Reynolds's "Notes on the Slipperiness of Ice," read before the Manchester Literary and Philosophical Society (NATURE, March 9, p. 455), the following extract from a brief paper, read by me before the Royal Dublin Society in 1886 (*Proc. R.D.S.*, vol. v. p. 453), may not be without interest.

"To the many phenomena which have found an explanation in Prof. J. Thomson's thermodynamic relation connecting melting-point with pressure, might be added those attending skating, *i.e.* the freedom of motion and, to a great extent, the 'biting' of the skate.

"The pressure under the edge of a skate is very great. The blade touches for a short length of the hog-back curve, and, in the case of smooth ice, along a line of indefinite thinness, so that until the skate has penetrated some distance into the ice the pressure obtaining is great; in the first instance, theoretically infinite. But this pressure involves the liquefaction, *to some extent,* of the ice beneath the skate, and penetration or 'bite' follows as a matter of course. As the blade sinks, an area is reached at which the pressure is inoperative, *i.e.* inadequate to reduce the melting-point below the temperature of the surroundings. Thus, estimating the pressure for that position of the edge when the bearing area has become 1/50 of a square inch, and assuming the weight of the skater as 140 lbs., and also that no other forces act to urge the blade, we find a pressure of 7000 lbs. to the square inch, sufficient to ensure the melting of the ice at − 3·5° C. With very cold ice, the pressure will rapidly attain the inoperative intensity, so that it will be found difficult to obtain 'bite'—a state of things skaters are familiar with. But it would appear that *some* penetration must ensue. On very cold ice, 'hollow-ground' skates will

have the advantage.

"This explanation of the phenomena attending skating assumes that the skater, in fact, glides about on a narrow film of water, the solid turning to water wherever the pressure is most intense, and this water, continually forming under the skate, resuming the solid form when relieved of pressure."

J. JOLY.

Geological Laboratory, Trinity College, Dublin.

VOL 59, 485 1899

What conclusions were drawn from these extraordinary observations or how they influenced educational theories in the United States is not revealed.

DR. A. MACDONALD, of the U.S. Bureau of Education, has sent us a number of statistics, showing the sensibility to pain, by pressure, in hands of individuals of different classes, sexes, and nationalities. So far as they go, the results indicate that the majority of people are more sensitive to pain in their left hand than in the right. Women appear to be more sensitive to pain than men, but of course it does not necessarily follow that women cannot endure more pain than men. American professional men are more sensitive to pain than American business men, and also than English or German professional men. The labouring classes are much less sensitive to pain than the non-labouring classes, and the women of the lower classes are much less sensitive to pain than those of the better classes. The general conclusion is that the more developed the nervous system, the more sensitive it is to pain. It is worth remark that, while the thickness of tissue on the hand has some influence, it has by no means so much as one might suppose, *à priori;* for many with thin hands require much pressure before experiencing any pain.

VOL 51, 299 1895

Professor O'Dea's resounding dignities do not shield him from the wrath of the reviewer:

AN IRISH ALGEBRA.

The New Explicit Algebra in Theory and Practice: for Teachers and Intermediate and University Students. By James J. O'Dea, M.A., formerly Professor of Mathematics, Natural Philosophy, and English Literature in St. Francis' College, Brooklyn, New York, and St. Jarlath's College, Tuam. Parts I. and II. Pp. x + 616, liv. (Longmans, Green, and Co., 1897, 1898.)

* * *

Like the poor Irish schoolboys, he is the victim of a most iniquitous system: that "payment by results" which warps and corrodes every branch of primary and intermediate education in Ireland.

... it puts a premium upon wrong methods, it encourages quackery and cruelty, it destroys sympathy between master and pupils, and the "results" which it produces are a delusion and a sham. It is heart-breaking to think of whole generations of clever, docile Irish lads condemned to the soul-destroying slavery which this rotten system perpetuates.

The reviewer's indignation mounts as he concludes:

Meanwhile My Lords the Commissioners of National Education in Ireland refuse to budge, in spite of the overwhelming verdict of competent opinion, nay in defiance of the unanimous protest of their own inspectors (see the *Manchester Guardian* for September 19, p. 7). No doubt their precious system works smoothly enough from their point of view ; the papers are set on traditional lines, the marks obtained are neatly tabulated, and the grants and scholarships impartially distributed accordingly ; how can any one, they may ask, reasonably object to such an obviously fair and practical procedure ? And so the costly, wasteful, and inefficient machinery continues to grind ; for all the world like a mill devised to scatter the flour and preserve the husk and bran.

G. B. M.

Vol 59, 25 1898

This sounds suspiciously like an attempt to account in terms of aberrant anatomy for the mesmeric effect that Gladstone seems to have exerted on some of his opponents.

A correspondent has called our attention to a statement which has appeared in various newspapers as to a peculiar characteristic of Mr. Gladstone's eyes. There is no doubt that Mr. Gladstone had striking and powerful eyes, but, according to the statement referred to, he also possessed nictitating membranes, which he occasionally used to paralyse his opponents in argument. We have asked the opinion of a distinguished authority upon the story, and he expresses the conviction that it is "all nonsense." He adds : "The nictitating membrane is not present, either in human eyes or in those of apes, except as a rudimentary crescentic fold at the inner corner, too small to cover the eye ; and the muscles which, in birds and some mammalia, cause the membrane to advance, are wholly wanting in men and apes. In birds the whole mechanism is very elaborate : in mammalia it is comparatively simple. If Mr. Gladstone possessed a nictitating membrane, and a power of moving it, he must have thrown back behind the hypothetical " missing link " ancestry of the human race. Moreover, the nictitating membrane, when present, as may be seen in five minutes in any fowl-house, does not cover the eye during waking life, and is not transparent. It is only drawn across the surface momentarily, from time to time, as a means of cleansing it. Mr. Nettleship, who operated on Mr. Gladstone for cataract, would, of course, be able to speak positively as to the suggested malformation."

Vol 59, 376 1899

This account, again by Grace Frankland, shows that women were attempting to assert themselves in science.

The subject of research work was also discussed, and stress was laid upon the fact that, inasmuch as the majority of students who take up science do so either as an avenue to a degree or with the idea of earning a livelihood by teaching later on, their training was as a rule insufficient and quite inadequate to permit them to undertake independent original work ; whilst on the other hand the demands upon their time made by teaching was so great as to leave practically no leisure for higher work, even when they were qualified to do it. Until this condition of things is altered, and until more women are attracted towards science for its own sake, and not as a means to an end, the contribution of women in the shape of original work must necessarily be limited. It was highly satisfactory to find that, in the open discussion which followed, an attempt on the part of two speakers to introduce the question of vivisection

from the anti-vivisectionist point of view was not tolerated by the audience, these speakers being refused a hearing. It is not too much to say that the papers contributed were worthy both of their subjects and their authors, and that there was a refreshing absence of the hackneyed comparison of the relative position and intellectual powers of men and women, which has been such a favourite theme with so many speakers at this Congress. The next International Congress of Women will be held five years hence in Berlin.

Vol 60, 228 1899

Here is a physicist's observation on the smug complacency that still pervaded the ancient seats of learning.

The Position of Science at Oxford.

Your correspondent " W. E. P." shows a curious ability for injuring his own side. He says that "Oxford collectively has done her best to remove any inferiority she may have had in the past " in respect of her scientific school, and further, " it would be difficult to name a body better qualified to decide what is a good general education than Convocation itself." And yet the whole tone of his letter is a practical confession that Oxford has failed in her best attempt, and that her view of general education has resulted in a practical failure to forward an essential branch of general education. The fact is Oxford's best is bad, and her ideal education is one-sided. The most serious cause of complaint of modern society against the old universities is that they have so controlled the education of the wealthy classes of the community that the landed and professional classes have been educated apart from the commercial and industrial classes to the very great injury of both. One might as well consult a committee of clergy as to the best education for a doctor, as advise with university dons as to the best education for the general community. The influence of a Pagan civilisation has created in them an ideal of life founded on contemplative learning, rather than on a Christian benevolent activity.

Geo. Fras. Fitzgerald.

Trinity College, Dublin, August 19.

Vol 54, 391 1896

Gout, which features so prominently in eighteenth-century literature, had now become demoded but spas were still in fashion.

Strawberry Cure for Gout.

The season of strawberries is at hand, but doctors are full of fads, and for the most part forbid them to the gouty. Let me put heart into those unfortunate persons to withstand a cruel medical tyranny by quoting the experience of the great Linnæus. It will be found in the biographical notes, written by himself in excellent dog-latin, and published in the Life of him by Dr. H. Stoever, translated from German into English by Joseph Trapp, 1794. Linnæus describes the goutiness of his constitution in p. 416 (*cf.* p. 415), and says that in 1750 he was attacked so severely by sciatica that he could hardly make his way home. The pain kept him awake during a whole week. He asked for opium, but a friend dissuaded it. Then his wife suggested, " Won't you eat strawberries ? " It was the season for them. Linnæus, in the spirit of an experimental philosopher, replied, " *tentabo*—I will make the trial." He did so, and quickly fell into a sweet sleep that lasted two hours, and when he awoke the pain had sensibly diminished. He asked whether any strawberries were left : there were some, and he eat them all. Then he slept right away till morning. On the next day he devoured as many strawberries as he could, and on the subsequent morning the pain was wholly gone, and he was able to leave his bed. Gouty pains returned at the same date in the next year, but were again wholly driven off by the delicious fruit ; similarly in the third year. Linnæus died soon after, so the experiment ceased.

What lucrative schemes are suggested by this narrative. Why should gouty persons drink nasty waters, at stuffy foreign Spas, when strawberry gardens abound in England ? Let enthusiastic young doctors throw heart and soul into the new system. Let a company be run to build a Curhaus in Kent, and let them offer me board and lodging gratis in return for my valuable hints.

F. G.

Vol 60, 125 1899

chapter eight
1900-1906

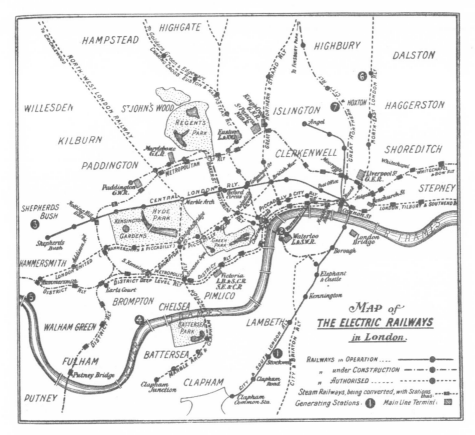

H. G. WELLS PREDICTS THE DEMISE OF HORSE TRAFFIC. C. S. SHERRINGTON EXPLAINS EMOTION IN DOGS. MAGIC SQUARES AND THE HOLY SHROUD OF TURIN MAKE THEIR FIRST, BUT BY NO MEANS LAST, APPEARANCE IN *NATURE*. SAMUEL PIERPOINT LANGLEY DEBUNKS FIRE-WALKING. RENÉ-PROSPER BLONDLOT'S *N*-RAYS COME AND GO. OLIVER LODGE OBSERVES THE BEHAVIOUR OF GLOW-WORMS IN THUNDERSTORMS. GEORGE DARWIN RESOLVES IN HIS FATHER'S FAVOUR THE ARGUMENT WITH LORD KELVIN ABOUT THE AGE OF THE EARTH. KELVIN LOSES AGAIN ON RADIOACTIVITY AND W. G. GRACE WIELDS THE WILLOW.

1900–1906

Ladysmith and Mafeking were relieved, and for the Empire the Boer War was brought to a satisfactory conclusion. In China, the Boxer Rising was brutally suppressed. In the United States, yet another president, William McKinley, was assassinated. Three years later, Theodore Roosevelt was elected. The Russo-Japanese war ended in humiliation for the Russians at Port Arthur. Earthquake and fire devastated San Francisco. Dreyfus returned from Devil's Island and was reinstated in the French Army. Guglielmo Marconi transmitted morse signals by the Atlantic cable, the Wright brothers flew at Kitty Hawk and so did the first zeppelin. HMS *Dreadnought* was launched.

Vivid descriptions survive of Tycho Brahe's strange ménage at Uraniborg, in which cavorted the malicious dwarf Jepp, who was his intimate. Tycho died of uraemia, which, he believed, resulted from his determination not to breach etiquette by leaving the table of his royal host to seek easement. He spoke his own epitaph: "He lived like a prince and died like a fool." The discovery in the coffin of his false nose of silver — he was parted from the original in a duel — must have afforded the seekers a romantic shiver.

A CORRESPONDENT sends us the following translation of an article which appeared in the *Neue Freie Presse* of Vienna, and was translated in the Copenhagen Journal *Dannebrog* on June 28, upon the removal of Tycho Brahe's remains from his tomb. This is the first report we have seen of the event :—" On the occasion of the 300th anniversary of Tycho Brahe's death the Prague Town Council decided to gather together the remains of the celebrated astronomer, which were in the Teyn Church, and bury them anew. Under the guidance of Mr. Herlein this operation was commenced yesterday. After having lifted the stone block on the monument, which is situated near the first column in the nave and which bears a full-length effigy of the great astronomer, a semi-collapsed arch was found, and on removing the stones two mouldering coffins were seen. On the following day a committee met to determine whether these bodies were those of Tycho Brahe and his wife. Two workmen with candles descended into the vault and removed the débris which covered the coffins, the wood of which was quite rotten and fell to pieces at every rough touch. About 10 a.m. the lid of the first coffin was free to be removed. It was a surprising sight that met the eye ; the body in the coffin was a wonderful likeness of the effigy on the monument. The head was slightly turned to one side, the bones of the face and the peaked Spanish beard being well preserved. The head was covered with a skull cap, and the neck was surrounded by a Spanish ruff which, like the remainder of the clothing, had suffered little during the 300 years since Tycho Brahe was laid in his last resting place. The feet were shod in long cavalry boots reaching up over the knee. That the body was Tycho Brahe's was also seen from the absence of the nose ; Tycho lost this organ in a duel and wore a silver one in its place. Amongst the rubbish was found a silver wreath and spray of flowers. The construction of the grave was rather remarkable, the stones being laid loosely over one another. This is all the more astonishing seeing Tycho Brahe was buried with great pomp and honours, but it is supposed that the vault broke down during the restoration of the church in 1721."

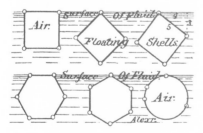

R. W. Wood, whom we shall meet again, was an American physicist and a pioneer of molecular spectroscopy. He was also a noted farceur, *remembered for such original fancies as training his cat to walk through his spectrograph to clean out dust and cobwebs. When Wood was staying in a rooming house in Paris, he was seen at dinner to sprinkle a white powder on*

the chicken bones left on his plate. The next night he brought a spirit lamp to the table and poured a drop of soup into the flame. There was a red flash and Wood was able to proclaim to the other diners that yesterday's bones had gone into the soup, for he had tagged them with lithium chloride. Wood's collection of clever verses (such as "The carrot and the parrot"), with the title How to Tell the Birds from the Flowers, *is still in print.*

Pseudoscopic Vision without a Pseudoscope: A New Optical Illusion.

A METHOD of securing an illusion of binocular vision wholly without instrumental aid occurred to me recently, which is interesting in connection with the study of pseudoscopic vision. It is fully as startling as any of the results obtained with the lenticular pseudoscope, which I showed at the Royal Institution in February, 1900, and which I shall speak of presently, and, requiring the aid of no optical instrument, is much more impressive.

A lead pencil is held point-up an inch or two in front of a wire window screen, with a sky background. If the eyes are converged upon the pencil point, the wire gauze becomes somewhat blurred, and of course doubled. Inasmuch, however, as the gauze has a regularly recurring pattern, the two images can be united, and with a little effort the eyes can be accommodated for distinct vision of the combined images of the mesh. To accommodate for a greater distance than the point upon which the eyes are converged requires practice, but the trick is very much easier in this case than in the case of viewing stereoscopic pictures without a stereoscope.

As soon as accommodation is secured, the mesh becomes perfectly sharp and appears to lie nearly in the plane of the pencil point, which still appears single and perfectly sharp. If now the pencil is moved away from the eyes which are to be kept fixed on the screen, it *passes through the mesh and becomes doubled*, the distance between the images increasing until the point brings up against the screen. If now the pencil be removed it will be found that the sharp images of the combined images of the gauze persists, even though the eyes be moved nearer to, or farther away from, the screen. Bring the eyes up to within six or eight inches of the plane in which the mesh appears to lie and attempt to touch it with the finger. *It is not there :* the finger falls upon empty space, the screen being in reality a couple of inches further off. This is by all means the most startling illusion that I have ever seen, for we apparently see something occupying a perfectly definite position in space before our eyes, and yet if we attempt to put our finger on it we find that there is nothing there.

It is best to begin by holding the pencil an inch or less in front of the screen. As the eyes become accustomed to the unusual accommodation, the distance can be increased. I have succeeded in bringing up the apparent plane of the mesh, five or six inches, but this requires as great a control over the eyes as is necessary in viewing stereoscopic pictures without an instrument.

The pseudoscope, which I have alluded to above, I have described in *Science* (about November, 1899), but inasmuch as the description of it which I sent to NATURE, the editor informs me, was never received, a brief account of it may not be out of place. Two lenses of about three inches focus are mounted in front of a pair of stereoscope lenses in such a way that the real inverted images formed by them in space can be combined by the stereoscope. The lenses should be mounted in slide tubes attached to the frame of the stereoscope, so that proper focussing can be accomplished. This instrument has been named the lenticular pseudoscope by the psychologists, and gives results far superior to those obtained by the Wheatstone and other forms of mirror pseudoscopes. Viewed through the instrument, a hollow bowl appears as a beautifully convex dome, and if a marble be dropped into it we witness the astounding phenomenon of a ball rolling up hill, crossing the top, descending part way down the other side and then returning to the summit, in defiance of the law of gravitation.

Johns Hopkins University.   R. W. WOOD.

The premature death of G. F. FitzGerald dismayed British physicists, for he was a popular figure. His obituary in NATURE *was by his fellow Irishman Joseph Larmor. It quotes this letter from Oliver Heaviside, a close friend of FitzGerald's, even though they met only a few times, because Heaviside, who was self-taught and became famous for his work on telegraphy (giving his name of course to the Heaviside layer), was something of a recluse.*

In a private letter, in response to a hurried intimation of FitzGerald's death, Mr. O. Heaviside writes as follows :— " I only saw him twice knowingly, once for two hours, and then again for six hours, after a long interval ; yet we had a good deal of correspondence at one time, and I seemed to have quite an affection for him. A mutual understanding had something to do with that. You know that in the pre-Hertzian days he had done a good deal of work, not large in bulk but very choice and original, in relation to the possibilities of Maxwell's theory, then considerably undeveloped and little understood ; and his way of looking at things was more like my own than anybody's. Well, he found that I had done a lot of work in the same line, and he was most generous in recognising and emphasising it. Too generous, of course. You remember that review of my ' Electrical Papers ' that he wrote ? No one knew better than myself how to allow for his temperament and desire to help me. He used to write to me a good deal about electromagnetic problems, and I laid down the law to him like—like myself, in fact. He took it all very pleasantly. But I knew all the while that he had a wider field than myself, and no time to specialise much. He had, undoubtedly, the quickest and most original brain of anybody. That was a great distinction ; but it was, I think, a misfortune as regards his scientific fame. He saw too many openings. His brain was too fertile and inventive. I think it would have been better for him if he had been a little stupid—I mean not so quick and versatile, but more plodding. He would have been better appreciated, save by a few."

VOL 63, 446 1901

Jacques Loeb expresses the pain of the misrepresented scientist. He was a German who spent most of his working life in the United States and was the model for Professor Gottlieb in Sinclair Lewis's classic novel of science, Martin Arrowsmith.

Sensational Newspaper Reports as to Physiological Action of Common Salt.

IN the interest of the dignity of scientific research I venture to hope you will print the following statement. Some American papers have recently published sensational and absurd reports of physiological theories and experiments whose authorship they attributed to me. These reports, which in America nobody takes seriously, were reprinted and discussed in European papers. I hardly need to state that I am in no way responsible for the journalistic idiosyncrasies of newspaper reporters and that for the publication of my experiments or views I choose scientific journals and not the daily Press.   JACQUES LOEB.
The University of Chicago, Physiological
    Laboratory, January 16.

VOL 63, 372 1901

Here the celebrated American astronomer, physicist and inventor Samuel Pierpoint Langley sets out to debunk a triumph of mind over matter.

The Fire Walk Ceremony in Tahiti.

THE very remarkable description of the " Fire Walk " collected by Mr. Andrew Lang and others had aroused a curiosity in me to witness the original ceremony, which I have lately been able to gratify in a visit to Tahiti.

Among these notable accounts is one by Colonel Gudgeon, British Resident at Raratonga, describing the experiment by a man from Raiatea, and also a like account of the Fiji fire ceremony from Dr. T. M. Hocken, whose article is also quoted in Mr. Lang's paper on the " Fire Walk," in the *Proceedings* of the Society for Psychical Research, February, 1900. This extraordinary rite is also described by Mr. Fraser in the " Golden Bough," and by others.

I had heard that it was performed in Tahiti in 1897, and several persons there assured me of their having seen it, and one of them of his having walked through the fire himself under the guidance of the priest, Papa-Ita, who is said to be one of the last remnants of a certain order of the priesthood of Raiatea, and who had also performed the rite at the island of Hawaii some time in the present year, of which circumstantial newspaper accounts were given, agreeing in all essential particulars with those in the accounts already cited. According to these, a pit was dug in which large stones were heated *red hot* by a fire which had been burning many hours. The upper stones were pushed away just before the ceremony, so as to leave the lower stones to tread upon, and over these, " glowing red hot " (according to the newspaper accounts), Papa-Ita had walked with naked feet, exciting such enthusiasm that he was treated with great consideration by the whites, and by the natives as a God. I found it commonly believed in Tahiti that anyone who chose to walk after him, European or native, could do so in safety, secure in the magic which he exercises, if his instructions were exactly followed. Here in Tahiti, where he had " walked " four years before, it was generally believed among the natives, and even among the Europeans present who had seen the ceremony, that if anyone turned around to look back he immediately was burned, and I was told that all those who followed him through the fire were expected not to turn until they had reached the other side in safety, when he again entered the fire and led them back by the path by which he had come. I was further told by several who had tried it that the heat was not felt upon the feet, and that when shoes were worn the soles were not burned (for those who followed the priest's directions), but it was added by all that much heat was felt about the head.

* * *

The *mise en scène* was certainly noteworthy. The site, near the great ocean breaking on the barrier reefs, the excited crowd, talking about the " red-hot " stones, the actual sight of the hierophant and his acolytes making the passage along the ridge where the occasional tongues of flame were seen at the centre, with all the attendant circumstances, made up a scene in no way lacking in interest. Still, the essential question as to the actual heat of these stones had not yet been answered, and after the fourth passage I secured Papa-Ita's permission to remove, from the middle of the pile, one stone which from its size and position every foot had rested upon in crossing, and which was undoubtedly at least as hot as any one of those trodden on. It was pulled out by my assistants with difficulty, as it proved to be larger than I had expected, it being of ovoid shape with the lower end in the hottest part of the fire. I had brought over the largest wooden bucket which the ship had, and which was half-filled with water, expecting that this would cover the stone, but it proved to be hardly enough. The stone caused the water to rise nearly to the top of the bucket, and it was thrown into such violent ebullition that a great deal of it boiled over and escaped weighing. The stone was an exceedingly bad conductor of heat, for it continued to boil the water for about twelve minutes, when, the ebullition being nearly over, it was removed to the ship and the amount of evaporated water measured.

*   *   *

The real question is, I repeat, how hot were those trodden on? and the answer to this I was to try to obtain after measuring the amount of water boiled away.

On returning to the ship this was estimated from the water which was left in the bucket (after allowing for that spilled over) at about ten pounds. The stone, which it will be remembered was one of the hottest, if not the hottest, in the pile, was found to weigh sixty-five pounds, and to have evaporated this quantity of water. It was, as I have said, a volcanic stone, and on minuter examination proved to be a vesicular basalt, the most distinctive feature of which was its porosity and non-conductibility, for it was subsequently found that it could have been heated red hot at one end, while remaining comparatively cool at the top. I brought a piece of it to Washington with me and there determined its specific gravity to be 0·39, its specific heat 0·19 and its conductivity to be so extremely small that one end of a small fragment could be held in the hand while the other was heated indefinitely in the flame of a blow-pipe.

VOL 64, 397 1901

The spider-silk industry seems not to have achieved prosperity.

THE manufacture of silk cord from spiders' web seems likely to attain commercial importance, for we learn through the *Board of Trade Journal* that one of the most novel exhibits in the Paris Exposition will be a complete set of bed-hangings manufactured in Madagascar from the silk obtained from the halabe, an enormous spider that is found in great numbers in certain districts of the island. The matter has been taken up by M. Nogue, the head of the Antananarivo Technical School. The results he has already achieved show that the production of spider silk should quickly become a highly important industry. Each spider yields from three to four hundred yards of silk. After the thread has been taken from the spiders they are set free, and ten days afterwards they are again ready to undergo the operation. The silk of these spiders, which is of the most extraordinary brilliant golden colour, is finer than that of the silkworm, but its tenacity is remarkable, and it can be woven without the least difficulty.

VOL 62, 17 1900

Lord Kelvin invented an analogue engine for solving simultaneous equations, made like this one:

AN ingenious machine for solving any algebraic equation of the form $px^n + px^{n_1} + p_2x^{n_2} + \&c. = A$, by an application of the principle of Archimedes, is described by M. Georges Meslin in the *Journal de Physique* for June. It consists of a beam balanced on a knife-blade from any point of which may be suspended a solid of revolution, and a series of such solids is provided, constructed in such a manner that in the solid of order n the volume cut off by a horizontal plane is proportional to the nth power of the distance of the horizontal plane from the lowest point. Thus for orders 1, 2, 3, the forms of the solids are a cylinder ; a paraboloid of revolution, a cone. If the solid of order n is suspended at a distance p from the knife-blade, then when it is immersed to a depth x in liquid, the moment of the resultant upward thrust of the fluid about the knife-edge is proportional to px^n. The operation of solving the equation consists in adjusting the weights at suitable distances, p, p_1, p_2 from the axis, and balancing them, then running water into a trough containing the solids until the fluid thrusts balance a weight A fixed at unit distance from the axis of the beam ; when this is done the equation of moments takes the form of the given algebraic equation and x, the root of the equation is equal to the depth of immersion of the solids.

VOL 62, 253 1900

Here is another example of Francis Galton's ingenuity. The smile of the Cheshire cat (product of the fancy of another of NATURE's *contributors (pp.56 and 79)) would today be extracted from digitised images and referred to as a difference map.*

AN excellent article dealing with the photographic side of the suggestions as to analytical portraiture made by Mr. F. Galton in NATURE of August 2 appears in *Photography* of August 9. Illustrations are given of results obtained by combining two portraits of a single person in the same pose, but having different expressions during the two exposures. In one picture the sitter has a normal expression ; in the other he is smiling. A transparency was made from the normal negative ; and when this positive and its negative were superimposed they neutralised one another. But by placing the positive of the normal expression of face upon the negative of the smiling expression, the two do not, of course, exactly obliterate one another. Certain parts of the features are common to both, and these disappear when the different positive and negative are superimposed, leaving only portions which represent the smile of the sitter's features. In a similar way, by superimposing the positive of a glum portrait upon the negative of a normal expression, it is possible to obtain differences representing an individual's glumness. Readers of " Alice in Wonderland " will remember that the Cheshire cat gradually disappeared and left only its grin behind. This facetious idea has now been realised, for as our enterprising contemporary points out, Mr. Galton's analytical portraiture shows how the factors of a grin or a scowl can actually be discriminated, so that a grin can be obtained without the face upon which it appeared.

VOL 62, 374 1900

H. G. Wells was often on target in his predictions, though whether the pollution by horses was more to be feared than that from the internal combustion engine must have seemed questionable. Alarm about the contribution to the greenhouse effect of methane in bovine flatus came later.

MR. H. G. WELLS commences, in the current number of the *Fortnightly Review*, a series of speculative papers upon some changes of civilised life and conditions of living likely to occur in the new century. To construct a prehistoric animal from one or two fossil bones is a much easier task than the prediction of future developments from the point of view of the present ; but Mr. Wells attempts to do this, and even if his prophetic visions do not materialise they will convince the conservative mind that there is some virtue in dissatisfaction at many of the methods of to-day. The subject of the first article is land locomotion in the twentieth century, and it scarcely requires a prophetic afflatus to know that the present systems will be largely superseded or modified. Horse traffic, with its cruelty and filth, while the animals exhaust and pollute the air, must give place to motor carriages in a few years. The railways will then develop in order to save themselves. There will be continuous trains, working perhaps upon a plan like that of the moving platform of the Paris Exhibition, or utilising the principle of the rotating platform outlined by Prof. Perry in these columns (vol. lxii. p. 412, 1900). Nothing is said about the possibilities of aëronautics, not because of any doubt as to its final practicability, but because " I do not think it at all probable that aëronautics will ever come into play as a serious modification of transport and communication." It is, of course, impossible to project ourselves into the future so as to say exactly what will or will not come to pass ; for an estimate of future performances can only be made with the material now available, and it leaves out of account the completely novel discoveries which often revolutionise the whole conditions. Nevertheless, it is not unprofitable to meditate upon the promise of progress.

VOL 63, 546 1901

This kind of verbal lapse had evidently not yet become associated with the Reverend Dr Spooner (p.96), who was in actuality noted less for what are now called Spoonerisms ("The bean is dizzy" or "the Lord is a shoving leopard") than for hybrids resembling the specimens given here. "It's 50 miles to Malvern as the cock crows", Spooner would say, and he was once heard to make reference to a journey from "Land's End to John of Gaunt".

It has been said that every person is mentally a little un-balanced, and that education from this point of view is simply the attempt to secure and maintain mental equilibrium, which, however, is never actually attained. Lapses of thought, inad-vertencies in expression, and other slips in speaking or writing (*lapsus linguae* and *lapsus calami*) are thus of interest to the psychologist as useful guides to the understanding of mental processes. Every one has experienced unaccountable lapses of this kind, and the lapse often comes as a surprise to the speaker or writer himself. During a lecture, a professor inadvertently referred to the "tropic of Cancercorn," intending to say "the tropics of Capricorn and of Cancer." Many similar instances might be cited, for example, the man who was going for a walk to "get a breash of freth air," the person who inquired for the "portar and mestle," and another who said "the pastor cut the shermon sort." A physicist is recorded to have said that he feared he should "get the instrument out of needle," when he intended to say he feared he would "get the instrument out of level and deflect the needle." This is curious, but it is not so amusing as the order of "beggs and acon" for breakfast, or the remark of a nervous churchman to a stranger in his seat, "Excuse me, but you are occupuing my pie." Mr. H. Heath Bawden has made a detailed study of similar mental lapses, both oral and graphic, and his results are described in a monograph of the *Psychological Review*. It is suggested that the aberrations dealt with are due to incipient aphasia or agraphia, and the similarity between them is held to show that our ordinary experience borders at every point on what is called the abnormal or pathological condition.

The great physiologist C. S. Sherrington was a fine writer. His popular work Man on His Nature *was something of a classic and he also published poetry, to critical acclaim. Here, in a long review article, he touches on a number of somewhat gruesome phenomena, of which the following is one example.*

EXPERIMENTATION ON EMOTION.

OF points where physiology and psychology touch, the place of one lies at the phenomenon "emotion." Built upon sense-feeling much as cognition is built upon sense-perception, emotion may be regarded almost *as* a "feeling"—a "feeling" excited, not by a simple unelaborated sensation, but by a group or train of ideas. To such compound ideas it holds relation much as does "feeling" to certain species of simple sense-perceptions. It has a special physiological interest in that certain visceral reactions are peculiarly concomitant with it. Heart, blood-vessels, respiratory muscles and secretory glands play special and characteristic *rôles* in the various emotions. These viscera, though otherwise remote from the general play of psychical process, are affected vividly by the emotional. Hence many a picturesque metaphor of proverb and phrase and name—"the heart is better than the head," anger "swells within the breast," "Richard Cœur de Lion." It was Descartes who first relegated the emotions to the brain. Even this century Bichat wrote, "The brain is the seat of cognition, and is never affected by the emotions, whose sole seat lies in the viscera." But brain is now admittedly a factor necessary in all higher animal forms to every mechanism whose working has consciousness adjunct.

What is the meaning of the intimate linkage of visceral actions to psychical states emotional? To the ordinary day's consciousness of the healthy individual the life of the viscera contributes little at all, except under emotion. The perceptions of the normal consciousness are rather those of outlook upon the circumambent universe than inlook into the microcosm of the "material me." Yet heightened beating of the heart, blanching or flushing of the blood-vessels, the pallor of fear, the blush of shame, the Rabelaisian effect of fright upon the bowel, the action of the lacrymal gland in grief, all these are prominent characters in the pantomime of natural emotion. Visceral disturbance is evidently a part of the corporeal expression of emotion. The explanation is a particular case in that of movements of expression in general. The hypothesis of Evolution afforded a new vantage point

The Board of Visitors of the Royal Observatory, Greenwich, photographed at the Observatory on Visitation Day, June 5, 1897. Among the party are several astronomers who delivered discourses at the Royal Institution. From left to right : R. B. Clifton (Royal Astronomical Society), A. W. Rücker (Royal Society), H. H. Turner (Savilian Professor of Astronomy, Oxford), Sir George Stokes (Royal Society), W. D. Barber (Secretary), Lord Rayleigh (Royal Society), J. W. L. Glaisher (Royal Astronomical Society), G. H. Darwin (Plumian Professor of Astronomy, Cambridge), Earl of Rosse (Royal Society), A. A. Common (Royal Astronomical Society), W. Huggins (Royal Astronomical Society), and F. W. Dyson (Chief Assistant, Royal Observatory). (Photo: Royal Greenwich Observatory.)

for study of that question. Fixed bodily expressions of emotion are hereditary. They are, especially in the "coarser or animal emotions," largely common to man and higher animals. The point of view is exemplified by Darwin's argument concerning the contraction of the muscles round the eyes during screaming. "Children when wanting food or suffering in any way cry out loudly like the young of most animals, partly as a call to their parents for aid, and partly from any great exertion serving as relief. Prolonged screaming inevitably leads to the engorging of the blood-vessels of the eye ; and this will have led at first consciously and at last habitually to the contraction of the muscles round the eyes in order to protect them." Mr. Spencer writes : "Fear, when strong, expresses itself in cries, in efforts to hide or escape, in palpitations and tremblings ; and these are just the manifestations which would accompany an actual experience of the evil feared. The destructive passions are shown in a general tension of the muscular system, in gnashing of the teeth and protrusion of the claws, in dilated eyes and nostrils, in growls : and these are weaker forms of the actions that accompany the killing of prey." In a word, expression of emotion is instinctive action.

* * *

Few dogs even when very hungry can be prevailed on to touch dog's-flesh as food. Almost all turn from it with signs of repugnance and dislike. I had strictly refrained from testing this animal previously with regard to disgust at dog's-flesh offered in her food. Flesh was given her daily in a bowl of milk, and this she took with relish. The meat was cut into pieces rather larger than the lumps of sugar usual for the breakfast table. It was generally horse-flesh, sometimes ox-flesh. We proceeded to the observation thus : the bowl was placed by the attendant in the corner of the stall, with milk and meat in every way as usual ; but the meat was flesh from a dog killed on the previous day. Our animal eagerly drew itself toward the food ; it had seen the other dogs fed, and evidently itself was hungry. Its muzzle had almost dipped into the milk before it suddenly seemed to find something there amiss. It hesitated, moved its muzzle about above the milk, made a venture to take a piece of the meat, but before actually seizing it stopped short and withdrew again from it. Finally, after some further examination of the contents of the bowl (it usually commenced by taking out and eating the pieces of meat), without touching them, the creature turned away from the bowl and withdrew itself to the opposite side of the cage. Some minutes later, in result of encouragement from us to try the food again, it returned to the bowl. The same hesitant display of conflicting desire and disgust was once more gone through. The bowl was then removed by the attendant, emptied, washed, and horse-flesh similarly prepared and placed in a fresh quantity of milk was offered in it to the animal. The animal once more drew itself toward the bowl, and this time began to eat the meat, soon emptying the dish. To press the flesh upon our animal was of no real avail on any occasion ; the coaxing only succeeded in getting her to, as it were, re-examine but not to touch the morsels. The impression made on all of us by the dog's behaviour was that something in the dog's-flesh was repulsive to her, and excited disgust unconquerable by ordinary hunger. Some odour attaching to the flesh seemed the source of its recognition.

Fear appeared clearly elicitable. The attendant, approaching from another room of which the door was open, chid the dog in high scolding tones. The creature's head sank, her gaze turned away from her advancing master, and her face seemed to betray dejection and anxiety. The respiration became unquiet, but the pulse never changed its rate.

In the face of these observations the vasomotor theory of the production of emotion becomes, I think, untenable : also that visceral sensations or presentations are *necessary* to emotion. A mere remnant of all the non-projecting or affective senses was left, and yet emotion persisted. If I understand it aright, Prof. James and Lange's theory lays stress on organic and visceral presentations, but re-presentations of the same species might no doubt be put forward in their place. That would be a somewhat different matter. To exclude the latter hypothesis, the deprivation of vascular and organic sensation might have to date from a very early period of the individual life. Experience early acquires its emotional data. If after that all fresh presentation were precluded, re-presentation might still be possible on the basis of already gained experience. But it is

noteworthy that one of the dogs under observation had been deprived of its sensation when only nine weeks old. Disgust for dog's flesh could hardly have genesis in the experience of nine weeks of puppy life in the kennel of the laboratory.

Sir Walter Raleigh, as this correspondent says, was no Buffon; nor yet was he a Gilbert White, but he makes a sage observation or two about what might constitute a species.

Sir Walter Raleigh and Evolution.

I HAVE recently come across a passage in Sir Walter Raleigh's "History of the World" which seems to me sufficiently remarkable for the author to deserve a notable place among those early naturalists who anticipated in some measure the modern views on evolution. In the historical sketch at the beginning of the "Origin of Species" Darwin quotes Buffon, who was born a century and a half later than Raleigh, as "the first author who in modern times has treated the subject in a scientific spirit " ; but although, scientifically, Raleigh cannot be compared with Buffon, the fact of his having penned at such an early date the words I am about to quote possesses some interest. The passage I refer to is to be found in the 1621 edition (part i., book i., chap vii., § 9, p. 94). Speaking of the days of the Flood, he says : "But it is manifest, and undoubtedly true, that many of the *Species*, which now seeme differing, and of severall kindes, were not then *in rerum natura*. For those Beasts which are of mixt natures, eyther they were not in that age, or else it was not needfull to preserve them, seeing they might be generated againe by others : as the Mules, the *Hyæna's*, and the like ; the one begotten by Asses and Mares, the other by Foxes and Wolves. And whereas by discovering of strange Lands, wherein there are found divers Beasts and Birds differing in colour or stature from those of these Northerne parts ; it may be supposed by a superficiall consideration, that all those which weare red and pyed Skinnes, or Feathers, are differing from those that are lesse painted, and were plaine russet or blacke ; they are much mistaken that so thinke. And for my owne opinion, I find no difference, but onely in magnitude, betweene the Cat of *Europe*, and the Ownce of *India* ; and even those Dogges which are become wilde in *Hispagniola*, with which the *Spaniards* used to devoure the naked *Indians*, are now changed to Wolves, and begin to destroy the breed of their Cattell, and doe also oftentimes teare asunder their owne Children. The common Crow and Rooke of *India* is full of red feathers in the drown'd and low Islands of *Caribana* ; and the Black-bird and Thrush hath his feathers mixt with blacke and carnation, in the North parts of *Virginia*. The Dog-fish of *England* is the Sharke of the South Ocean : For if colour or magnitude made a difference of *Species*, then were the *Negro's*, which wee call the Blacke Mores, *non animalia rationalia*, not Men, but some kind of strange Beasts : and so the Giants of the South *America* should be of another kind, than the people of this part of the World. We also see it dayly, that the natures of Fruits are changed by transplantation, some to better, some to worse, especially with the change of Clymate. Crabs may be made good Fruit by often grafting, and the best Melons will change in a yeere or two to common Cowcummers, by being set in a barren Soyle."   AGNES ROBERTSON.

The Old Hall, Newnham College, Cambridge, January 13.

Another overhead system for letter projectiles:

ACCORDING to a Reuter telegram from Rome, the Italian postal authorities have examined a scheme submitted by an engineer, named Piscicelli, for the establishment of an electric postal service. It is proposed, by means of this system, to transmit letters in aluminium boxes, travelling along overhead wires at the rate of 400 kilometres an hour. A letter could thus be sent from Rome to Naples in twenty-five minutes and from Rome to Paris in five hours. A technical commission has been appointed to report on the system before instituting a series of experiments between Rome and Naples.

Magic squares clearly had wide allure and, to judge by the communications to Nature, *whipped up considerable animation among the readers. The article by Major P. A. MacMahon, a fellow of the Royal Society, runs to six pages. It begins:*

MAGIC SQUARES AND OTHER PROBLEMS UPON A CHESS-BOARD.

THE construction of magic squares is an amusement of great antiquity ; we hear of them being constructed in India and in China before the Christian era, whilst they appear to have been introduced into Europe by Moschopulus, who flourished at Constantinople early in the fifteenth century. On the diagram you see a simple example of a magic square, one celebrated as being drawn by Albert Dürer in his picture of "Melancholy," painted about the year 1500 (Fig. 1). It is one of the fourth order, involving 16 compartments or cells. In describing such squares, the horizontal lines of cells are called "rows," the vertical lines "columns," and the oblique lines going from corner to corner

1	15	14	4
12	6	7	9
8	10	11	5
13	3	2	16

FIG. 1.

"diagonals." In the 16 compartments are placed the first 16 numbers, 1, 2, 3, . . . 16, and the magic property consists in this, that the numbers are placed in such wise that the sum of the numbers in every row, column and diagonal is the same, viz., in this case, 34.

It is probable that magic squares were so called because the properties they possessed seemed to be extraordinary and wonderful ; they were, indeed, regarded with superstitious reverence and employed as talismans. Cornelius Agrippa constructed magic squares of orders 3, 4, 5, 6, 7, 8, 9, and associated them with the seven heavenly bodies, Saturn, Jupiter, Mars, the Sun, Venus, Mercury and the Moon. A magic square engraved on a silver plate was regarded as a charm against the plague, and to this day such charms are worn in the east.

However, what was at first merely a practice of magicians and talisman makers has now for a long time become a serious

17	24	1	8	15
23	5	7	14	16
4	6	13	20	22
10	12	19	21	3
11	18	25	2	9

FIG. 2.

study for mathematicians. Not that they have imagined that it would lead them to anything of solid advantage, but because the theory of such squares was seen to be fraught with difficulty, and it was considered possible that some new properties of numbers might be discovered which mathematicians could turn to account. This has, in fact, proved to be the case ; for from a certain point of view the subject has been found to be algebraical rather than arithmetical, and to be intimately connected with great departments of science, such as the "infinitesimal calculus," "the calculus of operations" and the "theory of groups."

In the next diagram (Fig. 2) I show you a magic square of order 5, the sum of the numbers in each row, column and diagonal being 65. This number 65 is obtained by multiplying 25, the number of cells, by the next higher number, 26, and then dividing by twice the order of the square, viz., 10. A similar rule applies in the case of a magic square of any order. The formation of these squares has a fascination for many persons, and, as a consequence, a large amount of ingenuity has been expended in forming particular examples and in discovering general principles of formation. As an example of the amount of labour that some have expended on this matter, it may be mentioned that in 1693 Frénicle, a Frenchman, published a work of more than 500 pages upon magic squares. In this work he showed that 880 magic squares of the fourth order could be constructed, and in an appendix he gave the actual diagrams of the whole of them. The number of magic squares of the order 5 has not been exactly determined, but it has been shown that the number certainly exceeds 60,000.

As a consequence it is not very difficult to compose particular specimens and, for the most part, the fascinated individuals, to whom I have alluded, have devoted their energies to the discovery of principles of formation.

The Major then explains these principles of construction and gives an account of Euler's work on the subject, including the following poser:

He commences by remarking that a curious problem had been exercising the wits of many persons. He describes it as follows :—There

aα	aβ	aγ	aδ	aε	aθ
bα	bβ	bγ	bδ	bε	bθ
cα	cβ	cγ	cδ	cε	cθ
dα	dβ	dγ	dδ	dε	dθ
eα	eβ	eγ	eδ	eε	eθ
fα	fβ	fγ	fδ	fε	fθ

FIG. 7.

are 36 officers of six different ranks drawn from six different regiments, and the problem is to arrange them in a square of order 6, one officer in each compartment, in such wise that in each row, as well as in each column, there appears an officer of each rank and also an officer of each regiment. Of a single regiment we have, suppose, a colonel, lieutenant-colonel, major, captain, first lieutenant and second lieutenant, and similarly for five other regiments, so that there are in all 36 officers who must be so placed that in each row and in each column each rank is represented, and also each regiment. Euler denotes the six regiments by the Latin letters a, b, c, d, e, f, and the six ranks by the Greek letters α, β, γ, δ, ϵ, θ, and observes that the character of an officer is determined by a combination of two letters, the one Latin and the other Greek ; there are 36 such combinations, and the problem consists in placing these combinations in the 36 compartments in such wise that every row and every column contains the 6 Latin letters and also the 6 Greek letters (Fig. 7). Euler found no solution of this problem in the case of a square of order 6, and since Euler's time no one has succeeded either in finding a solution or in proving that no solution exists.

The general problem, it turns out, was indeed insoluble.
Here finally is the parting curiosity:

In conclusion, I bring before you an interesting example of magic arrangement that I found whilst engaged in rummaging amongst the books and documents of the old Mathematical Society of Spitalfields (1717–1845) for the purpose of extracting something which might interest or amuse, if it might not instruct, the audience I addressed in Section A of the British Association for the Advancement of Science at Glasgow last autumn. It is an arrangement of the first eighteen numbers on five connected triangles; the magical property consists in the circumstance that the numbers 19, 38 and 57 appear as sums in a variety of ways. The number 19 appears nine times, 38 twelve times and 57 fourteen times (Fig. 15).

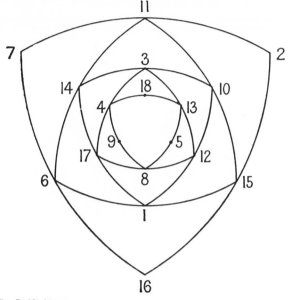

$9=$ 7+12=14+ 5= 4+15
  = 6+13=17+ 2= 9+10
  =16+ 3= 1+18= 8+11
$38=$ 7+11+14+ 6=11+ 2+15+10=15+16+ 6+ 1
  =11+10+ 3+14=10+15+ 1+12= 1+ 6+14+17
  =14+ 3+ 4+17= 3+10+12+13=12+ 1+17+ 8
  = 3+13+18+ 4=13+12+ 8+ 5= 8+17+ 4+ 9
$57=$ 7+14+ 4+ 5+12+15= 6+17+ 9+13+10+ 2=16+ 1+ 8+18+ 3+11
  = 7+11+ 2+15+16+ 6=11+10+15+ 1+ 6+14=14+ 3+10+12+ 1+17
  = 3+13+12+ 8+17+ 4= 4+18+13+ 5+ 8+ 9
  = 9+ 4+ 3+10+15+16=18+13+12+ 1+ 6+ 7= 5+ 8+17+14+11+ 2
  = 9+ 8+12+10+11+ 7=18+ 4+17+ 1+15+ 2= 5+13+ 3+14+ 6+16

FIG. 15.

VOL 65, 447 1902

In the years around the turn of the century, NATURE ran a series of articles each describing the work and denizens of a notable laboratory. The account of physics at the University of Heidelberg, where Robert Bunsen had held sway, began with a good anecdote. Lord Kelvin had given the composition of the Sun as an exemplar of a problem that could never be solved.

SOME SCIENTIFIC CENTRES.

THE HEIDELBERG PHYSICAL LABORATORY.

MOST travelled Englishmen are doubtless acquainted with the ancient town of Heidelberg, so famous for the beauty of its situation and the grandeur of its ruined castle. But far fewer know the charms of the long and romantic valley of the Neckar, at the almost sensational exit of which, from the Odenwald into the level plain of the Upper Rhine, Heidelberg stands. So also it is true that while most educated people connect Heidelberg with the great names of Kirchhoff and Bunsen and their

epoch-making discoveries in spectrum analysis, it is only the special students who know how large in extent and how important in result and example is the work which has steadily gone on for many years in the physical laboratory in the Friedrichsbau.

Its small beginnings in the middle of the last century are marked by the name of Kirchhoff scratched on the window of what is now the private room of the senior assistant. From this window one may look out over the Rhine plain towards busy Mannheim, as Bunsen and Kirchhoff did one night when a fire was raging there, and they were able by spectroscopic examination of the flames to ascertain that barium and strontium were present in the burning mass. But the same window also looks across the Neckar to the Heiligenberg, along the slopes of which runs the " Philosophers' Walk," the chief of the many paths among the wooded hills around the town, which the two friends were wont to traverse in their daily " constitutionals." Bunsen is known to have said that it was during such walks that his best ideas came to him. One day the thought occurred, " If we could determine the nature of the substances burning at Mannheim, why should we not do the same with regard to the sun ? But people would say we must have gone mad to dream of such a thing." All the world knows now what the result was, but it must have been a great moment when Kirchhoff could say, " Bunsen, I *have* gone mad," and Bunsen, grasping what it all meant, replied, " So have I, Kirchhoff ! "

VOL 65, 587 1902

Nikola Tesla, born in Croatia, emigrated to the United States as a young man and in due course (among many other remarkable achievements, including the invention of the induction coil) brought power to New York from the Niagara Falls. Here NATURE illustrates two of Tesla's spectacular electrical experiments.

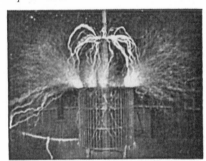

FIG. 1.—Combustion of atmospheric nitrogen by the discharge of an electrical oscillator giving twelve million volts and alternating 100,000 times per second. The flame-like discharge shown in the photograph measured 65 feet across.

FIG. 2.—The coil, partly shown in the photograph, creates an alternating current of electricity at the rate of 100,000 alternations per second. The discharge escapes with a deafening noise, striking an unconnected coil 22 feet away, and creating such an electrical disturbance that sparks an inch long can be drawn from a water-main at a distance of 300 feet from the laboratory.

VOL 62, 116 1900

Debates about the authenticity of the Turin shroud have erupted at fairly regular intervals up to our own time. Raphael Meldola despatches the arguments in its favour with aplomb:

THE HOLY SHROUD OF TURIN.

Le Linceul du Christ; Étude scientifique. By Paul Vignon, Dr. è Sci. Nat. Pp. 207 and 9 photogravures. (Paris : Masson et Cie, 1902.)

The Shroud of Christ. By Paul Vignon, D.Sc. (Fr.). Translated from the French. Pp. 170; 9 photogravures and collotype plates and 38 illustrations. (Westminster : Archibald Constable and Co., Ltd., 1902). Price 12*s.* 6*d.* net.

WHETHER the relic described, figured and discussed in this handsomely got up volume is the veritable shroud which enwrapped the body of Christ is a question which need not be seriously considered in the columns of a scientific publication. Dr. Vignon seems to have convinced himself that the relic is genuine, and his object in publishing this work is (presumably) to convince his readers, or at any rate to place before them the evidence on which his conclusions are based. So far as the antiquarian evidence goes, it will suffice to remind readers of NATURE that during the recent controversy—which appears to have been the last of a series of controversies concerning the authenticity of the relic in question—Father Herbert Thurston, S.J., communicated a letter to the *Times* of April 28, from which we make a few extracts :—

"The Abbé Ulysse Chevalier claims to have proved to demonstration that the linen winding-sheet exhibited at Turin is a spurious relic manufactured in the fourteenth century, and, as the writer believes, with fraudulent intent."

"We are not, of course, in any way bound to believe that those responsible for the subsequent veneration of this alleged relic have been guilty of conscious fraud. It may even in the first instance have been fabricated without intent to deceive. . . . Just as in the case of so many facsimiles of the Holy Vails, what was in the first instance a mere copy for devotional purposes has come in time to figure as an original, the wish, no doubt, being father to the thought, but probably without any deliberate insincerity."

Thus, out of the seven chapters composing this work, there are but two which come within our province, viz., chapter vi., in which the author deals with the scientific evidence, and chapter vii. more particularly, in which he puts forward an explanation of the image which is to be seen on the shroud. The antiquarian lore of the preceding chapters has no particular interest for us, and we may add, further, that the question whether the shroud is the real article or whether it was "faked" in the fourteenth century is a point which in no way affects the discussion of Dr. Vignon's scientific evidence, because the explanation with which we have to deal is equally miraculous whether the image is some twenty centuries old or whether it is only six hundred years old.

He then demolishes the chemical evidence step by step in nearly three pages and dusts his hands:

If by ammoniacal or any other vaporous emanation Dr. Vignon can succeed in producing an impression as distinctly recognisable as a likeness as the image on the shroud in all its details, we will waive the question of twenty centuries' permanence and go so far as to admit that there is at any rate some justification for "vaporographic" portraiture. As the "explanation" stands now, it is purely in the region of hypothesis, and pending that rigorous verification required by science, we consider that the author's case is "not proven." If there are any scientific readers who are convinced that the conclusions in this work are satisfactorily established, we shall be disposed to credit the shroud with having wrought a greater miracle than was ever ascribed to it by the Chapter of Lirey in the fourteenth century.

R. MELDOLA.

VOL 67, 241 1903

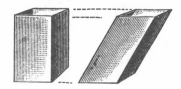

In the next issue the geologist and Anglican priest T. G. Bonney (p.81) also puts a boot into the unhappy Vignon's ribs:

The Holy Shroud.

PROF. MELDOLA'S notice from a truly scientific standpoint of Dr. Vignon's book, entitled "The Shroud of Christ," is not less interesting than valuable, but I think two difficulties which hardly fell within the scope of his article may also be raised. One struck me at once in examining the facsimile of the photographic negative plate of the Holy Shroud (facing p. 17). The body had been lying, of course, face upwards. I presume that if a corpse were thus placed on a stone slab, within a very few hours of death, the nates would be slightly flattened by pressure, but their normal roundness—as in a nude standing figure—caught my eye at once when examining the plate.

But a still more serious difficulty awaits Dr. Vignon. The shroud in shape has a general resemblance to an elongated bath-towel ; on one half, smoothed out, the body was laid, and the other was neatly doubled over the head and brought down so as completely to cover the feet. This mode of burial, so far as I know, was not usual among the Jews at that date (the corpse being more or less wrapped up, as described in the raising of Lazarus). But passing over this point, for Dr. Vignon pleads that the arrangement was a temporary one (though, by the way, it would make the preservative myrrh and aloes much less effective), we find the authors of the four Gospels all use language which excludes any such arrangement of the so-called shroud. Matthew and Luke both write ἐνετύλιξεν αὐτὸ ἐν σινδόνι ; Mark in a nearly identical sentence substitutes the verb ἐνείλησεν. But both these words mean to wrap or to roll up, not to lay a sheet over (and under). John, in a rather more minute description, says, ἔδησαν αὐτὸ ἐν ὀθονίοις μετὰ τῶν ἀρωμάτων, adding "as is the custom of the Jews in burial." He also mentions bandages or body-cloths a second time, and a napkin bound about the head—which would have interfered with the photographic process. Dr. Vignon endeavours to elude the plain meaning of these passages, but, as it seems to me, he can only prove the genuineness of the shroud by rejecting the four principal witnesses to the facts of which it is supposed to be a record, a process which has a suspicious resemblance to sawing off the branch on which you are sitting.   T. G. BONNEY.

VOL 67, 296 1903

Alfred Russel Wallace, reminiscing in old age on his relations with Charles Darwin, displayed a saintly generosity.

THE article which Dr. A. R. Wallace contributes to *Black and White* of January 17, on his relations with Darwin in connection with the theory of natural selection, is a historical document of great scientific interest. Dr. Wallace was introduced to Darwin in the insect-room of the British Museum in 1854. While living in Borneo in 1854, Dr. Wallace wrote a paper "On the Law which has Regulated the Introduction of new Species," which was published in the *Annals of Natural History* in the following year. Hearing that Darwin was preparing some work on varieties and species, Dr. Wallace sent him a copy of his paper and received a long letter in reply, but no hint was given by Darwin of his having arrived at the theory of natural selection. Darwin had, however, actually written out a sketch of his theory in 1842, and in 1844 this sketch was enlarged to 230 folio pages, giving a complete presentation of the arguments afterwards set forth in the "Origin of Species." Dr. Wallace arrived at the idea of the survival of the fittest as the operating cause in evolution in 1858, and immediately sent the outlines of this theory to Darwin, who brought the communication before Sir C. Lyell and Sir Joseph Hooker, and urged that it should be printed at once. Upon their advice, however, he consented to let an extract from his sketch of 1844 be presented to the Linnean Society with Dr. Wallace's paper on July 1, 1858. "In conclusion," Dr. Wallace says, "I would only wish to add that my connection with Darwin and his great work has helped to secure for my own writings on the same questions a full recognition by the Press and the public; while my share in the origination and establishment of the theory of natural selection has usually been exaggerated. The one great result which I claim for my paper of 1858 is that it compelled Darwin to write and publish his 'Origin of Species' without further delay." The story reflects great credit upon both Dr. Wallace and Darwin, and many naturalists will be glad to read it. We congratulate Dr. Wallace upon having presented the world with such an interesting record after attaining his eightieth birthday.

Here is a minuscule contribution from Oliver Lodge (the second of James Clerk Maxwell's "two Olivers"), a thoroughbred physicist whose reputation suffered from his interest in the paranormal. Lodge was also famous as the loser in one of the best-known examples of simultaneous discovery. He recognised electromagnetic radiation and measured its wavelength, but instead of publishing his revelations he departed on a climbing holiday in the Alps. Returning in time for the annual meeting of the British Association for the Advancement of Science, he heard the announcement of Heinrich Hertz's observations on the propagation of electromagnetic waves. Lodge seems to have felt no rancour.

Glow-worm and Thunderstorm ; also Milk.

IN the *Daily News* of July 14 is printed an observation by a Mr. Haswell, of Handsworth, which bears the marks of genuineness, that during a thunderstorm a glow-worm extinguished its light for a second or a second and a half before each flash, relighting at an equal interval after the flash. May I ask if this has been noticed by anyone else?

It may also be worth while for someone to examine whether radium can assist milk to turn sour, or can otherwise influence organic processes of that kind.

OLIVER LODGE.

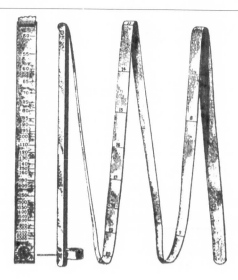

The affair of the n-rays, of which more presently, began in 1903 with a paper read to the French Academy of Sciences in Paris. Such proceedings at academies around the world were monitored in NATURE from the beginning.

On a new kind of light, by M. R. **Blondlot.** It has been shown in previous papers that the radiation from a focus tube, filtered from light rays by passing through a thin sheet of aluminium or black paper, proves to be polarised when examined with a small spark, and the plane of polarisation is rotated by quartz or sugar. It has now been found that a rotation of the plane is also produced when the rays are passed through a Reusch mica pile. A single sheet of mica produces elliptical polarisation, thus indicating that these rays are liable to double refraction. But if this is the case, there should also be simple refraction. Using a small spark as detector, the refraction of these rays by a prism was clearly made out, and an attempt to concentrate the rays by means of a quartz lens was also successful. These effects cannot be due to the X-rays, since the latter undergo neither refraction nor reflection. These results indicate the existence of a new set of radiations emitted by a Röntgen tube; these rays pass through aluminium, paper, wood, are rectilinearly polarised on their emission, are susceptible of both rotatory and elliptical polarisation, can be reflected and refracted, but produce neither fluorescence nor photographic action.

Next one of the odder theories of disease:

The concluding address was by Dr. Jonathan Hutchinson, F.R.S., the retiring president, who at two successive congresses has delighted his audience by a finely-argued discussion of a subject not at the first blush very attractive. His theme was leprosy. His theory is now well known, that this disease is caused by the consumption of badly cured fish, or occasionally by the eating of food which has been handled by lepers. During the last two years he has visited Africa and India, everywhere seeking out lepers and leprous communities, especially in places where he had been told that a fish diet was out of the question. Everywhere he found that in that particular his informants had been misinformed. A quotation from Erasmus sent to Dr. Hutchinson by a classical friend represented the Pope himself as proposing to proscribe the use of salt fish on account of its supposed tendency to spread leprosy, though it is not salt fish in itself that lies under any evil imputation. Erasmus often makes ironical statements, but on the foul effects produced in his day by the consumption of putrid fish his dialogue "Ichthyophagia" speaks with no ambiguity.

Frederick Soddy made many important contributions to the study of radioactive processes and discovered isotopes (indeed he coined the term). His great work was done early in his career, some of it in association with Ernest Rutherford. His scientific productivity ceased with his appointment to the chair of physical chemistry at Oxford. Thereafter he devoted his energies to the formulation and propagation of economic theories and schemes for world currencies, but he also foresaw that atomic energy would one day cause dilemmas for mankind. He became interested in the thorium mines in the East, and radiation from thorium and radium is the subject of a paper, which terminates alarmingly:

The emanation from a kilogram or more of thorium salt could be effectively employed on the lungs of a single patient. Thorium nitrate, a very soluble salt, is the most suitable compound to employ, but the free nitric acid present should be neutralised after the salt has been dissolved in water by cautious addition of ammonia with stirring, until precipitation is about to take place. A gas washing bottle, with outlet and inlet tubes ground in, could be used as the inhaler, and this should be filled as full as possible with the moderately concentrated solution. There is not much fear that an hour's daily inhalation of the emanation from 100 grams of dissolved thorium nitrate would produce any ill effect, and both the quantity employed and the time of inhalation could, after due trial, be increased indefinitely. For use with the radium emanation the inlet and outlet tubes should be provided with taps. A few milligrams of the salt, radium bromide, for example, should be placed in the dry bottle, and water drawn in to dissolve it, the taps being then closed. For the first trials, a few bubbles only of the total gas contained in a fairly large bottle should be drawn into the lungs with a deep breath of air, and retained as long as possible before being exhaled. The dose should be only very gradually increased, and the effect on the system very carefully watched, for the radium emanation is an exceedingly powerful agent. Mixed with air it glows brightly in a dark room, and exerts a very rapid oxidising action on carbonaceous matter, and even on mercury. The maximum possible dose for any one quantity of radium solution would be obtained by inhaling the whole gaseous contents of the bottle, a few bubbles at every breath, once every twenty-four hours.

The immunity of these processes from external interference, the simple nature of the treatment proposed, the infinitesimal quantity of the active agents employed, the manner in which the emanations may be inhaled to do their work at the very seat of the disease, leaving behind in their place the excited activity to continue the work in a gentle manner after they have been exhaled, make out a strong case why the attention of medical men should be directed to these new weapons which physics and chemistry have placed at their disposal. Indeed, if nature had designed these phenomena for the purpose proposed, it is difficult to see in what way they could be improved upon.

FREDERICK SODDY.

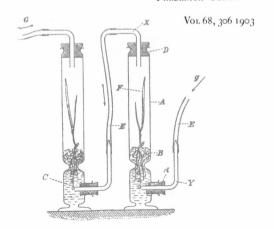

VOL 68, 306 1903

The grief that Lord Kelvin caused Charles Darwin with his assertions about the age of the Earth — based on calculations of the rate of cooling of a sphere — is well known. A mere 100 million years from the solidification of the crust could not begin to accommodate evolution. Kelvin must be wrong, Darwin wrote, "else my views would be wrong, which is impossible — QED." Kelvin and his associates showered scorn on Darwin and T. H. Huxley of the kind that physicists have ever harboured for biologists. Darwin asked his mathematician son, George, to check Kelvin's calculations and was dismayed when George could not fault them. Thirty years later George found the answer, and it must have given him no small satisfaction.

Radio-activity and the Age of the Sun.

IN the Appendix E of Thomson and Tait's "Natural Philosophy," Lord Kelvin has computed the energy lost in the concentration of the sun from a condition of infinite dispersion, and argues thence that it seems "on the whole probable that the sun has not illuminated the earth for 100,000,000 years, and almost certain that he has not done so for 500,000,000 years. As for the future, we may say, with equal certainty, that inhabitants of the earth cannot continue to enjoy the light and heat essential to their life for many million years longer, unless sources now unknown to us are prepared in the great storehouse of creation."

The object of the present note is to point out that we have recently learnt the existence of another source of energy, and that the amount of energy available is so great as to render it impossible to say how long the sun's heat has already existed, or how long it will last in the future.

The lost energy of concentration of the sun, supposed to be a homogeneous sphere of mass M and radius a, is $\frac{3}{5}\mu M^2/a$, where μ is the constant of gravitation. On introducing numerical values for the symbols in this formula I find the lost energy to be 2.7×10^7 M calories, where M is expressed in grammes. If we adopt Langley's value of the solar constant this heat suffices to give a supply for 12 million years. Lord Kelvin used Pouillet's value for that constant, but if he had been able to use Langley's his 100 million would have been reduced to 60 million. The discrepancy between my result of 12 million and his of 60 million is explained by a conjectural augmentation of the lost energy to allow for the concentration of the solar mass towards its central parts. I should have thought the augmentation somewhat too liberal, but for the present argument it is immaterial whether it is so or not.

Now Prof. Rutherford has recently shown that a gramme of radium is capable of giving forth 10^9 calories. If, then, the sun were made of such a radio-active material it would be capable of emitting 10^9 M calories without reference to gravitation. This energy is nearly forty times as much as the gravitational lost energy of the homogeneous sun, and eight times as much as Lord Kelvin's conjecturally concentrated sun.

Knowing, as we now do, that an atom of matter is capable of containing an enormous store of energy in itself, I think we have no right to assume that the sun is incapable of liberating atomic energy to a degree at least comparable with that which it would do if made of radium. Accordingly, I see no reason for doubting the possibility of augmenting the estimate of solar heat as derived from the theory of gravitation by some such factor as ten or twenty.

In an address to Section A of the British Association in 1886 I discussed the various estimates which have been made of geological time, and I said, "Although speculations as to the future course of science are usually of little avail, yet it seems likely that meteorology and geology will pass the word of command to cosmical physics as the converse." I think the recent extraordinary discoveries show that this forecast was reasonable.

It is probable that the bearing of radio-activity on the cosmical time-scale has occurred to others, but I do not happen to have seen any such statement.

Cambridge, September 20.   G. H. DARWIN.

A topic beloved of Nature *readers returns — animals with a sense of humour:*

Reason in Dogs.

Apropos of " thinking cats," perhaps the following story of a practical joke played by a dog will interest your readers.

A friend of mine, Mr. W., owns a Manchester terrier of which he is very fond, and for that reason receives rather more than doggy attention. The dog passes most of his time in the library, where a basket and rug are provided for him, but he prefers, when it is possible, to take possession of his master's easy chair. A short time ago I had occasion to call on Mr. W., and the dog was, as usual, occupying the chair, from which he was removed to his basket. He showed his resentment of this disturbance of his slumbers by becoming very restless. Presently he trotted over to the door, which he rattled by pushing with his nose, his usual method of attracting attention when he wished to go out. His master immediately rose and opened the door, but instead of the dog going out he rushed back and jumped into the chair his master had just vacated ! The rapid wagging of his tail and the expression on his face showed the dog to be very pleased with the result of his ruse. The dog has repeated the same joke once or twice since, with much evident delight to himself.   Arthur J. Hawkes.

Bournemouth.

Vol 71, 54 1904

Named after his native city of Nancy, n-*rays were discovered by René-Prosper Blondlot, professor of physics at the university there. They were an electromagnetic radiation given off by vacuum discharge tubes, hot filaments and various bodily tissues. Many attempts were made to reproduce Blondlot's experiments but — except in France — negative results were almost invariably reported. Blondlot had a considerable reputation as a physicist and many were willing to grant him the benefit of the doubt. Here is a suggested physiological explanation. In France it was widely held that Gallic sensibilities were more acute than those of the Germans or the Anglo-Saxons and that possession of the* esprit de finesse, *which was uniquely theirs, enabled French experimenters to detect the small differences in luminous intensity that betrayed the presence of* n-*rays.*

M. Blondlot's *n*-Ray Experiments.

It would be interesting to know whether anyone has succeeded in confirming the above, as described in your columns and elsewhere.

Personally, I have repeated most of M. Blondlot's experiments, but I have not been able to discern the slightest trace of any of the remarkable phenomena that he describes. This is also the case with Mr. J. C. M. Stanton and Mr. R. C. Pierce, who have assisted me in the investigations.

In order to get away from personal physiological idiosyncrasies we have also applied delicate photographic methods of observation, but without result, and as a general conclusion I am inclined to think that M. Blondlot's observations must be due, not to physical, but to physiological processes, and further, that these are not operative in the case of all persons.

Perhaps others may have tried the experiments and may have met with greater success.

A. A. Campbell Swinton.

66 Victoria Street, London, S.W., January 19.

Vol 69, 272 1904

Soon thereafter the bubble burst. R. W. Wood's account of what he witnessed in Blondlot's laboratory is a classic of scientific demolition: n-*rays were a figment, as Wood proved when he managed to secrete the dispersing prism in his trouser pocket without interrupting the process of data collection. The French scientific community was appalled and mortified, and the flood of papers on* n-*rays abruptly ceased. Blondlot himself never conceded and could not forgive what he saw as a breach of trust by the guest he had invited into his laboratory.*

The *n*-Rays.

The inability of a large number of skilful experimental physicists to obtain any evidence whatever of the existence of the *n*-rays, and the continued publication of papers announcing new and still more remarkable properties of the rays, prompted me to pay a visit to one of the laboratories in which the apparently peculiar conditions necessary for the manifestation of this most elusive form of radiation appear to exist. I went, I must confess, in a doubting frame of mind, but with the hope that I might be convinced of the reality of the phenomena, the accounts of which have been read with so much scepticism.

After spending three hours or more in witnessing various experiments, I am not only unable to report a single observation which appeared to indicate the existence of the rays, but left with a very firm conviction that the few experimenters who have obtained positive results have been in some way deluded.

A somewhat detailed report of the experiments which were shown to me, together with my own observations, may be of interest to the many physicists who have spent days and weeks in fruitless efforts to repeat the remarkable experiments which have been described in the scientific journals of the past year.

The first experiment which it was my privilege to witness was the supposed brightening of a small electric spark when the *n*-rays were concentrated on it by means of an aluminium lens. The spark was placed behind a small screen of ground glass to diffuse the light, the luminosity of which was supposed to change when the hand was interposed between the spark and the source of the *n*-rays.

It was claimed that this was most distinctly noticeable, yet I was unable to detect the slightest change. This was explained as due to a lack of sensitiveness of my eyes, and to test the matter I suggested that the attempt be made to announce the exact moments *at which I introduced my hand* into the path of the rays, by observing the screen. In no case was a correct answer given, the screen being announced as bright and dark in alternation when my hand was held motionless in the path of the rays, while the fluctuations observed when I moved my hand bore no relation whatever to its movements.

I was shown a number of photographs which showed the brightening of the image, and a plate was exposed in my presence, but they were made, it seems to me, under conditions which admit of many sources of error. In the first place, the brilliancy of the spark fluctuates all the time by an amount which I estimated at 25 per cent., which alone would make accurate work impossible.

Secondly, the two images (with *n*-rays and without) are built of " instalment exposures " of five seconds each, the plate holder being shifted back and forth by hand every five seconds. It appears to me that it is quite possible that the difference in the brilliancy of the images is due to a cumulative favouring of the exposure of one of the images, which may be quite unconscious, but may be governed by the previous knowledge of the disposition of the apparatus. The claim is made that all accidents of this nature are made impossible by changing the conditions, *i.e.* by shifting the positions of the screens; but it must be remembered that the experimenter is aware of the change, and may be unconsciously influenced to hold the plate holder a fraction of a second longer on one side than on the other. I feel very sure that if a series of experiments were made jointly in this laboratory by the originator of the photographic experiments and Profs. Rubens and Lummer, whose failure to repeat them is well known, the source of the error would be found.

I was next shown the experiment of the deviation of the rays by an aluminium prism. The aluminium lens was removed, and a screen of wet cardboard furnished with a vertical slit about 3 mm. wide put in its place. In front of the slit stood the prism, which was supposed not only to bend the sheet of rays, but to spread it out into a spectrum. The positions of the deviated rays were located by a narrow vertical line of phosphorescent paint, perhaps 0.5 mm. wide, on a piece of dry cardboard, which was moved along by means of a small dividing engine. It was claimed that a movement of the screw corresponding to a motion of less than 0.1 of a millimetre was sufficient to cause the phosphorescent line to change in luminosity when it was moved across the *n*-ray spectrum, and this with a slit 2 or 3 mm. wide. I expressed surprise that a ray bundle 3 mm. in width could be split up into a spectrum with maxima and minima less than 0.1 of a millimetre apart, and was told that this was one of the inexplicable and astounding properties of the rays. I was unable to see any change whatever in the brilliancy of the phosphorescent line as I moved it along, and I subsequently found that the removal of the prism (we were in a dark room) did not seem to interfere in any way with the location of the maxima and minima in the deviated (!) ray bundle.

I then suggested that an attempt be made to determine by means of the phosphorescent screen whether I had placed the prism with its refracting edge to the right or the left, but neither the experimenter nor his assistant determined the position correctly in a single case (three trials were made). This failure was attributed to fatigue.

I was next shown an experiment of a different nature. A small screen on which a number of circles had been painted with luminous paint was placed on the table in the dark room. The approach of a large steel file was supposed to alter the appearance of the spots, causing them to appear more distinct and less nebulous. I could see no change myself, though the phenomenon was described as open to no question, the change being *very* marked. Holding the file behind my back, I moved my arm slightly towards and away from the screen. The same changes were described by my colleague. A clock face in a dimly lighted room was believed to become much more distinct and brighter when the file was held before the eyes, owing to some peculiar effect which the rays emitted by the file exerted on the retina. I was unable to see the slightest change, though my colleague said that he could see the hands distinctly when he held the file near his eyes, while they were quite invisible when the file was removed. The room was dimly lighted by a gas jet turned down low, which made blank experiments impossible. My colleague could see the change just as well when I held the file before his face, and the substitution of a piece of wood of the same size and shape as the file in no way interfered with the experiment. The substitution was of course unknown to the observer.

I am obliged to confess that I left the laboratory with a distinct feeling of depression, not only having failed to see a single experiment of a convincing nature, but with the almost certain conviction that all the changes in the luminosity or distinctness of sparks and phosphorescent screens (which furnish the only evidence of *n*-rays) are purely imaginary. It seems strange that after a year's work on the subject not a single experiment has been devised which can in any way convince a critical observer that the rays exist at all. To be sure the photographs are offered as an objective proof of the effect of the rays upon the luminosity of the spark. The spark, however, varies greatly in intensity from moment to moment, and the manner in which the exposures are made appears to me to be especially favourable to the introduction of errors in the total time of exposure which each image receives. I am unwilling also to believe that a change of intensity which the average eye cannot detect when the *n*-rays are flashed " on " and " off " will be brought out as distinctly in photographs as is the case on the plates exhibited.

Experiments could be easily devised which would settle the matter beyond all doubt ; for example, the following :—
Let two screens be prepared, one composed of two sheets of thin aluminium with a few sheets of wet paper between, the whole hermetically sealed with wax along the edges. The other screen to be exactly similar, containing, however, dry paper.

Let a dozen or more photographs be taken with the two screens, the person exposing the plates being ignorant of which screen was used in each case. One of the screens being opaque to the *n*-rays, the other transparent, the resulting photographs would tell the story. Two observers would be required, one to change the screens and keep a record of the one used in each case, the other to expose the plates.

The same screen should be used for two or three successive exposures, in one or more cases, and it should be made impossible for the person exposing the plates to know in any way whether a change had been made or not.

I feel very sure that a day spent on some such experiment as this would show that the variations in the density on the photographic plate had no connection with the screen used.

Why cannot the experimenters who obtain results with *n*-rays and those who do not try a series of experiments together, as was done only last year by Cremieu and Pender, when doubt had been expressed about the reality of the Rowland effect?   R. W. Wood.
Brussels, September 22.

Vol 70, 530 1904

Clever Hans's arithmetical feats, as is well known, stemmed from his sensitivity to his master's unconscious movements. Informed opinion must, however, have been curiously receptive to the idea of intellectual animals.

Scientific critics in Berlin are now much exercised with regard to the remarkable performances of " Clever Hans," the thinking horse. According to the daily Press, a representative committee, which included the director of the Berlin Zoological Gardens, a veterinary surgeon, and a professor of the Physiological Institute of the Berlin University, witnessed these performances with the view of ascertaining whether they were the result of a trick, or whether they were due to the mental powers of the animal. Their verdict, it is reported, was unanimous in favour of the latter view. It is stated that when told that the day was Tuesday, and asked which day of the week this represented, the horse would give the correct answer by taps. Similarly he will tell not only the hour, but the minutes indicated by a watch ; while he is also reported to be able to record the number of men and of women among a row of visitors, and to indicate the tallest and the shortest members of the party.

Vol 70, 510 1904

Alas, Clever Hans's moment of fame was all too brief, but it was sensational while it lasted.

The performances of an intelligent horse—" Clever Hans "—at Berlin two or three months ago attracted much attention. In a letter which appeared in Nature of October 20 (vol. lxx. p. 602) the Rev. J. Meehan pointed out that the performances of the horse were much the same as those of the horse " Mahomet " shown at the Royal Aquarium twelve or thirteen years ago, and depended entirely upon the animal's observation of movements of the trainer or the tones of his voice. Much the same opinion has been reached by a commission of psychological experts, headed by Prof. Stumpf, of Berlin University, that has subjected " Clever Hans " to a scientific examination. The conclusion arrived at is that the horse is not capable of independent thought. According to the Berlin correspondent of the *Daily Chronicle*, Prof. Stumpf found that this horse is gifted with remarkable powers of observation, which four years of patient and skilful treatment have developed. When asked a question " Hans " knows he has to beat with his hoof in reply, but he does not know when to cease beating until he detects some movement on the part of the person questioning him. The commission expresses the opinion that, so far as Herr von Osten, the owner, is concerned, these movements are given involuntarily, and are sometimes of so imperceptible a nature as to be undetected, save by highly trained human observers. There has been no trickery, says Prof. Stumpf, but, on the other hand, there have been no reasoning powers on the horse's part. The whole secret is in von Osten's skill, patience, and judicious reward, and, on " Hans's " part, in keen powers of observation.

VOL 71, 156 1904

Dmitri Ivanovich Mendeleev (the youngest of 17 children, born into grinding poverty in Siberia) imposed order on the chemical elements but could not divine that the periodic table reflected differences in atomic structure. Yet his views, as represented in this article by Kropotkin, were perceptive, though the luminiferous ether defeated him. As to Prince Peter Alekseevich Kropotkin, he was a man of wide learning — a geographer and geologist who had made extensive studies of the terrain of Siberia. A committed anarchist, he served time in prison, escaped and eventually settled in Britain, though he returned to Russia at the time of the revolution.

The researches concerning the double stars prove that the masses of the stars which we know do not exceed the mass of our sun more than thirty-two times, while in other cases they are equal to it ; therefore, if we attribute to the ether the properties of gases, we must admit, on the basis of the kinetic theory of gases, that its specific gravity must be very much smaller than the specific gravity of hydrogen. In order that the ether may escape from the sphere of attraction of stars the mass of which is fifty times greater than the mass of the sun, it must, while it chemically resembles argon and helium, have an atomic weight not more than 0·000 000 000 053 (and a density, in relation to hydrogen, half as large, as I have proved in the above mentioned article on ether). The very small value of this figure already explains why there is little hope of isolating the substance of the ether in the near future, as it also explains why it penetrates all substances, and why it is condensed in a small degree, or collects in a physicomechanical way, round ponderable substances—being mostly condensed round such immense masses as that of the sun or of stars.[1] "

In conclusion, Mendeléeff indicates that while the conception of the chemical elements is connected in the most intimate way with the generally received teachings of Galileo and Newton about the mass and the ponderability of matter, as also with the teaching of Lavoisier concerning the indestructibility of matter, " the conception of the ether originates exclusively from the study of phenomena and the need of reducing them to simpler conceptions. Amongst such conceptions we held for a long time the conception of imponderable substances (such as phlogiston, luminous matter, the substance of the positive and negative electricity, heat, &c.), but gradually this has disappeared, and now we can say with certainty that the luminiferous ether, if it be real, is ponderable, although it cannot be weighed, just as air cannot be weighed in air, or water in water. We cannot exclude the ether from any space ; it is everywhere and penetrates everything, owing to its extreme lightness and the rapidity of motion of its molecules. Therefore such conceptions as that of the ether remain abstract, or conceptions of the intellect, like the one which also leads us to the very teaching about a limited number of chemical elements out of which all substances in nature are composed."

VOL 71, 66 1904

After Mendeleev's death in 1907, Sir Edward Thorpe wrote his obituary, which ended with this pleasant reminiscence.

On the occasion of his delivering the Faraday lecture it fell to the writer's duty, as treasurer of the Chemical Society, to hand him the honorarium which the regulations of the society prescribe, in a small silken purse worked in the Russian national colours. He was pleased with the purse, especially when he learned that it was the handiwork of a lady among his audience, and declared that he would ever afterwards use it, but he tumbled the sovereigns out on the table, declaring that nothing would induce him to accept money from a society which had paid him the high compliment of inviting him to do honour to the memory of Faraday in a place made sacred by his labours.

T. E. THORPE.

VOL 75, 373 1907

F. G. is presumably Francis Galton again, occupying his restless mind while sitting for his portrait.

Number of Strokes of the Brush in a Picture.

THE number of strokes of the paint brush that go to making a picture is of some scientific interest, so I venture to record two personal experiences. Some years ago I was painted by Graef, a well known German artist, when, finding it very tedious to sit doing nothing, I amused myself by counting the number of strokes per minute that he bestowed on the portrait. He was methodical, and it was easy to calculate their average number, and as I knew only too well the hours, and therefore the number of minutes, I sat to him, the product of the two numbers gave what I wanted to learn. It was 20,000. A year and a half ago I was again painted by the late lamented artist Charles Furse, whose method was totally different from that of Graef. He looked hard at me, mixing his colours the while, then, dashing at the portrait, made his dabs so fast that I had to estimate rather than count them. Proceeding as before, the result, to my great surprise, was the same, 20,000. Large as this number is, it is less than the number of stitches in an ordinary pair of knitted socks. In mine there are 100 rows to each 7 inches of length, and 102 stitches in each row at the widest part. Two such cylinders, each 7 inches long, would require 20,000 stitches, so the socks, though they are only approximately cylinders, but much more than 7 inches long, would require more than that number.

The following point impressed me strongly. Graef had a humorous phrase for the very last stage of his portrait, which was " painting the buttons." Thus, he said, " in five days' time I shall come to the buttons." Four days passed, and the hours and minutes of the last day, when he suddenly and joyfully exclaimed, " I am come to the buttons." I watched at first with amused surprise, followed by an admiration not far from awe. He poised his brush for a moment, made three rapid twists with it, and three well painted buttons were thereby created. The rule of three seemed to show that if so much could be done with three strokes, what an enormous amount of skilled work must go to the painting of a portrait which required 20,000 of them. At the same time, it made me wonder whether painters had mastered the art of getting the maximum result from their labour. I make this remark as a confessed Philistine. Anyhow, I hope that future sitters will beguile their tedium in the same way that I did, and tell the results.

F. G.

VOL 72, 198 1905

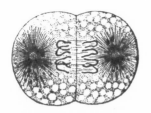

Here is the editor's classic quandary: a communication from Henry Charlton Bastian must have been as welcome as a boil on the neck. Bastian was a monomaniac, though by no means an outsider, for he became a professor and a fellow of the Royal Society. Like all his kind, he felt victimised by the guardians of established opinion, among whom were included all editors of learned journals and their referees. Norman Lockyer accepted the challenge — but of course the referees of the two journals that rejected Bastian's offering were right.

Archebiosis and Heterogenesis.

THE columns of the daily papers have during the last two weeks contained many references to the question of the origin of life. One of the most recent utterances has been that of Lord Kelvin, who has roundly declared himself an unbeliever in the natural origin of living matter either in the present or in the past. We must suppose, therefore, that in reference to this question he is content to believe in miracles.

Prof. Ray Lankester and Dr. Chalmers Mitchell, however, proclaim themselves, as followers of Huxley, believers in evolution generally, and in the natural origin of living matter in the past. They, like many others, refuse to believe that it takes place at the present time, because undoubted proof of its occurrence cannot be produced by laboratory experiments. The uniformity of natural phenomena would certainly lead us to believe, as Sir Oliver Lodge has intimated, that if such a process occurred in the past, it should have been continually occurring ever since—so long as there is no evidence to show cause for a break in the great law of Continuity. Certainly no such evidence has ever been produced, and if the origin of living matter takes place by the generation in suitable fluids of the minutest particles gradually appearing from the region of the invisible, such a process may be occurring everywhere in nature's laboratories, though altogether beyond the ken of man.

My point may be illustrated thus. Bacteriologists all over Europe and elsewhere have been working for the last thirty years by strict laboratory methods, and notwithstanding all that they have made out and the good that has thereby accrued to suffering humanity, they have apparently never yet seen the development from Zoogloea aggregates of Fungus-germs, of flagellate Monads, or of Amœbæ. If, however, they would only examine what goes on in nature's laboratory when a mixed bacterial scum forms on suitable fluids, they would have no difficulty in satisfying themselves as to the reality of these processes. I described such processes in your columns in 1870, more fully in the *Proceedings* of the Royal Society in 1872, and finally in my " Studies in Heterogenesis " (pp. 65–84, pls. vi. and vii., Figs. 53–71). Even during the last week I have again obtained photo-micrographs demonstrating the origin of flagellate Monads from Zoogloea aggregates forming in a bacterial scum, and if you will admit an illustrated communication on this subject to your columns, proving by such a test case my position as to the reality of heterogenesis, I shall be happy to present it, and to show that something beyond the recognised strict laboratory methods of the day is needed if we are to fathom some of nature's deepest secrets.

The councils of the Royal and Linnean Societies are guided in the acceptance of papers by referees who are wedded, on biological questions, to laboratory methods. It is useless for me, therefore, again to attempt to submit such a communication to them. Their referees (probably not having worked at such subjects themselves) would not advise the acceptance of the paper, and my communication might simply be consigned to their archives. The Royal Society " for the Promotion of Natural Knowledge " on two occasions would not even allow me to submit my views to the consideration of, and discussion by, its fellows. In these circumstances, Sir, I appeal to you, in the interests of science, to allow me to send you an illustrated paper proving, so far as such proof can go, the heterogenetic origin of flagellate Monads and of Fungus-germs.

H. CHARLTON BASTIAN.

Manchester Square, October 31.

[IN reply to Dr. Bastian's appeal we will print his communication, and also any important replies from competent workers on the subject which may be sent to us.—ED.]

The success of identification by fingerprint was a triumph for Francis Galton. Its application was not always free from controversy. As in our time, an expert witness could seemingly always be found to represent any view, no matter how absurd. This law is sometimes stated in the form: to every PhD there is an equal and opposite PhD.

IN a murder trial concluded last week, a finger mark left by one of the prisoners upon a cash-box tray at the shop where the crime was committed was used for purposes of identification. An inspector gave evidence that there were 80,000 or 90,000 sets of finger prints in the finger print department of Scotland Yard, and that he had never found two such impressions to correspond. The right thumb print of one of the prisoners agreed in twelve characteristics with an impression made with perspiration upon the cash-box tray, and therefore gave corroborative evidence of identity. It is probable, as Mr. Galton pointed out some years ago, that no two finger-prints in the whole world are so alike that an expert would fail to distinguish between them. The system was largely used in India by Sir William Herschel nearly fifty years ago, and was found by him to be most successful in preventing personation, and in putting an end to disputes about the authenticity of deeds. He described his methods in these pages in 1880 (vol. xxiii. p. 76); and in the previous volume (vol. xxii. p. 605) Mr. Henry Faulds referred to the use of finger-marks for the identification of criminals. There is no doubt as to the value of this system of identification, which was described in the pages of NATURE long before its practical applications had been realised, and we regret that anything should have occurred to throw discredit upon it. It appears from the reports of the trial referred to that a person who professed to be properly qualified wrote to the Director of Public Prosecutions, and also to the solicitors for the defence, offering to give evidence as an expert on the finger impressions, although he had not seen the impressions. It is not to be wondered at that Mr. Justice Channell should denounce such action in strong language, and whether the jury agreed with him or not—that the witness was " absolutely untrustworthy "—they no doubt considered that evidence which could be given on either side could not be of much importance. From the scientific point of view, we regret that a method which is associated with the names of men of such scientific eminence as Sir William Herschel and Mr. Francis Galton should be brought into disrepute. Finger prints are not only of value for personal identification, but also for hereditary investigations, and any action which produces comments like those made by Mr. Justice Channell is to be deplored, because it tends to shake the confidence of men in methods which rest on secure scientific foundations.

The triumph of Japanese arms in the war against Russia, culminating in the annihilation of the Russian fleet at Tsushima, excited admiration, not unmixed with gloating, in Britain. NATURE's *comment conveys the hint of a dangerous new giant growing to maturity in the East.*

WHY JAPAN IS VICTORIOUS.

TEN years ago, after the conclusion of the war between Japan and China, it was remarked that the sound of the Japanese cannon at the mouth of the Yalu River awoke the nations of the world to the fact that a new Power had arisen in the Far East which in future would require to be taken into account when any political problems arose. It is, of course, recognised by all who know modern Japan that the most important factor in the making of new Japan has been the applications of science to the arts both of peace and war. Without these, even the spirit of the samurai would have been as powerless before the attacks of Western Powers armed with all the latest warlike appliances, as were the dervishes at the battle of Omdurman. Spectators speak with admiration of the bravery of these men and with pity that their lives were thrown away in a vain resistance. Without the help of science and its applications it is very certain that, before this time, Japan would have been overrun by a European Power after immense slaughter, for the last man would have died, fighting with his primitive weapons, rather than recognise a foreign domination.

A careful study of the evolution of modern Japan shows plans founded on enlightened principles and carried out in every detail. In fact, one of the secrets of the success of the Japanese in the present war is that nothing is left to chance; every detail is worked out and carefully provided for. They soon recognised that their national ideals would never be realised without a system of education, complete in every department, which would supply the men who were required to guide the nation under the new conditions which had emerged. Elementary education was organised all over the country, secondary education in central districts, and technical education wherever it seemed to be required. Above all, there are two national universities which in equipment and quality of work done will bear favourable comparison with similar institutions in any other country in the world.

The educational work of the country was directed not simply to personal or sectional purposes, as is unfortunately too often the case in the West; it was also consciously directed to the attainment of great national ends. Every department of the national life was organised in a rational manner, and, therefore, on scientific principles. In many departments there is still much to be done, but past achievements promise well for the future.

Special attention has been paid by the Government to the applications of science. Without the railways, the telegraphs and telephones, the dockyards, the shipbuilding yards, the mines, and the engineering establishments, the existence of the army and navy would have been impossible; at least, if they did exist they would have been nearly powerless. The operations of the present war with Russia have clearly demonstrated the importance of the introduction of the scientific spirit into all the national activities. The railways which have been built in Japan have been fully utilised to convey men and materials and the ships to transport them oversea. The telegraphs have been used to communicate instructions and to keep the authorities informed regarding movements and requirements. The dockyards and shipbuilding yards have been ready to undertake repairs, and the arsenals and machine shops to turn out war material of all kinds, as well as appliances which aid operations in the field. Light railways have been laid down on the way to battlefields, and wireless telegraphy and telephones to convey instructions to the soldiers; in short, all the latest applications of mechanical, electrical, and chemical science have been freely and intelligently used.

The Japanese have not only modified Western appliances to suit their conditions, but they have also made numerous distinct advances. The ships of their navy are probably the best illustration of the Japanese method of procedure. In naval matters they accepted all the guidance the Western world could give them, but at the same time they struck out a line of their own, and the fleet which they have created is unique in the character of its units. British designs have in many respects been improved upon, with the result that they have obtained in their latest ships many features which have won the admiration of the world. The training of Japanese naval officers is very complete in every way, and in some respects offers an example to the British authorities, and the men are devoted to their profession Japan now sends her picked men to Europe to complete their studies, so that in every department of national life they are kept up with the latest developments. The siege of Port Arthur, the battle of Mukden and the other battles in Manchuria, and the exploits of the Japanese Navy prove most distinctly that they have profited by their experience.

VOL 72, 128 1905

W. R., the author of these verses, was Sir William Ramsay, discoverer of the noble gases.

THE DEATH-KNELL OF THE ATOM.[1]

Old Time is a-flying; the atoms are dying;
  Come, list to their parting oration :—
" We'll soon disappear to a heavenly sphere
  On account of our disintegration.

" Our action's spontaneous in atoms uranious
  Or radious, actinious or thorious :
But for others, the gleam of a heaven-sent beam
  Must encourage their efforts laborious.

" For many a day we've been slipping away
  While the savants still dozed in their slumbers;
Till at last came a man with gold-leaf and tin can
  And detected our infinite numbers."

Thus the atoms in turn, we now clearly discern,
  Fly to bits with the utmost facility;
They wend on their way, and in splitting, display
  An absolute lack of stability.

'Tis clear they should halt on the grave of old Dalton
  On their path to celestial spheres;
And a few thousand million—let's say a quadrillion—
  Should bedew it with reverent tears.

There's nothing facetious in the way that Lucretius
  Imagined the Chaos to quiver;
And electrons to blunder, together, asunder,
  In building up atoms for ever!

                              W. R.

[1] Sung at the Chemical Laboratory dinner at University College, November 17.

VOL 73, 132 1905

Scientific aspects of manly pursuits were occasionally aired in Nature. *Here is the beginning of a review of a cricket book by G. W. Beldam and C. B. Fry, with dramatic photographs. Fry was every schoolboy's hero — a dashing cricketer and captain of England, an athlete and a classical scholar. He was offered, but rejected, the crown of Albania and had political views not dissimilar to those of Tamburlaine the Great.*

SCIENCE AND ART OF CRICKET.

THE golfing world already owes a debt of gratitude to Mr. Beldam for his "Great Golfers." This companion volume, setting forth on the same lines the styles of play of our greatest cricketers, cannot fail to appeal strongly to all lovers of the most English of our national games. The method adopted here is identical with that of the earlier book. Each of the many batsmen pictured has been photographed in one or more characteristic attitudes before, during, or after the striking of the ball, and after a careful study of every picture Mr. Fry has set down his own interpretation for the guidance of the reader. No better guide could have been got, for among the great cricketers of our day Mr. Fry stands conspicuous as one who has studied the art of cricket with phenomenal success.

The book is divided into two parts. In part i. (individualities) close on 300 photographs are given of eighteen of our best known batsmen, including Grace, Ranjitsinhji, Trumper, Fry, Hill, Jackson, Duff, MacLaren, and so on. In part ii. (strokes illustrated) the various kinds of recognised strokes are systematically discussed and illustrated by photographs of other great batsmen. There is, of course, a good deal of repetition of the same ideas in the letterpress of these two parts, but each has its own value. In the one case it is the individual batsman whose pose and actions are being studied; in the other it is the kind of stroke which is the object of discussion, and this is helped out by an appeal to the example of a number of different cricketers. The volume ends with a short but very practical and interesting chapter upon the art of timing with the camera. We learn that the operator sometimes used a finger release of the shutter and sometimes an electric. The latter method enabled Mr. Beldam to act, in same cases, both as bowler and photographer. The requisites for good work of this kind are complete knowledge of the mechanism and capabilities of the camera, thorough acquaintance with the game itself, and a delicacy of judgment which must be partly inborn and strongly developed by practice. We are not told what proportion of photographs taken were failures, but the beauty and clearness of the 600 here shown prove that Mr. Beldam is a master hand in the art of taking action-photographs.

Where almost every picture is admirable, and illustrates some essential part of a particular stroke, it is not possible to choose for reproduction any that might be regarded as representative. W. G. Grace, for example, is shown in twenty-six different attitudes, and all have some lesson to tell. In the photograph reproduced we have the finish of an on-drive, in which the turn of the body has aided powerfully in giving full effect to the stroke. The eyes are

Fig. 1.—W. G. Grace—Finish of an On-drive.

still looking at the spot where the ball was when it was struck. The whole series of photographs proves that all great batsmen follow the ball with their eye right up to the moment of striking. It is this which gives precision, just as in golf.

Here is a useful mnemonic for anyone wishing to remember the value of π to 30 decimal places.

IN a recent issue (August 5) the *Academy* directs attention to a curious poetical tribute—composed by a French mathematician—to Archimedes, referring to the evaluation of π, which, set out in thirty places of decimals, is 3·141592653589793238462643383279. It will be observed that each of the thirty-one words in this quatrain contains the number of letters corresponding with the successive numbers in the numerical expression :—

3 1 4 1 5 9 2 6 5 3 5
Que j'aime à faire apprendre un nombre utile aux sages
8 9 7 9
Immortel Archimède, artiste ingénieur !
3 2 3 8 4 6 2 6
Qui de ton jugement peut priser la valeur ?
4 3 3 8 3 2 7 9
Pour moi ton problème eut de pareils avantages.

The *Frankfurter Zeitung* reproduces the French verse, and adds a similar effort emanating from a German poet and geometrician :—

3 1 4 1 5 9 2 6 5
Dir, o Held, o alter Philosoph, Du Riesen-Genie !
3 5 8 9 7
Wie viele Tausende bewundern Geister,
9 3 2 3 8
himmlisch wie Du und göttlich !—
4 6 2 6
Noch reiner in Aeonen
4 3 3 8
wird das uns strahlen,
3 2 7 9
wie im lichten Morgenrot !

The *Academy* asks for English parallels to these efforts.

VOL 72, 385 1905

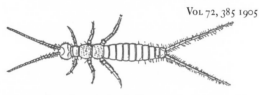

A NATURE reader takes up the challenge:

Rhymes on the Value of π.

THE following rhyme is in imitation of the French and German verses given in NATURE (August 17) in which the number of letters in each word correspond to a numeral in the value of π. The three concluding lines are somewhat obscure ; it seems to have occurred to the author that the method is a misuse of language, and he expresses the hope that NATURE will take a more lenient view than Dr. Johnson might be imagined to express.

To the Editor of NATURE.

Sir,—I send a rhyme excelling
3 1 4 1 5 9
In sacred truth and rigid spelling.
2 6 5 3 5 8
Numerical sprites elucidate
9 7 9
For me the lexicon's dull weight.
3 2 3 8 4 6
If " Nature " gain,
2 6 4
Not you complain,
3 3 8
Tho' Dr. Johnson fulminate.
3 2 7 9

F.R.S.

VOL 72, 558 1905

Readers of NATURE were still assumed to possess classical and modern linguistic learning. Sentences such as the one below litter the pages — this from an article entitled "The application of scientific methods to the study of history".

Through history mankind attains to self-consciousness. As Droysen puts it :—" Die Geschichte ist das γνῶθι σαυτὸν der Menschheit, ihr Gewissen."

VOL 73, 308 1906

The principles of radioactivity were not embraced by the crustier section of the physics establishment. Lord Kelvin, for instance, did not apparently care to relinquish his conclusions concerning the age of the Earth. Frederick Soddy, like the other followers of Ernest Rutherford, was plainly exasperated. Here is the beginning of his article and the concluding paragraph. Professor Armstrong, who aligned himself so incautiously with Kelvin, was an organic chemist and a kind of hot-air balloon, hovering for some 70 years over the scientific scene in England. Many years later Rutherford told of how, as he was beginning a lecture on aspects of radioactivity, which was to include its implications for the age of the Earth, he spied Kelvin in the audience, albeit asleep. As he approached the sensitive topic, he "saw the old bird sit up, open an eye and cock a baleful glance at me." Rutherford adroitly introduced his argument: "... a sudden inspiration came", he recalled, "and I said Lord Kelvin had limited the age of the earth, provided no new source of heat was discovered. That prophetic utterance refers to what we are now considering tonight, radium! Behold! the old boy beamed upon me."

THE RECENT CONTROVERSY ON RADIUM.

THE recent correspondence on the subject of radium, started in the *Times* by Lord Kelvin, has, after lasting nearly a month and causing widespread interest, apparently closed without any very definite conclusion being reached. Whatever opinion may be formed of the merits of the controversy, all must unite in admiration for the boldness with which Lord Kelvin initiated his campaign, and the intellectual keenness with which he conducted, almost single-handed, what appeared to many from the first almost a forlorn hope against the transmutational and evolutionary doctrines framed to account for the properties of radium. The weight of years and the almost unanimous opinion of his younger colleagues against him have not deterred him from leading a lost cause, if not to a victorious termination, at least to one from which no one will grudge him the honours of war. If peace and tranquility now result, and a measure of agreement is arrived at between conflicting views, it will be a result which all concerned will heartily welcome. The most ardent believer in the truth of the new doctrines cannot be other than satisfied that every feature and assumption that is admittedly speculative should be clearly recognised as such and separated from that which is not, if thereby the experimental foundations of the science of radio-activity are freed from further wordy and unprofitable controversy. There seems now to be a reasonable prospect that this has been secured.

Lord Kelvin's opening challenge (August 9) was broad and sweeping. He took exception to the statement, made by the writer in opening the discussion on the evolution of the elements at the British Association at York, that the production of helium from radium has established the fact of the gradual evolution of one element into others, and denied that this discovery affected the atomic doctrine any more than the original discovery of helium in cleveite. The obvious conclusion was that both cleveite and radium contained helium. He also stated that there was no experimental foundation for the hypothesis that the heat of the sun was due to radium, and ascribed it to gravitation.

The challenge was taken up on the other side successively by Sir Oliver Lodge, the Hon. Mr. Strutt, and other well-known authorities, and it soon became apparent that for argument at least Lord Kelvin on his side had to rely practically on himself alone. Prof. Armstrong, it is true, immediately enrolled under Lord Kelvin's banner, and entered the lists with an embracing criticism of physicists in general, whom, he declared, are strangely innocent workers under the all-potent influence of formula and fashion. He made the statement that no one had handled radium in such quantity or in such manner that we can say precisely what it is, and throughout put the word *radium* in inverted commas.

Whether or no his opponents are all as innocent and ignorant as Prof. Armstrong imagines, the fact remains that, except for this *ex cathedra* utterance and a leading article, argument against the accepted view there was little or none except that contributed by Lord Kelvin himself. Prof. Armstrong's letter merely served to provide Sir Oliver Lodge with justification for his favourite theme, which appears to be that whereas chemists have an instinct of their own for arriving at their results, reason is the monopoly of the physicist, whose results the chemist usually manages to absorb in the end. No better argument against the unfairness of this could be provided than by the history of radio-activity itself, which owes at least as much to the chemist as to the physicist. Prof. Armstrong is almost alone among chemists, as Lord Kelvin is among physicists, in his hostility to the new doctrines.

* * *

It would be a pity if the public were misled into supposing that those who have not worked with radio-active bodies are as entitled to as weighty an opinion as those who have. The latter are talking of facts they know, the former frequently of terms they have read of. If, as a result of the recent controversy, it has been made clear that atomic disintegration is based on experimental evidence, which even its most hostile opponents are unable to shake or explain in any other way, the best ends of science will have been served. The sooner this is understood the better, for in radio-activity we have but a foretaste of a fountain of new knowledge, destined to overflow the boundaries of science and to impregnate with teeming thought many a high and arid plateau of philosophy.

F. SODDY.

Vol 74, 516 1906

Here is a mathematical problem that has lain dormant pending the invention of the computer. The discoverer of the manuscript was Gotthold Ephraim Lessing, poet, dramatist and librarian.

THE *Popular Science Monthly* for November contains a note by Prof. Mansfield Merriman on the " cattle problem " of Archimedes. This problem occurs in the form of a poem of forty-four lines in a manuscript in the library of Wolfenbüttel, and it was brought into notice by Lessing shortly after his appointment as librarian there in 1769. The problem consists, in the first place, in determining the total number of cattle grazing on the plain of Sicily,

divided into white, black, dappled, and yellow bulls and cows, from seven equations of condition connecting the numbers in the eight various categories. The problem in this form is easy, but a further rider imposes the additional conditions that the number of white and black bulls shall be a square number, and the number of dappled and yellow bulls a triangular number. Amthor showed in 1880 that numbers satisfying these conditions could be found, but instead of the total number representing a possible herd of cattle, it would consist of no less than 206,545 digits. Finally, in 1889 Mr. A. H. Bell, in conjunction with two other mathematicians, began the work of solution, and in the course of four years determined the first thirty or thirty-one and the last twelve digits of the actual numbers. It is, however, pointed out that to determine all the 206,545 digits would occupy a thousand men for a thousand years.

Vol 73, 86 1905

Here is another dissertation on cerebrating insects.

The Intelligence of Animals.

IN his review of Father Wassmann's book (NATURE, February 1, p. 351) Lord Avebury dissents from Father Wassmann's conclusion that the sagacity of ants is " instinctive and essentially different from intelligence and reflection," and repeats the opinion which he has held for many years, that " it is difficult altogether to deny to them the gift of reason." The following incidents, which I observed on a footpath in the Donetz Coalfield, in Russia, in the summer of 1898, appear to me to show that the insects here referred to possess both intelligence and the gift of reason, and, therefore, to lend a general support of Lord Avebury's views.

* * *

... a solitary beetle rolling a comparatively new ball had reached a distance of nine or ten inches from the heap when a second unoccupied beetle coming from the opposite direction stood up in front of the rolling ball as if with the intention of pulling it forward and assisting the first. Instead of doing so, however, it brought the ball to a dead stop. In vain the first tried to move the ball; the second held it fast. The first then got down and peered round the side of the ball, apparently with the object of ascertaining the nature of the obstacle. While this examination was proceeding, the second, with its fore-feet still resting on the upper part of the ball, neither pushed nor moved in any way. The first then stood up again behind the ball and pushed it as before, but still the ball did not move. For the second time the beetle got down, made an examination as before, then, crouching with its back well under the lower curve of the ball, heaved with all its might—in the same way as a workman does in similar circumstances—but the ball remained stationary. The first beetle then came out from under the ball, and was proceeding round its right-hand side, with some new intention, when the two seemed to catch sight of each other. The second beetle threw itself on the ground with the quickness of thought, and fled pursued by the other, both running at their utmost speed. Fear, and a sense of guilt, seemed to spur the flight of the one, resentment and anger the pursuit of the other. In a chase which was continued for a distance of six inches, the fleeing beetle, which had started with an advantage of about an inch and a half, increased the distance between its pursuer and itself to more than two inches, when the former, seeing the futility of further pursuit, stopped, returned to the ball, and resumed its occupation of rolling it.

The reason why the second beetle stopped the ball, remained absolutely motionless when the other got down to reconnoitre, and ran away when it saw it was discovered is not apparent. Dare we suppose that it was simply amusing itself at the expense of the other? This was the impression left on my mind at the time.

W. GALLOWAY.

Vol 73, 440 1906

chapter nine
1907-1913

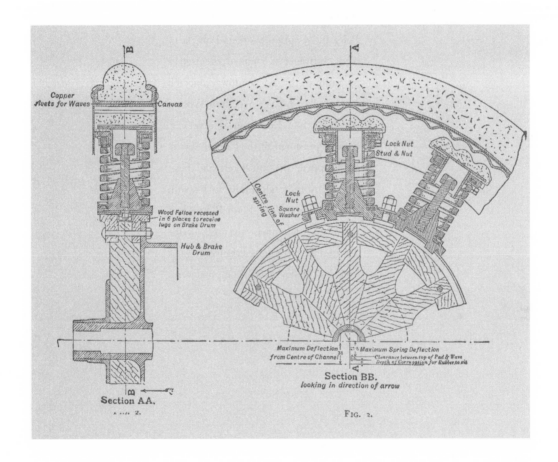

Copper
Rivets for Waves   Canvas

Wood Felloe recessed
in 6 places to receive
lugs on Brake Drum

Hub & Brake
Drum

Lock Nut
Stud & Nut

Centre line of
spring

Lock
Nut
Square
Washer

Maximum Deflection   Maximum Spring Deflection
from Centre of Channel   Clearance between top of Pad & Wave
                            Depth of Corrugation for Rubber to sit

Section BB.
looking in direction of arrow

Section AA.

Fig. 2.

J. J. Thomson discourses on the physics of golf, A. E. Tutton on crystal structure analysis
by diffraction of X-rays and D'Arcy Wentworth Thompson on sunshine and fleas. Edwin
Ray Lankester examines the motion of galloping horses and dogs. New shapes of aeroplanes
appear in the skies. The Piltdown skull occasions debate. The law is exposed to
mathematical analysis. Lord Lister is eulogised.

1907-1913

Roald Amundsen reached the South Pole, and Captain Scott and his party died. Suffragettes chained themselves to railings, broke windows and fell under horses' hooves. Woodrow Wilson, president of Princeton University, became president of the United States. Dr Crippen, the murderer, had been caught by radiotelegraphy as his ship docked in New York, Louis Blériot had flown the Channel, the unsinkable *Titanic* had encountered the iceberg and sunk, and Halley's comet came and went.

The scientific mind here takes a literal view of poetry. Lord Tennyson had himself fallen foul of the remorseless quest for precision not many years before, when Charles Babbage had taken exception to his lines —

Every minute dies a man,
Every minute one is born

—on the grounds that, as everyone knew, the world's population was expanding. The poem then should go:

Every minute dies a man,
And one and a sixteenth is born.

And Babbage added that the exact figure was in actuality 1.167, "but something must, of course, be conceded to the laws of metre". Tennyson compromised with:

Every moment dies a man,
Every moment one is born.

The Rainbow.

In " Poems by Two Brothers," written by the Tennysons, and published in 1827, is a poem called " Phrenology." The following lines occur :—

" Shall we, with Glasgow's learned Watt, maintain
That yon bright bow is not produced by rain?
Or deem the theory but ill surmised,
And call it light (as Brewster) polarised? "

Can any of your readers kindly tell me (1) what view was held by James Watt about the rainbow? (2) If Brewster was the first to point out that its light is polarised? Brewster states that he observed the fact in 1812. (3) Having regard to the date (1827), what were the most probable sources of information to which the writer of the poem was indebted?
Lord Tennyson kindly informs me that the poem was probably written by Charles Tennyson.

Chas. T. Whitmell.

Invermay, Hyde Park, Leeds, June 5.

Vol 76, 174 1907

In 1907 we are still in the great era of the mad inventor.

In the Transactions of the Institution of Engineers and Shipbuilders in Scotland, Dr. Victor Cremieu describes his proposed apparatus for extinguishing the rolling of ships, some references to which have appeared in the daily Press. One method involves the use of a heavy sphere rolling in viscous liquid in a curved tube at the bottom of the ship; in the second form the moving weight takes the form of a pendulum swinging in a chamber in the form of a sector of a circle, again filled with viscous liquid. The paper contains no reference to what would happen in the event of the weight striking the boundaries of the chamber in a heavy sea or in a disaster.

Vol 77, 114 1907

Here are more poetic allusions to tricks of light.

Reflections in Water.

In the coloured photographs from Egypt, printed in *The Illustrated London News* on February 25, one picture has white clouds and blue sky with their reflection in still water. The image has the appearance of being stronger than the original. The fact is that the blue sky has much more polarised light than the clouds : the cross-polarisation by reflection at the water darkens the sky and scarcely alters the clouds. At the various incidences, by which the different points of the sky reach us, the conditions are altered. Thus the reflected scene is one of greater variety and stronger contrasts. The effect is not due to anything in the photographic process; I was surprised to see such a correct presentation of what I have sometimes observed.

No one could surpass the late Lord Tennyson in his love for noting various moods of nature, but perhaps the habit is more frequent with great masters of language in France than with us. Pierre Loti abounds with such passages as " Avec cette sonorité particulière que les cloches prennent pendant les nuits tranquilles des printemps." Rostand devotes the opening verses of " Chantecler " to the varied powers of sunlight : the sinking sun, for instance, which chooses

L'humble vitre d'une fenêtre
Pour lancer son dernier adieu.

There are two marked forms in which we often see this. Sometimes it is difficult to believe that there is not a fierce red fire raging in a distant house. At other times, with a higher and a whiter sun, a house on the hill may reflect the sun to us, surrounded with brilliant coloured halos. I suppose that dust upon the window diffracts the light, as do the dust plates of Fraunhofer.
The College, Winchester.

W. B. Croft.

Vol 86, 45 1911

The Japanese author of this letter wrote frequently in Nature *on Japanese topics, usually historical.*

Polypus Vinegar—Sea-blubber Arrack.

(1) Although I am afraid it is now much too late to reply to Mrs. Hoskyns-Abrahall's inquiry anent the so-called *Polype vinaigre* (Nature, August 9, 1906, vol. lxxiv., p. 351), to which hitherto no answer has appeared in your columns, I may be allowed to quote the following passage as a probably important clue to its scientific elucidation :—
" Amongst the greatest curiosities of the Yellow Sea there is a wonderful polypus, only recently discovered. This curious zoophyte is known on the coast of Newchwang by the name of *Chang-yu*, and possesses the property of turning into vinegar the fresh water in which it is placed. This fact was noticed for the first time in Huc's travels in China and Thibet, but our savants at home were rather sceptical on the point, and refused to believe in its existence till it was lately sent to Paris by another missionary, Mr. Pernys, and the specimens, one alive and one dead, being put in tank at the aquarium of the Société d'Acclimatisation, they both turned into vinegar the fresh water in which they were placed " (A. Fauvil, " The Province of Shantung," in the *China Review*, vol. ii., No. 6, 1875, pp. 366-7).
So far as my limited reading goes, not a single Chinese work mentions or describes this remarkable creature. But I may hazard a remark that peradventure by *polype* Huc really meant a cephalopod, for the " Pen-tsao " applies the name Chang-yü (not *yu*) to the octopus, which formed a member of the classic authors' Polypi, as is manifest in Pliny's " Natural History," bk. ix., ch. 48 (see also the " Encyc. Brit.," ninth edition, vol. xix., p. 428).
(2) In " A New Account of East India and Persia in Eight Letters, being Nine Years' Travels, begun 1672 and finished 1681," by Dr. John Fryer, F.R.S., published London, 1698, pp. 68-9, the writer, recounting the causes of the bad health of the inhabitants of Bombaim, an island situated sixty leagues south of Surat, and the same distance north of Goa, says, " Among the worst of these, Fool Rack (Brandy made of *Blubber*, or *Carvil*, by the Portugals, because it swims always in a blubber, as if there were nothing in it; but touch it, and it stings like nettles; the latter, because sailing on the Waves it bears up like a Portugal Carvil; it is, being taken, a Gelly, and distilled causes that take it to be Fools), and Foul Women may be reckoned."
It is well known that certain species of jelly-fishes are eaten with gusto by the Japanese and the Chinese, but we have never heard, except the above instance, of any acaleph capable of yielding a spirituous liquor. Will any of your readers kindly tell whether it is fiction or truth?

Kumagusu Minakata.

Tanabe, Kii, Japan, August 6.

Vol 79, 8 1908

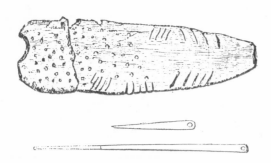

The botanist W. T. Thiselton-Dyer misjudged H. G. Wells's critical faculties.

Mulattos.

Mr. H. G. WELLS, in his interesting book " The Future in America " (1906), tells (pp. 269–270) a story at second-hand which apparently, however, he accepts as accurate in perfect good faith. I transcribe the facts as they were given to him :—

" A few years ago a young fellow came to Boston from New Orleans. Looked all right. Dark—but he explained that by an Italian grandmother. Touch of French in him too. Popular. Well, he made advances to a Boston girl—good family. Gave a fairly straight account of himself. Married."

The offspring of the marriage was a son :—

" Black as your hat. Absolutely negroid. Projecting jaw, thick lips, frizzy hair, flat nose—everything."

In this case Mr. Wells observes :—" The taint in the blood surges up so powerfully as to blacken the child at birth beyond even the habit of the pure-blooded negro."

This is, at any rate, ultra-Mendelian. Such a story would hardly be told and repeated unless it corresponded to popular belief. What one would like to have is precise evidence that such cases actually occur. If verifiable, it would be of great importance both on scientific and political grounds. I find, however, nothing resembling it in such authorities as I am able to consult. No such case is mentioned by either Darwin or Delage, though neither would have been likely to pass over such a striking instance of reversion had it been known to him. Sir William Lawrence, in his " Lectures on Physiology, Zoology, and the Natural History of Man " (1822), a book still worth consulting, has industriously collected (pp. 472–484) all the facts available at the time about mulattos, but has no instance of the kind.

The problem involved is thus stated by Galton (" Natural Inheritance," p. 13) :—" A solitary peculiarity that blended freely with the characteristics of the parent stock, would disappear in hereditary transmission." He then discusses the case of a European mating in a black population :— " If the whiteness refused to blend with the blackness, some of the offspring of the white man would be wholly white and the rest wholly black. The same event would occur in the grandchildren, mostly, but not exclusively, in the children of the white offspring, and so on in subsequent generations. Therefore, unless the white stock became wholly extinct, some undiluted specimens of it would make their appearance during an indefinite time, giving it repeated chances of holding its own in the struggle for existence." *Mutatis mutandis*, the same law would hold for a black mating in a white population.

Lawrence quotes a single case (p. 279) in which a refusal to blend certainly existed :—" A negress had twins by an Englishman : one was perfectly black, with short, woolly curled hair ; the other was light, with long hair." He also points out that in " mixed breeds " " children may be seen like their grandsires, and unlike the father and mother," a fact observed by Lucretius.

" Fit quoque, ut interdum similes existere avorum
  Possint, et referant proavorum sæpe figuras."

On the other hand, according to Lawrence, there was a legal process in the Spanish colonies of South America by which a mulatto could claim a declaration that he was,

at any rate politically, free from any taint of black blood. Of Quinterons, who were one-sixteenth black, he says :— " It is not credible that any trace of mixed origin can remain in this case," and even of Tercerons, who were one-quarter black, " in colour and habit of body they cannot be distinguished from their European progenitors." He says (p. 274) that Jamaica Quadroons " are not to be distinguished from whites." But " there is still a contamination of dark blood, although no longer visible. It is said to betray itself sometimes in a relic of the peculiar strong smell of the great-grandmother." If these statements can be relied upon, Galton's hypothetical law does not appear to apply to mulattos, and some doubt is thrown on the case cited by Wells. On the other hand, Lawrence quotes from the Philosophical Transactions (" v., 55 ") a case of two negroes who had a white child, the paternal grandfather being white. This seems purely Mendelian.

November 25.   W. T. THISELTON-DYER.

VOL 77, 126 1907

And Wells's riposte:

MAY I have a line to correct Sir William Thiselton-Dyer's impression (p. 126) that the tragic story of The Pure White Mother and the Coal-black Babe was accepted by me " as accurate and in perfect good faith "? I suppose I ought to have underlined the gentle sneer at a blackness transcending the natural blackness of a negro baby. At any rate, I told the anecdote simply to illustrate the nonsense people will talk under the influence of race mania, and I hope it will not be added too hastily to the accumulation of evidence on the Mendelian side.

H. G. WELLS.

VOL 77, 149 1907

Shelford Bidwell's invention — a sort of proto-fax machine (p.65) of 26 years before — was brought to perfection by Professor Korn of Munich. The description is Bidwell's.

PRACTICAL TELEPHOTOGRAPHY.

EARLY in 1881 I described in NATURE (vol. xxiii., p. 334) an experimental apparatus for the electrical transmission of pictures to a distance, in which use was made of one of the sensitive selenium cells devised a few months previously (*ibid.*, p. 58). Fig. 1 shows the arrangement diagrammatically. The transmitting cylinder T is mounted upon a screwed spindle, which moves it laterally through 1/64 inch at each revolution ; a selenium cell S is fixed behind the pinhole H, 1/20 inch in diameter, and is electrically connected through the spindle with the line wires L, E ; the picture to be transmitted—about two inches square—is projected upon the front surface of the cylinder by the lens *l*. The brass receiving cylinder R is of the same dimensions as T, and is similarly mounted ; F is a platinum stylus, which is pressed vertically against the metal by the flat spring G ; W is a variable resistance, and B₁, B₂ are batteries at the transmitting and receiving stations respectively. A piece of paper moistened with a solution of potassium iodide is wrapped round R, and, the pinhole H having first been brought to the brightest part of the focussed picture (thereby reducing the resistance of S to its minimum value), the resistance W is adjusted so that no current passes along the " bridge " C D, which, assuming the two batteries to be equal, will be the case when the resistance of W is the same as that of S. If now the Se cell is darkened, its resistance will be increased and a current will pass through the receiver in the direction C D, liberating iodine at the point of the stylus F.

To transmit a picture, the two cylinders are caused to rotate synchronously, at the same time moving from end to end of their traverses ; in the course of its spiral path the pinhole H covers successively every point of the focussed image, the illumination of the Se cell being proportional at any moment to the brightness of the spot occupied by the pinhole; the consequent variation in the resistance of the cell causes the stylus F to trace upon the paper a brown line which is lighter or darker in correspondence with the illumination of the Se. The close spiral line with breaks in its uniformity constitutes a picture, which should be a counterpart of that projected upon T. The earliest achievement of the apparatus consisted in the reproduction of the image of a hole cut in a piece of black paper; after some improvements simple black and white pictures painted upon glass were very perfectly transmitted, as was demonstrated upon several occasions when the apparatus was exhibited in operation.[1] It was, however, unable to cope with half-tones, and owing to pressure of work the experiments were shortly afterwards discontinued.

The problem of telegraphic photography has recently been attacked with conspicuous success by Prof. A. Korn, of Munich, whose work is described in a little book entitled "Elektrische Fernphotographie und Ähnliches" (Leipzig, 1907). His latest

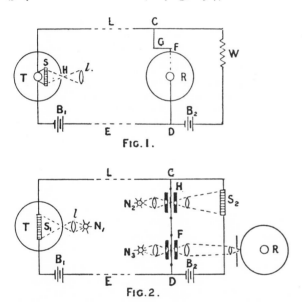

FIG. I.

FIG. 2.

method is indicated in Fig. 2. The transmitting and receiving cylinders T, R turn synchronously on screwed axes, the regulating mechanism of the receiver is situated in the bridge C D, and a suitable resistance is placed at S_2. A celluloid film negative of the picture to be transmitted is wrapped round the cylinder T, which is made of glass. The light of a Nernst lamp N_1 is concentrated by a lens upon an element of the film, through which it passes more or less freely according to the translucency of the film at the spot, to the Se cell S_1, which is fixed in position, and does not, like mine, move with the cylinder; thus the resistance of the Se is varied in correspondence with the lights and shades of the picture. The receiving cylinder R is covered with a sensitised photographic film or paper, upon a point of which light from a lamp N_3 is concentrated. Before reaching the paper the light passes through perforations in two iron plates at F, which are, in fact, the pole-pieces of a strong electromagnet; between these is a shutter of aluminium leaf, which is attached to two parallel wires or thin strips forming the

[1] Among others, at the Telegraph Engineers' *soirée* in 1881 (*see* NATURE, vol. xxiii., p. 563).

bridge C D. When there is no current through C D, the opening is covered by the shutter; when a current traverses the wires, they are depressed by electromagnetic action, carrying the shutter with them, and a quantity of light proportional to the strength of the current is admitted through the perforations. By means of this "light-relay," as it is termed, the intensity of the light acting at any moment upon the sensitised paper is made proportional to the illumination of the selenium in the transmitter.

It remains to mention a device of admirable ingenuity which has rendered it possible to transmit half-tones with fidelity. In its response to changes of illumination selenium exhibits a peculiar kind of sluggishness, to which reference was made in my old article: "Some alteration takes place almost instantaneously with a variation of the light, but for the greater part of the change an appreciable period of time is required." Prof. Korn has succeeded in eliminating the effects of the sluggish component by substituting for my box of resistance coils R a second

FIG. 3.

Se cell S_2, which is as nearly as possible similar to S_1, and which, by means of a second light-relay H, placed in series with the first, is subjected to similar changes of illumination. Thus any subpermanent fall in the resistance of S_1 due to the action of light is compensated by an equal fall in that of S_2, and only such changes as respond immediately to the varying illumination of S_1 are utilised for regulating the transmission current.

Such is in brief outline the nature of the new process. As regards the many carefully considered details which have made it a practical success, those interested will find ample information in the pamphlet mentioned above. The apparatus has been worked with excellent effect over long distances; a specimen of its performance, for which I am indebted to the kindness of Prof. Korn, is given in Fig. 3. The parallel lines traced by the point of concentrated light —in this case about 50 to the inch—are easily recognisable.

SHELFORD BIDWELL.

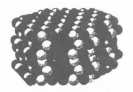

Elie Metchnikoff, the discoverer of phagocytosis, developed in later life the engagingly dotty theory that ageing is a pathological process caused by rampaging gut bacilli. These generate toxins that provoke the phagocytes into eating up the tissues. The proposed cure is set out in his book, reviewed here with restrained scepticism.

LIFE AND DEATH.

The Prolongation of Life. By Élie Metchnikoff. The English translation edited by P. Chalmers Mitchell. Pp. xx + 343. (London : W. Heinemann, 1907.) Price 12s. 6d. net.

* * *

Prof. Metchnikoff is of opinion that when old age approaches, the phagocytes, which have hitherto been man's friends, become his enemies, and hasten death by devouring the essential cells of the vital organs of the body, especially those of the nervous system. These cells are rendered particularly vulnerable to phagocytes by the action of poisons manufactured by the bacteria of the large intestine, and Prof. Metchnikoff suggests that this might to a large extent be prevented by taking skimmed milk which has been boiled and rapidly cooled, and on which pure cultures of the Bulgarian bacillus have been sown. This produces a pleasant, sour, curdled milk containing about 10 grams of lactic acid per litre, the lactic acid of which prevents intestinal putrefaction.

The author is dependent mainly upon two kinds of evidence, experimental and numerical, and therefore his difficulties are chiefly two. Many experiments which might bear upon the prolongation of life must necessarily be observed for many years. For example, he devotes much space to the uselessness of the large intestine; so far as his facts go there is nothing to be said against them—indeed, from them and others we are probably justified in thinking poorly of the large intestine—but before we can certainly know much about this numbers of human beings who have been deprived of their large intestine will have to be observed for many years.

As the question is the prolongation of life, the numerical evidence as to how long certain animals and plants live is of the greatest importance, but the author has to depend largely upon hearsay. Very few of his statements are evidence in the technical sense of the word. We are more likely to be correct in our knowledge of very old human beings than very old animals, but even with regard to human beings the evidence of extreme old age—say over 100 years—often breaks down when carefully examined. Those in doubt on this point should read T. E. Young " On Centenarians." Sometimes the age is accepted because it is on the tombstone, but, as Johnson says, " In lapidary inscriptions a man is not upon oath." Prof. Metchnikoff is inclined to accept the commonly stated age of Parr, but there is no real evidence as to his age at

death. Still, when we remember the extreme difficulty of getting suitable facts to support his views it must be admitted that the author has shown marvellous skill in the presentation of his case. No one can put down the book without feeling that it makes us think, will well repay careful critical reading, and induces gratitude to Dr. Chalmers Mitchell for his translation and excellent introduction.

Vol 77, 289 1908

This is an early shot fired in the nature-versus-nurture campaign, which continues to this day in unchanged terms, reflecting more often the political persuasions of the contestants than the factual evidence. The author of the report was clearly taken aback by her results.

The Francis Galton Eugenics Laboratory has published a lecture by its research scholar, Miss Ethel M. Elderton, entitled " The Relative Strength of Nurture and Nature," which was recently delivered in a course of lectures on national eugenics at the laboratory. By the method of correlation used by the lecturer and her colleagues, she claims to establish the fact that " overcrowding, bad economic conditions, bad physical and moral conditions of the parents, have practically no effect on the intelligence, eyesight, glands and hearing of the children." The results, indeed, show that the children of drunken parents are somewhat healthier and more intelligent than those of sober parents, and generally that the influence of environment is almost negligible compared with that of heredity. As the author admits, some of " these results are certainly startling and rather upset one's preconceived ideas."

Vol 81, 463 1909

The preoccupation with bowel movements was more a European than a British phenomenon and accounted for the rise of the spas with their aperient waters. (Baden-Baden was said to have been built on undelivered faeces.) It was, however, in Britain that the most drastic treatment for constipation was devised. Mr (later Sir William) Arbuthnot Lane, who voices his alarming opinions in this report, was the model for Cutler Walpole in George Bernard Shaw's The Doctor's Dilemma. *Lane believed that constipation was the source of most ills, including cancer, and was a consequence of our (unnatural) upright posture and of social inhibition. He discovered the efficacy of castor oil as a cure-all, but not before he had performed many heroic operations to ease the habitually costive.*

The discussion on chronic constipation turned very largely on the Röntgen-ray examination of the large bowel after the patient had taken a meal containing an insoluble salt of bismuth. This discussion was opened by Dr. J. F. Goodhart, who pointed out that constipation in old persons was frequently due to failure of voluntary effort. He said he held a brief for the importance and utility of the large bowel in opposition to those who, following the teaching of Metchnikoff, have come to regard the large bowel as a mere place of storage for the waste material of the food, in which poisons were generated which were very apt to be injurious. The large bowel, he stated, is meant to be full, not empty. Mr. Arbuthnot Lane said that in certain cases poisons were actually generated in the large bowel to such an extent that the patient's life was intolerable. In such cases he had removed a part or the whole of the large bowel with great benefit to his patient.

Vol 84, 153 1910

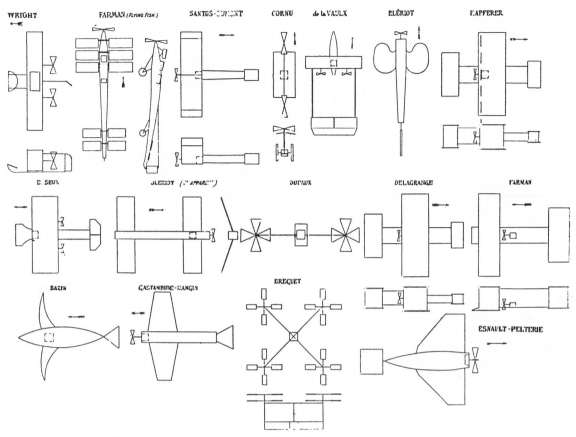

Plans of the principal Aëroplanes. From " Le Problème de l'Aviation," with slight modification.

This meeting of science with classical scholarship is more characteristic of the Victorian than the Edwardian period, during which the dilettante and the polymath yielded to the specialist.

The Gallop of the Horse and the Dog.

IN a note in NATURE of October 28 (p. 526) it is stated that Mr. Francis Ram, in a recent book, says I am in error (in an article lately published by me) in regard to the position of the legs and feet in a running dog.

I have not seen Mr. Ram's book, but I should be glad if you will print the enclosed outline figure of a running dog taken from a series of instantaneous photographs of a running dog by Mr. Edward Muybridge.

The horizontal line AB gives the actual level of the ground below the dog. The figure is one drawn for a book which I have in preparation, and I think has considerable value, since it serves to establish my suggestion that the Mycenæans (who were the originators of the pose of the galloping horse, which was never used by Greeks, Egyptians, Assyrians, Romans, or Europeans, but travelled, as Salomon Reinach has shown, across Tartary to China and Japan, and came from Japan to England at the end of the eighteenth century) did not *invent* the well-known conventional pose, but *observed* it in the dog, and very reasonably, but incorrectly, *applied* it to representations of the horse and other animals which do not really assume that pose. The pose in question satisfies the artist's judgment even when applied to the horse, because the outstretched position of the hind legs, with upturned hoofs and the forward-reaching position of the fore-legs, do *succeed* one another in the galloping horse so rapidly as to cause, not a continuity of the retinal impressions, but

a continuity of the more slowly formed mental appreciations of the positions of the legs.

It is an important fact that the late Prof. Marey, of Paris, did not succeed in photographing the dog with all the feet " off " the ground and the legs in the position shown in Muybridge's photographs, and consequently archæologists have supposed that the Mycenæans imagined the pose as an artistic expression of rapid galloping. It seems to me, on the contrary, certain that they constantly saw and admired this pose in their hunting dogs.

E. RAY LANKESTER.
29 Thurloe Place, South Kensington, October 29.

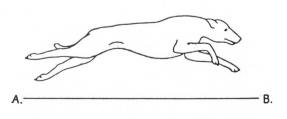

The telegraphic reproduction of pictures with high fidelity made heavy demands on Professor Korn's technology, but the results were remarkable.

THE TELEGRAPHY OF PHOTOGRAPHS, WIRELESS AND BY WIRE.

IT frequently happens that when two alternate processes are available for certain work, and one of them is considerably less practical than the other, the less practical one is possessed of much higher scientific interest. This may certainly be said of the telegraphy of pictures and photographs. The whole of the methods of transmission can be classed as either purely mechanical or dependent on the physical properties of some substance which, like selenium, is sensitive to light.

Fig. 1.—Photograph showing a Portion of the Photo-telegraphic Apparatus.

The latter methods are of no little scientific interest, and, although very delicate and for the moment obsolete, there is every likelihood of their coming into more extended use later on.

The telegraphy of pictures differs only from the transmission of ordinary messages in that the telegraphed signals, recorded by a marker on paper, must essentially occupy a fixed position. In the case of an ordinary telegram, it matters little whether the received message occupy two, three, or more lines when written out on paper, but when a picture is telegraphed every component part of it must be recorded in a definite position on the paper.

Suppose you greatly enlarge a portrait, and divide it up by ruled lines into a thousand square parts. Suppose, also, that the photograph is printed on celluloid, so that it is transparent. If, now, the portrait be held in front of an even source of illumination, it will be seen that each square—each thousandth part—is of different density. The light parts of the photograph will consist of squares of little density, the dark parts of squares of greater density, and so on. In this way the photograph is analysed into composite sections, each section corresponding precisely to a letter in a message; letters and spaces recombined form words and messages; squares of different densities recombined, in correct position, form a photograph.

I propose to deal with the more practical system first, which, as already pointed out, is perhaps the less interesting from the theoretical point of view. The telectrograph system has been employed by the *Daily Mirror* for the transmission of photographs since July, 1909, and has been worked very regularly between Paris and London and Manchester and London.

Instances of its use may be recognised in the publication of photographs taken in court in the recent Steinheil case at Paris, when photographs of witnesses or prisoners were sometimes received in London actually before the court rose at which they were taken, a clear day being gained in the time of publication.

The method of telegraphing photographs that has been employed on a large scale by the *Daily Mirror* may be called a practical modification of several early attempts. The effect of an electric current to discolour certain suitable electrolytes or to set free an element or ion that can be used to form with a second substance a coloured product was employed in many early forms of instruments for telegraphing writing, &c. If we break up a photographic image in the way already described into lines which interrupt the current for periods depending on their width, these interrupted currents can be used at the receiving station to form coloured marks, which join up *en masse* to form a new image. My telectrographic process is thus briefly as follows:—

At the sending station we have a metal drum revolving under an iridium stylus, to the drum being attached a half-tone photograph printed on lead foil. Current flows through the photographic image to the line, and thence to the receiver. The receiver consists of a similar revolving metal drum, over which a platinum stylus traces. Every time the transmitter style comes in contact with a clear part of the metal foil, current flows to the receiver, and a black or coloured dot or mark appears on the chemical paper. But you will readily understand that if our reproduction—built up of these little marks, which have to be made at the rate of some two hundred per second—is to be accurate, each mark must be only exactly as long, in proportion, as the clear metal space traversed by the stylus.

Fig. 2.—Fashion Plate Transmitted by Prof. Korn's Telautograph.

Fig. 3.—Photograph Wired from Paris to London by the Author's Telectrograph.

This picture shows some of the curious objects that darkened the skies (although sometimes not for long) in the years before World War I.

Fig. 2.—8. Blériot monoplane after accident. 9. Luyties American helicopter. 10. Bonnet Labrauche biplane. 11. Vuia helicopter. 12. Goupy triplane. 13. Curtiss's American biplane. 14. Zeus aëroplane. 15. Ferber biplane. 16. Santos Dumont's Demoiselle. 17. Gastambide Mangin biplane. 18. Farman's biplane which travelled from Bouy to Rheims (October 30, 1908). 19. Wright's machine which made the record flight with two passengers (1h. 9m. 45s.). 20. The Antoinette V., which made the record for monoplanes (1h. 7m. 35s.) and attempted to cross the Channel. 21. The Blériot XI., which crossed the Channel on July 25, 1909.

Erasmus Darwin's vision long pre-dates Lord Tennyson's airy navies, grappling in the central blue —

Writing in the *Times* of Friday last, Prof. R. Meldola says that it appears to have been overlooked that Erasmus Darwin, the grandfather of Charles Darwin, besides prophesying the introduction of steam as a motive power, foretold, in the following lines, the advent of aërial navigation :—

" Soon shall thy arm, unconquered steam, afar,
Drag the slow barge and drive the rapid car ;
Or on wide waving wings expanded bear
The flying chariot through the streams of air ;
Fair crews triumphant leaning from above
Shall wave their fluttering kerchiefs as they move ;
Or warrior bands alarm the gaping crowd,
And armies shrink beneath the shadowy cloud."

— and is amplified in this observation:

Erasmus Darwin on Flying Machines.

Prof. Meldola's reference to Erasmus Darwin's prophecy of flying machines (p. 370) omits the most remarkable proof, as it seems to me, of his insight into the future. The verses which he quotes are from Canto I., lines 289-96, of the " Botanic Garden " ; on line 254 there is a note in which occurs the following passage (the italics are mine) :—

" As the specific levity of air is too great for the support of great burthens by balloons, there seems no probable method of flying conveniently but by the power of steam, *or some other explosive material*, which another half-century may probably discover."

University College, London.   Arthur Platt.

The physics of golf evidently engrossed J. J. Thomson as much as it had P. G. Tait in the previous generation (p.80). This long paper was based on a Royal Institution lecture, with demonstration experiments. Here is how J. J. introduces his subject.

THE DYNAMICS OF A GOLF BALL.

THERE are so many dynamical problems connected with golf that a discussion of the whole of them would occupy far more time than is at my disposal this evening. I shall not attempt to deal with the many important questions which arise when we consider the impact of the club with the ball, but confine myself to the consideration of the flight of the ball after it has left the club. This problem is in any case a very interesting one; it would be even more interesting if we could accept the explanations of the behaviour of the ball given by many contributors to the very voluminous literature which has collected round the game; if these were correct, I should have to bring before you this evening a new dynamics, and announce that matter, when made up into golf balls, obeys laws of an entirely different character from those governing its action when in any other condition.

If we could send off the ball from the club, as we might from a catapult, without spin, its behaviour would be regular, but uninteresting; in the absence of wind its path would keep in a vertical plane; it would not deviate

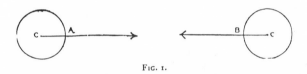

FIG. 1.

either to the right or to the left, and would fall to the ground after a comparatively short carry.

But a golf ball when it leaves the club is only in rare cases devoid of spin, and it is spin which gives the interest, variety, and vivacity to the flight of the ball. It is spin which accounts for the behaviour of a sliced or pulled ball, it is spin which makes the ball soar or "douk," or execute those wild flourishes which give the impression that the ball is endowed with an artistic temperament, and performs these eccentricities as an acrobat might throw in an extra somersault or two for the fun of the thing. This view, however, gives an entirely wrong impression of the temperament of a golf ball, which is, in reality, the most prosaic of things, knowing while in the air only one rule of conduct, which it obeys with unintelligent conscientiousness, that of always following its nose. This rule is the key to the behaviour of all balls when in the air, whether they are golf balls, base balls, cricket balls, or tennis balls. Let us, before entering into the reason for this rule, trace out some of its consequences. By the nose of the ball we mean the point on the ball furthest in front. Thus if, as in Fig. 1, C the centre of the ball is moving horizontally to the right, A will be the nose of the ball; if it is moving horizontally to the left, B will be the nose. If it is moving in an inclined direction CP, as in Fig. 2, then A will be the nose.

Now let the ball have a spin on it about a horizontal axis, and suppose the ball is travelling horizontally as in Fig. 3, and that the direction of the spin is as in the figure, then the nose A of the ball is moving upwards, and

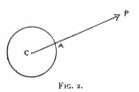

FIG. 2.

since by our rule the ball tries to follow its nose, the ball will rise and the path of the ball will be curved as in the dotted line. If the spin on the ball, still about a horizontal axis, were in the opposite direction, as in

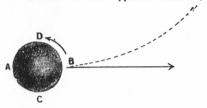

FIG. 3.

Fig. 4, then the nose A of the ball would be moving downwards, and as the ball tries to follow its nose it will duck downwards, and its path will be like the dotted line in Fig. 4.

Let us now suppose that the ball is spinning about a

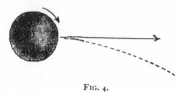

FIG. 4.

vertical axis, then if the spin is as in Fig. 5, as we look along the direction of the flight of the ball the nose is moving to the right; hence by our rule the ball will move off to the right, and its path will resemble the dotted line in Fig. 5; in fact, the ball will behave like a sliced ball.

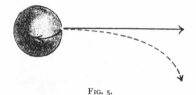

FIG. 5.

Such a ball, as a matter of fact, has spin of this kind about a vertical axis.

If the ball spins about a vertical axis in the opposite direction, as in Fig. 6, then, looking along the line of flight, the nose is moving to the left, hence the ball moves

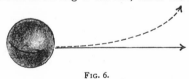

FIG. 6.

off to the left, describing the path indicated by the dotted line; this is the spin possessed by a "pulled" ball.

If the ball were spinning about an axis along the line of flight, the axis of spin would pass through the nose of the ball, and the spin would not affect the motion of the nose; the ball, following its nose, would thus move on without deviation.

Thus, if a cricket ball were spinning about an axis parallel to the line joining the wickets, it would not swerve in the air; it would, however, break in one way or the other after striking the ground; if, on the other hand, the ball were spinning about a vertical axis, it would swerve while in the air, but would not break on hitting the ground. If the ball were spinning about an axis intermediate between these directions it would both swerve and break.

Excellent examples of the effect of spin on the flight of a ball in the air are afforded in the game of base ball; an expert pitcher, by putting on the appropriate spins, can make the ball curve either to the right or to the left, upwards or downwards; for the sideway curves the spin must be about a vertical axis, for the upward or downward ones about a horizontal axis.

A lawn-tennis player avails himself of the effect of spin when he puts "top spin" on his drives, *i.e.* hits the ball on the top so as to make it spin about a horizontal axis, the nose of the ball travelling downwards, as in Fig. 4; this makes the ball fall more quickly than it otherwise would, and thus tends to prevent it going out of the court.

And on to the demonstrations, of which this is only a small sample. As the lecture proceeds, the complexity increases, and Thomson demonstrates with the aid of electrified particles of metal exposed to a magnetic field how types of spin are generated. We leave him before he gets to this firework display and while he is still blowing air over captive golf balls.

...when a golf ball is moving through the air, spinning in the direction shown in Fig. 10, the pressure

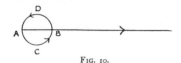

FIG. 10.

on the side ABC, where the velocity due to the spin conspires with that of translation, is greater than that on the side ADB, where the velocity due to the spin is in the opposite direction to that due to the translatory motion of the ball through the air.

I will now try to show you an experiment which proves that this is the case, and also that the difference between the pressure on the two sides of the golf ball depends upon the roughness of the ball.

In this instrument, Fig. 11, two golf balls, one smooth and the other having the ordinary bramble markings, are mounted on an axis, and can be set in rapid rotation by an electric motor. An air-blast produced by a fan comes through the pipe B, and can be directed against the balls; the instrument is provided with an arrangement by which the supports of the axis carrying the balls can be raised or lowered so as to bring either the smooth or the bramble-marked ball opposite to the blast. The pressure is measured in the following way:—LM are two tubes connected with the pressure-gauge PQ; L and M are placed so that the golf balls can just fit in between them; if the pressure of the air on the side M of the balls is greater than that of the side L, the liquid on the right-hand side Q of the pressure-gauge will be depressed; if, on the other hand, the pressure at L is greater than that at M, the left-hand side P of the gauge will be depressed.

I first show that when the golf balls are not rotating there is no difference in the pressure on the two sides when the blast is directed against the balls; you see there is no motion of the liquid in the gauge. Next I stop the blast and make the golf balls rotate; again there is no motion in the gauge. Now when the golf balls are spinning in the direction indicated in Fig. 11 I turn on the blast, the liquid falls on the side Q of the gauge, rises on the other side. Now I reverse the direction of rotation of the balls, and you see the motion of the liquid in the gauge is reversed, indicating that the high pressure has gone from one side to the other. You see that the pressure is higher on the side M, where the spin carries this side of the ball into the blast, than on L, where the spin tends to carry the ball away from the blast. If we could imagine ourselves on the golf ball, the wind would be stronger on the side M than on L, and it is on the side of the strong wind that the pressure is greatest. The case when the ball is still and the air moving from right to left is the same from the dynamical point of view as when the air is still and the ball moves from left to right; hence we see that the pressure is greatest on the side where the spin makes the velocity through the air greater than it would be without spin.

Thus, if the golf ball is moving as in Fig. 12, the spin increases the pressure on the right of the ball and diminishes the pressure on the left.

To show the difference between the smooth ball and the rough one, I bring the smooth ball opposite the blast; you

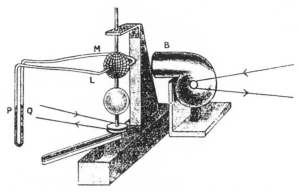

FIG. 11.

observe the difference between the levels of the liquid in the two arms of the gauge. I now move the rough ball into the place previously occupied by the smooth one, and you see that the difference of the levels is more than doubled, showing that with the same spin and speed of air blast the difference of pressure for the rough ball is more than twice that for the smooth.

We must now go on to consider why the pressure of the air on the two sides of the rotating ball should be different. The gist of the explanation was given by Newton nearly 250 years ago. Writing to Oldenburg in 1671 about the dispersion of light, he says, in the course of his letter :— "I remembered that I had often seen a tennis ball struck with an oblique racket describe such a curved line. For a circular as well as progressive motion being communicated to it by that stroke, its parts on that side where the motions conspire must press and beat the contiguous air more violently, and there excite a reluctancy and reaction of the air proportionately greater." This letter has more than a scientific interest—it shows that Newton set an excellent precedent to succeeding mathematicians and physicists by taking an interest in games. The same explanation was given by Magnus, and the mathematical theory of the effect is given by Lord Rayleigh in his paper on "The Irregular Flight of a Tennis Ball," published in the *Messenger of Mathematics*, vol. vi., p. 14, 1877. Lord Rayleigh shows that the force on the ball resulting from

FIG. 12.

this pressure difference is at right angles to the direction of motion of the ball, and also to the axis of spin, and that the magnitude of the force is proportional to the velocity of the ball multiplied by the velocity of spin, multiplied by the sine of the angle between the direction of motion of the ball and the axis of spin. The analytical investigation of the effects which a force of this type would produce on the movement of a golf ball has been discussed very fully by Prof. Tait, who also made a very interesting series of experiments on the velocities and spin of golf balls when driven from the tee, and the resistance they experience when moving through the air.

As I am afraid I cannot assume that all my hearers are expert mathematicians, I must endeavour to give a general explanation, without using symbols, of how this difference of pressure is established.

Vol 85, 251 1910

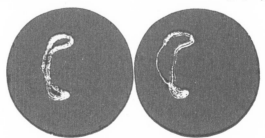

A problem agitating the mind of, among others,
F. G. Donnan, physical chemist and statesman of science,
was how to overcome the barriers of language that
separated scientists from different countries. The
necessity for a universal language, fashioned for the use of
scientists, was mooted. Was it to be Latin, Esperanto,
Volapük or, as Donnan hazarded, Ido?

Internaciona Matematikal Lexiko en Ido, Germana
Angla, Franca e Italiana. by Dr. Louis Couturat.
Pp. 36. (Jena : Gustav Fischer, 1910.) Price 1.50
marks.

* * *   The "Inter-
naciona Matematikal Lexiko," by Dr. Louis Couturat,
contains all the technical terms commonly used in
mathematics. The language of the International
Commission constitutes in many respects a great advance on its predecessors. If there is one feature that
possibly calls for improvement, it is that the new
language is not based on Latin as much as it might
be, in view of the fact that Latin is taught in schools
in every civilised country. By adopting the Latin
vocabulary free from all unnecessary grammatical
technicalities, the need of a new language could have
largely been obviated. It is true that a large proportion of the words are taken from Latin, but there are
exceptions, such as "lasta" for ultimate, "sam-
centra, sam-foka," and so forth, for concentric and
confocal, "ringo" for annulus, and "helpanta" for
auxiliary.

VOL 85, 269 1910

Studies by geneticists such as Karl Pearson on crime,
alcoholism and vice did not always produce the answers
that the moralists desired, and the traditional arguments
against iniquity and in favour of punishment were dusted
off and rolled out.

ALCOHOL AND EUGENICS.

DURING the course of the year 1910 there issued
from the Eugenics Laboratory of London
University a memoir, entitled "A First Study of the
Influence of Parental Alcoholism on the Physique and
Ability of the Offspring." The conclusion arrived at
by the authors (Prof. Karl Pearson and Miss Elderton) was, broadly speaking, that parental alcoholism
has no such influence. A result so sensational and
so opposed to the opinions of many social workers
was bound to arouse a storm of hostile criticism. It
weakened one of the arguments against the excessive
use of alcohol, and was interpreted as being a direct
encouragement of vice.

(1) Prof. Pearson divides his critics into three
classes :—(1) Paid officials and platform orators of
various temperance organisations; (2) economists
(already answered in a supplement to the original
memoir); (3) men with medical training who have
written on the subject of alcohol. It is the last class
who are dealt with in the first of the two papers now
under consideration. Their attacks—for one can
hardly apply the term criticism to much that they have
written—are repulsed with considerable losses.
It is shown that many of the errors attributed by
them to Prof. Pearson and his fellow-author may be
found in an aggravated form in the investigations
quoted as evidence rebutting their conclusions. A
sample of this evidence is itself examined and its
complete worthlessness exposed. It consists of data
obtained by Dr. MacNicholl in America, by Prof.
Laitinen in Helsingfors, by Demme in Berne, also a
curious piece of statistical work by Bezzola. The
defence and counter-attack are admirably conducted,

the writing is clear, so concise as to make
a summary impossible, and as entertaining as some of the controversial essays
of Huxley. Yet while according this high
praise to the memoir, we regret the necessity which compelled its production and
thus diverted from its proper channel of
original investigation any part of the
energies of the Eugenics Laboratory
staff.

It is with the greater satisfaction that
we turn to (2), in which the relations
between extreme alcoholism, mental
capacity, education, occupation, and religious profession are discussed. The
material on which the discussion is based
consists of the published reports of the
Langho, or Lancashire Reformatory, for
the years 1905-10, supplemented by
special information from Dr. F. A. Gill.
Particulars as to the age, number of convictions, religion, and education of 333
female inebriates were obtained this way,
and of the mental condition, physical
state, and conduct of 207 among them.

As the authors point out, results based
on numbers so small are not in any way
final; they may, however, suggest a solution of the
problems, or at any rate indicate methods by which
they can be profitably attacked. They certainly emphasise the need for the publication of good records
of individual cases.

Perhaps the most pressing of the problems referred
to is the relation of alcoholism to mental defect. The
closeness of the association between the two is shown
very clearly in the memoir. In table x. 223 female
inebriates are classified with regard to their mental
state. Of these only 37 per cent. were of normal
intelligence; 53 per cent. were defective mentally; 6
per cent. very defective; and 3 per cent. actually insane.
It is of the utmost importance therefore to determine
whether it is the intellectual deficiency which leads to
the alcoholism or the alcoholism which causes the
deficiency. Light can be thrown on this point by
measuring the correlations between education age and
mental capacity, among the alcoholists. If it is the
abuse of alcohol which causes a progressive degenera-
tion of the intellect one would expect to find a sensible
negative correlation between mental capacity and age
—mental capacity diminishing as age increases. No
such relation has been found. Allowing for differences
of education the correlation between mental capacity
and age is found to be 0·006±0·047, or quite
negligible.

Questions of great interest are also raised in the
dicussion of the relation between alcoholism and religion. Of the female inebriates in Langho Asylum
quite one-half are Roman Catholics, while of the
populations from which they are drawn not more than
one-third are of this denomination in Liverpool or
one-sixth in Manchester. These facts indicate that
the Roman Catholics in Manchester and Liverpool are
more given to alcoholic excess than the Protestants,
and it is suggested that a reason for this may be
found in a racial difference. The Roman Catholics
are largely Irish immigrants, and the Irish immigrants in the industrial towns of England are not the
most desirable specimens of their race. In this connection it is noted that "the Irish district of Liverpool
. . . is one of the few instances in which during the
last twenty years there has not been a fall in the birth-
rate." Thus if alcoholism is due to an hereditary
deficiency the differential birth-rate in Liverpool (and
Liverpool is probably not exceptional in this respect),
must lead to its propagation to a disproportionate
extent.

That prostitution is in intimate association with

alcoholism and mental defect is shown also in the tables of this paper. More than one-third of the whole number of women dealt with were prostitutes, but among these no greater proportion of mental defectives was found than among the remaining women. The Roman Catholic inmates of the asylum included a relatively smaller proportion of prostitutes than the Protestants, but this is due to the fact that the total proportion of alcoholists among the Roman Catholic community is greater, and not that the proportion of inebriate prostitutes is less.

VOL 85, 479 1911

Now the botanist W. T. Thiselton-Dyer on a topic of classical interest:

Origin of Incense.

IT is natural that incense should interest a botanist. For at least 4000 years mankind has used for this purpose the product of several species of Boswellia, natives of S.E. Arabia and Somaliland (the land of Punt). The English name Frankincense, borrowed from old French, substantially means incense *par excellence*, and represents the fact that, except amongst the Hebrews, it has been the substance exclusively employed in ritual. At last Epiphany frankincense and myrrh, in accordance with custom, were offered at the altar of the Chapel Royal, St. James's, on behalf of the King.

The use of incense might have originated in two different ways, and it is not perhaps always easy to distinguish these developments. Fumigation with fragrant or pungent herbs would easily arise as a sanitary expedient. The Greeks called this θυμίαμα, which connects with *fumus*; the plant name, thyme, derives from the same root. This, as there is evidence it did, would develop into the notion of ceremonial purification and then of consecration and honour. For such purposes it would be natural to burn frankincense on a fire-pan or censer. This was the Egyptian practice. Mr. Arthur Evans has discovered in Crete censers of Minoan age with lumps of some undetermined incense still adhering. Much of the use of incense in modern religious ceremonies has only a sanitary significance. Thus, at the coronation of George III., an official held a fire-pan on which frankincense was burnt, and this appears to have had no ritualistic meaning. It was not until the seventh century B.C. that frankincense was exported to Mediterranean countries. It doubtless carried with it is religious significance, and from this period dates the use of incense both by the Greeks and the Hebrews. That incense was of exotic origin is shown by the fact that the Hebrews called it *lebōnāh* and the Greeks λιβανωτός, names which, like the Arabic *lubān*, probably all derive from some local name at the place of production.

The sacrificial use of incense developed gradually and from a different source from the sanitary. Sacrifices were primarily offerings of food to the gods. It was a later development to burn them so as to present them in an ethereal form. Starting from the idea that the gods were to be propitiated through the sense of smell, frankincense was sprinkled on the burnt offerings to make them more fragrant. The latest refinement was to burn incense on the altar alone. The former the Greeks called λιβανωτὸν ἐπιτιθέναι, the latter λιβανωτὸν καθαγίζειν. Aristophanes in the fifth century B.C. carefully distinguishes (Clouds, 426) the three sacrificial acts: the sacrifice proper (θύος), the libation, and the addition of incense.

The use of frankincense spread to Italy, where it was used much as in Greece. The Romans called it *tus*, which is the equivalent of θύος. The substitution of the letter *r* in the oblique case, *tus, tur-is*, shows that θύος could not have found its way into Latin later than the fourth century B.C. In Greece θύος was always a sacrificial offering. Mr. Christopher Cookson, who has taken much kind trouble for me in this matter, informs me: " I can find no passage where θύος need mean ' incense ' and many where it cannot." Now, the Romans had their own word for a sacrifice, *sacrificium*. When they began to use frankincense, instead of borrowing its Greek name, they used *tus*, the latinised form of θύος, substituting the name of the whole rite for that of a mere incident in it.

The confusion so produced has existed for some 2000 years. There have been several notices in NATURE of the so-called " Incense Altar of Aphrodite " at Paphos. This is apparently based on the passage in the Odyssey (8.363), where Homer calls it βωμὸς θυήεις. But this is merely one of his common forms. He uses it of the altar of Jupiter on Mount Ida (Iliad, 8, 48), and (Il., 23, 148) of the altar of Sperchius, on which Peleus had vowed that Achilles should offer fifty rams. It is quite true that θυήεις has been translated " smelling with incense "; it really has its obvious and simple meaning of " reeking with sacrifice." Virgil was, however, misled, and paraphrases the passage in the Odyssey (Æneid, 1, 416) with his usual amplification into: " centumque Sabaeo)ture calent arae." But it is evident that this was not accepted at the time. The elder Pliny more than once discusses the question and asserts emphatically " Iliacis temporibus . . . nec ture supplicabatur " (N. H., 13, 1, 1). Whatever, therefore, may have been the development in later times, the Homeric altar of Aphrodite at Paphos could not have been an incense-altar. It is true that it has been contended that sacrifices of blood were not offered to Aphrodite. But this is not sustainable. Victims were offered to the Paphian Venus in the time of Horace.

W. T. THISELTON-DYER.

VOL 85, 507 1911

The time balls of London have sadly vanished, except for one at Greenwich.

IN connection with the subject of the synchronisation of public clocks, it is of interest to record that a time ball 4 feet in diameter has been provided on the summit of the dome of Messrs. S. H. Benson's building on the west side of Kingsway, and the ball is dropped at each hour by electric current. Unlike time balls which only work once a day, and require to be set up by hand daily before their fall, this one is wound up quite automatically by an electric motor shortly before each hour of daylight, and is released precisely at every hour by the Greenwich time signal. It was laid down as a condition by the architects that there should be no shock or jar occasioned by the fall, and this has been overcome by a system of counterbalancing, whereby the acceleration due to gravity is neutralised just before the ball reaches the bottom. The installation was designed by Mr. Hope-Jones, and carried out by the Synchronome Company, of 32–34 Clerkenwell Road, E.C.

VOL 85, 483 1911

The French electrical engineer F. Ducretet evidently invented the telephone answering machine in 1911.

H. **Lioret**, F. **Ducretet**, and E. **Roger**: A self-recording telephone. A combination of a loud-speaking telephone and phonograph is described.

VOL 86, 507 1911

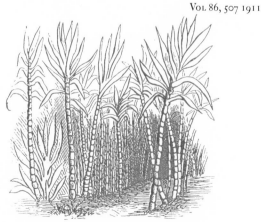

D'Arcy Wentworth Thompson, who spent most of his life as professor of biology at the University of St Andrews, was a colourful and intimidating figure. His most influential book, still read today for the sweep of its scholarship and its magisterial style, was On Growth and Form. *He was, besides, a classical scholar, and translated and edited Aristotle's* Historia Animalium, *whence (presumably) the following. Ten years earlier the Greek would have needed (or at all events had) no translation.*

Sunshine and Fleas.

ARISTOTLE (H.A. viii. 605*b*) makes the following curious and perplexing statement:—πάντα δὲ τὰ ἔντομα ἀποθνήσκει ἐλαιούμενα· τάχιστα δ', ἄν τις τὴν κεφαλὴν ἀλείψας ἐν τῷ ἡλίῳ θῇ. That is to say: " All insects die if they be smeared over with oil; and they die all the more rapidly if you smear their head with the oil and lay them out in the sun." So Pliny, Albertus Magnus, and recent commentators read and interpret the passage. But in the former half of the sentence, for ἐλαιούμενα, several MSS. read ἡλιούμενα: *i.e.* not " if they be smeared with oil," but simply " if they be exposed to the sun"; while in the latter half there is an obvious ambiguity, which inclines me to think that τὰ ἔντομα is used *sensu restricto*, and that τὴν κεφαλὴν refers, not to the insect's head, but to the experimenter's.

I take it, in short, that the heat of the sun was the main agent recommended for the destruction of the insects, and it is interesting to find this agency again coming into practical use for a very similar purpose. One of the latest of the Indian Medical Department's "Scientific Memoirs," by Capt. J. Cunningham, is entitled " On the Destruction of Fleas by Exposure to the Sun." The writer recommends the wholesale disinfection of clothing and baggage, for the special purpose of destroying plague-carrying fleas, by the simple process of laying out the garments or bedding on a sandy floor, exposed to the full rays of the sun. The author has made many careful and elaborate experiments, and has succeeded in showing that in less than an hour's time, under an Indian sun, the fleas are all dead.

D'ARCY W. THOMPSON.

VOL. 87, 77 1911

Another intelligent animal, or possibly pair of animals:

Animal Intelligence.

THE following incident may be of interest to readers of NATURE.

We have a black retriever dog, very well trained. She is kept chained in a kennel in the yard, to which a number of fowls have access. During the last few days a black hen nearly every day lays an egg in the kennel, the dog meanwhile sitting outside. Unless someone takes the egg out directly afterwards, the dog takes possession of it and eats it.

This curious proceeding raises the question whether the hen lays the egg in the kennel for the dog's benefit, and whether the dog for her own advantage allows the hen to enter the kennel without molestation.

M. N. W.

Frankland, St. Leonards, near Tring, April 19.

VOL. 89, 192 1912

Here, from an article about new domestic gadgets, is a rather modern toaster.

Fig. 3 is a " table toaster " shown open. The heating element is a ribbon of high-resistance alloy wound spirally on strips of mica. This glows with a dull-red heat when current is turned on. The bread is placed in the removable sides (one only is seen in the illustration), a piece on either side of the heater, and these are closed up so that the bread is held vertically, and parallel to the surface of the heater, with only a very short space between. As a result

FIG. 3.—Electric " Table Toaster." (Simplex Conduits, Ltd.)

the bread is toasted absolutely evenly, and with no loss of heat at all. Fig. 4 is a section of the lower part of an electric kettle, and it is seen that the heating element projects into the bottom of the kettle, and is practically surrounded by the water.

VOL. 87, 521 1911

The death of Lord Lister was marked by an extraordinary effusion of grief, veneration and affection. Here is one reminiscence from one of the many appreciations of the man.

From another of Lister's Glasgow students, and one who was his house-surgeon in the Royal Infirmary, Dr. J. Coats (now Colonel Coats), who was among the first to practise antiseptic surgery in private, an interesting letter of reminiscences has been received, from which the following is culled:—

" One day when Lister was visiting his wards in the Glasgow Royal Infirmary, there was a little girl whose elbow-joint had been excised, and this had to be dressed daily. Lister undertook this dressing himself. The little creature bore the pain without complaint, and when finished she suddenly produced from under the clothes a dilapidated doll, one leg of which had burst, allowing the sawdust to escape. She handed the doll to Lister, who gravely examined it, then, asking for a needle and thread, he sat down and stitched the rent, and then returned the dolly to its gratified owner."

VOL. 90, 504 1913

An effort to bridge the culture gap comes, unusually, from a man of letters rather than of science.

POETRY AND SCIENCE.

THE Professor of Poetry at the University of Oxford, Dr. T. Herbert Warren, President of Magdalen College, gave a public lecture on March 2 on the subject of "Poetry and Science." He began by quoting his predecessor Matthew Arnold, who wrote on New Year's Day, 1882 : " If I live to be eighty, I shall probably be the only person left in England who reads anything but newspapers and scientific publications."

Has Matthew Arnold's gloomy prophecy been fulfilled? Have newspapers and science killed real literature? In particular, are the interests of science hostile to the interests of literature?

Where science has dominated, has poetry languished? This is a very burning question, for science has certainly made great advances. It impresses the man in the street, chiefly by its usefulness. It is the poet and the poetic person who are impressed by the marvel, the magic, and the mystery of science. Matthew Arnold inherited the tradition of Wordsworth, who was a great poet of Nature, but not a poet of Natural Science. He strove hard to do justice to it, both in his prose prefaces and in his poetry, but with imperfect success. Wordsworth's poem " The Poet's Epitaph " contains a most beautiful and memorable description of the poet, but is scarcely fair to the man of science, who is generally a man also of natural affections. The man of science may be as fond of his mother as the poet, who is often one of the most selfish of beings, and if he would not " botanise upon his mother's grave " because he knows no botany might be quite capable of turning her into copy.

Further, the poet is not "contented to enjoy the things that others understand." He must synthesise in his own way. Wordsworth himself was for ever philosophising and moralising.

Keats, again, is often cited as complaining that Newton had destroyed the beauty of the rainbow by reducing it to prismatic colours, but Keats was perhaps not serious in this charge.

Goethe, on the other hand, did not object to Newton for reducing the rainbow to prismatic colours, but only for doing so wrongly.

Matthew Arnold " poked fun " at science as he did at religion, and was even less willing to treat it seriously than religion. He was often exceedingly amusing, and his famous description of a scientific education in " Friendship's Garland " was highly so.

Darwin, who began by being a great lover of poetry, thought that in later days he had lost the power through atrophy, but in point of fact the atrophy was by no means complete. He remained a most poetical writer. The closing paragraphs of the " Origin of Species " were worthy of Lucretius, which they strongly resembled.

History shows that poetry, philosophy, and science had all begun life together as children of one family. The early Greek poets, like the authors of the Books of Genesis and Job, dealt with the origin of things and the Story of Creation. The early thinkers who succeeded them expressed their thoughts in verse, and were often highly poetical. What could be more poetical than the " dark " science of Heraclitus? The same relation was maintained through Greek literature. The greatest astronomer of antiquity, the inventor of the Ptolemaic system, was the author of a beautiful epigram which was truly poetic. From Greece and Alexandria, science and poetry passed together to Rome, and might be found combined in Lucretius and Virgil. The greatest singers of antiquity were the most alive to science. Modern literature shows the same phenomenon in Dante and in Milton and in Tennyson. This is specially well brought out in a book by a living man of science, Sir Norman Lockyer's " Tennyson as a Student of Nature." On the last of the three poets Sir Oliver Lodge has also written briefly, but with rare force, in the recent volume " Tennyson and his Friends."

As time has gone on, the scientific spirit has increasingly made itself felt in poetry, and may be seen in the works of F. W. H. Myers and his brother, in the late Duke of Argyll, in George Romanes, in Richard Watson Dixon, and still better in his friend and editor, Mr. Robert Bridges. And others of the earlier poets had also been acquainted with science, notably Gray and Shelley.

With regard to the greatest of all, if Bacon wrote Shakespeare it is odd that Bacon's science does not appear more often in the plays, but in any case it may be remembered that Bacon wrote poetry of his own and had a place in the " Golden Treasury."

Other lands and literatures too have had their scientific poets, the most famous being Goethe, of whom the best account is to be found in the popular lectures of a most poetical man of science, Helmholtz. I can speak at length only of one, the French poet of the last century, Sully Prudhomme, who combined science, philosophy, and poetry. The best account of him is to be found in the study by M. Zyromski. " Poetry," said Sully Prudhomme, " is not only the lyrical outburst of our sentiments. The great poetry has noble destinies, and will sing the conquests of science and the synthesis of thought."

The average man does not care for " great poetry," or only for that part of it which appeals directly to his own feelings. Just now, what Sully Prudhomme calls *lyrisme,* that is, personal poetry, holds the field, but that has not always been so, and will not always be so. Science has not destroyed poetry. Cambridge, the University of Science, has been the University of Poetry, and with the revival of Science at Oxford in the last century, beginning in Shelley's time, poetry revived too. The really great poet must respond to the main and moving interests and influences of his day. The old facts and factors, the old *motifs,* do not change. Rebekah at the Well, David's lament over Saul and Jonathan, Hector and Andromache, Catullus at his brother's grave, still move us. But while these remain, our outlook on the world does gradually change, as Sully Prudhomme foretold in his fine sonnet to " The Poets of the Future." Science will certainly go on, and scholarship and poetry will go on at its side and beneath its ægis. The " scientific use of the imagination " on which Tyndall, that most poetic man of science, discoursed so finely forty years ago will be balanced more and more by the imaginative use of science.

The famous epigram by Ptolemy, the author of the Ptolemaic system, with the Professor's version of it, may conclude the address :—

ΠΤΟΛΕΜΑΙΟΥ.

Οἶδ' ὅτι θνατὸς ἐγὼ καὶ ἐφάμερος· ἀλλ' ὅταν ἄστρων
῾μαστεύω πυκινὰς ἀμφιδρόμους ἕλικας
οὐκέτ' ἐπιψαύω γαίης ποσίν, ἀλλὰ παρ' αὐτῷ
Ζανὶ θεοτρεφέος πίμπλαμαι ἀμβροσίης.

I know that I am mortal, and doomed to fleeting days,
  But when I track the circling stars in myriad-orbèd maze,
I tread the earth no more, but sit beside the Lord of
    Heaven,
  And taste the ambrosial food whereby the life of Gods is
    given.

*The most sensational catastrophe of the decade was the loss of
the* Titanic. Nature *reported in great detail the proceedings
of the various investigations into the cause. There was a
feeling abroad that hubris about the unsinkable liner played
some part in the disaster.*

The " Titanic."

Your article (Nature, August 29, 1912) on the
report of the Advisory Committee having emphasised
the contention from the first of some of us (students
of science and old naval commanders) as to the insanity of high speed " at night in the known vicinity
of ice," it behoves surely men of science to ask the
question whether we have not reached the imperative
limits of that false security which the "practical
man" is wont to feel in his contempt for scientific
"theory"; and, further, whether the time has not
therefore come for legislation requiring commanders
of the largest ocean-going steamers to hold a diploma,
guaranteeing such a systematic course of study (say in
a class at Greenwich or Kensington) in marine physio-
graphy and the elementary laws of mechanics as
would quicken their imagination as to the uncertainty and the magnitude of the risks to be run in
an abnormally ice-drifted sea. Lord Mersey's report
may whitewash the facts, but the facts *en évidence*
remain; and the chain of cause and effect in the
lamentable and tragic loss of the *Titanic* leads us in
the last resort to the notorious contempt for scientific
acquaintance with the facts and laws of nature on the
part of the "practical man."   A. Irving.

Hockerill, Bishop's Stortford, September 2.

Vol 90, 38 1912

*This was to be one of the last generous references to a
German for some years.*

It is with great regret that we record the disappearance of Dr. Rudolph Diesel from the G.E.R.
steamer *Dresden* on her voyage from Antwerp to
Harwich on the night of September 29; the circumstances are such as to leave no hope of his being
alive. Dr. Diesel will be remembered as the inventor
of the oil engine which bears his name. Born in
Paris in 1858, of German parentage, his training included courses at the Augsburg technical schools and
at the Munich Technical College. His first published
description of the Diesel engine appeared in 1893;
aided financially by Messrs. Krupp and others, the
next few years were spent in arduous efforts to realise
the principle of his engine in a commercially successful machine. The difficulties to be overcome were
very great. In the earliest attempt, compression of
the air was effected in the motor cylinder and the
fuel injected direct. This engine exploded with its
first charge and nearly killed the inventor. The
modern Diesel engine compresses the air in the motor
cylinder to a pressure above 400 lb. per square inch,
during which process the air becomes hot enough to
ignite the fuel. At the end of compression, the fuel is
injected by means of a separate air supply at a pressure higher than that in the cylinder. Nothing of
the nature of an explosion occurs in the cylinder; the
oil burns as it is injected, and, as the piston is moving
outwards at the same time, the pressure does not
rise to any extent. The fuel consumption of these
engines is remarkable, being roughly one-half of any
other type of oil motor. Engines both of a two-

stroke cycle and of a four-stroke cycle are now being
developed by many firms both on the Continent and
in Britain. In Dr. Diesel's opinion the two-stroke
engine would probably be the standard type for marine
purposes. Marine Diesel engines of very large power
have not yet been constructed, but many important
experiments in this direction are being made. Dr.
Diesel's loss will be regretted by men of science on
account of his efforts to interpret practically the Carnot
ideal cycle, and by engineers on account of the immense strides which his untiring energy and indomit-
able pluck have made possible.

Vol 92, 173 1913

*The study of molecular structure began with the Braggs'
realisation that the wavelength of X-rays would allow
diffraction from crystal lattices. This article by the
crystallographer A. E. Tutton shows the astonishing
imagination and percipience of the early practitioners.*

THE CRYSTAL SPACE-LATTICE REVEALED BY RÖNTGEN RAYS.

During a visit to Munich at the beginning of
August last the writer was deeply interested
in some extraordinary photographs which were
shown to him by Prof. von Groth, the *doyen* of
the crystallographic world, and professor of
mineralogy at the university of that city. They
had been obtained by Dr. M. Laue, assisted in the
experiments by Herren W. Friedrich and P. Knip-
ping, in the laboratory of Prof. A. Sommerfeld in
Munich, by passing a narrow cylindrical beam of
Röntgen rays through a crystal of zinc blende,
the cubic form of naturally occurring sulphide of
zinc, and receiving the transmitted rays upon a
photographic plate. They consisted of black
spots arranged in a geometrical pattern, in which
a square predominated, exactly in accordance with
the holohedral cubic symmetry of the space-
lattice attributed by crystallographers to zinc
blende.

Prof. von Groth expressed the opinion, in agreement with Herr Laue, that owing to the exceedingly short wave length of the Röntgen rays
(assuming them to be of electromagnetic wave
character), they had been able to penetrate the
crystal structure and to form an interference
(diffraction) photograph of the Bravais space-
lattice. This latter is the structural foundation of
the more complicated regular point-system according to which the crystal is homogeneously built
up, and the points of which (the point-system)
represent the chemical elementary atoms. The
space-lattice, in fact, was conceived to play the
same function with the short-wave Röntgen rays
that the diffraction grating does to the longer
electromagnetic waves of light.

The details of this work were laid before the
Bavarian Academy of Sciences at Munich in two
memoirs, on June 8 and July 6 last, and the two
memoirs are now duly published in the *Sitzungs-
berichte* of the Academy.[1] Besides a diagram of
the apparatus, which is reproduced in Fig. 1,
they are illustrated by reproductions of a dozen

of these photographs, one of which is also reproduced in Fig. 2. There can be no doubt that they are of supreme interest, and that they do in reality afford a visual proof of the modern theory of crystal structure built up by the combined labours of Bravais, Sohncke, Schönflies, von Fedorow, and Barlow. Moreover, they emphasise in a remarkable manner the importance of the space-lattice, so strongly insisted on from theoretical considerations by Bravais, Lord Kelvin, and von Groth, and from experimental considerations by Miers and the writer.

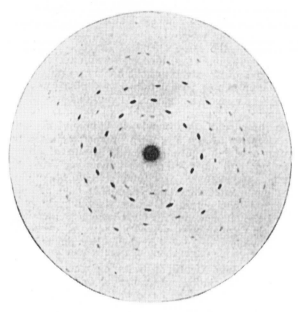

FIG. 2.—Photographic effect of passage of Röntgen rays through zinc blende.

* * *

The tetragonal nature of the axis of symmetry along which the Röntgen rays were travelling through the crystal is most strikingly apparent in the photograph. One recognises at once also the presence of two perpendicular planes of symmetry in the arrangement of the spots. In fact, *the figure corresponds to the holohedral or full symmetry (class 32) of the cubic system,* in spite of the fact that zinc blende belongs to the hexakistetrahedral class 31 (one of the so-called hemihedral classes) of cubic symmetry. Now this interesting fact affords the most beautiful and perfect proof that it is the space-lattice (Raumgitter) of the crystal structure which is affording the figure, and that no other property than this space-lattice is concerned. For space-lattices alone always possess holohedral symmetry, and they determine the crystal system and angles and obedience with the law of rational indices.

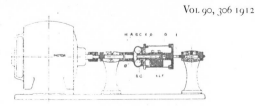

Francis Galton notwithstanding, this may be the only attempt ever to apply mathematical rigour to that most unscientific of all human activities, the law.

An Application of Mathematics to Law.

I HAVE attempted to apply mathematical symbolism to some of the difficult problems of patent law. The question to be decided by the Court in a patent law suit is usually this: assuming that the alleged invention deals with "a manner of manufacture" (*i.e.* is, or yields, something concrete), was there ingenuity and utility in the step from what was already known? Ingenuity means inventive or creative ingenuity as apart from the normal dexterity of the craftsman, which of itself is insufficient to support a patent, as otherwise patents would unduly hamper industry. It will be seen at once that it is a most subtle question for any court to determine whether a given act, the selection of one out of many alternatives, the assemblage of various old elements, the adaptation of old elements to new uses—whether such an act is one which calls for ingenuity as apart from the expected skill of the craftsman.

To express the problem symbolically I will start from an admirable dictum of Lord Justice Fletcher Moulton (Hickton Pat. Syn. *v.* Patents Improvements). He stated that invention might reside in the idea, or in the way of carrying it out, or in both; but if there was invention in the idea plus the way of carrying it out, then there was good subject-matter for a patent. I express this by representing any idea as a functional operator, and the way of carrying it out (*i.e.* the concrete materials adopted) as a variable. Calling result I:

$$I = f(x).$$

Here I represents what the Germans call the "technical effect" of the invention, or what Frost calls the manufacturing "art," and we see at once that a patent cannot be obtained for a mere principle or idea (f, which is not concrete) unless some way of carrying it out (x) is also given. But the invention may reside either in f or in x.

Let us express in general terms a manufacture (M) which is not an invention. We will use f to represent a known operator or idea, ϕ to represent a new operator or idea. a, b ... will represent known variables, ways of carrying out an invention (*e.g.*, valves, chemical substances, &c.), and x, y, new variables.

It is obvious that $f(a)$ is not an invention, nor will it normally be an invention to add $f(b)$ to it. Moreover, the craftsman is not to be tied down to this. He is at perfect liberty, within limits, to make variations in his variables, to alter the size of a crank, to substitute one alkali for another, and so on; in other words, he can take $f(a + \delta a)$.

Generalising, we may say:

$$M = \Sigma f(a + \delta a).$$

Developing this by Taylor's theorem, and proceeding from an infinitesimal to a finite change, we have, neglecting quantities of the second order:

$$M = \Sigma f(a) + \Sigma \delta f(a).$$

This is the general equation for a manufacture which is not an invention. To be an invention, ingenuity (i) must be involved.

$$I = M + i \text{ or } I = \psi(M),$$

thus:

$$I = \psi[\Sigma f(a) + \Sigma \delta f(a)] = \Sigma f(a) + \Sigma \delta f(a) + i.$$

I will now consider in various actual examples the nature of ψ, the inventive function, and of i, the inventive increment.

One of the commonest cases in which a decision is necessary is that of a combination. Suppose that

$f(a)$ and $f(b)$ are old; will there be invention in combining a and b?

The answer is this :

(1) $I = f(a, b)$
(2) $M = f(a) + f(b)$
(3) $I = M + i$.

If the result of the combination is given by (1), there is an invention; this is termed a "combination." If the result is given by (2), there is no invention; this is termed an "aggregation." It is interesting to compare this definition with one given by Lord Justice Buckley (Brit. United Shoe Mach. Co. v. Fussell) of a "combination" as "a collocation of inter-communicating parts, with a view to obtaining a simple result."

An example of a true "combination" is found in the case of Cannington v. Nuttall, in which a patent was upheld for a glass furnace, although each and every part (a, b, c) had been employed before in glass furnaces (employment $= f$). But, owing to the combination, and the co-operation of the parts, a new result was obtained.

$$I = f(a, b, c) = f(a) + f(b) + f(c) + i.$$

On the other hand Bridge's case is an example of an aggregation; in fact, a patent was refused by the Law Officer, showing that the case was considered absolutely devoid of invention. The alleged invention consisted in the employment in a shutter for dividing-up rooms (f) of means (a) to guide the shutters along the floor, and cogs (b) to hold the shutters against the wall. $f(a)$ and $f(b)$ were both old, and no new result flowed from their juxtaposition. Hence $M = f(a) + f(b)$: there was no invention; each part simply played its own rôle, and there was no inter-action.

Another type of invention is that of varying proportions in a known combination. Here, if $M = f(a, b, c)$, and if there is a maximum at one value or range of values of c, invention may be involved. The maximum relates to the technical effect, and may be with respect to efficiency, economy, &c.

Thus if $\dfrac{\partial f(a, b, c_1)}{\partial c} = 0$ at the value c_1, the function will be a maximum or a minimum and there may be an invention. This will not be the case if $\dfrac{\partial f(a, b, c_1)}{\partial c} \neq 0$.

Other singular points may be inventions, *e.g.* where $\dfrac{\partial f(a, b, c_1)}{\partial c} = \infty$ (discontinuity), or where $\dfrac{\partial^2 f(a, b, c_1)}{\partial c^2} = \infty$ (kink in the curve). This also holds for a range of values from c_1 to c_2.

Examples of the application of this equation are to be found in the cases of Edison v. Woodhouse, and Jandus Arc Lamp Co. v. Arc Lamp Co. In Edison's case f represented the employment in an incandescent lamp of an exhausted glass vessel (a), leading-in wires (b), and a carbon filament (c_1). $f(a, b, c)$ was known, but it had never been proposed to use a very thin carbon conductor or "filament." Here, owing to the high resistance and flexibility of the filament, the efficiency was a maximum :—

$$\frac{\partial f(a, b, c_1)}{\partial c} = 0,$$ and the choice of this value c_1, which

made the difference between failure and success, was held to be an invention.

In the Jandus Arc Lamp case, f represented the employment in an arc lamp of carbons (a), a tightly fitting sleeve (b), and an envelope of glass, &c. (c), inside the outer globe. By making the glass envelope 3 in. in diameter a maximum efficiency was obtained, and on this ground the patent was upheld, although envelopes had previously been made 9 in. in diameter.

Here again :

$$\frac{\partial f(a, b, c_1)}{\partial c} = 0 \text{ when } c_1 = 3 \text{ in.}$$

A further example is an old case (Muntz v. Foster) in which a sheathing for ships was made of sixty parts of copper (a) and forty of zinc (b).

Alloys of copper and zinc had been used before in about the same proportions, but in this case the same result would not have been attained, because Muntz specified the best selected copper and highly purified zinc. The impurities (δx) were of great and unsuspected importance. Moreover, other alloys of copper and zinc (probably even of purified metals) had been made. We may consider the two points separately.

(1) Impurities :—
$f(a + \delta x, b + \delta x)$ was old, where δx represents impurities. Muntz's alloy was $f(a, b) = f(a + \delta x, b + \delta x) + i$, hence there was an invention.

(2) Selection of 60 : 40 percentage :—
$\dfrac{\partial f(a_{60}, b_{40})}{\partial a} = 0$ since at this percentage the efficiency

was a maximum, because the alloy oxidised just fast enough to prevent barnacles adhering to the ship, but not fast enough to waste away excessively.

On the contrary, the case of Savage v. Harris was one in which there was held to be no invention in changing the size of part of a device for retaining ladies' hats in place. There was a back portion (a) and teeth (b), and the size of the back was altered :—
$\dfrac{\partial f(a_1, b)}{\partial a} \neq 0$, and there was no invention.

A known device or material (a) may be employed for a new purpose (ϕ). If $f(a)$ is the old use, and $\phi(a)$ the new use, we have for an invention $I = \phi(a) = f(a) + i$. But if $M = \phi(a) = f(a)$, there is no invention. The oft-quoted case of Harwood v. Great Northern Railway Company was one of the latter type. Fishplates (a) had been used for connecting (f) logs of timber, and it was held there was no invention in applying them (ϕ) to rails in which they acted in the same manner :—

$$\phi(a) = f(a).$$

But in Penn v. Bibby, wood (a) was employed (ϕ) for the bearings of propellers in order to allow the water to pass round the friction surfaces. Wood had previously been employed (f) in water-wheels, but $\phi(a) = f(a) + i$, and it was held that there was invention.

A similar type of invention is that in which different materials are employed in the same process. Here $f(a)$ is old, and $f(x)$ is new. If $f(x) = f(a)$ there is no invention. If $f(x) = f(a) + i$ there is invention. In the recent case, Osram Lamp Works v. Z Lamp Works, a patent was upheld for the use (f) in incandescent filament lamps of tungsten (x), though osmium (a) was known. Tungsten was more efficient and cheaper :—
$f(x) = f(a) + \delta i$, where δi represents a small degree of invention. This in itself might not have been sufficient, but it was coupled with the fact that one particular process of removing the carbon from the filaments was selected out of three known processes. This may be considered to require an amount of ingenuity Δi. $\delta i + \Delta i = i$, and therefore $f(x) = f(a) + i$, and there is invention involved.

Another type is the omission of one step in a known process. In the case of Badische Anilin- und Soda-Fabrik v. Soc. Chim. des Usines du Rhône, it was held that there was subject-matter in such an omission. A process had been proposed for preparing dyes called anisolines (A) from rhodamines (r) by first forming a potassium salt (1st step $= f$), and then transforming this salt into anisoline (2nd step $= F$). Thus the known process was :—

$$A = F[f(r)].$$

Now it was shown that the potassium salt did not exist, *i.e.* $f(r)$ was imaginary; the patent in question obtained anisoline direct from rhodamine, $A = f(r)$, and this was held to be an invention.

I may note two final points. When a patent is granted, the criterion of ingenuity is not applied, as this is left for the Court to determine. However, if there is absolutely no ingenuity possible, the Law Officer may refuse to grant a patent. His criterion of rejection is, therefore, not $f(x) = f(a)$, as in the Court, but $f(x) \equiv f(a)$.

A patent is invalid for "insufficiency of description" if it casts on the public the burden of experiment beyond a certain point. This may be expressed by saying that in this case the equation $I = \phi(a)$ is indeterminate.   HAROLD E. POTTS.

University Club, Liverpool, April 2.

Vol 91, 187 1913

This received a dismissive rejoinder from a future Chancellor of the Exchequer. Stafford Cripps had begun his career as a lawyer with a science background, a specialist in the demolition of technical expert witnesses. Born into a highly political family, he entered politics relatively late in life and became in due course a wartime ambassador to the Soviet Union and a minister in the first postwar Labour government. He was an ascetic and widely seen as a killjoy. "Wherever Stafford has tried to increase the sum total of human happiness," one of his colleagues observed, "grass has never grown again."

An Application of Mathematics to Law.

I HAVE read Mr. Potts's letter in NATURE of April 24, but am at a loss to understand the use to which he would put his equations.

If it be his object to find some equation giving the validity of a patent or foretelling in any way the probability of its being upheld in a court of law, he has clearly failed to do anything of the sort.

If his equation $I = M + i$ is to be of any value, the quantity i must have a fixed value greater than zero. In fact, however, for any given patent, i may have an infinite number of values, including zero, since each person will have his own idea of the amount of ingenuity that must be shown in the particular case by the inventor. Thus the inventor will certainly put a high positive value upon i, while his opponent will as certainly say that the value of i is zero. It is clear that the value of i can only be finally settled when the validity of the patent has been settled by the House of Lords, and at this stage of a patent's career it is scarcely necessary to have an equation to test its validity. So far as the rest of his letter goes, he seems to have chosen a rather complex method of setting out a few of the chief principles of patent law.   R. STAFFORD CRIPPS.

Fulmer, Slough.

Vol 91, 270 1913

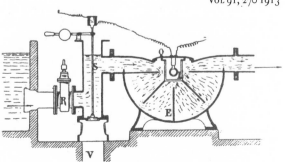

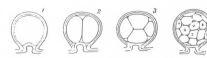

The Piltdown skull, or rather its ill-assorted fragments, was found by a respected amateur archaeologist, Charles Dawson. The human cranium on top of ape-like jaws became a topic of eager debate. The creature was named Eoanthropus, *or "man of the dawn", and was not exposed as a forgery until more than 40 years had passed. Earlier suspicions were dismissed with hauteur by the anthropological mandarins Arthur Smith Woodward, Arthur Keith and Grafton Elliot Smith. An authoritative analysis of the affair by J.S. Weiner concluded that the hoaxer was Dawson — playing perhaps the part of the amateur who had been snubbed or patronised by the professionals; but Stephen Jay Gould has explained why he believes that Teilhard de Chardin, whom Dawson had befriended, at least connived at the deception.*

THE PILTDOWN SKULL.

IN his evening lecture to the British Association at Birmingham on September 16, Dr. Smith Woodward took the opportunity of replying to Prof. Arthur Keith's recent criticisms on his reconstruction of the Piltdown skull. It will be remembered that Dr. Woodward regarded the mandible as essentially that of an ape, and restored it with ape-like front teeth, while he determined the brain-capacity of the skull to approach closely the lowest human limit. Prof. Keith, on the other hand, modified the curves of the mandible to accommodate typically human teeth, and reconstructed the skull with a brain-capacity exceeding that of the average civilised European.

Fortunately, Mr. Charles Dawson has continued his diggings at Piltdown this summer with some success, and on August 30, Father P. Teilhard, who was working with him, picked up the canine tooth which obviously belongs to the half of the mandible originally discovered. This tooth corresponds exactly in shape with the lower canine of an ape, and its worn face shows that it worked upon the upper canine in the true ape fashion. It only differs from the canine of Dr. Woodward's published restoration in being slightly smaller, more pointed, and a little more upright in the mouth. Hence, there seems now to be definite proof that the front teeth of Eoanthropus resembled those of an ape, and its recognition as a genus distinct from Homo is apparently justified.

The association of such a mandible with a skull of large brain-capacity is considered by Dr. Woodward most improbable, and he has made further studies of the brain-case with the help of Mr. W. P. Pycraft, who has attempted a careful reconstruction of the missing base. Dr. Woodward now concludes that the only alteration necessary in his original model is a very slight widening of the back of the parietal region to remedy a defect which was pointed out to him by Prof. Elliot Smith when he first studied the brain-cast. The capacity of the brain-case thus remains much the same as he originally stated, and he maintains that Prof. Keith has arrived at a different result by failing to recognise the mark of the superior longitudinal sinus on the frontal region and by unduly widening that on the parietal region.

It is understood that Mr. Dawson and Dr. Woodward will offer an account of the season's work to the Geological Society at an early meeting, and Prof. Elliot Smith will include a detailed study of the brain-cast of Eoanthropus in a memoir on primitive human brains which he is preparing for the Royal Society.

Vol 92, 110 1913

Henri Fabre was one of the most entertaining writers on natural history. The connection with Charles Darwin, in the face of Fabre's obdurate opposition to evolution, makes a pleasant footnote to history. Darwin referred to Fabre as "that inimitable observer".

THE October number of *The Fortnightly Review* contains an article by Henri Fabre, the veteran naturalist of Sérignan, on his relations with Charles Darwin. The article illustrates in an interesting way some of the leading characteristics of these two remarkable men—the combined fertility and caution in speculation shown by Darwin, with his determination to bring every hypothesis to the test of experiment; and the unrivalled powers of observation possessed by Fabre, his enthusiasm in the pursuit of his favourite study, and the charm of his literary style. Darwin, being interested in the homing instincts of the mason-bees, suggested to Fabre the making of experiments to determine, if possible, whether this instinct was at all dependent on a perception by the insects of the direction in which they were first carried away from their nests. A whole series of trials was carried out by Fabre, the essential feature in which was the enclosure of marked bees in a dark box, the carrying of the box with its inmates in a direction opposite to that from which the release was to take place, and the repeated rotation of the box at different points of the route, in order to ensure that the captive bees should lose their bearings during the journey. The experiment was repeated, with variations, many times over, the almost uniform result being that from 30 to 40 per cent. of the liberated bees found their way home without difficulty. This was contrary to the expectation of both inquirers, and Darwin next proposed to try the effect of placing the insects within an induction coil, "a curious notion," as Fabre observes. The experiment was performed, with amusing results. But in the end the experimenter was fain to confess that the homing instinct of his bees remained a mystery.

VOL 92, 237 1913

Mr Sleeper's supposed publication bears the marks of a hoax. Was the question ever resolved?

A REMARKABLE ANTICIPATION OF DARWIN.

THE presidential address to the Linnean Society of London, delivered last May by Prof. E. B. Poulton, F.R.S., and recently published in separate form, deals with a truly astonishing work by G. W. Sleeper, printed, apparently, in Boston, U.S.A., in the year 1849, and containing an anticipation of modern views on evolution and the causes and transmission of disease, which, considering all the circumstances, is extraordinary.

The work, which is a small pamphlet of some thirty-six pages, was sent by an American gentleman, Mr. R. B. Miller, to the late Dr. Alfred Russel Wallace, who forwarded it to Prof. Poulton with an interesting letter quoted in the latter's address. Dr. Wallace justly observed that the author's "anticipation of diverging lines of descent from a common ancestor, and of the transmission of disease germs by means of insects, are perfectly clear and very striking."

It is well known that the idea of the derivation of species by descent, and even of the operation of natural selection, had occurred to other thinkers before Darwin. The passage cited by Darwin himself from the "Physicæ Auscultationes" of Aristotle shows, though its import has often been misunderstood, that the Greek philosopher had before his mind the doctrine of natural selection. The medieval schoolmen were by no means wedded to the theory of special creation, and in the eighteenth and early nineteenth centuries the transformist view was freely canvassed, without, however, making much way among scientific thinkers. The "Historical Introduction" prefixed to the later editions of the "Origin of Species" gives an account of several anticipations, more or less exact, of the Darwinian theory.

But the present treatise goes far beyond most, if not all, previous attempts at solving the problem of evolution. The clear grasp shown by the author of the Darwinian principles of the struggle for life, and origin of fresh species by the preservation of those forms best adapted for their environment, his advocacy of the persistence of germinal characters, and the very terminology that he uses, might well suggest a doubt as to whether the pamphlet is really what it professes to be, or whether it is not, in fact, a cleverly devised fabrication with a falsified date. We find, for example, such expressions as the following :—"Life owes its faint beginning to primal germs . . . pervading the entire terrestrial atmosphere; and, perhaps, the entity of the Cosmos"; "everywhere about us we see waged the pitiless battle for life . . . the useless perish, the useful live and improve"; "Man and the Ape are co-descended from some primary type"; "The life germ resident in Man transmitted to his descendants goes on existing indefinitely." Here are anticipations, not only of Darwin, but also of Arrhenius, Galton and Weismann. Not less surprising are his enunciation of the germ-theory of disease, his experiments on the cultivation of streptococci from a sore throat, with the use as a germ-filter of cotton wool sterilised by heat, his suggestion of the action of phagocytes, and his recommendation of metal gauze protective frames for doors and windows in order to ward off infection carried by insects.

The question of the genuineness and authenticity of the pamphlet is carefully discussed by Prof. Poulton. The evidence on the point is perhaps not absolutely conclusive; but it may fairly be said that after weighing the interesting information brought together by Prof. Poulton respecting the book and its author, few will doubt that Mr. Sleeper's work was really printed and published at the time stated, and that it contains one of the most remarkable anticipations of modern views and forms of expression respecting evolution and the germ-theory of disease that have yet come to light.   F. A. D.

VOL 92, 588 1914

chapter ten
1913-1919

N IELS B OHR AND E RNEST R UTHERFORD EXPLORE THE ATOM AND SUPERCONDUCTIVITY IS DISCOVERED BY
H EIKE K AMERLINGH O NNES. T HE TWO-WHEELED MOTORCAR IS SEEN IN R EGENT'S P ARK. A RTHUR
S TANLEY E DDINGTON EXPATIATES ON RELATIVITY AND C HANDRASEKHARA V ENKATA R AMAN ON THE
"WOLF-TONE" OF VIOLINS AND CELLOS. P ATRIOTIC OUTRAGE TAKES POSSESSION OF EMINENT SCIENTISTS.
T HE VOICE OF EUGENICISTS IS HEARD. A CCUSATIONS AND COUNTER-ACCUSATIONS ARE TRADED ON THE USE
OF WAR GASES. T HE TELEPHONE FINDS A USE IN SURGERY IN THE FIELD. C HEMISTS TACKLE THE NITROGEN
PROBLEM CAUSED BY THE A TLANTIC BLOCKADE. H. G. W ELLS AND E DWARD S CHÄFER PROCLAIM THE
ASCENDANCY OF SCIENCE OVER CLASSICS. A DOLF VON B AEYER IS REMEMBERED WITH AFFECTION.

1913-1919

Archduke Franz Ferdinand and his duchess were immolated at Sarajevo, and the armies of Europe mobilised and clashed in the opening battles of World War I. The engines of war continued to grind relentlessly with the battles of Ypres and Loos, Verdun and the Somme, Passchendaele and Caporetto. A German submarine sank the *Lusitania* and General John Pershing's army landed in France. The Panama Canal opened for business. In Britain, the Franchise Act was passed, giving women the vote. The statesmen of Europe concocted the Treaty of Versailles. In Russia, the Allies tried unsuccessfully to thwart the Bolsheviks. The League of Nations was founded. John William Alcock and Arthur Whitten Brown flew the Atlantic and the great influenza epidemic took 20 million lives.

Niels Bohr's observations on line spectra formed the basis for the Bohr-Rutherford model of the atom. His letter represents the beginning of one of the great scientific revolutions of the century. This is how Bohr introduces the subject, before launching into the quantitative analysis:

The Spectra of Helium and Hydrogen.

RECENTLY Prof. Fowler (Month. Not. Roy. Astr. Soc., December, 1912) has observed a number of new lines by passing a condensed discharge through mixtures of hydrogen and helium. Some of these lines coincide closely with lines of the series observed by Pickering in the spectrum of the star ζ Puppis, and attributed to hydrogen in consequence of its simple numerical relation to the ordinary Balmer series. Other lines coincide closely with the series predicted by Rydberg and denoted as the principal series of the hydrogen spectrum. The rest of the new lines show a very simple relation to those of the latter series, but apparently have no place in Rydberg's theory.

From a theory of spectra (*Phil. Mag.*, July, 1913) based on Rutherford's theory of the structure of atoms and Planck's theory of black-radiation, I have been led to the assumption that the new lines observed by Fowler are not due to hydrogen, but that all the lines are due to helium and form a secondary helium spectrum exactly analogous to the ordinary hydrogen spectrum.

VOL 92, 231 1913

And here a few weeks later is Ernest Rutherford:

The Structure of the Atom.

IN a letter to this journal last week, Mr. Soddy has discussed the bearing of my theory of the nucleus atom on radio-active phenomena, and seems to be under the impression that I hold the view that the nucleus must consist entirely of positive electricity. As a matter of fact, I have not discussed in any detail the question of the constitution of the nucleus beyond the statement that it must have a resultant positive charge. There appears to me no doubt that the α particle does arise from the nucleus, and I have thought for some time that the evidence points to the conclusion that the β particle has a similar origin. This point has been discussed in some detail in a recent paper by Bohr (*Phil. Mag.*, September, 1913). The strongest evidence in support of this view is, to my mind, (1) that the β ray, like the α ray, transformations are independent of physical and chemical conditions, and (2) that the energy emitted in the form of β and γ rays by the transformation of an atom of radium C is much greater than could be expected to be stored up in the external electronic system. At the same time, I think it very likely that a considerable fraction of the β rays which are expelled from radioactive substances arise from the external electrons. This, however, is probably a secondary effect resulting from the primary expulsion of a β particle from the nucleus.

The original suggestion of van der Broek that the charge on the nucleus is equal to the atomic number and not to half the atomic weight seems to me very promising. This idea has already been used by Bohr in his theory of the constitution of atoms. The strongest and most convincing evidence in support of this hypothesis will be found in a paper by Moseley in *The Philosophical Magazine* of this month. He there shows that the frequency of the X radiations from a number of elements can be simply explained if the number of unit charges on the nucleus is equal to the atomic number. It would appear that the charge on the nucleus is the fundamental constant which determines the physical and chemical properties of the atom, while the atomic weight, although it approximately follows the order of the nucleus charge, is probably a complicated function of the latter depending on the detailed structure of the nucleus.   E. RUTHERFORD.

Manchester, December 6, 1913.

VOL 92, 423 1913

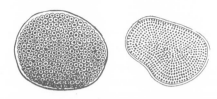

Like so many great physicists, Lord Rayleigh could not observe any physical phenomenon without experiencing an irresistible itch to explain it. Here is a characteristic contribution, typical also of his style.

Reflection of Light at the Confines of a Diffusing Medium.

I SUPPOSE that everyone is familiar with the beautifully graded illumination of a paraffin candle, extending downwards from the flame to a distance of several inches. The thing is seen at its best when there is but one candle in an otherwise dark room, and when the eye is protected from the direct light of the flame. And it must often be noticed when a candle is broken across, so that the two portions are held together merely by the wick, that the part below the fracture is much darker than it would otherwise be, and the part above brighter, the contrast between the two being very marked. This effect is naturally attributed to reflection, but it does not at first appear that the cause is adequate, seeing that at perpendicular incidence the reflection at the common surface of wax and air is only about 4 per cent.

A little consideration shows that the efficacy of the reflection depends upon the incidence not being limited to the neighbourhood of the perpendicular. In consequence of diffusion [1] the propagation of light within the wax is not specially along the length of the candle, but somewhat approximately equal in *all* directions. Accordingly at a fracture there is a good deal of "total reflection." The general attenuation downwards is doubtless partly due to defect of transparency, but also, and perhaps more, to the lateral escape of light at the surface of the candle, thereby rendered visible. By hindering this escape the brightly illuminated length may be much increased.

The experiment may be tried by enclosing the candle in a reflecting tubular envelope. I used a square tube composed of four rectangular pieces of mirror glass, 1 in. wide, and 4 or 5 in. long, held together by strips of pasted paper. The tube should be lowered over the candle until the whole of the flame projects, when it will be apparent that the illumination of the candle extends decidedly lower down than before.

There follows a rigorous mathematical analysis, culminating in this explanation:

The increased reflection due to the diffusion of the light is thus abundantly explained, by far the greater part being due to the total reflection which ensues when the incidence in the denser medium is somewhat oblique.   RAYLEIGH.

VOL 92, 450 1913

The dangers that lay in the fashionable use of radium for therapeutic or even prophylactic purposes were at last starting to be appreciated. The doses to which patients were exposed must often have been enormous. Marie Curie herself of course was one of the martyrs; indeed a plan a few years ago to put her laboratory notebooks on public display was abandoned when it was found that they were still radioactive enough to constitute a hazard to the public. In the United States, radium decoctions were being marketed as a general restorative, and a scandal eventually broke when a member of New York high society, celebrated as a sportsman, died in appalling circumstances from ingestion of radium.

RADIUM AND QUACK MEDICINES.

In view of the fact that a large number of drugs, earths, and waters, said to be radio-active, are being offered for sale to the general public for the treatment of certain diseases, the medical committee of the British Science Guild recently instituted an inquiry into the question of radium and its therapeutic uses.

The result of the inquiry indicates the urgent necessity for legislation in order to safeguard the interests of the community in the sale of these substances, by compelling a written guarantee to be given as to the quantity of radium present in the substances offered for sale.

The use of radium in cases of cancer is now widely known, but it is necessary to warn the public that no definite evidence that cancer is permanently curable by radium is yet forthcoming. The immediate effect of the treatment of cancer by radium is often highly satisfactory, but it must not be forgotten that agents other than radium are known to give equally good results. It is only by keeping under observation for at least five years patients who have been so treated that a definite decision can be come to as to the place radium-therapy shall take in the treatment of malignant diseases.

The great strides that have been made in recent years in the use of radium for the treatment of disease, and the results obtained, encourage the medical profession to persevere with this therapeutic agent. However, radium in its application to disease is still but little understood, and until more experimental, pathological and clinical data have been collected to show the effect of this agent upon, not only the diseased but also the healthy tissues of the body, dogmatic statements as to its therapeutic value cannot be made.

In these circumstances of uncertainty the public is warned that there is danger that the claims which have been advanced for radium as a curative agent may lead to frauds on the credulous section of the public, which may be imposed upon by the sale of substances or waters in which radium does not exist, or may be harmfully treated by persons with no medical qualifications.

The inclusion of radium in the Pharmacopœia would be of material benefit to the public, and it is proposed to take the steps necessary to secure this end. It has also been suggested that radium should be scheduled as a poison under the Foods and Drugs Act, which would be an additional safeguard against the victimisation of the public.

The report of the medical committee of the British Science Guild contains further valuable and important information concerning the sources, etc., of radioactive substances, the price of radium, and diseases which are treated with radium, and this will be published in full in the annual report of the guild, to be issued in May next.

Vol 93, 225 1914

A stylish mnemonic, following earlier efforts (p. 123), for the value of π.

In reading a review in the Bulletin of the American Mathematical Society (xix., 10) of Dr. Gerhard Kowalewski's recent Calculus, we find quoted some interesting French verses from which, by counting the letters of the words, the ratio of the circumference to the diameter may be written down to thirty decimals. They are as follows :—

"Que j'aime à faire apprendre un nombre utile aux sages !
Immortel Archimède artiste ingénieur
Qui de ton jugement peut priser la valeur !
Pour moi ton problème eut de pareils avantages."

It is much easier to remember these verses than the numbers, derived from counting the letters, namely—

$$3\cdot14159265358979323846264643383279.$$

Vol 92, 458 1913

The two-wheeled motorcar must have made a brave sight in Regent's Park.

THE SCHILOWSKY GYROSCOPIC TWO-WHEELED MOTOR-CAR.

A LARGE two-wheeled motor-car, constructed from the design of Dr. Schilowsky, by the Wolseley Tool and Motor Company, Ltd., was given a trial run in London last week. The car is a six-seated car, and it carried six people as it slowly made a circuit of Regent's Park. The gyroscopic mechanism is placed in the cupboard under the middle four seats. This consists of a heavy gyrostat rotating at the moderate speed of 1100 revolutions a minute, and driven by an electric motor of $1\frac{1}{2}$ horse-power. The axis is vertical, and it is mounted in a ring supported on transverse trunnions, so that it may tilt in a fore and aft plane. As the car is necessarily unstable on its two wheels, the gyrostatic ring must also be carried unstably for it to have corrective influence. If, as a ship, the car could have been carried stably, then the gyrostatic ring would also have to be stably mounted. If one is stable and the other unstable then the gyrostat operates in the opposite sense to that intended.

The unstably mounted gyrostat will not maintain the car in its upright position for long, as the precessional oscillations increase in amplitude. Dr. Schilowsky counteracts this by an ingenious piece of mechanism. Driven by worm-gearing from the gyrostat axle are two spur wheels, each just out of gear with a segmental rack, but capable of being brought into gear by a heavy pendulum which feels any tilting of the car away from the dynamical vertical. This is only allowed to engage at such times as the gyrostat ring is approaching the neutral position. During this time the engagement causes a hurrying of the precession and a consequent steadying of the motion. At the moment the neutral position is reached the pinion and rack are disconnected by a snap mechanism reminding one of that used for closing the valves of a Corliss engine. One pendulum controls the engagement when the gyrostatic ring is approaching the neutral position from one side, while the other effects the control on the other side of the neutral position. Either alone might be used, but the two alternate with one another and maintain a more continuous control. It is a curious fact that the controlling mechanism is more easily adjusted so as to maintain the equilibrium of the car when it is turning in the opposite direction to the rotation of the wheel.

For turning in the same direction more exact adjustment is necessary. A working model railway on this system has been presented by Dr. Schilowsky to the South Kensington Museum, where it may be seen by anyone interested.

The car weighed three tons, having been designed for running on a rail, while the engine was one of the maker's standard 16-h.p. engines. This was insufficient in power to drive the heavy car, as well as the motor of the flywheel, more than about four miles an hour. At this speed and at rest or moving backwards the car maintained its position with passengers jumping on or off. When a new load was applied to one side the car moved almost imperceptibly so as to raise it and maintain the centre of gravity over the line of support as has already been made familiar by Mr. Brennan with his monorail.

It will be interesting to see how the car behaves when a more powerful engine is fitted and higher speeds are possible. The inventor is, of course, aware of the very great couple, ordinarily resisted by the four-wheel support of the motor-car when ordinary curves and speeds are negotiated together, which he will have to contend with in like circumstances. The demonstration in the Regent's Park did not show that the gyrostatic control then existing would be sufficient for this, but it did show, and that perfectly, that the first step has been successfully accomplished. It may be worth while to add that the bicycle balance is not used, the gyrostatic control being independent of speed or direction of motion.

C. V. Boys.

Vol 93, 251 1914

Heike Kamerlingh Onnes was the great pioneer of low-temperature physics. Below is the beginning of a report describing the phenomenon of superconductivity.

EXPERIMENTAL DEMONSTRATION OF AN AMPERE MOLECULAR CURRENT IN A NEARLY PERFECT CONDUCTOR.

IT has long been known that the electrical resistance of metals falls with a reduction of temperature in an approximately straight line law, indicating that, in the neighbourhood of absolute zero, there would be no resistance whatever. Prof. H. Kamerlingh Onnes, of Leyden, has carried experiments on this subject down to extremely low temperatures, and has found that it is at a point a few degrees above absolute zero that the resistance of certain pure metals practically vanishes. His later experiments illustrate the properties of these almost resistanceless bodies, or, as he terms them, "super-conductors," in a very striking way. Taking a closed coil of lead wire, he cooled it down by immersion in liquid helium to a temperature at which its resistance is of the order of 2×10^{-10} that at normal temperatures. He then induced a current in the coil, which, instead of ceasing with the E.M.F., was shown to persist with scarcely sensible diminution for as long a period as the coil could be kept cold. As there was practically no resistance, there was practically no dissipation of energy, and the system behaved like the imagined molecular currents of Ampère, and realised the conception of Maxwell as to a conductor without resistance.

Vol 93, 481 1914

The implications of the theory of relativity were still not palatable to all. Sir Oliver Lodge objects:

SPACE AND TIME.

"FROM this time forth space and time apart from each other are become mere shadows, and only a kind of compound of the two can have any reality." So spoke Herrmann Minkowski in 1908. But his statement has not yet been realised. It is still the elect to whom it is given to escape from the bondage of their own consciousness so completely that they can think of time as nothing more than the most convenient means of ordering events. Sir Oliver Lodge was voicing the feeling of the man in the street when, at Birmingham, he said: "Surely, we must admit that space and time are unchangeable: they are not at the disposal even of mathematicians."

Vol 93, 532 1914

With World War I now in full swing, chauvinistic excesses began to manifest themselves in Nature, *as indeed everywhere else. This was the period when anyone with a German name was persecuted in England; the First Sea Lord, the passionately loyal Lord Louis Battenberg, was made to resign (he changed his name to Mountbatten, but too late) and patriots kicked dachshunds in the street. Sir Arthur Schuster, professor of physics at the University of Manchester and president of the British Association for the Advancement of Science in 1915, was harried by the press on account of his German name while his son was fighting (and being wounded) in the Dardanelles. Here is a small voice, pleading for sanity.*

Renunciation of Honorary Degrees.

I HAPPEN to see in one of the Dutch journals that a number of German men of science have divested themselves of honours bestowed upon them by British universities and learned societies, on account of the war between England and Germany.

Will you allow me to express the hope, by means of this letter, that my British friends will not reciprocate this action by a similar one?

To my mind, worse than the young lives sacrificed, worse than the destruction of ancient monuments of arts and science, is the almost inevitable consequence of this terrible war: the sowing of hatred and distrust between different nations.

Now it is my firm belief that it is the duty and the privilege of scientific men all the world over to do all in their power gradually to allay these feelings of hatred and distrust.

For this reason especially I regret greatly the action of the German "savants," and earnestly pray my British friends to abstain from similar action.

J. P. Lotsy.

Perpetual Secretary of the Dutch Society of Sciences, Haarlem, September 12.

Vol 94, 88 1914

A leader with the title "Germany's Aims and Ambitions", written by Sir William Ramsay (the admired friend and patron of, for instance, Otto Hahn), displays the ugly temper of the time. One passage from the middle will suffice.

The nation, in the elegant words of one of its distinguished representatives, must be "bled white."

Will the progress of science be thereby retarded? I think not. The greatest advances in scientific thought have not been made by members of the German race; nor have the earlier applications of science had Germany for their origin. So far as we can see at present, the restriction of the Teutons will relieve the world from a deluge of mediocrity. Much of their previous reputation has been due to Hebrews resident among them; and we may safely trust that race to persist in vitality and intellectual activity.

Vᴏʟ 94, 138 1914

The war was already taking its toll of the younger generation of scientists.

A CENTRAL NEWS telegram from Amsterdam, published in the *Times* of December 21, announced that Prof. Otto Sackur was blown to pieces by an explosion which occurred in the laboratory of the Kaiser-Wilhelm Institute at Dahlem, near Berlin, where experiments in high explosives were being conducted. The Kaiser-Wilhelm Institute was founded in 1911 by the Kaiser-Wilhelm Association for the Advancement of Science, and the first volume of results of researches issued by it was described in NATURE of May 28 (vol. xciii., p. 322). Prof. Sackur was one of the ablest of the younger physical chemists in Germany. A pupil of Richard Abegg, he became a Privat-dozent in the University of Breslau, and on Abegg's appointment to the Technische Hochschule there, succeeded him as extraordinary professor in the University. He later received an appointment in the Kaiser-Wilhelm Institute. Sackur was more distinguished as a theorist than as a practical worker. His papers cover a wide range, and are characterised by considerable independence of thought. In 1912 he published a "Lehrbuch der Thermochemie und Thermodynamik," which is an admirable modern introduction to the subject, and is, we understand, being translated into English.

Vᴏʟ 94, 486 1914

The utterances of the eugenicists about the war make queasy reading.

IN an editorial article on eugenics and war in the October issue of the *Eugenics Review,* it is pointed out that "the British Empire, by reason of maintaining her army on a voluntary basis, must inevitably suffer racially more than other nations. The battle death-rate must strike her unevenly and reduce the number of her males amongst the class from whom it is most desirable that she should produce the stock of the future. In the countries with universal compulsory service, the reduction in effective males will be spread over the entire population; good and bad will alike be reduced. In this country the types which are physically and mentally superior will volunteer for active service. . . . The sample of those killed will not be the average of the race, but the best type of the race. . . . Although the system may give victory and national prestige, the racial effect must be injurious."

Vᴏʟ 94, 208 1914

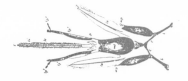

The war also brought the proponents of social Darwinism out of the woodwork.

WAR AND THE RACE.

THE Manchester Statistical Society has printed an eloquent address by Dr. C. W. Saleeby, "The Longest Price of War." The thesis is the old, but politically ignored, result of war in "reversed selection." Quoting Michelet's epigram that the campaigns of Napoleon lopped a cubit from the stature of the French, and Prof. J. A. Thomson's observation that not even the discoveries of Pasteur could restore the physique which the victories of Napoleon's armies had destroyed, Dr. Saleeby notes the small size of the present-day French soldier, as remarked by many observers. To-day, for our own forces, "the brave, the vigorous, the healthy, the patriotic are taken, and the others left. . . . The rejected recruits recruit the race." The whole question is one which statisticians should investigate in special reference to the present war. Dr. Starr Jordan's study, "The Human Harvest," and the late J. Novikow's "Darwinisme Sociale," are the best of a meagre list of popular expositions of the thesis, of which the decay of the Roman Empire is the classic type. Speeck estimated that of every hundred thousand Romans, eighty thousand were slain. "Vir" thus gave place to "homo"; "the Roman Empire perished," says Seeley, "for want of men."

No scientific mind wishes to eulogise war, in the German fashion, which depends for its argument on the primitive athletic form of war, whereas war of to-day is simply peace riddled with casualties. Darwin's famous sentences refer only to a more or less imaginary conscript army in a country which is always at war—"in every country in which a large standing army is kept up, the finest young men are taken by the conscription or are enlisted. They are thus exposed to early death during war, are often tempted into vice, and are prevented from marrying during the prime of life. On the other hand, the shorter and feebler men, with poor constitutions, are left at home, and consequently have a much better chance of marrying and propagating their kind." It seems a fairly obvious inference that the dysgenic results of modern warfare remain to be proved. The deliberate sacrifice of life by exploiting the mass-formation is a special case needing investigation. The whole subject calls for investigation; until this is carried out, nothing is at all clear either for or against the biological effects of war.

Vᴏʟ 94, 544 1915

Here is Sir William Ramsay again, not even pausing to wipe the foam from his lips.

THE WAR.

THE war drags on; and we are learning to understand the mentality of the German race more completely. It is being revealed in various forms. The policy always adopted by the bully, of attempting to terrorise by attack on defenceless persons, is shown by the shelling of the watering-places of Yorkshire, resulting in the murder (for that is the only word which fits the case) of 103 harmless people. "Murder" should certainly have been the verdict, although it was disallowed by the coroner; for although the commanders of the German vessels may not yet be known by name, a verdict of murder would have rendered them subject, when captured, to trial by a British jury. The cowardly and murderous onslaught has led, we are told, to rejoicings in Berlin. It is as we feared; the German nation has lost its moral perspective. They may rest assured, however, that there will be no similar reprisals on the side of the Allies. We do not revenge ourselves on innocent women and children.

It was scarcely worth while, perhaps, for the French universities and British men of letters and science to have replied to the self-named "intellectuals" of Germany. Neutral countries have already made up their minds from the perusal of official documents, not the least important being those from German sources, that the war is one of pure aggression on the part of the Germans. We hear from Switzerland, from America, and from Scandinavia that the public in these countries now pay no attention to German polemic literature. If they had conceivably had any case, they have given it away by their inhuman acts, which have raised a sentiment of disgust in every civilised mind.

We look with contemptuous amusement at the childish renunciation of foreign honours by our Teutonic colleagues in science. That is even the attitude of some of their own countrymen; Prof. Verworn, of Bonn, writing in the *Berliner Tageblatt*, describes it as unworthy of German men of science, and Profs. Waldeyer, Martin, and Orth have protested against the foolish conduct of their countrymen. We can only shrug our shoulders and say that the loss is theirs, not ours.

We have also been disillusioned by the words of the well-known Celtic scholar, Prof. Kuno Meyer, late of Liverpool University, now of Berlin, who has acted, and is acting, as an agent of the Prussian Government in attempting to excite the feelings of Nationalist Ireland and of American Irish in favour of Germany. Here is a man, eminent in his own subject, speaking English without an accent, who has spent thirty years of his life in an English university, a man who has (or had) many intimate friends in this country and has been received in many English households as a friend, turning out to be a dastardly enemy. Savages have a code that, after breaking bread in a man's house, it is treacherous to war against him; not so Prof. Kuno Meyer. This is evidently another instance of "Kultur." It behoves us to treat with suspicion all naturalised aliens of Teutonic extraction; and yet we know, alas! that in doing this, we are acting unjustly in some cases. But the individual, in these days, must suffer for the crimes of his countrymen. It is such instances as these which make the Allies determined that such a race must be deprived of power to do further mischief, whatever be the cost in life and money.

Some correspondence has appeared in the Press as to the relative merits of German contributions to science, as compared with the achievements of members of other races. The discussion is perhaps a useful one; for there is little doubt that the German estimate of the scientific ability of their own people is a much exaggerated one. The statement made in a previous issue of NATURE (October 8) that German science has not been remarkable for originality appears to meet with general assent. We in England have been always more intent on welcoming a discovery than in inquiring into the nationality of the discoverer; indeed, it is a common saying that science is international. But we are beginning to revise our verdict. Prof. Karl Pearson, Prof. Sayce, and Sir E. Ray Lankester have shown that Germany has played only a small part in inception of scientific truths, although by organisation she has greatly extended their application. Huxley and Bywater held this view, each as regards his own subject; and it appears to be shared by geologists, physicists, and chemists. "Ausarbeiten" is the goal of the Germans; the inventive faculty has not been their strong point. Perhaps a mixed race gains in original ability; both flint and steel are necessary to produce a spark. But one thing the German man of science knows how to do well —to exalt the achievements of his nation, often by ignoring that of others. This has probably been done in many cases without intention; it appears to be one way in which German patriotism manifests itself.

The authors of the first report on the use of poison gas in the war had little trouble in identifying it as chlorine. The system for releasing the gas was devised by Fritz Haber. The Haldanes, father and son, were both active in the war and applying their scientific skills.

ASPHYXIATING GASES IN WARFARE.

DR. J. S. HALDANE'S report on his investigation of the nature and effects of the asphyxiating gases, used by the Germans in their attack last week on the French and British lines near Ypres, leaves but little doubt that chlorine or bromine was the chief agent employed, whilst shells containing other irritant poisons were also used.

Prof. H. B. Baker, who accompanied Dr. Haldane, is carrying out an investigation as to the chemical side of the question, and until his report is available, surmises as to the nature of the poisonous gases and the methods adopted for their use would be premature, but the evidence seems to point clearly to the fumes floated by the wind on to the Allies' lines being chlorine, as at ordinary pressure bromine is a liquid below 59° C., and at ordinary temperatures would not give off its vapour with sufficient rapidity to cause the seven-foot bank of vapour that drifted on to the Allies' trenches, whilst the colour of the cloud would have been a rich brown and not the "greenish" or "yellowish-green" colour so frequently described, which undoubtedly points to chlorine.

Chlorine gas is 2·45 times heavier than air, and if discharged "down wind" would only slowly rise, so that at a distance of one hundred yards from its point of disengagement the bank of fume might be expected to be six or seven feet deep, but with bromine vapour, which is more than five times the weight of air, the thickness of the layer of vapour would, under the same conditions, be much less. Liquid chlorine has, for many years, been a commercial article : the gas is liquefied by a pressure of six atmospheres at 0° C., and is stored in lead-lined steel cylinders, being largely exported for use in the extraction of gold in localities where, from difficulties of transport, plant and materials for making the gas *in situ* would be more expensive.

It is said that such cylinders, 4 ft. 6 in. long, were sunk in the German trenches and were connected to pipes six feet long pointing towards the Allies' lines : under these conditions, intense cold would be produced at the point where the cylinders discharged into the delivery pipes by gasification of the liquid and expansion ; this would soon check the rapid production of gas, and the white smoke seen behind the greenish cloud of gas may well have been caused by brushwood fires lighted above the delivery pipes to warm them and prevent stoppage.

Although all the evidence and the symptoms found in the unfortunate victims overcome at this particular section of the line point to chlorine as the gas employed, there seems every probability that liquid bromine has also been used in shells or grenades, which, bursting in the air, would scatter the liquid under conditions that would rapidly gasify it, when the weight of the vapour would cause it to descend on the troops below.

Both chlorine gas and bromine vapour, when present to the extent of 5 per cent. in air, rapidly cause death by suffocation, by acting on the mucous linings of the nose, throat, and lungs, so causing acute inflammation ; but bromine poisoning is generally distinguishable by the skin of the victim being stained yellow, and the intense action on the eyes, which is much greater than with chlorine.

The Germans have an unfailing source of bromine in the crude carnallite, worked at Stassfurt for the production of potassium chloride, but when full particulars are available it will probably be found that, besides such obvious asphyxiants as chlorine, bromine, and sulphur dioxide, they have also employed compounds of a more complex character.

Vol 95, 267 1915

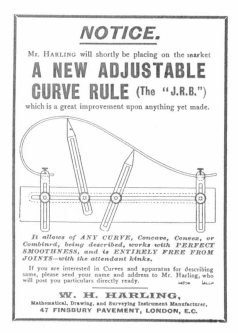
Gas warfare quickly became a subject of assertive but uninformed comment.

THE *Daily Telegraph* published a telegram "From our own correspondent at Copenhagen," on May 26, reporting that "A Danish surgeon and scientist of the highest reputation has succeeded in discovering what the German soldiers use to protect themselves against the asphyxiating gases which they employ against the enemy." The "discovery" is that the Germans make use of solutions of hyposulphite and bicarbonate of soda to moisten their respirators. The announcement reminds us, however, of the belated discovery of the lamented death of Queen Anne ! The use of such solutions is well known to all workers with chlorine gas, and was mentioned in daily papers a day or two after the Germans commenced to discharge the gas upon our troops.

Vol 95, 380 1915

With the war raging, NATURE *fulminates against the ignorance, officiousness and incompetence that ties the hands of loyal scientists. The government was exacting excise duty on laboratory alcohol and derivatives made from it, and also on the use of stills.*

...chloroform and ether made from ethyl alcohol pay duty, whereas that from methylated spirit (methylated ether) in one case and acetone in the other do not, although the products are practically identical. Again, methyl and ethyl alcohol used for research are exempt from duty, whereas ethyl acetate and butyrate, ethyl chloride, bromide, iodide, and chloral hydrate, in all of which ethyl alcohol is used, are not exempt. The corresponding methyl derivatives which are obtained in precisely the same way from methyl alcohol are not scheduled and, we presume, are free to all consumers.

The article then gets to the nub of the matter — scientists are undervalued:

But it is not the duty on alcohol which has been the main factor in crippling the colour industry during the last thirty years. Nor is it defective training, equipment, or ability of the young chemists turned out from our universities, whose scientific work stands second to none. It is that the manufacturing world is only beginning to realise at this time of crisis in the chemical industry the true value of the research chemist. We say "beginning to realise," for it was only a few days ago that a professor of chemistry in one of our provincial universities received a request from a large and wealthy corporation to recommend a first-rate chemist, to whom the handsome salary of thirty shillings a week was offered, or about a third of the earnings of a coal-miner working full time !

And so to a rousing conclusion:

The conclusions are obvious. Our highly-educated Government officials and princes of industry are more or less ignorant of science. It is for many of them an unknown and mysterious region into which they would prefer not to penetrate. That Nemesis now confronts our industries may be a blessing in disguise. It is only in a struggle that the weak points in one's armour are disclosed. We are learning a lesson, which might have been learnt years ago had we not been so inexorably bound by tradition, and the sooner we profit by it the better.

According to the physicist Silvanus P. Thompson, Sir Mackenzie Davidson, the surgeon in this description, had success with the telephonic technique in distinguishing between a fragmented bullet and pieces of bone. Alexander Graham Bell's failure in the case of President James Abram Garfield was resounding: moving an induction coil over the president's body and listening with an earphone, he discovered (according to T. P. McMahon's story) a series of bullets distributed in a regular pattern. They were, it transpired, the nodes in the bed-spring lattice, and the president died of septicaemia before Bell could grapple with the problem again.

THE TELEPHONE IN SURGERY.

IN the *Lancet* of January 30 is published an address by Sir James Mackenzie Davidson, delivered before the Medical Society of London, on the telephone attachment in surgery. By this phrase the author refers to the attachment of a telephone receiver to a probe, or lancet, or other metallic instrument used by a surgeon when exploring a wound containing a bullet or other piece of extraneous metallic matter, in such a way that the sound heard in the telephone when the probe comes into contact with the bullet enables the surgeon to make certain of the position of the bullet in the wound.

As this matter appears to be of real importance at the moment to surgeons in the field hospitals of our armies abroad, we make no apologies for giving our readers a summary of the more salient features of Sir Mackenzie Davidson's address. His attention was first directed to the use of the telephone as an auxiliary in surgery thirty-two years ago, by the accounts of the attempts made by Graham Bell, to determine, by means of the induction balance, the position of the bullet in the body of President Garfield when he was assassinated in 1881. Speaking afterwards of these attempts, and of the difficulties attending the method—which had failed in that notable case to yield satisfactory indications—Graham Bell outlined another and simpler electrical method for the detection of bullets, as follows :—

It consists of a telephone, to one terminal of which a fine needle is fixed, and to the other a plate of metal of the same nature as the needle. The plate is placed on the limb to be examined, and the needle is thrust in where the bullet is believed to be; and when it strikes the ball a galvanic battery is formed within the body. . . . *This will cause a click to be heard in the telephone each time the bullet is struck.* This is a far simpler apparatus than the induction balance, and one far more easily procured.

*The war cut down several outstanding young scientists,
among whom H. J. G. Moseley — felled by a Turkish bullet
at Suvla Bay — was the most remarkable. Ernest Rutherford
had sought to get Moseley transferred to a position in which
his talents could be used, rather than aimlessly extinguished,
and he wrote the obituary, beginning like this—*

HENRY GWYN JEFFREYS MOSELEY.

SCIENTIFIC men of this country have viewed
with mingled feelings of pride and apprehension the enlistment in the new armies of so
many of our most promising young men of science
—with pride for their ready and ungrudging response to their country's call, and with apprehension of irreparable losses to science. These
forebodings have been only too promptly realised
by the death in action at the Dardanelles, on
August 10, of Henry Gwyn Jeffreys Moseley,
2nd Lieut. in the Royal Engineers, at the age of
twenty-seven.

— and ending:

Moseley's fame securely rests on this fine series
of investigations, and his remarkable record of four
brief years' investigation led those who knew him
best to prophesy for him a brilliant scientific
career. There can be no doubt that his proof
that the properties of an element are defined
by its atomic number is a discovery of great and
far-reaching importance, both on the theoretical
and the experimental side, and is likely to stand
out as one of the great landmarks in the growth
of our knowledge of the constitution of atoms.

It is a national tragedy that our military organisation at the start was so inelastic as to be
unable, with few exceptions, to utilise the offers
of services of our scientific men except as combatants in the firing line. Our regret for the
untimely end of Moseley is all the more poignant
that we cannot but recognise that his services
would have been far more useful to his country
in one of the numerous fields of scientific inquiry
rendered necessary by the war than by exposure
to the chances of a Turkish bullet.

E. RUTHERFORD.

VOL 96, 33 1915

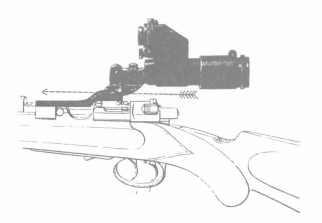

*Losses were equally calamitous on the German side.
Rutherford received letters from the front — sent through a
neutral country — from a former protégé, Hans Geiger, that
included a roll-call of the fallen. The young Austrian physicist
Fritz Hasenöhrl, whose death is announced here, was adopted
by the Nazis as a totem of Germanic science: they claimed that
it was from him that Albert Einstein had stolen his discoveries.*

PROF. KAMERLINGH ONNES announces in the *Nieuwe
Rotterdamsche Courant* that he has received news
from Vienna of the death of Prof. F. Hasenöhrl, professor of physics in the University of Vienna, who was
killed in action on the Italian front. The deceased,
who was a pupil of Boltzmann, began in 1899 an investigation on the dielectric constants of liquefied
gases, in the cryogenic laboratory at Leyden. Having
returned to Vienna, he became privat-dozent, and later
succeeded Boltzmann, whose collected papers he edited
with much care. Earlier in the war in Galicia, Prof.
Hasenöhrl had been wounded in the shoulder, but
after a complete recovery he again went to the front,
where his lamented death occurred.

VOL 96, 267 1915

*Reviewing this book of Darwin letters, W. T. Thiselton-Dyer
laments the passing of a golden age and the arrival on the
scientific landscape of the joyless professional.*

EMMA DARWIN AND HER CIRCLE.

*Emma Darwin: A Century of Family Letters,
1792–1896.* Edited by her daughter Henrietta
Litchfield. Vol. i., pp. xxxi + 289. Vol. ii.,
pp. xxv + 326. (London: John Murray, 1915.)
Price 21s. net. Two vols.

* * *

... though well-to-do the middle class of the
early nineteenth century was simple and
unaffected in its mode of life and content with
intellectual pleasures. With leisure and freedom
from anxiety it could turn to science and recruited
the Royal Society. This swept that body into the
social life of the day, which is now inevitably
ebbing away from it. Darwin speculates in a
letter to a son as to "what makes a man a discoverer of undiscovered things," and remarks that
"many men who are very clever—much cleverer
than the discoverers—never originate anything."
He conjectures that "the art consists in habitually
searching for the causes and meaning of everything that occurs." Perhaps the explanation lies
in the difference between the inductive and
deductive temperament. Hereditary aptitude must
also count for something. The clan was clever
enough but never failed to throw up originality.
Tom Wedgwood was "the first discoverer of
photography," Hensleigh was a mathematician
and philologist, John with Sir Joseph Banks
founded the Horticultural Society at Hatchard's
shop, Sir Henry Holland and the Galtons were
cousins. The amateur has been the glory of

English science; there is now little place for him. The ground to be traversed before the fighting line is reached is too vast, and each worker must be content to "nibble" at his own little section with small knowledge of what the rest are doing. And so, rather unkindly, Prof. Armstrong describes the Royal Society as a "rabble." Science must now be content to be professional, if not professorial. In the last century it was not so. Leading men of science were in touch with one another and the larger life and influenced it. It seems strange to read that in 1842 at the Athenæum "they have *soirées* every Monday evening, and as all the literary and scientific men in London are in the club, they must be very pleasant."

Vol 95, 503 1915

The uncertain supplies of "Chile saltpetre" brought a crisis in the chemical industries of the combatant nations. Chemists (most notably Fritz Haber in Germany and Chaim Weizmann in England) bent their minds to the "nitrogen problem". Here the illustrious American physical chemist A. A. Noyes worries about it.

THE NITROGEN PROBLEM IN RELATION TO THE WAR.

PROF. ARTHUR A. NOYES, of the Massachusetts Institute of Technology, who is chairman of the Committee on Nitrate Investigations of the National Research Council of America —a body which owes its existence to the war— recently delivered a lecture before a joint meeting of the Washington Academy of Sciences and the Chemical Society of Washington, a report of which, under the above title, is published in the Journal of the Washington Academy of Sciences for June 19. The lecture dealt with the vital importance of an adequate supply of nitrogen compounds, particularly of nitric acid and ammonia, in connection with the war, and gave a brief description of the various efforts America was making in order to meet the demand. Nitric acid enters, directly or indirectly, into the composition of all the more important explosives, such as smokeless powder, picric acid, ordinary black powder, dynamite, trinitrotoluol, and ammonium nitrate. The last-named substance is now used on so enormous a scale that the demand for ammonia is scarcely less urgent than that for nitric acid.

The main sources of these two nitrogen compounds are: (1) Chile saltpetre; (2) by-product gas from coke-ovens; (3) atmospheric nitrogen, which is "fixed" by (a) the cyanamide process, (b) the cyanide process, (c) the arc process, and (d) the synthetic process.

For its supply of nitric acid the United States, like ourselves, has hitherto mainly depended upon imported sodium nitrate (Chile saltpetre), which is now recognised as a rather precarious source, as it depends upon adequate shipping, and is liable to be affected by enemy machinations in interfering with production, destroying plants, or blowing up the reservoirs of oil needed for fuel. Hitherto all attempts on the part of the enemy to establish a submarine base on the Pacific Coast have been foiled. But even if this source continues to be efficiently safeguarded, America realises that it is impracticable to get through imported nitrate the huge amount of nitric acid that will be needed for her Army, and that it will be necessary to supplement this supply by other means.

Vol 102, 26 1918

The patriotic frenzy that took possession of the articulate classes is reflected in the excesses of the book and the review by Sir Edward Thorpe.

THE BOOK OF FRANCE.

The Book of France. In Aid of the French Parliamentary Committee's Fund for the Relief of the Invaded Departments. Edited by Winifred Stephens. Pp. xvi + 272. (London: Macmillan and Co., Ltd.; Paris: E. Champion, 1915.) 5s. net.

* * *

The book opens with an appreciation of France by Mr. Henry James, written with the copiousness and *verve* which characterise all his work. It is followed by a short article by M. J. H. Rosny *aîné*, on British character and policy, translated by Mr. Thomas Hardy, in which, in a few pregnant paragraphs, our national excellences and shortcomings are dealt with in a manner as discriminating and tactful as it is just and true. An essay on the mentality of the Germans, by M. René Boylesve, translated by Dr. W. G. Hartog, is a keen and incisive psychological analysis of the Teutonic mind, written with detachment, and wholly dispassionate—an admirable example of the clear, penetrative insight of French criticism of the highest order. How true it all is Germany will yet come to realise in the awakening which is inevitably in store for her, no matter what the fortune of war may bring. Perhaps the most arresting and striking contribution to the work is the "Debout pour la Dernière Guerre!" by M. Anatole France, done into nervous, palpitating English by Mr. H. G. Wells. How true is all this!

"The prophetic nightmares of our scientific fantastics are being lamentably realised; they come about us monstrously alive, surpassing the horror of Dis, Malebolge, and all that the poet beheld in the Kingdom of Misery. But it is not Martians but German professors who accomplish these things. They have given this war a succession of forms that testify continually to their genius for grotesque evil, first the likeness of the waterspout and typhoon that brought them to the Marne and defeat irreparable, then the sullen warfare of the caverns, then the conflict of metals and chemicals. . . . A philosophical doctor, who sits beside me and reads as I write, interrupts : ' Be certain,' he says, ' that when they abandon that last method they will take to bacteriological war ; after the poison gas and the jet of fire they will fight as disease cultures. We shall have to create in every country a Ministry of Anti-Teutonic Serums.' And to this their science has brought them ! I recall the *mot* of our good Rabelais : ' Knowledge without conscience is damnation.' "

And how beautiful and how sublime is the invocation with which the whole ends !

"O Britain, Queen of the Seas and lover of justice ; Russia, giant of the subtle and tender heart ; beautiful Italy, whom my heart adores ; Belgium, heroic martyr ; proud Serbia ; and France, dear Fatherland, and all you nations who still arm to aid us, throttle and end for ever this hydra, and to-morrow you will smile and clasp hands across Europe delivered."

Vol 95, 667 1915

F. G. Donnan casts a look over his shoulder at the advancing behemoth of German science :

Science is standing Germany in good stead at present. It is known that the Badische Works, employing the process initiated by the scientific researches of Prof. Haber, had arranged for an enormous output of synthetic ammonia during the present year. About twelve years ago Prof. Ostwald, foreseeing (as he has himself publicly stated) a nitric acid famine in Germany during a period of war, investigated the conditions for the economical oxidation of ammonia to nitric acid. This process has been worked for several years at a factory near Vilvorde in Belgium. It is rumoured that Prof. Haber and the Badische Works have greatly improved the process, and that in conjunction with the synthetic ammonia process it now provides a large part of the nitric acid required by Germany for the manufacture of her explosives. *Ohne Phosphor kein Gedanke* said the materialist once upon a time. So might he now say, "No nitric acid, no war." Interesting notices have appeared from time to time in the *Chemiker Zeitung* relating to the activity of organised German science during the past twelve months. A new industry of zinc extraction has been developed, and it is reported that means have been found to replace the French bauxite required for the production of aluminium.

The shortage of copper has been discounted by the use of special alloys. It is also stated that processes have been developed for the manufacture of gasoline and lamp oil. Alcohol is being largely used in internal combustion engines.

We may feel sure that not only the universities and technical high schools, but also the splendid special laboratories of the *Kaiser Wilhelm Forschungsgesellschaft* are working at high pres-

sure in the service of their State.

It is necessary—urgently necessary—that we should do as much, if not more. Let Britain call, British science is ready. It is straining at the leash. All that is wanted on the part of our leaders are imagination and sympathy. A little more of these, and the good that has been done can be magnified a thousandfold.

F. G. Donnan.

Vol 95, 510 1915

This was not the only occasion on which Nature *made such an embarrassing mistake.*

NOTES.

We learn with much satisfaction that the announcement of the death of Prof. I. P. Pavlov is incorrect ; and we may hope, therefore, that the record of his work given in Nature of March 2 will be extended still further in the coming years. Prof. B. Menschutkin, of the Polytechnic Institute, Petrograd, writing on March 20, informs us that Prof. Pavlov is alive and well, and that the Prof. Pavlov who died in February was Eugeni Vasilievitch Pavlov, a celebrated surgeon. The name of Pavlov is common in Russia, there being no fewer than five professors of that name in Petrograd, so that the mistake in the *Times* of February 12 is quite comprehensible.

Vol 97, 185 1916

At this time, Henry Charlton Bastian died, secure in the conviction that he was right about spontaneous generation of life and the scientific world wrong. Here is the concluding passage from his obituary.

But it is particularly in connection with the "origins" of life that Bastian's name will be chiefly remembered. Contrary to generally accepted views, he denied that life always originates from pre-existing life, and maintained that, just as presumably in past ages life developed from non-living matter, so at the present time lowly living organisms are, under certain conditions, being generated from non-living elements. He was, in fact, an upholder of the doctrine of spontaneous generation, or, as he preferred to term it, of "archebiosis." By the use of solutions containing colloidal silica and iron, enclosed in sealed glass tubes and sterilised by heat, and maintained under particular conditions of light and temperature, he claimed that after a time microorganisms, such as bacteria and torulæ and even moulds, developed. His results have been detailed in several papers which have appeared in the pages of Nature during the last few years, and in "The Beginnings of Life" and "Nature and Origin of Living Matter." Few have cared to undertake the laborious investigations necessary to follow this work, which cannot be said to have been confirmed, though the MM. Mary, of Paris, and a correspondent writing only last week in the *English Mechanic,* state that they have observed the development of lowly organisms in culture tubes prepared according to his directions.

Dr. Bastian also supported the doctrine of "heterogenesis," the sudden appearance of one kind of organism as the offspring of another, *e.g.*, ciliates and flagellates descending from amœbæ. This work was published in book form under the title of "Studies in Heterogenesis." Dr. Bastian maintained his views against all opposition with a tenacity and ingenuity which won the respect of his bitterest opponents. A man of great personal charm and originality he literally died in harness, for up to three or four months ago he was continuing his investigations and planning new experiments with a vigour which showed little decline in spite of his four-score years, and to the last his interest in science he had served so well remained undimmed.   R. T. H.

Vol 96, 348 1915

Charles Richet, physiologist, Nobel prize-winner for his discovery of anaphylaxis, and the most gullible of parapsychologists, was also a poet in the magniloquent tradition.

A bust of Pasteur was installed in the place of honour, a prize was decreed to M. Schloesing, that veteran of the Académie des Sciences, who is now in his ninety-second year, and a most admirable address was given by M. Gaston Bonnier. It is true that English men of science, likewise, are well able to instal busts, decree prizes, and give addresses. But France does it better, for she is not afraid, as we are, of magniloquent oratory. And M. Bonnier not only gave his audience an address, but also read them a poem, "A la gloire de Pasteur"—a poem which won the Grand Prix of the Académie Française last year, the work of M. Charles Richet, professor of medicine in Paris, a man honoured by all physiologists in France and over here. This noble poem is published, with M. Bonnier's address, in the *Revue Scientifique*, March 11–18. The reference to Lister is delightful :—

> Honneur à toi, Lister, qui, seul dans cette foule,
> T'opposant aux clameurs des savants et des sots,
> Pendant qu'un vain torrent de critiques s'écoule,
> En admirant l'asteur, sus dompter nos fléaux.

But the whole poem deserves study. Truly, a pleasant little festival of gratitude, goodwill, and reverence; and while these quiet men of science were celebrating in Paris the glory of Pasteur, the batteries of Verdun were thundering out the everlasting glory of France.

Vol 97, 105 1916

Next another of the dismaying chauvinistic outbursts that the war provoked:

GERMANY AND RACIAL CHARACTERS.

The Germans: (1) The Teutonic Gospel of Race; (2) The Old Germany and the New. By J. M. Robertson. Pp. viii + 291. (London: Williams and Norgate, 1916.) Price 7s. 6d.

IN the first part of his book Mr. Robertson gives an admirable and timely exposition of the crude falsity of certain current doctrines of race. The much-used "Aryan," if understood ethnologically, is almost meaningless; all that we know is that certain peoples speak Aryan languages. We do not know that those peoples, *e.g.*, in Europe, are the descendants of the invaders who brought the original Aryan speech. Similarly with skull-measurement. Many writers have claimed a generic superiority for the long-headed type—

which, according to Gobineau, is that of the Teuton warrior—regardless of insuperable difficulties. For example, the Swedes are dolichocephalic, and they are not a leading nation; worse still, it is found that their best individuals are less dolichocephalic than the average. And dolichocephaly is characteristic of the negro, the Eskimo, and the gorilla. Equally fallacious is the Germans' claim that their ancestors were exceptional in their considerate treatment of women; Plutarch proves that the Ligurians excelled them, as the North American Indians did later on. Indeed, all talk about "Germanic" virtues is absurd if its aim is to glorify Germany; for East Germany is partly Slav, and Belgium and North-east France are ethnologically more Germanic than Bavaria.

Part ii. traces the process by which the Germany of Kant and Herder and Goethe became the Germany of Treitschke, Bernhardi, and the author of "The Hymn of Hate." Mr. Robertson gives an excellent historical survey, and, coming to recent times, quotes telling proofs of Germany's scheming for Britain's downfall from the writings of Prince von Bülow and other statesmen. It is clear enough now that only our supremacy at sea saved us from attack in 1900. The great blunder of Germany in 1914 was in supposing that Britain would not fulfil her treaty obligations to Belgium. Having no principles herself, no recognition of international morality, she expected a similar lack in others. Formerly few of us could believe in her criminal attitude. Now she has opened our eyes, and we see that her power must be crushed before stable peace in Europe can be hoped for.

Vol 97, 379 1916

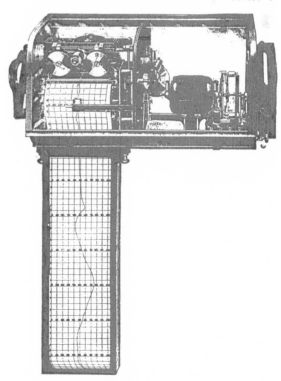

The war seems to have exacerbated the rancour between the representatives of science and the arts. Blackwood was the most venerable of the literary magazines (always called "Maga", from the time when its founder, in an easily guyed Edinburgh accent, was accustomed to refer to it as "mah mahgzine"). Sir Edward Schäfer (later Sharpey Schäfer, the discoverer of adrenaline) was a leading British physiologist.

Science versus Classics.

In "Musings without Method"—which might with equal alliteration be termed "Ravings without Reason"—the editor of *Blackwood* gives in the March number his views on the claim of science to occupy a more prominent position in general education than has hitherto been allotted to her. He calls this claim "a ferocious attack on the humanities," an evidence of "unbalanced minds" devoid of "the sense of humour and proportion." He gratuitously assumes that men of science desire "to kill all other learning than their own," and asserts that for all men there is a need of verbal expression "which is most easily satisfied by the study of Greek and Latin." He endeavours to pour scorn on the usefulness of scientific knowledge by the story —probably apocryphal—of a "commercial house in the East" which sent to Cambridge for a chemist, and when a chemist was forwarded to them, promptly returned him on the ground that although there was nothing wrong with him as a chemist, he had no knowledge of the world! One wonders what has become of "Maga's" "sense of humour"? Clearly the "commercial house in the East" did not want a chemist! Had they asked for what they really wanted they would have been sent a classical don; who doubtless would have proved more than a match for the heathen Chinee, which was probably the problem to be tackled!

It is essentially the cause of Oxford and Cambridge which our knight of the pen comes forward to champion—at least, it is what he conceives to be the cause of Oxford and Cambridge. But why should Oxford and Cambridge furnish an exception? They might, it is true, from their more ancient standing, claim to give a lead to the others, but it should surely be the aim of all the universities to provide the best system of education which the needs of the country require.

The question is: What is the best system? We others believe that it is to be found in the introduction of the study of natural science into the upbringing of everyone, whatever his ultimate aim in life may be. The prime object of education is, or should be, the attainment of a knowledge of ourselves and our surroundings: this knowledge can only be obtained through the study of natural science. That other branches of learning—mathematics, philosophy, history, language, and literature—may help, is not contested, but the basis of education in an age in which all our prosperity, present and prospective, depends upon proficiency in science must be scientific. If he who runs cannot read as much as this, he is either purblind or hopelessly slow of understanding!

We need not go outside the pages of "Maga" to prove the inadequacy of the classics. Of what is this Cabinet composed which the editor has denounced in unmeasured terms from month to month as patterns of imbecility, hesitation, and vacillation, unable to see beyond the ends of their noses, incompetent to manage any department of State? Are not the ranks of the "gallant twenty-two" (now twenty-three) recruited almost exclusively from the institutions on the system of education of which "Maga" sets so high a value? Is not the Prime Minister, against whom particularly the editorial fulminations of "Maga" have so often been directed, himself a notable example of

classical attainments? Far be it from scientific men to belittle these or any other accomplishments—philosophical, literary, or artistic. Our contention is that —along with the more advanced study of the natural sciences—these other branches of learning should be treated as subjects of special education: that they ought not to dominate the general education of the country. So far from having any wish to kill all other learning, we desire to promote *all* learning, but that desire does not prevent us from thinking that training in science will have to take the place in schools which is now occupied by Greek and Latin.

I am aware that our opponents may retort we have no right to assume that persons who have had training in science as an integral part of their education are more competent to manage the affairs of the nation or to carry on business or industrial operations than those who do not possess this advantage. We possess, however, an example of the influence of scientific training on efficiency in one of the largest of our public departments—the Navy. This is admitted, even, I believe, by "Maga," to be the best organised of those departments; it is certainly the one in which the public places the most confidence. But the men upon whom this efficiency depends are distinguished from those of all other public Services in the fact that their education is, from the beginning, purely scientific. They have had no opportunity for the acquisition of that knowledge of the classics which "Maga" appears to consider necessary for the making of men; yet even the boys of the Navy have again and again demonstrated by their actions that the scientific training which they have received has not prevented them from showing of what stuff they are made.

Nor has "Maga" the right to assume that it is only the classical members of our ancient universities who have come forward so splendidly in this crisis of our national life. For, side by side with those of their classical fellows, "stand imperishable upon the roll of honour" the names of hundreds of science students who have—whether "from their despised studies" or not I cannot say—also "learned and taught the habit of command," and many of whom have, alas! also made the supreme sacrifice. But to anyone with a "sense of proportion," it must be obvious that this can have nothing to do with the question at issue. For in showing their readiness to give their lives for their country the members of the universities are doing no more than is being done by millions of their fellow-subjects at home and abroad.

E. A. Schäfer.

Vol 97, 120 1916

Denunciations of all things German and assertions of Hunnish perfidy continued to flare up in the pages of NATURE. *Here is a small part of a particularly ferocious example. Who could the anonymous professor have been who urged the destruction of Shakespeare's tomb?*

As one of the conditions of their being granted peace by those who shall conquer them on their own selected arena of brute force backed by perverted machinery and prostituted chemistry, they should be made publicly to acknowledge the enormous benefits to science made initially, not by themselves, but by those whom they forced to become their enemies, the Italians, the British, and the French.

It would be a salutary humbling of their scientific pride to be made to confess that it was the Englishman, Newton, who discovered the law of universal gravitation; the Englishman, William Harvey, who discovered the circulation of the blood; the Englishman, Priestley, who first isolated oxygen; the Scotsman, Joseph Black, who discovered the chemical relations of carbon dioxide; and the Scotsman, Rutherford, who discovered nitrogen gas. They must be made to know that the Englishman, Stephen Hales, was the first to perceive the necessity of a mechanical system of ventilation, to estimate the magnitude of the blood pressure *in vivo* by an instrument which he had devised for estimating the pressure of sap in plants, and that he was the first to invent an apparatus for artificial respiration. Chemistry as a science was created by the Englishman, Dalton. They should be made to confess that the steam-engine was a British invention, as was also the steamboat; that the electric telegraph, the telephone, and the phonograph were all inventions of English-speaking people. The bicycle and the aeroplane were devised on the soil of Britain.

It was Faraday, they should be made to confess, who laid the basis of electromagnetics, and therefore the foundations of that amazing industrial application of electricity as a mode of motion. It was Davy who showed the elemental character of the alkaline metals —a discovery of the greatest moment. They must be made to realise that Boyle, Cavendish, Watt, Stephenson, Leslie, Hutton, and Lyell, as well as John Hunter, Jenner, Simpson, and Lister, were Britons who made discoveries of the first importance. They must be forced to confess the supreme character of the work of Napier, the Herschels, Adams, Clerk-Maxwell, and Kelvin. We, on our part, always acknowledge the indebtedness of science to such Germans as Mayer, Helmholtz, and Ehrlich; whereas our enemies systematically conceal their immense indebtedness for the enunciation of first principles to men of the English-speaking race.

* * *

Are the Germans grateful to us for what we have done in science? Do they realise, when they use railroads and steamers, dynamos and telephones, that they are all of British origination? They realise nothing of the kind. Not only are they not grateful for the benefits conferred on them by British science, but they have entered into a conspiracy of silence with regard to them.

Let us never forget that it was a German professor of physics who deliberately declared that German aircraft must destroy the tombs of Newton and of Faraday. He also included the tomb of Shakespeare, which was highly inconsistent with the widespread academic delusion that our and the world's greatest poet was a German.

D. FRASER HARRIS.
Halifax, Nova Scotia, September 30.

VOL 98, 168 1916

Sir Arthur Stanley Eddington (who famously replied, when asked whether it was true that only three men in the world understood the general theory of relativity, that he wondered who the third might be) was a writer with an exceptional gift for lucid exposition of difficult concepts (although his incoherence as a speaker became a legend; he was said never to have finished a sentence in any of his lectures). Here is how he treats the subject of relativity.

GRAVITATION AND THE PRINCIPLE OF RELATIVITY.

ACCORDING to the principle of relativity in its most extended sense, the space and time of physics are merely a mental scaffolding in which for our own convenience we locate the observable phenomena of Nature. Phenomena are conditioned by other phenomena according to certain laws, but not by the space-time scaffolding, which does not exist outside our brains. As usually expressed, the laws of motion and of electrodynamics presuppose some particular measurement of space and time; but, if the principle is true, the real laws connecting phenomena must be independent of our framework of reference—the same for all systems of co-ordinates. Of course, it may be that phenomena are conditioned by something outside observation—a substantial æther which plays the part of an absolute frame of reference. But the following considerations may show that the ideal of relativity is not unreasonable. Every observation consists of a determination of coincidence in space or time. This is sufficiently obvious in laboratory experiments; and even the crudest visual observation resolves itself into the coincidence of a light-wave with an element of the human retina. If, then, we trace the path of adventure of a material particle, it intersects in succession the paths of other particles or light-waves, and these intersections or coincidences constitute the observable phenomena. We can represent the course of Nature by drawing the paths of the different particles—on a sheet of paper in a two-dimensional case. The essential part of the diagram is the order of the intersections; the paths between the intersections are outside observation altogether, and are merely interpolated. The sequence of phenomena will not be altered if the paper is made elastic and deformed in any way, because the serial order of the intersections is preserved. This deformation of the paper corresponds to a mathematical transformation of the space in which for convenience we have located the phenomena.

Until recently the application of the principle of relativity was limited to one particular transformation, namely, a uniform translation of the axes. In this case there is a wide range of experimental evidence in support of the principle. In 1915 Prof. A. Einstein [1] finally succeeded in developing the complete theory by which the postulate of relativity can be satisfied for all transformations of the co-ordinates. Gravitation plays a part of great importance in the new theory, and therein lies much of the practical interest of Einstein's work.

No attempt is made to explain the cause of gravitation—as a kink in space or anything of that nature. But the extended law of gravitation is determined, to which Newton's law is an approximation under ordinary conditions. It has long been suspected that there must be some modification of the law when the bodies concerned are in rapid relative motion; moreover, the "mass" of a moving body no longer has a unique meaning, so that a further definition, if not extension, of Newton's law is clearly needed. Now, although we do not seek a *cause* of gravitation in the properties of space, it may well happen that the *law* of gravitation is determined by these properties. The inverse-square law represents the natural weakening of an effect through spreading out in three dimensions; we may say that it is determined by the properties of Euclidean space. There is, therefore, nothing unreasonable in proceeding, as Einstein does, to examine whether a more extended law is suggested by the properties of generalised space—that is, by geometry.

The way in which gravitation enters into the discussion may be seen from the following example. Suppose an observer is in a closed lift; let the supports break and the lift fall freely. To the observer everything in the lift will now appear to be without weight; gravity has been suddenly annihilated. The acceleration of his frame of reference (the lift) is equivalent to an alteration of the gravitational field. Now an acceleration of the axes is one of the transformations contemplated by the general principle of relativity, and it is therefore necessary to allow that the gravitational field depends on the choice of co-ordinates. There is a "local" gravity, just as there is a "local" time or magnetic field depending on the co-ordinates selected.

VOL 98, 328 1916

Chandrasekhara Venkata (later Sir Venkata) Raman, the first Asian to win a Nobel prize, is now remembered mainly for the effect that bears his name. He had, however, a lifelong interest, among many others, in the physics of musical instruments. The origin of the "wolf tone" — familiar to all players of stringed instruments — was a favoured topic in the columns of NATURE.

On the Alterations of Tone produced by a Violin-"Mute."

EXPERIMENTS on the "wolf-note" of the violin or 'cello (see NATURE, June 29, and September 14, 1916, and *Phil. Mag.*, October, 1916) suggest an explanation

of the well-known and striking alterations in the tone of the instrument produced by a "mute," which at first sight seems somewhat difficult of acceptance, viz. that they are due to the lowering of the pitch of the free modes of vibration of the *entire body of the instrument* produced by the added inertia. This view of the action of the mute (which was suggested by way of passing reference in my paper on the "wolf-note") has, I find, excited some incredulity, and its correctness has, in fact, been questioned in a note by Mr. J. W. Giltay in the *Phil. Mag.* for June, 1917. The following brief statement may therefore be of interest as establishing the correctness of my view of this important phenomenon :—

If N_1, N_2, N_3, etc., be the frequencies of the free vibrations of the body (in ascending order), the frequencies as altered by the addition of the "mute" are determined by equating to zero the expression (see Routh's "Advanced Rigid Dynamics," Sec. 76),

$$(N_1^2 - n^2)(N_2^2 - n^2) \times \text{etc.}, - \alpha n^2 (n_2^2 - n^2)(n_3^2 - n^2) \times \text{etc.},$$

where α is a positive quantity proportionate to the added inertia, and n_2, n_3, etc., are the limiting values of N_2, N_3, etc., attained when the load is increased indefinitely $|n_1 = 0$, and $n_2 < N_1$, $n_3 < N_2$, etc.]. The forced vibration due to a periodic excitation of frequency n is determined by the same expression, being inversely proportional to it except in the immediate neighbourhood of points of resonance. The sequence of the changes in the forced vibration produced by gradually increasing the load is sufficiently illustrated by considering a case in which n lies between N_1 and N_2. If $n_1 < n_2$, the load decreases the forced vibration throughout, but if $n > n_2$, the load at first *increases* the forced vibration until it becomes very large, when n coincides with one of the roots of the equation for free periods, subsequent additions of load decreasing it. The *increase* in the intensity of tone indicated by this theory has actually been observed experimentally by Edwards in the case of the graver tones and harmonics of the violin (*Physical Review*, January, 1911). Edwards's observation that the intensity of tones and harmonics of high pitch is *decreased* by "muting" is also fully explained on this view, as in the case of the higher modes of free vibration of the instrument a very small load would be sufficient to make the frequencies approximate to their limiting values.

Comparison of the effects of loading the bridge of the instrument at various points on the free periods and the tones of the instrument furnishes a further confirmation of the foregoing theory. For instance, on a 'cello tried by me, the lowering of the "wolf-note" pitch produced by a load fixed on *either* of the feet of the bridge was small compared with that obtained by fixing it on top of the bridge, and the observed "mute" effect was correspondingly smaller. In fact, the alterations of free period produced by loading furnish us with quantitative data regarding the relative motion of different parts of the instrument, and of their influence in determining the character of its tones.   C. V. RAMAN.

Calcutta, August 28.

Colonel Sir Almroth Wright was the head of the inoculation department at St Mary's Hospital in London, the patron of Alexander Fleming and the model for Sir Colenso Ridgeon in George Bernard Shaw's The Doctor's Dilemma *(see also p.130). He was much concerned with avoidance of septicaemia in war wounds. Here is the beginning of his long two-part article on the subject.*

THE TREATMENT OF WAR WOUNDS.

WE are wont to classify the patients in our military hospitals into sick and wounded. In reality all, or nearly all, are suffering from bacterial infections. And the essential difference between the sick and the wounded lies in this, that the sick are suffering from infections spontaneously contracted, the wounded from infections induced by mechanical injuries. My theme is the treatment of this latter class of infections. They are distinguished by certain quite special features.

In spontaneous infection we have to deal with microbes which have fought their way into the body, and generally only a single species of microbe will have done this. In wounds we have microbes mechanically driven in, and every sort of microbe which exists in external Nature may thus be introduced.

But let me, before embarking upon the question of their treatment, first tell you something about the

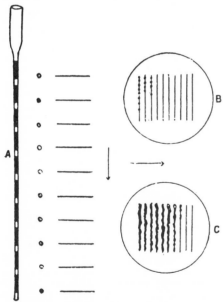

·FIG. 1.—Method of pyo-sero-culture. A, Pipette which has been implanted by the wet-wall method, and has then been filled in by the wash and after-wash procedure with unit-volumes of serum. By the side of the pipette to the right is ranged a series of drops representing the series of unit-volumes of serum blown out in order from the pipette, and, finally, to the right of the drops is a series of lines representing linear implantations made upon agar. B, Results of the series of linear implantations made with the unit-volumes of the patient's serum. C, Results of the series of linear implantations made with the unit volumes of the normal serum which was used as a control.

natural agencies by which the inroads of microbes are combated. You are, of course, aware that we are guarded against microbic infection by our blood fluids and our white blood corpuscles.

THE BODY FLUIDS.

Let me begin with the blood fluids, and let me take you directly to the following experiment. I call it the experiment of *pyo-sero-culture—i.e.* the experiment in which we implant pus into serum to see which of the microbes of the wound can grow in the blood fluids.

We procure for our experiment a suppurating wound. We take from it a specimen of pus containing a large variety of different organisms. At the same time we take from the patient's finger a sample of blood; and we take a specimen also of our own. When the serum has issued from the clot we take a capillary pipette, fit a rubber teat to the barrel, and inscribe a mark upon the stem at about, say, one-third of an inch from the tip. We now aspirate a little pus into the stem, drawing it up only so far as our fiducial mark, and, blowing it out again, leave a wash of pus upon the walls. This done, we sterilise the tip of the pipette, and then aspirate into the stem a series of unit-volumes of serum, dividing each volume off from the next by a bubble of air. The pipette when filled in this manner presents the appearance shown in Fig. 1, and we have in the proximal end our first and heaviest implantation of pus, and in the distal end our last and lightest implantation. The pipette is now placed in the incubator to allow every microbe which is capable of growing in serum to do so. After an interval of six or more hours we proceed to our examination. What we do is to blow out our series of unit-volumes of serum in separate drops and examine under the microscope; or, better, we plant out a sample of each drop upon a separate seed-bed. Here in B and C you have the results of such culture represented diagrammatically— the meagre crop in B being that obtained with the patient's serum, and the more copious crop in C being that obtained with normal serum.

And you have in the next figure (Fig. 2) a drawing of an agar tube implanted from a pyo-sero-culture made with the serum of a wounded man. In the upper part of the agar tube you see two seed-plots implanted from the distal portion of the capillary stem. These have remained sterile. In the middle of the tube you see four plots implanted from the unit-volumes of serum which occupied the middle region of the capillary stem. These have grown colonies of only one species of microbe—the streptococcus. At the bottom of the tube you see seed-plots implanted from the proximal end of the capillary stem. These are overgrown with colonies of staphylococcus. But no doubt interspersed with, and overgrown by, these are also colonies of streptococci. If, instead of cultures from the patient's serum, I had been showing you here cultures from normal serum, what you would have seen would have been a much larger number of fertile seed-plots, and the seed-plots implanted from the proximal end of the pipette would have shown a large assortment of different colonies.

FIG. 2.—A portion of a pyo-sero-culture planted out upon an agar slant divided up by furrows into a series of seed-beds.

We learn from such experiments three lessons : *first,* that in the uncorrupted serum in the distal region of the pipette only two species of microbes from the wound can grow and multiply; *secondly,* that in the corrupted serum in the proximal end of the pipette all the microbes of the wound can grow; and, *thirdly,* we learn from a comparison of the wounded man's serum with the normal serum that the former offers more resistance to microbic growth, and is less easily corrupted by the addition of pus.

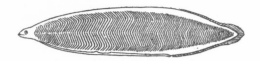

The second part shows how impressively microbiology could already be applied in the field.

Experimental Investigation of the Efficacy of Antiseptics.

Let me now try to indicate to you what sort of experiments should be undertaken before nourishing in connection with a particular antiseptic the expectation that it is going to be efficacious for sterilising and afterwards suppressing microbic growth in wounds. I can illustrate my points best if you let me show you here four tubes.

In tube No. 1 I have a suspension of microbes in water. I now add an equal volume of the antiseptic I wish to test and shake up thoroughly. These are, as you see, conditions which give every possible advantage to the antiseptic. It is applied in a non-albuminous medium and is intimately mixed with the microbes. To find out whether the microbes have been killed I draw off a sample and dilute with very many times its volume of nutrient medium. I then incubate to see whether I get any bacterial growth.

In tube 2 I make the conditions more favourable to the survival of the microbes—infinitely more favourable than if I left behind an antiseptic in a wound. I have here a mixture of staphylococci, streptococci, and gas-gangrene bacilli suspended in serum, and I now, as in tube 1, add an equal bulk of the antiseptic and shake up, and I then, following the technique of Prof. Beattie, pour on a little hot vaseline which will afterwards congeal. This, forming an air-tight seal, will allow the gangrene bacillus, if it survives, to grow out. It will also announce the growth of this microbe, for it will confine any gas which may be evolved from the culture.

Tube 3 is, as you see (Fig. 10), a tube which has

Fig. 10.—A test-tube standing on spike legs, representing a war wound with diverticula.

been drawn out into a number of hollow spikes to imitate the diverticula of the wound. My colleague, Dr. Alexander Fleming, its author and inventor, calls this form of tube the "artificial war wound." To imitate the conditions obtaining in the actual war wound we fill both the tube and its diverticula with an infected trypsinised serum. We now empty the tube, leaving behind of necessity in the diverticula a certain

amount of the original infected fluid. We then fill with an antiseptic; and the future of the infection will now depend on the penetrating power of the antiseptic. If the antiseptic penetrates into the infected fluid sterilisation will be obtained; if it fails to penetrate, microbes will survive. To test our result we empty out the antiseptic, refill with trypsinised serum, and incubate.

In the debate between scientists and humanists, the arrogance was fairly evenly divided, but seldom as crudely expressed as here by H. G. Wells. His adversary is a noted classicist, translator of Thucydides and head of an Oxford college.

Classical Education and Modern Needs.

A review by Mr. H. G. Wells in Nature of April 19 contains the following words: "This claim is pressed even more impudently by Mr. Livingstone in his recent 'Defence of Classical Education.' He insists that all our sons are to be muddled about with by the teachers of Greek up to at least the opening of the university stage."

This is a complete misrepresentation of my views, the more gratuitous because in several passages I insist on the importance of *not* teaching Greek and Latin to those boys who are unsuited for them—*e.g.* on p. 241: "It ought to be a first aim . . . to avoid diverting boys with mechanical or scientific tastes, who have no aptitude for linguistics, into studies that will be barren for them." (The context shows that the studies referred to are Latin and Greek.)

With regard to the present system of "compulsory Greek," after pointing out that it was an undesirable system, maintained on the ground that without it Greek teaching would, in present circumstances, disappear in many important educational areas, I remarked that "it would be possible, almost without opposition," to abolish it if such facilities were provided for the study of Greek as would put it within the reach of all boys in secondary schools *who wished to learn it*. This does not seem to me an impudent claim; it would be easy to satisfy; and I imagine that no one would take exception to it.

R. W. Livingstone.
Corpus Christi College, Oxford,
    April 23.

———

Mr. Livingstone's letter is satisfactory, so far as it goes, in promising to spare such boys as are unworthy of classical blessings, but I think many of the readers of Nature will see in its phrasing just that implicit claim to monopolise the best of the boys for the classical side of which I complain. We do not want the imbeciles, the calculating boys, the creatures all hands and no head, and so forth, for the modern side. We want boys for scientific work who may be not "unsuited," but eminently suited for Greek and Latin, in order that they may do something better and more important. I write with some personal experience in this matter. I am very much concerned in the welfare of two boys who have a great "aptitude for linguistics," and would make excellent classical scholars. I think I can do better with them than that, and that they can serve the world better with a different education. In each case I have had to interfere because they were being "muddled about with" by the classical side masters, and have got Russian substituted for the futile beginnings of Greek. The fact remains that Mr. Livingstone does, under existing conditions, wish to retain compulsory Greek.

H. G. Wells.

Sir Oliver Lodge's notorious interest in the paranormal was much intensified after the war had claimed his son, Raymond, with whom he communicated through a medium. Raymond, it appeared, was adapting to the mores of the Other Side, where the newly arrived "chaps" tended to ask for, and be given, cigarettes and whisky. Those who had been blown to pieces were slowly reassembled, but could also at a pinch grow back a missing limb. Sir Oliver's book Raymond *was one of the publishing successes of the war, repeatedly reprinted, though denounced by the Church and the* Daily Mail, *which called it "half a guinea's worth of rubbish".* NATURE *treated Lodge with forebearance.*

Raymond: or Life and Death. With Examples of the Evidence for Survival of Memory and Affection after Death. By Sir Oliver J. Lodge. Pp. xi + 403. (London: Methuen and Co., Ltd.) Price 10s. 6d. net.

LIEUT. RAYMOND LODGE was killed by shrapnel, near Ypres, in September, 1915. In this volume, which is at once scientifically important and humanly touching, we are given in part i. a selection of letters which show Raymond's fine and attractive personality; in part ii. some of the evidence which indicates his continued existence and occasional communication; and in part iii. Sir Oliver Lodge discusses the philosophy of the subject, with large tolerance but full conviction. Survival is reasonable enough. Life is not a form of energy. It guides or directs energy, but there is no sound reason to believe that it goes out of existence when it ceases to manifest through a particular body.

Of part ii., which more specially concerns a scientific journal, the most striking incident is the one referring to a photograph which was said—through two sensitives—to have been taken, though the family knew nothing about it, and learnt nothing for some months. This photograph was described very minutely. It was said to consist of a group of soldiers, numbering a dozen or more men, some standing and some sitting; Raymond would be found at the front, sitting down, with a stick, and a man was said to be leaning on him or trying to lean on him; vertical lines would be prominent at the back, and of the figures the most prominent would be that of a man whose name began with B. Ultimately the photograph (taken in Flanders) came to light, and all the details corresponded with the description received. As given in full in the book, this incident is very impressive, and it is supported by many others of varying degrees of evidential weight.

The volume is inevitably of an intimately personal nature, but a restrained and scientific temper is maintained throughout, and contentions are supported by facts. It will probably be considered the most important psychical book since Myers's great work on Human Personality; and it is unique in the sense that it is the first large book of its kind to be published by a man of science of the first rank.   J. A. H.

A tale of the amphibious louse:

Vitality of Lice.

I HAD occasion recently to examine microscopically some head-lice (*Pediculus capitis*) under water, and I noticed a phenomenon to which I have been unable to find any reference in standard works.

On first being placed into water contained in a watch-glass the lice struggled, but after a short time there was no activity visible, and life appeared to be extinct. After three-quarters of an hour I poured out the water from the watch-glass and dried the lice. In a few seconds they showed manifestations of life, and within a minute resumed their normal activity, internal metamorphosis and metabolism being visible. This led me to further experiments, and I find that after being submersed completely for fifteen hours in a beaker of distilled water free from air, they regain their normal activity within a quarter of an hour of their removal from water. I have not yet tried submersion for longer periods, but the subject is of great interest, and I should be obliged if any of your readers are able to throw any light upon it.

HENRY COHEN.

"Avenue House," Petworth Street, Cheetham, Manchester, September 2.

This letter represents what may be a vanished scholarly subdiscipline.

Folk-lore and Local Names of Woodlice.

AMONST the readers of NATURE there are many, I feel sure, who are interested in the folk-lore and local names of the members of our fauna. May I appeal to such for any information bearing upon the heading of this letter?

Already nearly one hundred local names, such as bibble-bug, chisel-hog, cud-worm, palmer, lock-chester, slater, tiggy-hog, etc., have been obtained, and the districts noted in which such are in use. Celtic and Gaelic names are particularly desired.

WALTER E. COLLINGE.

The University, St. Andrews, October 4.

Despite the unforgiving mood of the time, there seems to have been no animus against Adolf von Baeyer, whose death in 1917 was marked by a fulsome obituary by his most famous English pupil, W. H. Perkin, which contained the stories below. Baeyer was a widely revered figure whose many distinguished students were unswervingly loyal to him. He was one of the last of the old-style organic chemists for whom it was a point of pride to lead from the bench and perform the trickiest syntheses themselves. Baeyer retained throughout his long life a child-like pleasure in discovery.

Baeyer's immense power of work is shown by the fact that, until his eightieth birthday, he delivered his usual lectures on five mornings of each week and continued to experiment in his laboratory with his usual unflagging energy. Had the war not robbed him of his private assistant and laboratory staff, it is probable that he would have gone on even longer. He confided to one of his intimate friends that work in the laboratory gave him as much pleasure after fifty years' toil as at any time during his career, and to the last he took the greatest interest in any developments in the domain of natural science which were brought to his notice. It is well known that he viewed with disfavour and apprehension the growing domination of military power in Berlin and Prussia generally, and it was mainly, no doubt, for this reason that he refused to accept the invitation to Berlin on the death of Hofmann.

* * *

In his "Erinnerungen aus meinem Leben," which he wrote for the celebrations organised in connection with his seventieth birthday, he tells us that he converted a passage in the house into a small laboratory, and there carried out the usual dangerous and unpleasant experiments associated with early youth. It was during this time that he made his first discovery, that of the double salt, $CuCO_3, Na_2CO_3, H_2O$. The activity of the small laboratory does not seem to have been altogether appreciated, and the poet, Paul Heyse, who was a frequent visitor at the house, had reason to protest:

> Es stinkt in diesem Haus gar sehr
> Das kommt vom Adolf Baeyer her.

VOL 100, 189 1917

The radiological aeroplane may have been an idea before its time.

A "RADIOLOGICAL aeroplane" was described by Drs. Nemirowski and Tilmant before the Academy of Medicine of Paris at a meeting on September 3. It contains three places for the pilot, surgeon, and radiographer, and is provided with a generator for Röntgen-rays, one operating-table for operations performed with the aid of the rays, surgical instruments, and medicaments. The "Aerochir," as it is called, is intended to fly over the lines of action, ready to alight and render first aid to the wounded. The invention should be invaluable, provided, however, that it is not regarded by the enemy as a target for his fire.

VOL 102, 130 1918

The carnage on the Somme became associated in the imagery of the time with poppies—

> In Flanders fields the poppies blow
> Between the crosses, row on row

— but, as this poignant piece shows, there was plenty of scope for botanising among the mounds that marked the resting places of the fallen.

THE FLORA OF THE SOMME BATTLEFIELD.

THE ground over which the Battle of the Somme was fought in the late summer and autumn of 1916 rises gradually towards Bapaume, and at the same time is gently undulating, with some well-marked branching valleys initiating the drainage system of the area. Before the war the land was for the most part under cultivation, but on the highest levels there were large areas of woodland, such as High Wood and Delville Wood, now shattered and destroyed.

During last winter and spring all this country was a dreary waste of mud and water, the shell-holes being so well puddled that the water has remained in them, and even in the height of the summer there were innumerable ponds, more or less permanent, in every direction.

The underlying rock is everywhere chalk with a covering of loam of varying thickness. As a result of the bombardment the old surface soil has been scattered and the chalk partially exposed. One effect of the shelling, however, has been to disintegrate the underlying chalk and produce a weathering effect which has been accentuated by the winter rains, snow, and frost. A general mixing of chalk, subsoil, and scattered top soil and also a rounding of the sharp edges have taken place, so that instead of the new surface soil being sterile, the shelling and weathering have "cultivated" the land. That this is so is proved by the appearance of the Somme battlefield during the past summer.

Looking over the devastated country from the Bapaume Road, one saw only a vast expanse of weeds of cultivation which so completely covered the ground and dominated the landscape that all appeared to be a level surface. In July poppies predominated, and the sheet of colour, as far as the eye could see, was superb; a blaze of scarlet unbroken by tree or hedgerow. Here and there long stretches of chamomile (*Matricaria chamomilla*, L.) broke into the prevailing red and monopolised some acres, and large patches of yellow charlock were also conspicuous, but in the general effect no other plants were noticeable, though a closer inspection revealed the presence of most of the common weeds of cultivation, a list of which is given below.

Charlock not only occurred in broad patches, but was also fairly uniformly distributed, though masked by the taller poppies. Numerous small patches were, however, conspicuous, and these usually marked the more recently dug graves of men buried where they had fallen. No more moving sight can be imagined than this great expanse of open country gorgeous in its display of colour, dotted over with the half-hidden white crosses of the dead.

In all the woods where the fighting was most severe not a tree is left alive, and the trunks which still stand are riddled with shrapnel and bullets and torn by fragments of shell, while here and there unexploded shells may still be seen embedded in the stems. Aveluy Wood, however, affords another example of the effort being made by Nature to beautify the general scene of desolation. Here some of the trees are still alive, though badly broken, but the ground beneath is covered with a dense growth of the rose-bay willow herb (*Epilobium angustifolium*) extending over several acres. Seen from across the valley, this great sheet of rosy-pink was a most striking object, and the shattered and broken trees rising out of it looked less forlorn than elsewhere.

The innumerable shell-hole ponds present many interesting features to the biologist. In July they were half-full of water, and abounded in water beetles and other familiar pond creatures, with dragonflies flitting around. In nearly every shell-hole examined, just above the water-level, was a band of the annual rush (*Juncus bufonius,* var. *gracilis*), and this plant appeared to be confined to those zones where the ground was relatively moist, and to occur nowhere else. With the Juncus, and often growing out of the water, were stout plants of *Polygonum persicaria,* and water grasses, not in flower, were often seen spreading their leaves over the surface of the pools.

In the battlefield area not only were the common cornfield weeds to be seen, but here and there patches of oats and barley, and occasionally plants of wheat, sometimes apparently definitely sown, perhaps by the Germans, though more often the plants must have grown from self-sown seeds of crops that were on the land before the war. Here and there, too, could be seen opium poppies representing former cultivation and remnants of battered currant and other bushes which alone remained to show where once had been a cottage garden. Both weeds and corn afford good evidence that the soil has not been rendered sterile by the heavy shelling, but how and when the land can be brought into a fit state for cultivation are questions not easily answered.

On the banks and sides of the roads traces of the old permanent flora still remain, and perennial plants, such as *Scabiosa arvensis, Eryngium campestre, Galium verum,* chiccry, *Centaurea scabiosa, Cnicus acaulis,* and other characteristic chalk plants were occasionally seen.

The clothing of this large tract of country with such a mass of vegetation composed almost entirely of common annual cornfield weeds is remarkable when one remembers that it has been the seat of encampments, and for the most part out of cultivation since the autumn of 1914. It is well-nigh impossible that such masses of seed can have been carried by wind or birds to cover these thousands of acres, and the plants must therefore have grown from seed lying dormant in the ground. No doubt in the ordinary operations of ploughing and tilling of the ground in years before the war much seed was buried which has been brought to the surface by the shelling of the ground and subsequent weathering. In this connection the presence of charlock on the more recently dug graves, where the chalk now forms the actual surface, is of interest, since it adds further proof of the longevity of this seed when well buried in the soil.

VOL 100, 475 1918

As the denouement of the Great War approaches, resentments simmer. Here Lord Walsingham grumbles at the effrontery of the German biologists who, back in 1913, had pressed for international adoption of their preferred nomenclature. A devilish revenge is now in prospect: make them conform to English usage.

It may be remembered by those who were present at the meeting of the International Zoological Congress at Monaco in March, 1913, how persistently the representatives of German scientific societies endeavoured on that occasion to dominate the discussions, especially on the subject of the rules of nomenclature, and insisted that the names habitually employed in Germany should receive the sanction of long

usage, to the exclusion of all attempts to trace out the literary history of each species and to preserve for it the name bestowed by the first author who described or figured it.

* * *

Are American, English, French, or Italian naturalists to be expected to meet Germans and to join them in friendly discussion on the various questions that may arise? Considering that before the war every man, woman, and child in Germany, with scarcely an exception, was intent upon war, as has been amply demonstrated by the evidence of innumerable witnesses, it is impossible to dissociate the mental attitude of the population of that country, by no means excepting the highly educated and scientific classes, from the world-conquering aspirations of their rulers, or from the barbarous atrocities committed by them in pursuit of that national ideal. A conspicuous instance is that of a certain learned professor with whom I was on terms of friendship, who was honoured by the Universities of Liverpool and Dublin, and delivered lectures in London under the auspices of the London University, turning out eventually to be a German spy engaged in fomenting rebellion in Ireland and antagonism to England and her Allies in the United States. If an individual in any community commits murder or robbery, or is even plausibly suspected of swindling or cheating at cards, the unavoidable and universally recognised penalty is that no man with a grain of self-respect will ever again associate with him, shake his hand, or converse with him in friendship.

Let us trust that for the next twenty years at least all Germans will be relegated to the category of persons with whom honest men will decline to have any dealings.

* * *

To those Germans, if any there be, who are honestly well disposed, and who put the interests of science before the greed for world-domination, it can be no hardship to publish their descriptions in the English or French language, with which the great majority of their scientific workers are more or less intimately acquainted.

His Lordship boils over:

In Russia the paid agents of Germany have brought about, or at least connived at, the wanton destruction of treasures innumerable; some of the finest entomological collections in the world were in Russia, in Belgium, and in Rumania. I would urge that it is the plain duty of the Allies to insist not only that all objects removed shall be replaced, but also that equivalent value in kind shall be rendered for everything destroyed or damaged, and this should apply to specimens illustrating the study of natural history (best represented in value by original author's types), as well as to pictures, statues, and other objects of art or antiquity, for the selection of which from German museums special commissioners should be appointed.   WALSINGHAM.

6 Montagu Place, Portman Square, W.1,
August 29.

VOL 102, 4 1918

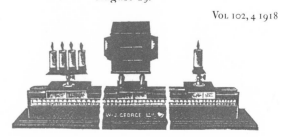

It has been an unceasing source of chagrin for British chemists to be taken by the ignorant for pharmacists. Under the title "The Profession of Chemistry" the writer begs "pharmacists" and "druggists" to keep off his patch. He patronisingly disclaims any snobbish intent and concludes like this (but he is of course quite wrong about the Huns, for whom a pharmacist is an Apotheker):

Pharmacy is an honourable occupation, and I cannot believe that the pharmacist would lose dignity or status by the change. Comparatively few pharmacists are chemists in the modern sense, and it is well known that in other great countries this confusion of titles does not prevail; in fact, this is one of the few points on which we are at variance with our Allies, whilst they are in complete harmony with the Hun.

M. O. FORSTER.

Savage Club, W.C.2, March 4.

VOL 103, 25 1919

Here is a further example of Sir Arthur Stanley Eddington's unsurpassed clarity of exposition. One needs to read the entire article, of which this is a sample, to appreciate fully the effortless sweep of the argument.

GRAVITATION AND THE PRINCIPLE OF RELATIVITY.

THERE were many difficulties to encounter in entering the room just now. To begin with, we had to bear the crushing load of the atmosphere, amounting to 14 lb. on every square inch. At each step forwards it was necessary to tread gingerly on a piece of ground moving at the rate of twenty miles a second on its way round the sun. We were poised precariously on a globe, apparently hanging by our feet, head outwards into space. And this acrobatic feat was performed in the face of a tremendous wind of æther, blowing at I do not know how many miles a second literally through us. We do not claim much credit for overcoming these difficulties—because we never noticed them. But I venture to remind you of them, because I am going to speak of some other extraordinary things that may be happening to us of which we are quite unconscious.

Not to go too far back in history, the present subject arises from a famous experiment performed in the year 1887, known as the Michelson-Morley experiment. The apparatus was elaborate, but the principle of the experiment is not very difficult. If you are in a river, which will be the quicker—to swim to a point fifty yards up stream and back again, or to a point fifty yards across stream and back again? Mathematically the answer is, perhaps, not immediately obvious, because the net effect of the current is a delay in both cases. But I think that anyone who has swum in a river will have no hesitation about the answer. The up-and-down journey takes longer. Now we *are* in a river—of æther. There is a swift current of æther flowing through this room; or, if we happen to be at rest in the æther at the present moment, six months hence the earth's orbital motion will be reversed, and then there must be a swift current. Michelson divided a beam of light into two parts; he sent one half swimming up the stream of æther for a certain distance, and then by a mirror back to the starting point; he sent the other half an equal distance (as he thought) across the stream and back. It was a race; and with his apparatus he could test very accurately which part got back first. To his surprise, it was a dead-heat. Clearly the two paths could not really have been equal, the along-stream path must have been a little shorter to compensate for the greater hindrance of the current. That objection was foreseen, and the apparatus, which was mounted on a stone pier floating in mercury, was rotated through a right angle, so that the arm which was formerly along the stream was now across the stream, and *vice versa*. Again the two portions of the beam arrived at the same moment; so this time the other arm had become the shorter—simply by altering its position. In fact, these supposedly rigid arms had contracted when placed in the up-and-down stream position by just the amount necessary to conceal the effect which was looked for.

That is the plain meaning of the experiment; but we might well hesitate to accept this straightforward interpretation, and try to evade it in some way, were it not for some theoretical discoveries made later. It has gradually appeared that matter is of an electrical nature, and the forces of cohesion between the particles, which give a solid its rigidity, are electrical forces. Larmor and Lorentz discovered that this property of contraction in the direction of the æther current was something actually inherent in the formulæ for electrical forces written down by Maxwell many years earlier and universally adopted; it only waited for some mathematician to recognise it. It would be going too far to say that Maxwell's equations actually prove that contraction must take place; but they are, as it were, designed to fall in line with the contraction phenomenon, and certain details left vague by Maxwell have since been found to correspond.

We are then faced with the result that a material body experiences a contraction in the direction of its motion through the æther. According both to theory and experiment the contraction is the same for all kinds of matter—a universal property. One reservation should be made; the experiment has only been tried with solids of laboratory dimensions, which are held together by *cohesion*. There is at present no experimental evidence that a body such as the earth the form of which is determined by *gravitation* will suffer the same contraction; we shall, however, assume that the contraction takes place in this case also.

I am going to ask you to suppose that we in this room are travelling through the æther at the rate of 161,000 miles a second, vertically upwards. Let us be bolder and say that that *is* our velocity through the æther—because no one will be able to contradict us. No experiment yet tried can detect or disprove that motion; because all such experiments give a null result, as the Michelson-Morley experiment did. With that speed the contraction is just one-half. This pointer, which I hold horizontally, is 8 ft. long. Now [turning it vertically] it is 4 ft. long. But, you may say, it is taller than I am, and I must be approaching 6 ft. No, if I lay down on the floor I should be, but as I am standing now I am under 3 ft. The contraction affects me just as it did the pointer. It is no use bringing a standard yard-measure to measure me, because that also will contract and represent only half a yard. "But we saw that the pointer did not change length when it turned." How did you tell that? What you perceived was an image of the pointer on the retina of your eye, and you thought the image occupied the same space of retina in both positions; but your retina has also contracted in the vertical direction without your knowing it, so that your estimates of length in that direction are double what they should be. And similarly with every test you could apply. If everything undergoes the same change, it is just as though there were no change at all.

We thus get a glimpse of what, from our present point of view, must be called the *real* world, strangely different from the world of appearance. In the real world, by changing position you extend yourself like a telescope; and the stoutest individual may regain slimness of figure by an appropriate orientation. It must be something like what we see in a distorting mirror; and you can almost see a living-picture of this real world reflected in a polished door-knob.

VOL 101, 15 1918

chapter eleven
1919–1923

PAUL KAMMERER'S EXPERIMENTS ON INHERITANCE OF ACQUIRED CHARACTERISTICS UNLEASH A STORM OF CONTROVERSY. THE DUC DE BROGLIE GIVES TONGUE ON WAVES AND QUANTA. RECRIMINATIONS ABOUT CHEMICAL WARFARE CONTINUE, WITH SIR EDWARD THORPE AND FRITZ HABER THE LEADING PROTAGONISTS. IN TENNESSEE, THE LEGALITY OF TEACHING EVOLUTION IN SCHOOLS IS TESTED IN THE "MONKEY TRIAL". AN ELECTRICAL LARYNX IS INVENTED. H. G. WELLS IS ELOQUENT ABOUT INTELLECTUAL FREEDOM IN POST-REVOLUTIONARY RUSSIA. SRINIVASA RAMANUJAN DIES IN INDIA AND IS EULOGISED BY HIS PATRON, G. H. HARDY.

1919–1923

The Russians were driven out of Warsaw, the Kronstadt mutiny was crushed and Stalin became General Secretary of the Soviet Communist Party. German reparations were fixed at an insupportable £6 billion. The Munich Putsch followed, Adolf Hitler scored a publishing success with *Mein Kampf* and inflation was stabilised by the *Rentenmark*. The University of Oxford caved in and admitted women to its degrees.

TROPISMS.

Forced Movements, Tropisms, and Animal Conduct. By Prof. Jacques Loeb. Pp. 209. (Philadelphia and London: J. B. Lippincott Co., 1918.) Price 10s. 6d. net.

A PROLIFIC investigator does a great service to his brethren when, without waiting to write an elaborate treatise, he collects the gist of some considerable portion of his work into a book; and if the book be a small one, so much the better. This Prof. Loeb has now done, and we are immensely obliged to him.

So begins D'Arcy Thompson's admiring review of Jacques Loeb's book (see p108). The concluding passage is in a characteristic vein.

I think it was Liebig who said, in one of his letters to Faraday, that (in those days) a man might be an eminent geologist in England who knew nothing of physics, nothing of chemistry, nothing even of mineralogy. Change the wording, and the biologist may (or once upon a time might) have begun to feel uneasy. It is something to be taught or reminded by Prof. Loeb, and by the whole brotherhood of experimental biologists, that the naturalist cannot live alone, but works in a field inextricably connected, for better for worse, with the whole range of the physical sciences.

Prof. Loeb has a boundless wealth of ideas. In this book and in his other books and papers we seem to see them tumbling one over another. He has enough and to spare for all his pupils and fellow-workers, so that all who come to him may eat and be filled. Moreover, his manifold experiments all have the hall-mark of simplicity, and this is surely one of the greatest things that can be said of any experimenter. There is no parade of elaborate apparatus, nor does it ever seem to be required. *Simplex sigillum veri!*

The book concludes with a bibliographical list of nearly six hundred titles—a catalogue of books and papers on experimental biology, in the sense in which Prof. Loeb himself deals with it. In the first hundred and fifty titles (and I have gone no further in my analysis) sixty-three are German, forty-three American, thirty-eight French, and four more are Dutch or Italian. I shrink from doing the addition and subtraction which would reveal our British share.

D'ARCY W. THOMPSON.

The dispute about Lamarckian inheritance that had riven the biological community in the years before the Great War was revived immediately after, though by then Paul Kammerer and his allies could have had few remaining supporters. One such, however, was E. W. MacBride, who appears to have been totally immune to self-doubt. Here William Bateson, clearly exasperated beyond the limits of patience, takes him on. He details his exchange of letters with Kammerer, and continues:

Later in the summer of 1910 I unexpectedly was able to attend the *Mendelfeier* at Brünn, and was for some time in Vienna, having the privilege of being the guest of my old friend Dr. Przibram. I was many times at the Versuchsanstalt, and inquired in vain for the Alytes. On one occasion especially, about October 3 or 4, I was there in company with Profs. E. Baur, Lotsy, Nilsson-Ehle, Dr. Hagedoorn, and the late M. Ph. de Vilmorin. Those who survive of that party will remember that, on conferring together, we all shared the same feeling of doubt. After seeing what Dr. Kammerer showed us we were entirely unconvinced, and in particular it seemed to us inexplicable that, if Alytes had existed with Brunftschwielen in July, one specimen of so great a curiosity should not have been preserved, if only for exhibition with the Salamanders at Dr. Kammerer's numerous lectures. I may add that I expressed my doubts categorically to Dr. Przibram, the head of the Anstalt, but I am glad to think that, though he defended Dr. Kammerer, our cordial intercourse continued unbroken up to the time of the war. Few, I imagine, will now consider that, on the evidence available, my scepticism was not justified.

He then describes the published photographs of the toads from which Kammerer drew his conclusions, all but accuses him of fraud (or at the very least delinquency) and finishes up:

Prof. MacBride urges that sceptics should repeat experiments on the inheritance of acquired characters. We, however, are likely to leave that task to those who regard it as a promising line of inquiry. Why do workers in that field so rarely follow up the claims of their predecessors? Each starts a new hare. Scarcely has one of their observations been repeated and confirmed in such a way that we could be sure of witnessing the alleged transmission if we were to try for ourselves. Brown-Séquard's observation on guinea-pigs is an exception. That has been repeated by various observers, until at length, by the work of Graham Brown, the mystery may be regarded as explained. The observation was true, but the interpretation was faulty. As I have often remarked, acquaintance with the normal course of heredity is an indispensable preliminary, without which no one can interpret the supposed effects of disturbance. This knowledge of normal genetic physiology is being slowly acquired, and already we have enough to show that several variations formerly attributed to changed conditions should not be so interpreted. Even in this case of Alytes, were a male with incontrovertible Brunftschwielen before our eyes, though confidence in Dr. Kammerer's statements would be greatly strengthened, the question of interpretation would remain, pending the acquisition of a knowledge of Batrachian genetics.   W. BATESON.

June 22.

The war is over; NATURE analyses the academic salaries offered in the job advertisement columns and does not like what it sees. Note the calm assurance that these men of intellect will pass their intelligence genes on to their children, who will therefore, it seems, require more expensive education.

The class of fixed salary earners, which comprises the brain-workers of the nation, the professional class, has borne the brunt of the rising prices without anything like an adjustment of salary corresponding with the rise. It is the hardest hit of all by the war, and yet this class, perhaps more than any other, has contributed to winning the war. Hitherto patriotism has kept it silent. Now, however, the time has come when the scale of the professional man's salary must be revised. Incomes such as those found above do not admit of the upbringing and education of a family as befits its inherited ability; of the expenses inevitable if a man is to keep abreast of his profession; and of saving and insurance against sickness, age, and death.

"In war-time," writes the *Economist* of July 12, in a "business note" on British and German science, "we make full use of our men of science. If we did so in peace they might be as useful for production as they have been for destruction." The first step is to see that they get what, for them, is a living salary, else there will be no men of science to use. The second step is to see that their teachers get adequate remuneration, else there will be no training to make men of science of them.   C.

Diver's Air Pipe disconnected.

Diver's Telephone Breast Rope disconnected.

Diver entering Decompression Chamber.

In the "land fit for heroes to live in" that politicians of Britain promised to build, education became an important topic. Frederick Soddy had been engaged in war work and by 1919, when he was appointed to the chair of physical chemistry at Oxford, he had already become preoccupied with political and economic theories. In this review he strikes a blow for the cause. In his assertion that a classical education would lead only to the breadline he was profoundly mistaken, for in Britain a degree in classics was and remained the gateway to success in public service and elsewhere — even if an Oxford classics don had with coy self-deprecation defined the advantage of a classical education as enabling a man to despise the success that it prevented him from achieving.

THE WASTE OF YOUTH.

Problems of National Education. By Twelve Scottish Educationists. With Prefatory Note by the Right Hon. Robert Munro. Edited by John Clarke. Pp. xxvi + 368. (London: Macmillan and Co., Ltd., 1919.) Price 12s. net.

THE extension of the school age from fourteen to fifteen, with compulsory education in continuation classes to eighteen years of age, which is the main provision of recent educational legislation, adds four additional years of schooling at the most critical and formative period of life. It is to be hoped rather than expected that better use may be made in the future than has been made in this country of the school period in the past.

* * *

So, the classics are still for the many and science for the few! Nothing is incredible, not even that this and much more like it should actually be written as a contribution to "Problems of National Education" at the close of the great war. If these are the people to whom their children's educational destinies are to be committed for four further years, the Labour Party will do well to expedite its attainment of a minimum State subsistence. For, be they turned out from school with their physique, morals, and manners, religious and æsthetic perceptions, civic ideals, and use of the subjunctive mood in subordinate clauses in the ancient languages never so perfect, it is difficult to see what else can save them from starvation in the hard times ahead. Until something more in keeping with the age is substituted for the intellectual training of the school, the words in the opening essay (p. 39) will continue to be true: "They begin their course with keen interest and lively curiosity. Then shades of the prison-house seem gradually to close upon the growing boy."

FREDERICK SODDY.

Fire-eating by the Indian toad:

Toads and Red-hot Charcoal.

TOADS are associated with some wonderful myths, and my scepticism was naturally great when my friend Mr. H. Martin Leake assured me, while on a visit to Cawnpore in October of 1915, that toads would eat red-hot charcoal. An after-dinner demonstration, however, soon dispelled my doubts. Small fragments of charcoal heated to a glowing red were thrown on the cement floor in front of several of the small toads (usually *Bufo stomaticus*) which so commonly invade bungalows at that time of year, and, to my surprise, the glowing fragments were eagerly snapped up and swallowed. The toads appeared to suffer no inconvenience, since not only did they not exhibit any signs of discomfort, but, on the contrary, several toads swallowed two or even three fragments in succession. A probable explanation of the picking-up is that the toads mistook the luminous pieces of charcoal for glow-worms or fireflies, the latter being numerous in the grounds of the Agricultural College at Cawnpore in October; but this does not account for the swallowing of the hot particles—the absence of any attempt to disgorge. I repeated the experiment at Allahabad in August, 1916, with the same results (the toads even attempting to pick up glowing cigarette-ends), though I have never observed glow-worms or fireflies in Allahabad at any time of year.

W. N. F. WOODLAND.
" Kismet," Lock Mead, Maidenhead.

VOL 106, 46 1920

In 1920 the peerless Indian mathematician Srinivasa Ramanujan died at the age of 32. The story of how he developed his powers in his parents' house in a small town in southern India, sent samples of his work in lined school exercise books to several mathematicians and found a patron in G. H. Hardy, who brought him to Cambridge, has been often told. His obituary was fittingly written by Hardy.

I first heard of Ramanujan in 1913. The first letter which he sent me was certainly the most remarkable that I have ever received. There was a short personal introduction written, as he told me later, by a friend. The body of the letter consisted of the enunciations of a hundred or more mathematical theorems. Some of the formulæ were familiar, and others seemed scarcely possible to believe. A few (concerning the distribution of primes) could be said to be definitely false. There were no proofs, and the explanations were often inadequate. In many cases, too, some curious specialisation of a constant or a parameter made the real meaning of a formula difficult to grasp. It was natural enough that Ramanujan should feel a little hesitation in giving away his secrets to a mathematician of an alien race. Whatever reservations had to be made, one thing was obvious, that the writer was a mathematician of the highest quality, a man of altogether exceptional originality and power.

It seemed plain, too, that Ramanujan ought to come to England. There was no difficulty in securing the necessary funds, his own University and Trinity College, Cambridge, meeting an unusual situation with admirable generosity and

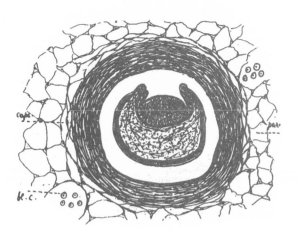

imagination. The difficulties of caste and religion were more serious; but, owing to the enterprise of Prof. E. H. Neville, who happened fortunately to be lecturing in Madras in the winter of 1913–14, these difficulties were ultimately overcome, and Ramanujan arrived in England in April, 1914.

The experiment has ended in disaster, for after three years in England Ramanujan contracted the illness from which he never recovered. But for these three years it was a triumphant success. In a really comfortable position for the first time in his life, with complete leisure assured to him, and in contact with mathematicians of the modern school, Ramanujan developed rapidly. He published some twenty papers, which, even in war-time, attracted wide attention. In the spring of 1918 he became the first Indian fellow of the Royal Society, and in the autumn the first Indian fellow of Trinity. Madras University endowed him with a research studentship in addition, and early in 1919, still unwell, but apparently considerably better, he returned to India. It was difficult to get news from him, but I heard at intervals. He appeared to be working actively again, and I was quite unprepared for the news of his death.

Ramanujan's activities lay primarily in fields known only to a small minority even among pure mathematicians—the applications of elliptic functions to the theory of numbers, the theory of continued fractions, and perhaps above all the theory of partitions. His insight into formulæ was quite amazing, and altogether beyond anything I have met with in any European mathematician. It is perhaps useless to speculate as to his history had he been introduced to modern ideas and methods at sixteen instead of at twenty-six. It is not extravagant to suppose that he might have become the greatest mathematician of his time. What he did actually is wonderful enough. Twenty years hence, when the researches which his work has suggested have been completed, it will probably seem a good deal more wonderful than it does to-day.   G. H. HARDY.

VOL. 105, 494 1920

H. G. Wells was enchanted to witness scientific activity in Russia after the ravages of the revolution.

Mr. H. G. Wells, who has recently been in Russia, describes in the *Sunday Express* of November 7 the position of some leading men of science whom he met at Petersburg, by which name, and not Petrograd, this city is now called. He saw Pavlov, the physiologist, Karpinsky, the geologist, Belopolsky, the astronomer, Oldenburg, the Orientalist, and Radlov, the ethnologist, among others who have survived the complete social disruption which Russia has undergone since the catastrophe of 1917–18. Such privileges as are possible in the country under existing conditions appear to be extended to scientific workers; for Mr. Wells mentions that the ancient palace of the Archduchess Marie Pavlova is now a House of Science, where a special rationing system "provides as well as it can for the needs of four thousand scientific workers and their dependents—in all, perhaps, for ten thousand people." In spite of this, however, there are much privation and misery, and unless food and clothing are provided few are likely to survive the coming winter. What struck Mr. Wells more than anything else was that even under the present disordered conditions, and with physical vitality reduced almost to its lowest limits, a certain amount of scientific work is still carried on, and there is a burning desire to know what has been done for the advancement of natural knowledge in other parts of the world since the Russian collapse. "The House of Literature and Art," we are told, "talked of want and miseries, but not the scientific men. What they were all keen about was the possibility of getting scientific publications; they value knowledge more than bread." There would, we are sure, be no difficulty in obtaining the books and publications needed by, or funds for providing warm clothing for, the great survivors of the Russian scientific world, if their colleagues here were assured that the parcels would reach their destination. This specific aid is, however, a different matter from general provision for the physical and mental needs of the "four thousand" scientific workers to whom Mr. Wells refers. We should scarcely have placed so many men in that category even before the war, and the ranks of scientific forces in Russia must have been greatly reduced by the revolution.

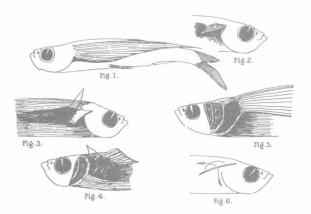

Fig. 1.   Fig. 2.   Fig. 3.   Fig. 4.   Fig. 5.   Fig. 6.

It has been stated as a general law that one can always find in some astronomy department in some major university a professor of astronomy who believes that the Earth is flat. Equally there are always to be found in the ranks of respectable scientists those who believe in the paranormal. One such was Fournier d'Albe, and Charles Richet (p.157) was another. Here they are seen off in fine style by the reviewer.

The Newer Spiritualism.

Phenomena of Materialisation: A Contribution to the Investigation of Mediumistic Teleplastics. By Baron von Schrenck-Notzing. Translated by Dr. E. E. Fournier d'Albe. Pp. xii + 340. (London: Kegan Paul, Trench, Trubner, and Co., Ltd.; New York: E. P. Dutton and Co., 1920.) Price 35s. net.

"OF making many books" on spiritualism "there is no end," and study thereof "is a weariness of the flesh." Certainly such is the effect of reading a ponderous and repellent volume of 200,000 words conveying the story of séances the details of which are as like one another as peas in a pod.

* * *

As already said, the record of the sittings, which extended from June, 1909, to July, 1913 (they were held chiefly in Paris), has a dreary uniformity. Mme. Bisson was always present; her watchful care over "Eva C." suggests more than friendship, and awakens suspicions as to collaboration in "materialisations." The appointments were as usual; the medium sat in a dark cabinet, Mme. Bisson hopping in and out, and then joining the other sitters, rarely more than three or four, in a room dimly lighted by a red lamp; a white light, as all spiritualists agree, "acting destructively on the pseudopods or psychic projections from the medium's body." Apart from M. Richet, a somewhat credulous savant, no prominent man of science was at the sittings, save one Dr. Specht, who, after three attendances, said that he had been "shown materialisations which do not exist." Baron von Schrenck-Notzing naïvely adds: "On account of this negative attitude, Dr. Specht was not invited to further sittings."

* * *

The reviewer can deal only with such statements as fill this book at their face value. The *onus probandi* lies on those who make them. As Faraday said in a lecture delivered before the Royal Institution in 1854: "I am not bound to explain how a table tilts any more than to indicate how, under the conjurer's hands, a pudding appears in a hat." Baron von Schrenck-Notzing and Dr. Fournier d'Albe have a clear course before them. Let them bring "Eva C." to London to exhibit her "materialisations" before a committee of which Sir Ray Lankester, Sir Bryan Donkin, and Mr. Nevil Maskelyne should be members. Then the matter would be put beyond doubt whether the so-called evidence, thus judicially sifted, is or is not based upon the collusive action of mediums and upon the bad, because prejudiced, observation of the sitters. The need to keep in mind what Hume says about occult phenomena was never more urgent than it is to-day. "As finite added to finite never approaches a hair's breadth nearer to infinite, so a fact [statement?] incredible in itself acquires not the smallest accession of probability by the accumulation of testimony."

Vol 106, 367 1920

The review of which the following is a part was signed "The Professor of Biology". The university in question too is anonymous, but was probably Oxford. What were the buried scandals at which the reviewer hints?

The Confidences of Men of Science.

The Purple Sapphire, and other Posthumous Papers. Selected from the Unofficial Records of the University of Cosmopoli by Christopher Blayre. Pp. x + 210. (London: Philip Allan and Co., 1921.) 7s. 6d. net.

THE author—or, to be more accurate, the editor—of this fascinating but blazingly indiscreet volume refers to Nature as "that admirable journal"—a compliment which ought perhaps to secure a benevolent review, but needless to say we shall not let it induce us to depart from our habitual detachment.

Mr. Blayre was for many years Registrar in a well-known university, and had certain manuscripts confided to him by more or less scientific members of the staff on the understanding that they should remain *in retentis*, as who should say, unless events occurred which rendered their publication desirable. In no case, however, were they to be published in the lifetime of the depositors, to whom the documentation served as a sort of Freudian relief. Now there is no doubt that the publication clears up many puzzling events, such as the ghastly damage that followed the acceptance of the so-called "purple sapphire" by the Mineralogical Museum, the mystery of Prof. Markwand's death, and the tragic case of Austin Black, who, if anyone, must be credited with laying the foundations of psychobiology.

To clear up these and other obscurities, more familiar to the older than to the younger readers of Nature, has seemed to Mr. Blayre sufficient warrant for publishing the deposited documents. He does not seem to be aware, however, that the Professor of Biology, the present reviewer, is still alive, and by no means so sure as he once was of Mr. Blayre's fiducial discretion. His feeling of relief when he found that his own document had been suppressed by the publishers enables him to sympathise at least with the relatives of the deceased gentlemen whose confidences are now blazoned abroad. It is true that names are sometimes suppressed or modified in the book, but in these days, when the study of the history of science is rife, it seems a cruelly thin disguise to refer to a professor by a pseudonym and then proceed to mention one of his well-known discoveries.

Vol 107, 677 1921

The Reparations Act, which required Germany to pay indemnity to the Allied powers, had some preposterous consequences.

EMPHATIC corroboration of recent correspondence in our columns upon the supply and cost of German publications is provided by a letter addressed to the *Times* signed by the Vice-Chancellors of the Universities of Liverpool, Sheffield, and Manchester and the Principals of Armstrong College, Newcastle, and Birmingham University. At each of these institutions the librarians have found it impossible to obtain current German scientific literature by reason of the operation of the Reparations Act. There has been a complete stoppage of delivery through the Customs of books of German origin, while books which have been ordered direct from agents in Germany are delayed for an indefinite period. Even when it has been proved that the order was placed before the present Act came into operation and the 50 per cent. Customs charge has been paid under protest, books are still undelivered. The writers of the letter emphasise the fact that it cannot be regarded as patriotic to cut off from this country all knowledge of scientific progress in Germany; on the contrary, it is to the advantage of our trade and ultimate prosperity to know without delay every addition to knowledge made in Germany as in other countries. German journals of science and other publications devoted to the advance of knowledge cannot be regarded as entering into competition with British journals and books, and vigorous protest is made against the interpretation of the Act by the Board of Trade to include such articles.

Vol 107, 630 1921

By 1921 the consequences of the revolution for science in Russia had become all too apparent. A Russian academic described in emotional terms the privations visited on the intelligentsia by the likes of Comrade Lunacharsky, the commissar for science. His article in NATURE drew a wide response. Here is a letter from an anonymous "well-known Russian professor of chemistry", forwarded to NATURE by the noted chemist J. W. Mellor. The professor's melancholy tale begins like this:

Communism and Science.

IN view of the article on "The Proletarisation of Science in Russia" in NATURE of September 1, the following extracts from a letter I have just received from a well-known Russian professor of chemistry may be of interest. I have omitted personal references and a few other matters.   J. W. MELLOR.

Pottery Laboratory, Stoke-on-Trent,
September 5.

"You doubtless know the old adage, 'Primo vivere, deinde philosophari.' I do this last, but the first part, 'vivere,' is more than uncertain for us who have the misfortune to be a little civilised, as one never knows what our wild, wild taskmasters are going to do next. The higher schools of Petrograd are under the control of a former apprentice of the dockyards of Cronstadt, who has learned to talk glibly and to sign his name with an appropriate flourish. He has not the remotest notions as to what is a seat of high learning; but that does not trouble him in the least, he just governs according to his lights, and actually does his best to destroy all culture, all real science, in our institutes. It is just the same everywhere, and the results are glaringly apparent in the utter failure of crops in the east and south of Russia, due not so much to exceptional climatic causes as to the countless requisitions of 'surplus' wheat, 'surplus' bullocks and horses, and other kindred measures of the reigning proletariat. The population of some twenty provinces, which supplied once upon a time almost all Russia with bread and exported thousands of tons of wheat to foreign lands, is now leaving their houses and fleeing to the east, the north, and the west of Russia, where there is still something to eat; they spread desolation wider and wider—also cholera and other diseases; tens of thousands perish daily. Almost nothing is left in the devastated provinces; they must now be colonised anew. We are fortunate for the nonce in being sufficiently far from these places, but the outlook for us is anything but reassuring.

"Up to now we receive a 'ration of scientists,' which during 1920 was comparatively good, but is now reduced to the following items, received, for instance, in June:—14 lb. of bread (made principally out of soya beans); 11 lb. of soya beans (it is not generally known that they contain poisonous constituents, and many were the cases of poisoning); 19 lb. of herrings; 4 lb. of tallow (the first fatty substance received since February); 9 lb. of wheat (we eat it boiled in the form of gruel); 3 lb. of macaroni made out of soya beans; 1 lb. of salt; 1¼ lb. of sugar; 3 lb. of lean pork—bones and hide, no lard, and very little meat; ¼ lb. of tea (surrogat); ¾ lb. of tobacco; some matches; and 1 lb. of washing soda (there is no soap). During the same month I received—only a few days ago—as salary for my lectures, etc., the stately sum of 21,000 roubles; but as bread costs about 4000 roubles and butter 30,000 roubles per lb., this sum is the equivalent of 5 lb. of bread, or some 20 kopeks (=5d.) of pre-war days. You will thus appreciate the munificence of my salary; the meanest mechanic or plumber gets from 250,000 to 500,000 roubles and more monthly, and it is nothing unusual to pay 1 lb.

of bread or 5000 roubles for one hour of manual work, whereas I, as a full-fledged professor and doctor of chemistry, receive for one hour of lecture 450 roubles. Consequently, to nourish the members of the family (I have, fortunately, neither wife nor children, but live with my old mother), I work in kitchen-gardens, sell the few things that are still left, etc. The prices for a new suit range up to 1,000,000 roubles; a pair of old high boots, which I could not wear and which cost originally some fifteen years ago 14 roubles, fetches now 700,000 roubles, as boots are very scarce; for a shirt you get 2 lb. of butter. During the first six months of 1921 my mother and I have eaten different foods to the value of about 6,000,000 roubles.

"Needless to say that, in spite of these millions, we are now paupers in the strict sense of the word. All my savings, made little by little during more than twenty-five years of professorship, were placed in State loans and annulled in 1917; our small landed estate not far from Petrograd was taken from us in 1918, and is now completely devastated, all the woods having been cut. It would be now utterly impossible to exist without the 'ration of scientists,' meagre as it is. I read in NATURE in 1920 a notice on the voyage of Mr. Wells to Petrograd and his opinion on the position of our men of science (NATURE, vol. 106, p. 352). He did not see much, as he was 'personally conducted,' and does not understand Russian. He estimates the number of men of science at four thousand; NATURE expresses some astonishment at this number—thinks there must be less, as many have died. Well, there is here an official commission, composed of Mr. Oldenburg and other learned gentlemen, who decide whether an applicant is a man of science or not. As it is often a question of life and death, the decision is usually in favour of the applicant, if certain formalities are fulfilled—if he has printed some learned articles, or even simply written one, etc. Professors are almost always included if they teach in higher schools. There are now very many proletariat "higher institutes"; for instance, a "Higher Institute of Anti-fire Technique" and a "Higher Institute of Plastic Arts"; the first turns out firemen for fire brigades, the second—dancers! "Professors" of these institutes are also "scientists"! Even among the professors of old high schools there are now some without scientific degrees called "red professors"; for now it is decreed that anybody can be a professor, just as anybody who is sixteen years of age can be a student. If he is quite ignorant he will attend a "preparatory course," but, like real students, will receive his ration and salary (students do not *pay* anything now, but receive salaries).

VOL 108, 113 1921

There quickly followed a letter that implied that the seeds of repression had been planted in Russia long before.

Following an invitation, I took part in the Congress of Naturalists which was held in Odessa in August 1883, and of which I am the only foreign survivor. I became acquainted with the most prominent professors of that University and found that they were first-rate men of science, without a trace of anything " backward and reactionary." And yet this reproach is correct, but it refers to the State director of the university. A man, unsympathetic, gloomy, reactionary, every inch a bureaucrat, and fairly old, inaugurated the first general meeting with the following severe words : " You came here to speak of science and I hope that you will speak *only* of science ! " After this rose Metchnikoff and gave a brilliant account of his recent and unpublished work on phagocytosis, which was received with enthusiastic applause by the whole assembly.

I congratulated my Russian colleagues and the University upon having such a professor, but they replied with regret that he no longer belonged to the University, and upon asking for reasons I was given the explanation : Metchnikoff as a professor of zoology announced a course of lectures " On the Theory of Evolution." And now the very reverse took place of what I described four months ago (see above). The director summoned Metchnikoff to his office and said to him : " It appears that you are going to lecture on Darwinism ? If it is so, then you must submit your written lectures to my censorship and I will tell you what I allow you to say to the students and what not ! " Metchnikoff did not accept this explanation of the ." Lehr- und Lernfreiheit," he did not submit his notes to the curator ; he resigned the professorship. Russia was not the soil for such a genius, and it was good fortune for him and for science that he left for Paris and for Pasteur.

BOHUSLAV BRAUNER.
Bohemian University, Prague, March 9.

Vol 109, 478 1922

Henry E. Armstrong, the organic chemist and educator (p.123), produced a stream of contributions to NATURE *over many decades, which proclaimed him a formidable windbag. Here his aim seems to be true.*

Is Scientific Inquiry a Criminal Occupation?

I ASK the question because, under the provisions of the Safeguarding of Industries Act, 1921, which came into operation on October 1, scientific workers and the public may be fined one-third of the value on all scientific appliances and on all chemicals—other than sulphate of quinine—imported into this country. Why this quinine salt alone of all chemicals should be free I do not understand, unless it be because it is largely used as a contraceptive and the philanthropic framers of the Act are alive to the fact, which of all others is the most important for us to recognise, that our country has double the population it can carry. Obviously, they are bent on discouraging and hindering scientific inquiry in every possible way ; the Act can have no other effect ; only a small proportion of the articles it covers are, or ever will be, made in this country. No more iniquitous measure was ever passed into law.

I have given notice that at the next meeting of the council of the Chemical Society I will move that action be taken forthwith to secure the repeal of the Act. If it be not annulled, scientific workers generally must agree to boycott all apparatus and materials of English manufacture. For once we must wake up

and show that we can both help ourselves and protect the interests of our country.

Sir William Pope, in a recent speech dealing with American conditions, pointed out that chemists at least were so organised in the U.S.A. that they could make their voice heard with effect in the legislature. Here the legislature, bureaucracy in general, does not care a rap for science. A request made by Sir William Pope several months ago to the Board of Trade, on behalf of the Federal Council, that the Council might be heard on the proposed Bill was never more than formally acknowledged.

If we believe in our craft and its national value we must be militant in its protection.

I shall be glad to receive names and addresses (written legibly on postcards, please) of those who are willing to join in a memorial to the Prime Minister. If we desire to gain a position for science in this country, it is our duty to show, for once, that we can do something—that we are not mere talkers.

HENRY E. ARMSTRONG.
55 Granville Park, Lewisham, London, S.E.13.

Vol 108, 241 1921

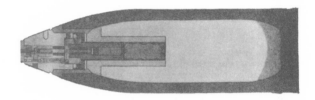

The decision over Hebrew as the language of the Jewish state was not made without a fight, for both English and German had their champions.

A FURTHER communication has reached us from Mr. Cyril Crossland (see NATURE, August 4, p. 733), who expresses the view that the use of Hebrew at the Jewish University in Jerusalem is an act of exclusiveness against non-Jews. He urges that Hebrew is a dead language, and only the Zionists are working for its revival, for purely racial and political ends. As a matter of fact, Hebrew is the real living language of the Jews in Palestine. It is the language of instruction of the Jewish schools in Palestine, both elementary and secondary, and is one of the official languages of the country. It would surely be an anomaly to have a Jewish university in Palestine without Hebrew as the language of instruction. Hebrew has already been used for scientific work with great success. To urge that the use of Hebrew means excluding non-Jews is the same as urging that the use of English at a British university means excluding Frenchmen or Germans. Mr. Crossland objects to the statement that Jews are opposed to clericalism, and asks, " Then how is it they remain Jews? " We suggest that he misunderstands the meaning of the word " clericalism," which signifies the usurpation of political power by the clergy, and to this Jews are opposed everywhere. Mr. Crossland refers to the college at Beirut, where instruction is given in English, and to the fact that there are in Palestine native qualified, energetic, and patriotic medical men who can deal with the public health of the country. These facts are irrelevant. Beirut is not in Palestine. Further, public health has been grossly neglected in Palestine, and to object to this being cared for by Jews is an inadmissible attitude.

Vol 108, 387 1921

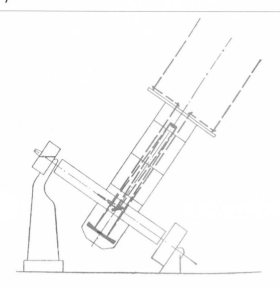

This excellent letter sparked off an acrimonious debate. It encapsulates the differences of perspective that so often separate scientists and philosophers.

Relativity and Materialism.

PROF. WILDON CARR has for a number of years been busily engaged in ringing the death-knell of materialism. I was therefore not a little surprised to read in NATURE (October 20) his statement that Einstein's theory was the "death-knell of materialism." I thought, from my previous acquaintance with Prof. Carr's writings, that Bergson, Croce, and others had already done all that was necessary in that direction. But no! Prof. Carr has resuscitated it for the express purpose of killing it once more. That unfortunate doctrine seems to exist mainly for the purpose of being periodically slaughtered by professors of metaphysics; and we are led to the conviction that materialism must have very singular properties to survive so many tragic executions.

Well, it does possess a property which must naturally appear singular to those steeped in metaphysics—it happens to be true. Scientific materialism, as now understood, does not profess to be a rounded or final system of philosophy: it is merely a name for a few general principles, laid down by science, and selected for emphasis on account of their high human significance. Science makes new knowledge; philosophy (rightly understood) does not; it simply collects together certain principles yielded by science, those principles being selected as having some bearing on the deep undying problems of most profound human interest.

Among the scientific principles thus selected and emphasised by materialism—and the only one among them still seriously controverted—is that which states that mind cannot exist apart from matter, or as I prefer to put it, that mind is a function of material organisms. Prof. Wildon Carr is of opinion that mind *can* and *does* exist apart from matter; and he is under the impression that this opinion is justified by the principle of relativity. So far as I can follow his argument, it amounts to this. Space and time are relative to the observer; therefore the existence of an observing mind must be antecedent to the existence of space and time. True; but space and time are not matter: they are not objective things; you cannot weigh them or touch them; they are part of the mental framework which we erect for our convenience in dealing with external nature. They are concepts; just as the number 10 is a concept; not a

thing, but a framework into which things can be fitted. "For the concept of relative space-time systems the existence of mind is essential." Prof. Carr might with equal profundity have said that for the presence of dew the existence of water is essential. Dew is aqueous; a concept is mental; but let me inform Prof. Carr that neither one nor the other of these propositions gives the slightest qualm to any scientific materialist, nor have they the least relevance to the question whether or not mind depends upon matter. We are not concerned with "concepts," which, of course, imply the previous existence of mind, but with objective *things*.

Now Prof. Carr argues that the "space-time system," involved by relativity, is conditional on the existence of mind. Very well then. It follows that if mind were to be extinguished throughout the universe, the laws at present ascribed to the universe would cease to operate, or perhaps the universe itself would cease to exist. Now that is an altogether incredible proposition. If Prof. Carr's mind were to be extinguished, the laws of nature would still remain as they are. If everybody else's mind were also to be extinguished the laws of nature would be unaltered. "Concepts" would vanish no doubt; but the validity of the principle of relativity itself does not depend on the existence of a mind which can testify to it. Prof. Carr exhibits that incurable confusion between concepts and objects which is common to all those who think that metaphysics is a rival method of science in the making of new knowledge.

Relativity of space and time no more conflicts with scientific materialism than does relativity of motion. But it is idle to argue with sentiment, and it is with sentiment alone that we have to do—sentiment unsupported by a fragment of evidence, and asserting itself in flat contradiction to every principle of logic. As a mere statement of truth, materialism will always reign, as it has reigned now for centuries as the basis of scientific experiment. But on a show of hands it will always be in a minority; its reign is that of an uncrowned king. There exists a wide and universal human sentiment which loathes materialism. That sentiment comes out in many different forms: in the vulgar superstitions of the uneducated, in spiritualism, in metaphysical dissertation. They are but the same deep sentiment on different intellectual grades, but as false and rotten in the higher grades as they are in the lower. Everywhere it comes out: in physiology we find it as vitalism; among the public at large it supports religion, the most powerful single factor that has moulded the destinies of civilised humanity. Materialism must always be unpopular; that is why it is so often being killed. But it is true; that is why it never dies; that is why it never will die; unless, indeed, it is one day drowned in the floods of oily sentimentalism.

Two hundred and fifty years ago the world of physics was fermenting as it is now. Newton was introducing a revolution of thought, comparable to the revolution of the last twenty years. Then, as now, the sudden upsetting of old ideas had in some sense a demoralising effect. There seems a real danger that metaphysics may take root, for a brief period, amid the general disorganisation consequent upon the revolution. A spectator does sometimes see most of the game, and I trust it may not be considered presumptuous in a spectator to sound an old note of warning at a time when many insidious invasions of science are being attempted by metaphysics: "Physics, beware of Metaphysics."

November 18.   HUGH ELLIOT.

Diatribes by eminent British men of science against the depravity of the beastly Hun meanwhile continued, Sir Edward Thorpe ever in the vanguard. Fritz Haber was a favoured target, and indeed he had carried the responsibility for the development of chemical weapons on the German side. In the course of his work, he had exposed himself to every danger of war and had suffered the suicide of his wife, an organic chemist, who had tried to dissuade him from pursuing what he saw as his patriotic duty. He was undoubtedly a passionate patriot and took a childlike pride in his promotion to the rank of a mere captain. But Haber was known to have believed that gas warfare would hasten the end of the war and save many thousands of lives. Here he attempts to defend his actions.

Chemical Warfare.

Sir Edward Thorpe, in his review of Victor Lefebure's book, "The Riddle of the Rhine," in Nature of November 10, p. 331, quotes a passage which deals with my own work during the initial stages of the war, and that of the Kaiser Wilhelm-Institut für Physikalische Chemie, of which I am the principal. The intention is to make the world believe that the materials for gas warfare were prepared by the German military authorities and chemical industry for the intended war, and that experiments with this end were carried out in my institution, if not previous to the war, at least from August, 1914, onwards.

It is always dangerous to attempt to form a correct estimate of the intentions of others from the traces of events they have left behind them. But the greatest errors must necessarily arise if an outsider tries to deduce from his own impressions the intentions of men whose ways of thinking he does not know and cannot understand.

Perhaps there might have been some ground for suspicion if Germany could have foreseen the trench warfare, and if we could have imagined that the German troops could ever be held up for weeks and months before the enemy's wire entanglements. But previous to the war, and up to the Battle of the Marne, everyone in Germany imagined that the course of the war would be a succession of rapid marches and great pitched battles, and what use would gas have been to a field army in such a war of movements? I think I may safely say that during the course of the war I became acquainted with every man of any importance in the army, in industry, and in science, who had anything to do with chemistry as applied to military offensive and defensive operations, and that I am well informed regarding the development and the course of chemical warfare. Yet among all these men I have never met one who, previous to the war or during the first two months of its course, had conceived the idea of providing the field army with gas, or had made experiments or preparations for such a purpose. We had actually first to read in the French, Italian, and English Press—as, for instance, in the *Pall Mall Gazette* of September 17, 1914—of the terrible things that were in preparation for us before we began to make similar preparations in view of the commencement of the war of position.

As regards my own institution and its work during the first months of the war, that intelligent person who, according to the passage in Lefebure quoted by Sir Edward Thorpe, observed my activities in my institute from behind a wall, lacked the gift of interpreting correctly what he saw and heard. Visitors in grey Headquarters motors did indeed come to my institution in August, 1914, though not to see me upon the subject of chemical means of warfare, but because Headquarters were very anxious to know how motor spirit could be made proof against the cold of a Russian winter without the addition of toluol. The question of gas as means of warfare did not begin to engage our attention until the first three months of war had passed.

In war men think otherwise than they do in peace, and many a German during the stress of war may have adopted the English maxim, "My country, right or wrong," but that German science and industry before the war made preparations with deliberate intent for gas warfare against other nations is an assertion that, in the interest of the necessary interdependence of the nations in the realms of science and industry, must not be allowed to go uncontradicted in so serious and respected a journal as Nature.

F. Haber.

Kaiser Wilhelm-Institut, Berlin-Dahlem,
     December 17.

Thorpe is not mollified: this is the beginning of his long response.

Herr Geheimrat Haber takes exception to the quotation I made from Major Lefebure's "Riddle of the Rhine," in the course of my notice of that book, on the ground that it implies that the German military authorities were prepared to ignore their undertaking, under the Hague Convention, to abstain from the use of asphyxiating or deleterious gases in war, if not for some time before, at least at its outbreak in the summer of 1914. I have, of course, no precise knowledge of the intentions of the German military authorities, but it was not unreasonable to surmise that these authorities, who deliberately intended to violate the treaty with Belgium, would not hesitate—as, indeed, the sequel showed—to disregard their promise under the Hague Convention if and when it suited their purpose to do so.

As regards their intentions, Field-Marshal Lord French, in his dispatch after the first German gas attack, with which Prof. Haber was concerned, wrote: "The brain-power and thought which has evidently been at work before this unworthy method of making war reached the pitch of efficiency which has been demonstrated in its practice shows that the Germans must have harboured these designs for a long time."

"It is an arresting thought," says Major Lefebure, "that even as early as 1887 Prof. Baeyer, the renowned organic chemist of Munich, in his lectures to advanced students, included a reference to the military value of these compounds"—*i.e.* to substances intended to produce temporary blindness.

Not long after this exchange, an article appeared based on a lecture on "Science and Gas Warfare" given by a Colonel C. H. Foulkes at the unveiling of a war memorial at Imperial College in London. Tucked away unobtrusively in the middle is this passage, which shows, as so often, that only losers have to answer for their transgressions.

Owing to the secrecy which it was necessary to maintain during the war, the general public has still, I believe, little idea of the prominent part which chemical warfare played on the field of battle on the Western Front. Between the Battle of Loos, in September 1915, and the armistice, the activity of the Special Brigade was almost incessant, and gas attacks were carried out on an average on two nights out of every three during the whole period. Some 800 separate attacks were made—against about 25 by the Germans against us—and nearly ten thousand tons of gas were liberated, quite apart from the work of the artillery: and many were the variations practised in the form of attack, as regards tactics, mechanical appliances, and meteorological conditions.

The enemy's casualties from these gas attacks probably numbered between 100,000 and 200,000, amongst whom the percentage of mortality was very high.

The scientists felt a moral necessity to share at least some of the dangers of the soldiers at the front.

In research work, Profs. Baker and Thorpe were very prominent throughout the whole period of the war. They, with other eminent men, gave themselves whole-heartedly to this work, to their own financial disadvantage, and without the prospects of reward which the successful soldier has in view. In science, at any rate, there was no profiteering. A lady of this College, Dr. Whiteley, introduced the use of "S.K." —symbolising South Kensington—a substance that was largely used against the Germans—as well as a new explosive.

In all the preliminary physiological tests, of course, animals were used; but volunteers were never wanting for the more important experiments in the lethal chamber: and at one time many of the experimental staff at Porton were in a constant state of ill-health owing to the trying nature of this work. One gallant action worthy of record was that of Mr. Barcroft of Cambridge, who, in order to confirm a theory which had an important bearing on our gas tactics, entered the lethal chamber together with a dog, both being entirely unprotected, and remained there while exposed to prussic acid gas until the dog died.

F. G. Donnan, as we have seen, had something of a bee in his bonnet about the need for a universal language for science. In 1922, NATURE allowed him full rein. Here is the crux of a long dissertation on the subject.

...the problem is a very pressing one. Those who have to do with science, industry, and commerce feel this very acutely. Before the war I attended several international scientific congresses. On these occasions it was open to any one to speak in English, French, German, or Italian. When the language of the speaker or lecturer changed, one half of the audience usually adjourned to the refreshment bar. I could follow German, but when it was a case of Italian or Parisian French I also used to get thirsty. I am going to an international scientific congress in June of this year. The representatives of at least thirteen different nations will be present, and I expect at least four languages will be used. As the language of the country where the congress is to be held is not one of these, one ought really to know five languages. I am glad to say that the civilised world is at last beginning to take a real interest in this problem. We may, indeed, say that, since the war, the whole question has entered on a new phase. Learned and scientific bodies of international influence and repute are beginning to study the matter seriously.

And the recommendations, which suggest we may all have had a narrow escape from compulsory courses in Ido:

The first national response to the appointment of the International Committee was by the British Association for the Advancement of Science, which, at its Bournemouth Meeting in September 1919, appointed a Committee "to study the practicability of an International Language." This British Committee has been very active, and at the Edinburgh meeting of the British Association in September last, presented its report. Its conclusions may be summarised very briefly as follows:

(1) Latin is too difficult to serve as an international auxiliary language.

(2) The adoption of any modern national language would confer undue advantages and excite jealousy.

(3) Therefore an invented language is best. Esperanto and Ido are suitable; but the Committee is not prepared to decide between them.

Two years after the death of its first editor, NATURE recorded the unveiling of a plaque in his honour at the observatory that bears his name.

FIG. 1.—Portrait Medallion of Sir Norman Lockyer.

17TH MAY, 1836 - 16TH AUGUST, 1920
TO THE ENDURING MEMORY OF THE GENIUS AND CONSTRUCTIVE IMAGINATION OF
SIR NORMAN LOCKYER, K.C.B., F.R.S.,
CORRESPONDANT DE L'INSTITUT DE FRANCE.
HON. LL.D., GLASGOW, ABERDEEN AND EDINBURGH; HON. SC.D., CAMBRIDGE AND SHEFFIELD;
HON. D.SC., OXFORD.
FOUNDER AND DIRECTOR OF THE SOLAR PHYSICS OBSERVATORY, SOUTH
KENSINGTON, 1885-1913 ; AND OF THIS OBSERVATORY, 1913-1920.
FOUNDER AND EDITOR OF "NATURE," 1869-1919.
PIONEER IN THE INVESTIGATION AN INTERPRETATION OF THE CHEMISTRY OF
THE SUN AND STARS AND IN THE SCIENCE OF ASTRONOMICAL PHYSICS.
DISCOVERER OF HELIUM IN THE SUN AND ORIGINATOR OF THE THEORY THAT
HELIUM AND HYDROGEN ARE THE ULTIMATE PRODUCTS OF THE DISSOCIATION
OF MOLECULES AND ATOMS.
FOUNDER OF THE SYSTEM OF STELLAR CLASSIFICATION BASED UPON ASCENDING
AND DESCENDING TEMPERATURES IN ORDERLY CELESTIAL EVOLUTION.
REVEALER OF THE ASTRONOMICAL SIGNIFICANCE OF STONEHENGE AND OTHER
ANCIENT MONUMENTS.
FOUNDER OF THE BRITISH SCIENCE GUILD FOR THE PROMOTION AND APPLICATION
OF SCIENTIFIC METHOD TO PUBLIC AFFAIRS.

THIS PORTRAIT WAS ERECTED BY RELATIVES AND FRIENDS.

VOL 110, 192 1922

Numerology exercising, as ever, an irrational fascination:

Some Curious Numerical Relations.

IN the course of a series of computations it was noticed that the ratio of the numerical values of the following pairs of quantities is in each case an integral power of ten. This curious relation is so surprisingly exact that it seems worthy of record :

$$(h/2\pi)^2 = 1\cdot08806 \times 10^{-54} \text{ erg}^2 \text{ sec.}^2$$
$$e^3/K_0 = 1\cdot08804 \times 10^{-28} \text{ erg cm. es.}$$

$$e = 4\cdot774 \times 10^{-10} \text{ es.}$$
$$h/k = 4\cdot777 \times 10^{-11} \text{ sec. deg.}$$

$$m_0 = 8\cdot9991 \times 10^{-28} \text{ gm.}$$
$$c^2 = 8\cdot9916 \times 10^{20} \text{ cm.}^2 \text{ sec.}^{-2}$$

$$r_1 = 5\cdot30507 \times 10^{-9} \text{ cm.}$$
$$e/m_0 = 5\cdot30500 \times 10^{17} \text{ es. gm.}^{-1}$$

The symbol *es* has been used to denote the electro-static unit of charge, r_1 the radius of the first Bohr ring in hydrogen, K_0 the dielectric constant of a vacuum, k the gas constant per molecule; the other symbols have their usual significance. The values that served as the basis of the computation were those just given for e, e/m_0, and c, and the following: h, $6\cdot554 \times 10^{-27}$ erg sec.; the faraday, $2\cdot89365 \times 10^{14}$ es. per equivalent; the volume of one gram-molecule of ideal gas at 0° C. and one standard atmosphere, $22411\cdot5$ cm.^{3} per mole; and 0° C., $273\cdot1$° K.

N. ERNEST DORSEY.

1410 H Street, N.W.,
Washington, D.C.

VOL 112, 505 1923

R. W. Wood, spectroscopist, farceur *and nemesis of René-Prosper Blondlot and his* n-rays *(p.115), put his skills to the service of the military. The consequences of his observations on fluorescence of the retina, excited by the ultraviolet mercury arc line, are not recorded.*

USES OF INVISIBLE LIGHT IN WARFARE.

PROF. R. W. WOOD, of Johns Hopkins University, Baltimore, gave to the Physical Society of London on March 14 a demonstration of the uses of invisible light in warfare. The first device shown was a signalling-lamp, consisting of a 6-volt electric lamp with a small curled-up filament at the focus of a lens of about 3 in. diameter and 12 in. focus. This gave a very narrow beam, only visible in the neighbourhood of the observation post to which the signals were directed. In order to direct the beam in the proper direction, an eyepiece was provided behind the filament. The instrument was thus converted into a telescope, of which the filament served as graticule. When directed so that the image of the observation post was covered by the filament, the lamp, when lit, threw a beam in the proper direction. In many circumstances the narrowness of the beam was sufficient to ensure secrecy; but sometimes it was not desirable to show any light whatever, and filters were employed to cut out the visible spectrum. By day a deep red filter, transmitting only the extreme red rays, was placed in front of the lamp. The light was invisible to an observer unless he was provided with a similar red screen to cut out the daylight, in which case he could see enough to read signals at six miles. By night a screen was used which transmitted only the ultra-violet rays. The observing telescope was provided with a fluorescent screen in its focal plane. The range with this was also about six miles. For naval convoy work lamps are required which radiate in all directions. Invisible lamps for this purpose were also designed. In these the radiator was a vertical Cooper-Hewitt mercury arc, surrounded by a chimney of the ultra-violet glass. This glass only transmits one of the mercury lines, viz. $\lambda = 3660$ A.U., which is quite beyond the visible spectrum. Nevertheless, the lamp is visible at close quarters, appearing of a violet colour, due to fluorescence of the retina. The lens of the eye is also fluorescent. This gives rise to an apparent haze, known as the "lavender fog," which appears to fill the whole field of view. Natural teeth also fluoresce quite brilliantly, but false teeth appear black.

Reverting to the use of the lamps at sea, they are picked up by means of a receiver consisting of a condensing lens in the focal plane of which is a barium-platino-cyanide screen the full diameter of the tube. An eyepiece is mounted on a metal strip across the end of the tube. When the fluorescent spot has once been found somewhere on the screen, it is readily brought to the central part and observed with the eyepiece. The range is about four miles, and the arrangement has proved invaluable for keeping the ships of a convoy together in their proper relative positions by night.

The author of this letter offers a genetically shaky explanation for his observations. The age of the children at the beginning of their first school year is a factor that seems not to have entered into his lucubrations.

Seasonal Incidence of the Births of Eminent People.

In order to find, if possible, the causes which underlie the production of increased numbers of eminent intellects at certain periods (as, for example, the year 1809 and a year or two before and after it), I collected statistics of the dates of birth of more than two hundred eminent persons. The list consists chiefly of creative intellects,—poets, literati, musicians, painters, architects, men of science, explorers, and inventors, with a few statesmen and military men. Analysis of the dates shows that the greater number of these persons were born in the colder months of the year; but the distribution of the numbers is somewhat erratic. February is distinctly the richest month, having produced a galaxy of eminent persons; December comes next; August and June are the richest among the warm months.

Sixty pre-eminent names, chosen for no reason but their pre-eminence, were found to be distributed as follows :—In warmer months : April, 4 ; May, 6 ; June, 7 ; July, 2 ; August, 5 ; September, 3 ; total, 27. In colder months : October, 4 ; November, 1 ; December, 9 ; January, 5 ; February, 9 ; March, 5 ; total, 33.

The difference is more evident when the months are taken in groups of three, as follows : December to February, 23 ; March to May, 15 ; June to August, 14 ; September to November, 8.

In order to find whether this distribution corresponds with the ordinary distribution of births through the twelve months, I compared the numbers with the average of twelve years taken at a venture from the Registrar General's Quarterly Returns, namely, the period 1844–55. The figures are too numerous for quotation, but it may suffice to say that I could find no correspondence between the ordinary distribution of births and the distribution of births of eminent persons. In the Registrar General's Returns the order of average frequency for the quarter-years was as follows : April to June, July to September, January to March, October to December.

Climate can scarcely explain the distribution, (See letter from Dr. Robert W. Lawson, Nature, June 3, p. 716.) Cold weather is not unhealthy for children, and in fact the diseases of the hot months are among the most fatal for them. I suggest that the reproductive organs, especially the germ cells, are more vigorous at certain seasons, producing offspring of higher quality. The many eminent persons born in the winter months, December to February, were conceived in the spring, the time of increased vigour of most living things; whereas the few born in the autumn months, September to November, were conceived in the winter.   F. J. Allen.

Cambridge, June 17, 1922.

The machine that speaks with a human voice was perhaps a premature invention, but it shows the vigour of AT&T from its inception.

An Electrical Analogue of the Vocal Organs.

In connexion with correspondence which recently has appeared in the columns of Nature relating to the physical characteristics of vowel-sounds, the following account may be of interest of an apparatus believed to be novel, which is conveniently capable of the artificial production of many speech-sounds. It is well known that Helmholtz succeeded in imitating vowels by combinations of tuning forks, and Miller by combinations of organ pipes. Others, notably Scripture, have constructed apparatus wherein the transient oscillations of air in resonant cavities were excited by series of puffs of air, in close physical imitation of the action of the human vocal organs. It seems hitherto to have been overlooked that a functional copy of the vocal organs can be devised which depends upon the production of audio-frequency oscillations in electrical circuits.

A schematic diagram of such an apparatus is given in Fig. 1. Periodic interruptions of the electric current, produced by a buzzer or a motor-driven circuit interrupter, corresponded to the periodic inter-

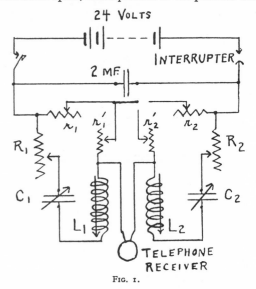

Fig. 1.

ruptions of the air current in the human throat by the vocal cords. The intermittent electric current thus produced excited the natural damped oscillations of the resonant circuits, 1 and 2. This was confirmed by observation with an oscillograph. In like manner, puffs of air from the vocal cords excite the natural damped oscillations of the air in the mouth cavities. The work of numerous investigators has indicated that the air in the mouth cavities possesses, as a rule, only one or two important modes of vibration. The oscillations of the electric current were transformed to sound-vibrations in the air by a loosely coupled telephone receiver. The distortion introduced by this telephone receiver appeared to be of little importance.

Appropriate adjustments of the resonant circuits 1 and 2 were observed to result in the production of all the various vowels and semi-vowels in turn. Alteration of the frequency or damping of either resonant circuit was observed to result in alteration of the vowel produced. The frequency of interruption, which was the group frequency of the recurrent damped oscillations, was observed to determine the pitch of the vowel; but it did not determine what vowel was produced. Similarly, in the case of the human voice the frequency of vibration of the vocal cords is known to determine the voice-pitch, while the adjustment of the mouth cavities is known to characterise the vowel.

The vowels and semi-vowels produced by the " electrical voice " with regular interruptions, in the manner just described, were equivalent to intoned or sung vowels, or, if the frequency of interruption was made to vary appropriately, to spoken vowels. It was found possible to produce the whispered vowels with interruptions that were non-periodic, which is in accordance with the idea that, in the human voice, whispered speech is due to irregular frictional modulation of the exhaled air. Whether the vowel was whispered, sung, or spoken depended upon the manner of making the interruptions ; while what particular vowel was produced depended upon the adjustment of the resonant circuits.

Diphthongs were produced by altering the circuit adjustments rapidly so as to shift from the initial to the final vowel-sound of the diphthong pair. Some of the fricative (hissing) consonants were approximated with irregular interruptions, provided the resonant circuits were set at somewhat higher frequencies than for the vowels and semi-vowels. None of the explosive consonants were satisfactorily imitated. It is believed that lack of success with the explosives was due to obvious difficulties of manipulation.

There was much room for improvement with respect to the naturalness of the " electrical voice." It was too monotonous, as was to have been anticipated. Contrary to expectation, alteration in the wave-form of the exciting current did not materially change the tone, provided the wave-form was sufficiently far from sinusoidal. The intoned vowels, semi-vowels, and diphthongs produced by the " electrical voice " were sufficiently natural to be recognised in at least fifty per cent. of the trials by eight or ten different observers. When arrangements, not indicated in Fig. 1, were made to give the appropriate circuit adjustments in rapid succession, simple words like " mama," " Anna," " wow-wow," yi-yi," were fairly well imitated. The whispered vowels and fricative consonants were not imitated so well, because (it is thought) of the lack of complete irregularity in the circuit interruptions.

* * *

The analytical expression for a single transient due to one resonant circuit when loosely coupled is, of course,

$$\text{Instantaneous displacement} = Ae^{-at} \sin 2\pi ft,$$

where e is the base of natural logarithms, a the damping constant, f the frequency, and A the amplitude. The displacement in the air-vibration is taken as proportional to the instantaneous current. Also, $f = 1/2\pi\sqrt{LC}$ nearly, and $a = R/2L$. The capacity C of each resonant circuit was variable in steps from 0·001 to 2 microfarads. The inductance L was continuously variable from about 0·3 to 0·7 henry. The resistance R was due largely to a dial box of range 1 to several thousand ohms, and included, in addition, the resistance of the inductometer and the (perhaps 100 ohms) small variable coupling resistances r and r' (Fig. 1).

The nature of the numerical results is indicated in Table I., which gives approximate values of the frequencies and dampings of the recurrent oscillations which characterise six of the more important vowels. Group frequencies (that is, voice pitches) were for each vowel varied over the range 75-300 per second. The first three vowels given in this table are each characterised by a single train of recurrent damped oscillations ; the remaining three are characterised by two trains of recurrent damped oscillations. The numerical values are approximate. Indeed, consider-

able changes in the circuit adjustments in some cases do not materially alter the vowel produced. The problem of determining the permissible range of variation for each speech-sound requires further study. For the latter three vowels the relative values of r_1 and r_2 are of some importance.

TABLE I.

Vowel.	Damping Oscillations. Frequency, f.	Damping constant, a. (Unit of Time, one second.)
rude . . .	320	small (<50)
law	650	100
father . . .	1000	500
mat	$\left\{\begin{matrix}750\\1500\end{matrix}\right.$	$\left.\begin{matrix}800\\800\end{matrix}\right\}$
pet	$\left\{\begin{matrix}420\\2300\end{matrix}\right.$	$\left.\begin{matrix}50\\50\end{matrix}\right\}$
cede . . .	$\left\{\begin{matrix}320\\2500\end{matrix}\right.$	$\left.\begin{matrix}50\\50\end{matrix}\right\}$

JOHN Q. STEWART.
Princeton University, Princeton, New Jersey, July 8.

VOL 110, 311 1922

Henry E. Armstrong, growing ever more crusty and intolerant, was never long absent from the pages of NATURE. *He had a good deal to say about newfangled ways with food, here under a stirring title.*

The Peril of Milk.

By Prof. HENRY E. ARMSTRONG.

A CONFERENCE of a most important and serious character was held in the Council Chamber of the Guildhall, London, on October 16-18, during the week of the Dairy Show, dealing with our milk supply in practically all its aspects—except the scientific ! Yet we speak of science as salvation, perpetually proclaim its importance, and deplore public apathy towards its priesthood. Our class was not invited to participate. I heard of the conference only casually and bought myself in, only at the very last moment ; consequently I was relegated to a place in the gallery behind the speaker's chair, where I could not hear a word. Being unobtrusive in my ways, I descended to the floor and trespassed into a vacant seat ; the platform was all but empty but no invitation to take a chair upon it came down to me. I do not wish to complain but merely point out the rewards of scientific service and the effusive way in which the man of affairs welcomes our aid.

And after a good deal more in this vein he gets to the nub:

Prof. Stenhouse Williams—dairy bacteriologist at Reading College—and I were the only speakers to sound the note of nutritional danger from Pasteurisation. We stood alone. Rothamsted, which claims to stand at the head of agricultural research, was unheard ; the Animal Nutrition station at Cambridge was voiceless. Sir W. Morley Fletcher, of the Medical Research Council, who took the chair at the discussion on Pasteurisation, had not a word to say by way of caution. The Medical Research Council, however, has never had a chemist among its members ; and yet medicine is nothing but applied chemistry.

Where, we may ask, are the Prophets ? Science is simply disgracing itself in this matter of milk : the call to wake up and defend the public health must go out everywhere.

VOL 110, 648 1922

Armstrong again, full sail hoisted:

A New Worship?

"Therefore no man that uttereth unrighteous things
    shall be unseen;
Neither shall Justice, when it convicteth, pass him by.
For in the midst of his counsels the ungodly shall be
    searched out;
And the sound of his words shall come unto the Lord
To bring to conviction his lawless deeds:
Because there is an ear of jealousy that listeneth to all
    things,
And the noise of murmurings is not hid.
Beware then of unprofitable murmuring."

AFTER a period of ennobling worship in that greatest of our English Cathedrals, the Scafell massif, on my return to town I chanced to enter that strange building, Burlington House, wherein be installed many altars to the great god, Science. Visiting that which ranketh first, I found an impassive figure, seated in a chair, at the High Altar, with a brass bauble before him: he needed but the peculiar head-dress to be an Egyptian Priest-King. Moreover, the service was apparently Graeco-Egyptian, if not Babylonian. The officiating young priest used many beautiful words clearly of Grecian origin, though at times an American phrase was noticeable, as when he spoke of Arrhenius doing chores, as I understood, for the god Isos. Most remarkable, however, was the way in which, at intervals, turning towards the altar, he solemnly gave utterance to the incantation—"See, Oh, Too!" My impression was that *Too* was the great king in the chair. The priest apparently was in fear of impending disaster, for at the close of his address he spoke much of concentration of the Hydrogen Ikons and their attack and repulse, often repeating the phrase "See, Oh, Too"—but Too seemed not to notice.

Two young acolytes then cast pictures of writing upon the wall as difficult to interpret as was that message expounded by Daniel in days long ago.

Most marvellous was the closing sermon, in which an account was given of the confusion wrought among a strange people, called "Lysodeiktics," by adding tears, nasal secretion, animal stews, turnip juice—seemingly muck of any kind—to their food: and how some of them were not killed. To one of an old faith, it seemed a strangely degenerate worship; indeed, that such service could be held worthy of attention amazed me.

In the evening, it chanced that I was led to peruse an article, in *The Times Literary Supplement*, on "Tradition and the French Academy," wherein is given Matthew Arnold's quotation, in his well-known essay, from the Academy's statutes:—

"The Academy's principal function shall be to work with all the care and all the diligence possible at giving sure rules to our language and rendering it *pure, eloquent and capable of treating the arts and sciences.*"

The whole article is worth reading; at the end is a quotation from a work by the late Pierre Duhem, the closing words being—

"*le respect de la tradition est une condition essentielle du progrès scientifique.*"

It is scarcely necessary to point out the application of these quotations; yet shall I ever pray: See to it, Oh, see to it, great Oh, Too!

HENRY E. ARMSTRONG.

The physicist E. N. da C. Andrade (always introduced with his full complement of initials), who became in later years something of a household figure as an all-purpose pundit on radio, reads a little homily to the credulous:

Occult Phenomena and After-images.

IF the hand be held against a dark background in a very subdued light, coming from behind the observer and falling on the hand, a diffuse glow will be observed round thumb and fingers, frequently uniting the finger tips. A little patience and a moderately clean hand are all that is required to observe the phenomenon

Further, however, if a hand be cut out of white cardboard (which is easily done by placing the hand, with thumb and fingers moderately spread, on the cardboard, tracing the outline in pencil, and cutting round with scissors) and feebly illuminated in the way described, a similar but somewhat stronger glow will be observed. In the case of both the flesh and the cardboard the shape of the glow can be modified by slow movement of the hand.

Such radiations are frequently described by writers on the occult sciences as being emitted by the human body. For example, in the chapter on magnetism in M. de Dubor's recently published "Mysteries of Hypnosis," I read of a doctor who was making magnetic passes over a lady. "The subject was wearing a black dress, and the doctor had his back to the light. Suddenly, in the semi-darkness which surrounded him, he observed a greyish vapour, like the fumes of a cigarette, issuing from the tips of his fingers, and, with especial clearness, from the index and the middle fingers. Moreover, the index fingers of the two hands seemed to be united by a luminous arc or semicircle. . . . Other persons, on the doctor's invitation, drew near and observed the same phenomenon. . . . Then the room was darkened. . . . In the darkness, twelve of the witnesses perceived nothing at all, and the remaining six perceived only very little."

M. de Dubor and the whole occult school explain the glow, or aura, seen round the hand as being due to magnetic emanations from the body (using the word magnetic in a superphysical sense). They appear to think that the phenomenon is more rare than it actually is, and do not treat the case of cardboard hands. For the phenomenon as observed with these, there would seem to be two possible alternative explanations. One is, that the cardboard is occult cardboard, and the scissors hypermagnetic scissors, and that I have unwittingly impregnated everything with induced ectoplasmic activity. The other is that the phenomenon is a retinal (and rational) one, which can be observed whenever a white, or whitish, surface is seen in a feeble light, the visual purple from the actual retinal image diffusing into the neighbouring parts of the retina. Accepting, for argument's sake, the latter explanation (which accounts at once for the fact that nothing is seen in the dark), the effect will be intensified by the restless movement of the eye, which undoubtedly takes place when objects are viewed in unfavourable circumstances.[1] The eye shifts the image into an unfatigued part of the retina, and the after image persists as a feeble glow. Such phenomena have been frequently described by Dr. Edridge-Green in a variety of forms, and I do not claim any particular originality for this prosaic explanation.

But a further very interesting phenomenon can be observed with the cardboard hand, which has not, I believe, been described. If it be looked at fixedly, the ends of the fingers will be seen to vanish intermittently, now one, now the other, while the extended thumb and little finger appear to move up and down, producing somewhat the appearance of a hand opening and shutting. The effect is very striking, and is pleasantly diversified by the complete disappearance of the hand at intervals. This is due either to retinal fatigue, combined with eye movement, or else to the ferro-forcificatory magnetism of the scissors, permeated as they must be with psychic influences and what not. I must leave it to the readers of NATURE to repeat the experiments, and judge for themselves.

Seeing that the festive season (I understand that this is the correct way to refer to Christmas) is upon us, I venture to describe a third occult phenomenon, somewhat analogous to that quoted by Dr. Edridge-Green in NATURE of December 9, p. 772. Two heads, facing one another, are cut out of white cardboard in profile, and observed in a very subdued light against a black background as before. (My heads are about two and a half inches in diameter, and the noses about half an inch apart.) By a delicate manipulation of the scissors one of the heads may be given a feminine character, largely by providing it with back hair. On careful observation the heads will be seen to approach and kiss repeatedly, separating with rapturous amaze after each contact. Like the other phenomena, including M. de Dubor's magnetic fluid, this cannot be observed in the dark, nor, I may add, even heard, in the case of the cardboard heads.

All the phenomena seem to be observed even more easily by myopic people than by myself. A morning's experiment has convinced me that with suitable illumination and white cardboard a very creditable *séance* can be arranged, including auræ, movements and levitations, magnetic emanations, and ectoplasm. This method involves no expense and no hymn singing. Even an atmosphere of reverence is not necessary for the production of the phenomena, although, I admit, the morning of my essay in the occult art was a Sunday morning, which may have had some favourable effect.   E. N. DA C. ANDRADE.

Artillery College, Woolwich,
    December 11.

VOL 110, 843 1922

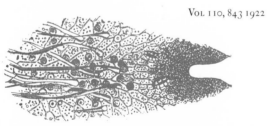

Julian Huxley was an exceedingly public figure, familiar in Britain during World War II as a member (sometimes in the company of Andrade) of the "Brains Trust", a radio panel that addressed itself each week to questions, some earnest, some frivolous, sent in by the numerous devotees. He was an author, together with H. G. Wells and Wells's son, G. P., of a three-volume encyclopaedia of popular biology — a work by which Wells set great store. Here Huxley reviews, in the form of a long essay, the latest literary effort by his friend (and the friend above all of the editor of NATURE, *Sir Richard Gregory).*

Biology in Utopia.

Men Like Gods. By H. G. Wells. Pp. viii + 304. (London, New York, Toronto and Melbourne: Cassell and Co., Ltd., 1923.) 7s. 6d. net.

THE columns of NATURE are not the place to discuss the literary merits of Mr. Wells's new book—although, for the matter of that, good style or artistic capacity and appreciation are qualities as natural as any others. Suffice it to say that he has achieved a Utopian tale which is not only interesting but also extremely readable. Most readable Utopias are in reality satires, such as "Gulliver's Travels," and the no less immortal "Erewhon." Mr. Wells has attempted the genuine or idealistic Utopia, after the example of Plato, Sir Thomas More, and William Morris; and, by the ingenious idea of introducing not a solitary visitor from the present, but a whole party of visitors (including some entertaining and not-at-all-disguised portraits of various living personages) has provided a good story to vivify his reflections.

However, since Mr. Wells is giving us not only a story, but his idea of what a properly-used human faculty might make of humanity in the space of a hundred generations, his romance has become a fit subject for biological dissection in these pages.

* * *

With the rediscovery of Mendel's laws and their recent working out, we are introduced to the theoretical possibility of an analysis of the hereditary constitution similar to the chemist's analysis of a compound; and so, presumably, in the long run to its control. There are great technical difficulties in higher organisms, and application to man presents yet further difficulties. Still, the fact remains that the theoretical possibility exists for us to-day, and did not exist twenty-five years ago. We must further remember that all discoveries concerning the history of man remind us that we must think, not in centuries as heretofore, but in ten-thousand-year periods when envisaging stages in human development.

We must further recall the lessons of evolutionary biology. These teach us that, however ignorant we may be regarding the details of the process, life is essentially plastic and has in the past been moulded into an extraordinary variety of forms. Further, that the attributes of living things have almost all been developed in relation to the environment—even their mental attributes. There is a causal relation between the absence of X-rays in the normal environment and the absence in organisms of sense-organs capable of detecting X-rays, between the habits of lions and their fierceness, of doves and their timidity. There is, thirdly, no reason whatever to suppose that the mind of man represents the highest development possible to mind, any more than there was to suppose it of the mind of monkeys when they were the highest organisms. We must squarely recognise that, in spite of proverbs to the contrary, it is probable that "human nature" could be considerably changed and improved.

* * *

Again, Mr. Wells, being a major prophet, perceives without difficulty that the substitution of some new dominant idea for the current ideas of commercialism, nationalism, and sectarianism (better not beg the question by saying *industry*, *patriotism*, and *religion*) is the most needed change of all. Here, again, he is in reality only adopting the method of Lyell and Darwin—uniformitarianism—and seeking the key of the future, as of the past, in the present. There is to-day a slowly growing minority of people who not only profoundly disbelieve in the current conceptions *and valuations* of the world and human life, but also, however gropingly, are trying to put scientifically-grounded ideas in their place.

Belief is the parent of action ; and so long as the majority of men refuse to believe that they need not remain the slave of the transcendental, whether in the shape of an imaginary Being, of the Absolute, or Transcendental Morality, they cannot reap the fruits of reason. If the minority became the majority, society and all its institutions and codes would be radically altered.

Take but one example, and a current one—birth-control. When Mr. Wells's "Father Amerton" finds that it is the basis of Utopian civilisation he exclaims in horror : "Refusing to create souls ! The *wickedness* of it ! Oh, my God ! "

.   .   .   .   .   .

This is the great enemy of true progress—this belief that things have been already settled for us, and the consequent result of considering proposals not on their merits, but in reference to a system of principles which is for the most part a survival from primitive civilisations.

Mr. Wells may often be disagreed with in detail : he is at least right in his premises. A perusal of his novel in conjunction with a commentary would be useful. "Men Like Gods" taken *en sandviche* with, say, Punnett's "Mendelism," Trotter's "Instincts of the Herd," Thouless's "Psychology of Religion," Carr-Saunders's "Population Problem," Whetham on eugenics, and a good compendium of recent psychology, would be a very wholesome employment of the scientific imagination.   J. S. H.

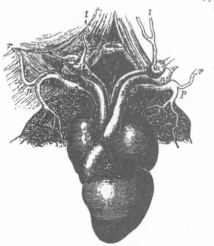

Paul Kammerer delivered in 1923 what turned out to be a fateful lecture in Cambridge on the experiments that purported to prove inheritance of acquired characteristics. These extracts from the resulting article give an indication of why his work had such wide appeal.

Breeding Experiments on the Inheritance of Acquired Characters.

ALMOST a quarter of a century has passed since I commenced to examine the inheritance of certain breeding- and colour-adaptations which I had obtained with amphibia and reptiles. I did not expect, in relatively so short a time, to obtain positive results, and, moreover, I was then well under the spell of Weismannism and Mendelism, which both agree that somatic characters are not inherited.

In the year 1909 I succeeded in ascertaining that *Salamandra atra* and *Salamandra maculosa* can be so bred as to produce a complete and hereditary interchange (of reproductive characters). The fact that *Salamandra atra*, which propagates itself in a highly differentiated manner, can be made to propagate itself in the manner of *Salamandra maculosa* need not necessarily be regarded as the acquisition of a new character, but may be an atavism. Since, however, the breeding habits of *Salamandra maculosa* can be changed to those of *Salamandra atra*, this objection is (in this case) excluded. I have hitherto always believed that no true inheritance underlay this phenomenon, but only the appearance of heredity (*Scheinvererbung*)—the external conditions applied (such as moisture) affect the germ plasm in the direct physical and not primarily physiological manner.

In view of my researches on the change of colour in *Salamandra maculosa* I could no longer entertain this belief. If the young animals are kept on a black background they lose much of their yellow marking and, after some years, appear mainly black. The offspring of these, if kept again in black surroundings, bear a row of small spots, chiefly in the middle line of the back. If the offspring, however, unlike their parents, are reared on a yellow background, these spots fuse to a band.

The yellow markings of the parent generation reared in yellow surroundings increase at the expense of the black colour of the Salamandra. If now the descendants of such strongly yellow individuals be kept on a yellow background, the yellow portions grow and appear as wide bilateral stripes. Descendants, however, which, unlike their parents, are now kept on a black background have less yellow, but proportionately far more than the background produces in the offspring of parents raised in black surroundings. The yellow markings are arranged symmetrically in rows of spots on both sides of the body.

*   *   *

... old and new characteristics can be distinguished, not only by means of crossing-experiments, but also by means of experiments on ovarian transplantation. If ovaries of spotted females are transplanted into the naturally striped ones, then the appearance of the young is determined by the origin of the ovaries—according to the *true* mother and not according to the foster - mother. They are always irregularly spotted. If, on the other hand, ovaries of spotted females are transplanted into artificially striped ones, then, if the father is spotted, the young are line-spotted ; if the father is striped, the young are wholly striped.

The ovary of the spotted female brings into the body of the naturally striped foster-mother only its own hereditary properties as effective in fertilisation. In the body of an artificially striped foster-mother this same ovary behaves as if it had been derived from the body of a striped female and as if the eggs of the striped female had been used in the crossings.

The objection cannot be raised that the operation was not thorough—that portions of the original ovaries may have been left behind in the foster-mother, as in Guthrie's experiments on fowls, which were afterwards tested by Davenport and found to be merely cases of regeneration of the original ovaries. Thanks to its enclosing membrane, the ovary of the Salamandra can be removed from the surrounding tissue as a whole. It is impossible that any remnants could have been left behind and that the descendants were derived from these remnants regenerated.

* * *

All existing objections, which rendered insoluble the inheritance of acquired characters, apply also to my breeding experiments on Alytes, and I myself would not have attached any special significance to this were it not that it is a result of just these experiments which has aroused the keenest interest in England— the development of a nuptial pad in the male Alytes. In male frogs, which pass their mating-time in water, there appears before mating, usually on the inner fingers, a rough, horny, glandular, dark-coloured pad. On the other hand, in Alytes, which mate on land, no trace of such a pad is to be seen. Yet it can be made to appear after several generations by compelling the Alytes to mate in water, like other European frogs and toads. This compulsion is brought about by raising the temperature, under which condition the mating animals stay longer in the water than usual, for if they did not do so they would run the risk of being dried up. Later in life compulsion becomes unnecessary. The stimulus of warmth produces an association through which henceforward the Alytes take to the water of their own accord when they wish to mate.

Of the many changes which gradually appear in this water breed during the various stages of development—egg, larva, and the metamorphosed animal, young and old—I will describe only one, the above-mentioned nuptial pad of the male. At first it is confined to the innermost fingers, but in subsequent breeding seasons it extends to the other fingers, to the balls of the thumb, even to the underside of the lower arm. After spreading, it exhibits an unexpected variability, both in the same individual and between one individual and another. The variability in the same individual is shown by the characters altering from year to year and in the absence of symmetry between the right hand and the left. In one specimen the dark pad extended to all the other fingers and almost over the whole of the left hand. On the right hand it was never so marked, and it was even less developed later, because the skin was stripped from this hand in the living animal for the purpose of histological investigation. The present skin and pad formation next to the inner finger is to be ascribed

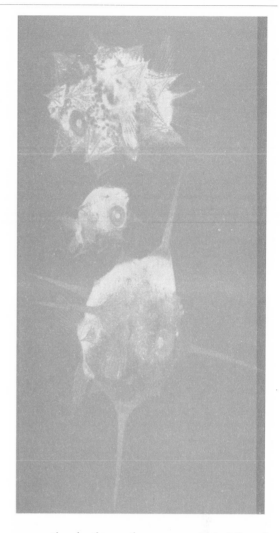

to regeneration in the mating season which followed. Microscopical preparations show the difference between the thumb skins of the mating male Alytes in the control breed and the padded skins of the water breed.

* * *

Not content with any of the previous experiments, I carried out, before 1914, what may really be an *experimentum crucis*. I have written a few words on it in my " Allgemeine Biologie." There has been no detailed publication as yet. The subject is the Ascidian, *Ciona intestinalis*. If one cuts off the two siphons (inhalent and exhalent tubes), they grow again and become somewhat larger than they were previously. Repeated amputations on each individual specimen give finally very long tubes in which the successive new growths produce a jointed appearance of the siphons. The offspring of these individuals have also siphons longer than usual, but the jointed appearance has now been smoothed out. When nodes are to be observed, they are due not to the operation but to interruptions in the period of growth, just as in the winter formation of rings in trees. That is to say, the particular character of the regeneration is not transferred to the progeny, but a locally increased intensity of growth is transferred.

One of the first reactions was from William Bateson, a most formidable adversary.

Dr. Kammerer's Alytes.

THOSE who have followed the discussion of Dr. Kammerer's claims will be aware that special interest has centred on the question whether he could produce for examination males of Alytes showing the modification alleged to occur in consequence of his treatment.

* * *

A photograph of the palm of a hand was thrown on the screen. This palm was pointed to as showing rugosities, but I saw none. In the specimen exhibited, the backs of the digits were not visible, nor were we shown any photograph of them.

I direct attention first to the fact that the structure shown did not look like a real *Brunftschwiele*. Next I lay stress on its extraordinary position *It was in the wrong place.* Commenting on the evidence, I pointed this out. In the embrace of Batrachians the palms of the hands of the male are not in contact with the female. Those who looked at the specimen naturally concluded that they must be. One speaker confidently told me in the discussion that I was wrong, and that in the common toad the rugosities *are* on the palmar surface! To show how the hands are placed I send a photograph (Fig. 1) of a pair of *Rana agilis* killed and preserved while coupled. The lower digits of the male's hands are the thumbs.

Clearly the rugosities, to be effective, must be on the backs and radial sides of the digits, round the base of the thumb, as in our common frog, on the inner sides of the forearms, or in certain other positions, but not on the palms of the hands. There are,

FIG. 1.

of course, minor variations, in correspondence with which the positions of the rugosities differ. The clasp of Alytes, for example, is first inguinal and afterwards round the base of the head (Boulenger). Minute thorns may be formed on the back of Bombinator and perhaps

in other places on the skins of Batrachians, where they cannot serve as *Brunftschwielen*; but on the palm of Alytes they would be as unexpected as a growth of hair on the palm of a man.

Dr. Kammerer's own reply was on different lines from that of the speaker I have mentioned, but curious and, as I thought, significant. He asked us to note that in his lecture he had refrained from using the word " Adaptation "—a defence sound perhaps, though surely disquieting to his disciples.

The discoveries claimed by Dr. Kammerer are many and extensive. To geneticists that regarding heredity and segregation in Alytes (*Verh. naturf. Ver. Brünn*, 1911) which I called in question at the Linnean meeting is the most astounding. But what I then heard and saw strengthens me in the opinion expressed in 1913, that until his alleged observations of *Brunftschwielen* in Alytes have been clearly demonstrated and confirmed, we are absolved from basing broad conclusions on his testimony.

W. BATESON.

VOL 111, 738 1923

E. W. MacBride, the unregenerate Lamarckian, weighed in for Kammerer and received a sharp slap on the wrist from J. T. Cunningham, a zoologist who himself leaned towards the Lamarckian view and some 30 years before had discharged a number of blasts against August Weismann in NATURE. *He believed, for instance, that the skewed cranium and asymmetrically placed eyes of flatfish were acquired characteristics. An innate intellectual rigour nevertheless turned him against Kammerer. Here is the conclusion of his argument.*

It would serve no useful purpose to reply to other points in Prof. MacBride's letter. He refers me to Dr. Kammerer's " long paper." But I was dealing with the lecture as delivered and printed, which in my opinion failed to show that Dr. Kammerer had an adequate conception of the range of knowledge, the completeness of evidence, and the validity of reasoning, required to establish the conclusions he asks us to accept. I am not, of course, suggesting any deception on Dr. Kammerer's part—except self-deception. Lamarckian doctrine has often suffered more from the indiscretion of its advocates than from the attacks of its enemies.

J. T. CUNNINGHAM.
East London College, Mile End, E.1,
June 26.

VOL 112, 133 1923

The fundamentalists were meanwhile strenuously active in the United States: this was the time of the Scopes or "monkey" trial, when a biology teacher was arraigned by the state of Tennessee for corrupting the young by teaching evolution and thus controverting Christian doctrine. William Jennings Bryan, sometime presidential candidate, was the prosecutor and was defeated and indeed broken by Clarence Darrow, who managed to goad him into the declaration that man was not a mammal. William Bateson in this admirable leader has Bryan's measure. Bryan died not long after, and H. L. Mencken wrote a classic memoir of the man, in which he gave this opinion of what it was that drove him: "What animated him from end to end of his grotesque career was simply ambition — the ambition of a common man to get his hand upon the collar of his superiors, or failing that, to get his thumb into their eye. He was born with a roaring voice, and it

*had the trick of inflaming half-wits. His whole career was
devoted to raising those half-wits against their betters, so
that he himself might shine." And as to the trial, Bryan
was not a bucolic fundamentalist, merely a hater of "the
city men who had laughed at him so long, and brought him
at last to so tatterdemalion an estate. He lusted for revenge
upon them. He yearned to lead the anthropoid rabble
against them, to punish them for their execution upon him
by attacking the very vitals of their civilization."*

The Revolt against the Teaching of Evolution in the United States.

THE movement in some of the Southern and
Western United States to suppress the teaching
of evolution in schools and universities is an interesting
and somewhat disconcerting phenomenon. As it was
I who, all unwittingly, dropped the spark which started
the fire, I welcome the invitation of the Editor of
NATURE to comment on the consequences.

First as to my personal share in the matter. At the
Toronto meeting of the American Association I was
addressing a scientific gathering, mainly professional.
The opportunity was unique inasmuch as the audience
included most of the American geneticists, a body
several hundreds strong, who have advanced that
science with such extraordinary success. I therefore
took occasion to emphasise the fact that though no
one doubts the truth of evolution, we have as yet no
satisfactory account of that particular part of the
theory which is concerned with the origin of *species* in
the strict sense. The purpose of my address was to
urge my colleagues to bear this part of the problem
constantly in mind, for to them the best chances of a
solution are likely to occur. This theme was of course
highly academic and technical. Nevertheless, to
guard against misrepresentation, I added the following
paragraph by the advice of a friend whose judgment
proved sound, though to me such an addition looked
superfluous.

" I have put before you very frankly the considerations which have made us agnostic as to the actual
mode and processes of evolution. When such confes-
sions are made the enemies of science see their chance.
If we cannot declare here and now how species arose,
they will obligingly offer us the solutions with which
obscurantism is satisfied. Let us then proclaim in
precise and unmistakable language that our faith in
evolution is unshaken. Every available line of argument converges on this inevitable conclusion. The
obscurantist has nothing to suggest which is worth a
moment's attention. The difficulties which weigh
upon the professional biologist need not trouble the
layman. Our doubts are not as to the reality or truth
of evolution, but as to the origin of *species*, a technical,
almost domestic, problem. Any day that mystery may
be solved. The discoveries of the last twenty-five years

enable us for the first time to discuss these questions
intelligently and on a basis of fact. That synthesis
will follow on analysis, we do not and cannot doubt."

The season must have been a dull one, for upon this
rather cold scent the more noisy newspapers went off
full cry, with scare-headings " Darwin Downed," and
the like.

All this seemed foolish enough, and that practical
consequences would follow was not to be expected.
Nevertheless, Mr. William Jennings Bryan, with a
profound knowledge of the electoral heart, saw that
something could be made of it and introduced the
topic into his campaign, which, though so far harmless
in the great cities, has worked on the minds of simpler
communities. In Kentucky a bill for suppressing all
evolutionary teaching passed the House of Representatives, and was only rejected, I believe, by one vote,
in the Senate of that State. In Arkansas the lower
house passed a bill to the same effect almost without
opposition, but the Senate threw it out. Oklahoma
followed a similar course. In Florida, the House of
Representatives has passed, by a two-thirds vote, a
resolution forbidding any instructor " to teach or
permit to be taught Atheism, agnosticism, Darwinism,
or any other hypothesis that links man in blood
relation to any form of life." This resolution was
lately expected to pass the Senate. A melancholy case
has been brought to my notice of a teacher in New
Mexico who has been actually dismissed from his
appointment for teaching evolution. This is said to
have been done at the instigation of a revivalist who
visited the district, selling Mr. Bryan's book.

The chief interest of these proceedings lies in the
indications they give of what is to be expected from a
genuine democracy which has thrown off authority
and has begun to judge for itself on questions beyond
its mental range. Those who have the capacity, let
alone the knowledge and the leisure, to form independent
judgments on such subjects have never been more than
a mere fraction of any population. We have been
passing through a period in which, for reasons not

altogether clear, this numerically insignificant fraction has been able to impose its authority on the primitive crowds by whom it is surrounded. There are signs that we may be soon about to see the consequences of the recognition of " equal rights," in a public recrudescence of earlier views. In Great Britain, for example, we may witness before long the results which overtake a democracy unable to tolerate the Vaccination Act, and protecting only some 38 per cent of its children.

As men of science we are happily not concerned to consider whether a return to Nature, as a policy, will make for collective happiness or not. Nor is it, perhaps, of prime importance that the people of Kentucky or even of " Main Street " should be rightly instructed in evolutionary philosophy. Mr. Bryan may have been quite right in telling them that it was better to know " Rock of Ages " than the ages of rocks. If we are allowed to gratify our abnormal instincts in the search for natural truth, we must be content, and we may be thankful if we are not all hanged like the Clerk of Chatham, with our ink-horns about our necks.

For the present we in Europe are fairly safe. A brief outbreak on the part of ecclesiastical authority did follow the publication of the " Origin of Species," but that is now perceived to have been a mistake. The convictions of the masses may be trusted to remain in essentials what they have always been; and I suppose that if science were to declare to-morrow that man descends from slugs or from centipedes, no episcopal lawn would be ruffled here. Unfortunately the American incidents suggest that our destinies may not much longer remain in the hands of that exalted tribunal, and that trouble may not be so far off as we have supposed.   W. BATESON.

Vol 112, 313 1923

Louis-Victor Pierre Raymond Broglie, the seventh Duc de Broglie, came to science late in his studies, deflected from a career as a historian by his elder brother, Maurice, a physicist of note who worked on the properties of X-rays in his laboratory in Paris. When World War I broke out, de Broglie was summoned to the colours and posted to the radiotelegraphy unit in the Eiffel Tower, where he remained for more than five years. Out of this practical experience of electromagnetism grew an absorbing interest in the nature of radiation. This celebrated paper, in which he makes the intuitive leap of treating light quanta as particles ("atoms of light"), formed the substance of his doctoral thesis.

Waves and Quanta.

THE quantum relation, energy = h × frequency, leads one to associate a periodical phenomenon with any isolated portion of matter or energy. An observer bound to the portion of matter will associate with it a frequency determined by its internal energy, namely, by its " mass at rest." An observer for whom a portion of matter is in steady motion with velocity βc, will see this frequency lower in con-

sequence of the Lorentz-Einstein time transformation. I have been able to show (*Comptes rendus*, September 10 and 24, of the Paris Academy of Sciences) that the fixed observer will constantly see the internal periodical phenomenon in phase with a wave the frequency of which $\nu = \dfrac{m_0 c^2}{h \sqrt{1 - \beta^2}}$ is determined by the quantum relation using the whole energy of the moving body—provided it is assumed that the wave spreads with the velocity c/β. This wave, the velocity of which is greater than c, cannot carry energy.

A radiation of frequency ν has to be considered as divided into atoms of light of very small internal mass ($< 10^{-50}$ gm.) which move with a velocity very nearly equal to c given by $\dfrac{m_0 c^2}{\sqrt{1 - \beta^2}} = h\nu$. The atom of light slides slowly upon the non-material wave the frequency of which is ν and velocity c/β, very little higher than c.

The " phase wave " has a very great importance in determining the motion of any moving body, and I have been able to show that the stability conditions of the trajectories in Bohr's atom express that the wave is tuned with the length of the closed path.

The path of a luminous atom is no longer straight when this atom crosses a narrow opening; that is, diffraction. It is then *necessary* to give up the inertia principle, and we must suppose that any moving body follows always the ray of its " phase wave "; its path will then bend by passing through a sufficiently small aperture. Dynamics must undergo the same evolution that optics has undergone when undulations took the place of purely geometrical optics. Hypotheses based upon those of the wave theory allowed us to explain interferences and diffraction fringes. By means of these new ideas, it will probably be possible to reconcile also diffusion and dispersion with the discontinuity of light, and to solve almost all the problems brought up by quanta.

    LOUIS DE BROGLIE.

Paris, September 12.

Vol 112, 540 1923

Hertha Ayrton, daughter of a Polish-Jewish immigrant family, friend of Marie Curie and one-time protégé of George Eliot (wife of Darwin's friend George Henry Lewes), was supposedly the model for Minah in George Eliot's last great novel, Daniel Deronda. Henry E. Armstrong's characteristically patronising obituary of this forgotten woman pioneer is calculated to incite the feminists.

MRS. HERTHA AYRTON.

APPEAL is made to me to give some account of Hertha Ayrton, the wife of my former colleague, who died last August.

" Is the study of heredity a science or a pure romance ? " asks Mrs. Trevelyan, in her biography of her mother, Mrs. Humphry Ward. I would set the question in another form : Is *das ewig Weibliche* to be suppressed by science ? Mrs. Ayrton was one of those who aspired to prove that woman can be as man as an original scientific inquirer. Did she succeed ? If we are to frame a psychology of the scientific mind, regarding this as a species apart, we must carefully note and analyse the doings of such as she. I have but small qualification for the office, yet as she was my colleague's wife and we often met and were in fair sympathy, I was able to take notice of her idiosyncrasies and of the conditions under which she was placed.

Ayrton and I met originally in the autumn of 1879, when we were appointed the first two professors of the City and Guilds Institute and set the ball of technical education rolling in London ; the ball rolled well and proved to be fissiparous but no one of the small band who gave it shape in the City and West End ever received the slightest recognition from the Guilds, their masters— and most of these have committed *hari-kari* as concerted workers in education. A strange world is ours and if we worked otherwise than for the sake of working, we should do little.

Ayrton had a peculiar experience : his then (first) wife—his cousin, Mathilda Chaplin—was a woman who had acquired merit in the cause of women's rights, as she was one of the three, I believe, over whom the fight first raged in Edinburgh whether women should be admitted to the study of medicine. When I met her, her health was more than failing. She was an ethereal being, a woman of infinite charm of manner but above the world—a mature Melisande ; indeed, when I first heard Debussy's opera her memory was recalled to me by the peculiar rhythm and tone of its melody. Her daughter, Mrs. Zangwill, has inherited not a few of her mother's characteristics—especially her charm of voice. Her chief occupation was novel-reading, from penny-dreadfuls upwards, in which she ran a caucus race with our erratic friend, John Perry.

Ayrton married his second wife in 1885. If I were to compose an opera with my scientific friends as the characters, I should associate the Melisande theme with the first Mrs. Ayrton ; I should not quite know where to place the second musically but it would be near to Brunhilde, as she had much of the vigour of Wotan's masterful daughter and, at least, aspired to be an active companion of scientific heroes—a race far above Wagner's dull and degenerate Teutonic gods, be it said.

Sarah Marks was the daughter of intelligent but poor Jewish parents in Portsmouth. She was a clever child and was early sent to a school in London kept by her paternal aunt, who became Mrs. Hartog ; Mr. Hartog was a teacher of French in London. Mrs. Hartog was the mother of Numa Hartog, Philip Hartog and the professor of botany in Cork ; also of two daughters, one very clever, a talented painter, who married Dr. Darmstadter of Paris ; the other earned her living as a musician. Numa Hartog died early, after a most brilliant university career and seems to have been unusually clever. Mrs. Marks had four undistinguished children, besides Sarah ; nothing is known of her parents. Mrs. Ayrton's ability, however, would seem to have been derived from the mother's side.

At about the age of fifteen, Sarah Marks became acquainted with Madame Bodichon, a well-to-do lady, strong on the women's rights question, who sent her young friend to Girton College, Cambridge. Apparently, she then changed her name to Hertha. She took honours in mathematics. She is credited with the invention, during the period, of a sphygmograph and also of an instrument for rapidly dividing up a line into a number of equal parts. Through Madame Bodichon, she became acquainted with George Eliot and several other people of distinction. In 1884 she entered the Finsbury Technical College. I remember her coming. She not only came but was seen and soon conquered—Ayrton ; and they married. As sole issue

they had a daughter, who has her father's gift of tongue ; she married a Christian, whilst his daughter by his first wife married a Jew. I often told him that he and his wife were an ill-assorted couple : being both enthusiastic and having cognate interests, they constantly worried each other about the work they were doing. He should have had a humdrum wife, "an active, useful sort of person," such as Lady Catherine recommended Mr. Collins to marry, who would have put him into carpet-slippers when he came home, fed him well and led him not to worry either himself or other people, especially other people ; then he would have lived a longer and a happier life and done far more effective work, I believe.

Under her husband's inspiration, Mrs. Ayrton soon entered upon the study of the electric arc. Her work is recorded in the book on the subject which she published in 1902, in part a reprint of papers submitted to the Royal and other Societies. She was an indefatigable and skilful worker. Whatever the absolute value of her observations, her husband and his good friend Perry were the last not to make the most of her achievement, so probably the scientific halo with which they and others who fancied that women could be as men surrounded her was overpainted. Most of us thought, at the time, that they were ill advised in preferring her claim to the Royal Society ; the nomination came to nothing on legal grounds. She was, however, elected into the Institution of Electrical Engineers and at her death was its only lady member. She also engaged in an inquiry into the formation of sand ripples and this led her, early in the War, when chlorine was first used as poison gas, to develop a fan-device for waving back the fumes. There is little doubt that she took too high a view of the practical value of the invention and was unwarrantably aggrieved at its rejection by the military authorities. She was awarded the Hughes Medal by the Royal Society in 1906.

Mrs. Ayrton was a very striking woman in appearance and of considerable personal charm, full of common sense ; this kept her from being a militant suffragist, though she promoted the cause in every possible way. I never saw reason to believe that she was original in any special degree ; indeed, I always thought that she was far more subject to her husband's lead than either he or she imagined. Probably she never had a thorough scientific equipment ; though a capable worker, she was a complete specialist and had neither the extent nor depth of knowledge, the penetrative faculty, required to give her entire grasp of her subject. Ayrton himself, though a genius, was in no slight measure partial in his interests : by heredity literary and artistic, educated intensively in the classical school, a born actor and therefore a good lecturer and public speaker, impelled into science through contact with Sir William Thomson, he was a worker chiefly at its technical and commercial fringe rather than in its depths : so he was not a good judge of his wife's scientific ability. His partner Perry was the solid member of the firm. In fine, my conclusion is, that *das ewig Weibliche* was in no way overcome in Mrs. Ayrton : nor could we wish that a thing so infinitely precious should be : she was a good woman, despite of her being tinged with the scientific afflatus.   HENRY E. ARMSTRONG.

The influenza epidemic, or Spanish flu, of 1919 claimed more lives in Europe than the Great War had done.

INFLUENZA has again further increased in severity over the British Isles, and the Registrar-General's return for the week ending February 15 shows the deaths in London (County) to be 273 due to the epidemic. Forty-eight per cent. of the deaths occurred at the ages from twenty to forty-five, so that the death incidence is similar to that when the present epidemic was most virulent at the commencement of last November, the complaint attacking most severely the strong and able-bodied. Influenza caused 13 per cent. of the total deaths during the week ending February 15, pneumonia 13 per cent., and bronchitis 16 per cent.; in the early part of November influenza caused 57 per cent. of the deaths from all causes, but deaths from pneumonia and bronchitis were not very different from those at present. In the ninety-six great towns of England and Wales, including London, there were 1363 deaths during the week from influenza, and since the commencement of the epidemic in October last there have been 48,736 deaths, whilst in London there have been 12,286 deaths. The total deaths in any previous epidemic in London have only amounted to about 2000. The present is the twentieth week of the epidemic, five of the previous epidemics having continued as long, and the epidemic from October 1904 to April 1905 continued for twenty-six weeks, but in London during the whole time the total deaths from influenza were only 707, and the maximum number in any week was only forty-five.

VOL 102, 510 1919

The Möbius band — the figure with only one surface — was described by the German mathematician August Ferdinand Möbius. Charles Vernon Boys characteristically finds a practical application for this topological oddity (curiously never referred to in his letter by its familiar name) in the interests of thrift.

A Puzzle Paper Band.

SOME thirty or forty years ago geometricians were much interested in the endless band of paper to which one half twist had been given before joining the ends. This gave the figure having only one surface and one edge. At that time those who studied this figure were so obsessed with the consequence of cutting down the middle line of such a band or of a band with two or more half twists that I believe no one noticed the result which I wish now to describe. It is the doubling up in a proper manner of an endless band to which four half twists have been given so as to produce the endless band first described but of double thickness. The first band is shown at A, as an endless belt connecting two crossed shafts

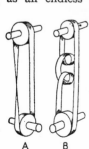

for which, as is well known, it is exactly fitted. B shows the band with the four half twists all on one side in the form of two complete loops, and to be uniform with the other it is shown as an open band connecting two parallel shafts. The only object of putting in the shafts and pulleys is to assist the perspective. They are not wanted in making the experiment. Now if B, which appears sufficiently uncompromising, is folded up properly it becomes A but of double thickness.

A   B

I used this doubled A for a time as a record sheet for my recording calorimeter, for, having a head-room to work in of four feet and a movement of paper of six inches a day, I was able in this way to obtain a continuous record for 32 days on one side of the paper only. This is superseded now by a more convenient arrangement.

It may be worth while to add that while lying awake one night I visualised A in two thicknesses and saw it to be what I wanted, and the next day I found it all right. What I did not visualise was the puzzle that it is to fold up B into an A of double thickness, and that it makes a first-class parlour puzzle game. It has this further advantage that a number may be made, some with right-hand and some with left-hand twists, so that any preliminary success gained on one may make the other seem the more difficult. The band should be not less than 50 times as long as it is wide.

Of the four half twists two are easily seen in the finished double thickness band, for each thickness has one half twist; it is amusing to find out where the other two have gone.   C. V. BOYS.

VOL 111, 774 1923

chapter twelve
1924-1930

PAUL KAMMERER, THE PROPONENT OF LAMARCKIAN INHERITANCE, TAKES HIS OWN LIFE, BUT IS DEFENDED BY BIOLOGICAL IDEOLOGUES, NOTABLY E. W. MacBRIDE. HENRY E. ARMSTRONG DENOUNCES AS FOOLISHNESS W. L. BRAGG'S CONCLUSIONS CONCERNING THE IONIC CHARACTER OF COMMON SALT. SIR WILLIAM ARBUTHNOT LANE GRAPPLES WITH CONSTIPATION. HILAIRE BELLOC INVEIGHS AGAINST DARWINISM. LORD RUSSELL IMPARTS HIS OBSERVATIONS ON KITTENS. TELEVISION COMES TO STAY AND MEN OF SCIENCE SHUDDER TO BE CALLED SCIENTISTS.

1924-1930

Lenin died. In Germany, Paul von Hindenburg became president; in England, the general strike was defeated; and in the United States people were shaken by the stock-market crash. Charles Lindbergh landed in France, a zeppelin circumnavigated the globe and the airship R101 crashed on its way to India, so establishing the supremacy of heavier-than-air flight.

In Russia, dire consequences for science continued to flow from the revolution.

The Position of Scientific Workers in Russia.

AFTER the Bolshevist revolution in 1917, Russian men of science adopted two courses of action. Those who engaged in active or passive opposition to the Soviet regime emigrated or were shot, and some were deported abroad by the government, so that there are now about 500 Russian scientific workers outside the borders of Russia. Others took up a strictly non-political attitude and decided to remain at their posts to safeguard and carry on Russian science. All teaching of moral sciences was abolished or "adapted to communist principles." Those professors who refused to comply were dismissed, and some died of starvation in spite of the existence of a government commission for improving the conditions of scientific workers, some of whom received extra food rations.

Russian men of science are now working under most difficult conditions. Their salary is far below a minimum living wage; some receive only about 9-12 per cent. of their pre-War salary, while the cost of living is far higher. They work in living-rooms and laboratories where the temperature is scarcely above freezing point; they chop wood, carry water, hunt for cheaper food, etc. There is acute shortage of gas, chemicals, implements, and books. Nevertheless, Russian science is not dead and has recorded many important achievements, but purely scientific works are difficult to publish. This is not so with works based on "Marxist principles," which are specially commended. Those men of science who have joined the communist party have been given special privileges.

An Institute of "Red" professors has been founded to supplant the old workers. The students are recruited from the communist party and lack proper scientific training. Admission to the universities and technical colleges is also officially confined to communists, raw uneducated peasants and workmen's youths. Lecturing to such an audience is particularly difficult.

This winter the government proposed to dismiss most of the old professors and replace them by "Red" ones. Faced by starvation or loath to hand over their beloved work to incompetent people, some of the former rallied to the Soviet platform. A congress was called in Moscow, at which an alliance was staged between science and the proletariat. All resistance was so far broken that the professors elected Zinoviev, the president of the International, whose activities are quite outside the realm of science, as honorary member of one of the sections of the congress.

VOL 113, 142 1924

G. I. Taylor was a physicist of legendary originality. Here he describes an experiment with a toasting-fork (an instrument that used to be considered essential for college life in England).

The Singing of Wires in a Wind.

THE following observation may be of interest in connexion with the singing of wires in a wind. A four-pronged toasting-fork was waved backwards and forwards through the air. It was found that if the plane of the prongs was in the direction of motion, a singing noise was produced, but that if the plane of the prongs was perpendicular to the direction of motion and the fork was moved at the same speed through the air, the singing was practically inaudible.

The prongs were 8 cm. long, 0·25 cm. diameter, and they were spaced 2·2 cm. apart.

The resistance of a wire when it is moved through air, and also the singing noise produced by its motion, are both due to eddies which are formed in its wake. It seems curious, therefore, that when several wires are arranged so that they shield one another and thus reduce their total resistance, the singing noise is thereby greatly increased instead of being reduced.

G. I. TAYLOR.

Trinity College, Cambridge,
March 15.

VOL 113, 536 1924

Sir Arthur Schuster wrote a warm obituary notice of Georg Hermann Quincke, best remembered now for his work on surface tension, who had been hard at work until a few weeks before his death at the age of 90. This paragraph seems to epitomise his character.

Quincke was an experimenter of the highest rank; for theories he had little affection. He looked with disfavour on many of the recent developments of physics, but his scepticism began so long ago that science has, at any rate in one case, justified it. The writer of this notice remembers more than one occasion when he asserted with some warmth his disbelief in the existence of the ether.

VOL 113, 281 1924

In 1925 there appeared a supplement devoted to the memory of T. H. Huxley. Here is a domestic vignette.

Without doubt some of Huxley's lecture jokes were annuals, like those for which Oliver Wendell Holmes became famous in his anatomical lectures at Harvard. Others were spontaneous and of the moment. He loved stories upon himself, as of his popular lecture on the brain and the one elderly dame whom he especially picked out to address as apparently the only intelligent member of an evening audience. As the lecture closed this dame advanced for a question: "Professor, there is one point you did not make quite clear to us: Is the cerebellum inside or outside of the skull?" This was a crusher. On his lack of orthodoxy, according to a story of youthful domestic experience which he told my wife, he was never rebuked so forcibly as in the early years of his married life by an intoxicated cook. After Mrs. Huxley had tried in vain to dislodge the cook from the kitchen floor, Huxley descended to the kitchen and with full assurance of masculine supremacy said: "Bridget, get up and go to your room, you ought to be ashamed of yourself." Whereupon Bridget gave a kick and replied: "I am not ashamed of myself, I am a good Christian woman, I am not an infidel like you."

VOL 115, 727 1925

The benign tolerance of the Medical Research Council paid dividends too numerous to list. A distinguished physiologist reflects on its record and sets out his scientific creed:

Discovery and Research.

In the leading article in Nature of April 5, on "Medical Research in Great Britain," dealing with the Report of the Medical Research Council for the year 1922–23, you seem to regard it as a sign of failure that the Report contains no record of a first-class medical discovery made in Great Britain during the year in question. The work of the Medical Research Council does not need support from me, but I am concerned that Nature should adopt the exaggerated idea of the merit of ' discovery ' which is held by the ' man in the street.' Every discovery, however important and apparently epoch-making, is but the natural and inevitable outcome of a vast mass of work, involving many failures, by a host of different observers, so that if it is not made by Brown this year it will fall into the lap of Jones, or of Jones and Robinson—simultaneously, next year or the year after.

One or two examples will illustrate my point. Bayliss and I once had the good fortune to make a ' discovery '! It was of no practical use to any one, but a source of much gratification to ourselves, since it seemed to open up a new chapter in our knowledge of the body. But there were at that time half a dozen workers skating along the edge of the discovery, and it is difficult to comprehend why, for example, Wertheimer and Lepage did not take the one further step which would have made them and not us the discoverers of *secretin*.

The same thing applies to other discoveries which have been of paramount importance for the welfare of mankind. Ross's discovery of the transmission of malaria by mosquitoes had been prepared by the work of many other men on the part taken by insects in conveying disease both to men and to the lower animals. If Ross had never lived, it is improbable that this great discovery would have been deferred by more than a few months or years. Similarly, the preparation of *insulin* by Banting and Best, an admirable piece of work, is but the last step of an arduous journey, in which hundreds of workers have taken part. In the history of insulin, the greatest achievement was probably that of Minkowski, when he showed that diabetes could be produced by extirpation of the pancreas. If tradition may be believed, this discovery was due to an accidental observation of the old laboratory servant, who tasted some crystals left on the floor of the laboratory by the evaporation of urine voided by the operated animals, and found them sweet. The actual discovery of insulin waited, however, forty years, until an easy and accurate method was devised for the estimation of sugar in small quantities of blood. A dozen workers were on the track. Banting and Best, coming in late, but with the freshness and ardour of youth, arrived first.

I believe that Nature has urged the desirability of a sum of money being set aside by Parliament for the reward of ' discovery.' I contend that any such sum should rather be used as consolation prizes for the many able scientific workers who have never made a discovery.

Discovery brings its own reward. Not only is there the joy of being first on a virgin peak—a joy that every mountaineer hopes to experience at least once in his lifetime—but, if the discovery is of a nature to be appreciated by the public, it brings in its train Royal medals, K.C.B.'s, Nobel prizes, and academic distinctions of all sorts, not to mention the plaudits of the daily press.

The Medical Research Council has no need to concern itself with ' discoveries.' All it has to do is to ensure that the growing tree of knowledge is dug round and pruned, dunged, and watered. The fruit will come in due season, and will fall to the lot of some one who may or may not have been assiduous in the labour of cultivation. The Report of the Council shows that this, its function, is being wisely and diligently carried out ; and we servants in the House of Science cannot but approve of its efforts to facilitate our work.   Ernest H. Starling.

University College, W.C.1,
     April 12.

Vol. 113, 606 1924

An audacious experiment on the US president's person, reminiscent of the unhappy attempt to treat the Duke of Wellington's deafness by pouring sulphuric acid into his ear:

It was reported in the *Times* of May 22 that President Coolidge had been treated for a cold by the inhalation of a chlorine gas mixture. Thus is put to good use a gas which was one of the first employed for its noxious effects during the War. The incident illustrates a principle which has a very general application, that substances which are poisonous in higher concentrations are useful in smaller amounts ; in fact, not only the substances classed as poisons, which are used beneficially every day in medical treatment, but also others, the presence of which in the body is essential to its normal functioning, are toxic in quantities above a certain level. The use of chlorine gas for the treatment of colds was suggested by certain observations upon men employed in its manufacture during the War ; it was found that these workers suffered less than others from influenza. Its mode of action is probably not a direct one upon the microbes in the nasal cavities, although this explanation has been suggested for the action of the fumes of nitrogen peroxide and sulphur dioxide in preventing influenza : these gases may make the secretions acid and so hinder bacterial growth. Chlorine, however, acts more probably as an irritant to the mucous membranes, producing an increased secretion from them, and this carries with it a host of white blood cells ; both the cells and the secretion attack the micro-organisms, while the latter also washes them away mechanically. Possibly also the microbes find it difficult to penetrate through these secretions into the mucosa itself. Thus it is seen that the gas acts more by stimulating the natural defensive processes of the body than by a direct effect upon the microbes themselves, showing once again the extreme importance of the defensive mechanisms of the body itself in resisting bacterial invasion.   Vol. 113, 796 1924

Dr Albert Abrams was a notorious mountebank whose diagnostic "black box" attracted the gullible in their hordes. NATURE *speaks:*

The Abrams' Cult in Medicine.

IN our issue of June 7 we published an article on "the Abrams' Cult in Medicine" and indicated that the claims made on behalf of the late Dr. Abrams were without scientific foundation. One of the many ridiculous diagnoses made by means of the so-called electronic dynamiser was quoted, and the methods employed by the "electronic reaction Abrams" (E.R.A.) were described. In the July, August, and September numbers of the *Scientific American* there is an extensive report of a committee of experts appointed by that journal to inquire into the actual facts of the case. The committee consisted of thoroughly trained medical and electrical investigators, including such well-known men as W. H. Park, W. C. Alvarez, R. C. Post, and J. M. Bird, with A. C. Lescarboura as secretary. These men have spent a year chasing the elusive methods of electronic medicine. They have interviewed and co-operated with electronic practitioners and experts in the new cult, and the result is disastrous to the reputation of Albert Abrams.

The report of this committee is in what may be described without offence as the characteristic American style—interesting to those on the eastern side of the Atlantic. The authors of the report state that the "Electronic Reactions of Abrams are not substantiated, and it is our belief that they have no basis in fact. In our opinion the so-called electronic reactions do not occur, and the so-called electronic treatments are without value." In another place the committee states that "the entire Abrams' electronic technique is not worthy of serious attention in any of its numerous variations. At best it is an illusion. At worst it is a colossal fraud."

The *Scientific American* committee gives a clear account of the genesis of the Abrams' myth, and shows how the Californian Cagliostro's claim that the basis of life is bound up with the movements of electrons is a question which does not admit of proof. On the electronic basis, however, Abrams invented a mysterious box called a dynamiser for collecting the electronic emanations, and elaborated a portentous system in which diagnoses of the most obscure ills of life were said to be possible, at a distance of thousands of miles, from the emanations given off by a drop of blood on a disc of blotting-paper. Not only the disease of the patient but also the height, habitus, colour, sex, and religion were said to be capable of revelation by the dynamiser—a childish toy which defies all the laws of electrical science. In one of Abrams' own demonstrations he placed in the dynamiser a photograph of a man and foretold his disease as insanity due to syphilis,

and on running the electrode across a map of America it stuck at Stockton, California, which was consequently proclaimed as the patient's address !

By judicious advertisements and newspaper "puffs" by credulous persons and those in quest of sensational material for the press, Abrams' claim to diagnose disease at a distance was soon noised abroad, and Abrams founded a school rich in financial possibilities. He then added to the diagnostic scheme one of electronic treatment, thus providing a wider field for commercial exploitation. New companies sprang up to manufacture the Abrams' dynamisers and oscilloclasts at fancy prices. Schools for training Abrams' practitioners were founded, and a motley array of medical men, osteopaths, chiropractors, and electronaturopaths hied them to sunny California to sit at the feet of Gamaliel and hear him expound the new science. Within a few years, more than three thousand practitioners of electronism were trained for their life work in the United States alone. They found that there was a rich harvest in dollars, in comparison with which ordinary medical practice was a poor profession.

When one reflects that we are living in the twentieth century, and that both medical and lay people can believe something which has no meaning, it is clear that there is no limit to credulity. There have been great humorists in history, but Dr. Albert Abrams of California must surely head the list. To invent a new form of energy which cannot be tested by any known method except by the five senses of Abrams and his pupils is something of an achievement. To determine the sex of a child *in utero* from a drop of the blood of the father a thousand miles away is asking one to believe a good deal. To explain errors in electronic diagnosis in California by wireless disturbances on the Eiffel tower in Paris or activities on the planet Mars is perhaps reasonable from the electronist's point of view. To expect, however, your fellow-creatures to believe all this nonsense, and to earn vast sums of money by the delusion, may be classed as a form of humour to which our psychologists and mental philosophers have not paid due attention. Although the *Scientific American* committee has exposed the absurdity of Abrams' system, it does not follow that its labours will be crowned with a universal success, for in all probability there will still remain hundreds of genial persons who do not mind paying their dollars to hear the joyous tick-tack of Abrams' oscillating metronome.

We live in an age of science, but it is as easy now as ever it was for the charlatan to impose on the credulity of the public. The penalty of great scientific achievement seems to be that people are ready to accept the most astounding claims, provided they are put forward in the name of science.   W. B.

Paul Kammerer was not alone in extravagant claims
about the success of improbable animal experiments.

Chimæras Dire.

DR. FINKLER'S experiments on the transplantation of the heads of insects have attracted both scientific and popular attention to a degree which was marked on the one hand by an exhibit last year at a Royal Society soirée and, on the other, by mention in the pages of *Punch*. It is desirable, therefore, to direct attention to an emphatic repudiation of his claims which has just been published in the *Zeitschrift für wissenschaftliche Zoologie* (vol. cxxiii. pp. 157-208) by Hans Blunck and Walter Speyer.

It will be recalled that Finkler stated that the heads of adult insects could be successfully grafted on to bodies of the other sex, and even on to bodies of distinct species belonging to widely different genera. He inferred, rather than observed, the union of tissues following the operation and hastened on to describe its remarkable results, physiological and psychological. The head of an herbivorous water-beetle persuaded a carnivorous body to be content with, and seemingly to digest, a vegetable diet; a male head led a female body into unwonted perversities; and a Dytiscus strove to moderate the colouring of its wing-cases to suit the sober tastes of its new Hydrophilus brain. Experiment was added to experiment, and water-boatmen abnormally coloured by inverted illumination transferred the abnormal coloration with their heads to other individuals not so illuminated.

It was to be expected that a field of work offering such remarkable possibilities would speedily be occupied by other investigators. The living material is easy to obtain, the technique is simple (" an Roheit schwerlich zu übertreffen," say Blunck and Speyer), and the results are got in a brief space of time; and yet no one, with the solitary exception of Dr. Kammerer, appears to have claimed success in repeating even the less startling of Dr. Finkler's experiments.

Now Dr. Blunck and Dr. Speyer (already known by a long series of anatomical and biological researches on the very water-beetles that were among the chief of Finkler's *corpora vilia*) come forward with a detailed and documented confession of their failure, after persistent attempts, to confirm the simplest of Finkler's results. They followed carefully his instructions as to procedure, and only after this failed did they try to refine upon his technique. The severed heads certainly adhered to the bodies on which they had been placed, cemented, in fact, by the coagulated blood which dried to a chitin-like hardness, but in no case did the insects survive longer than others in which, after decapitation, the wounds were stanched with a little melted wax; in no case did the chimæras behave differently from the headless trunks; in no case did stimulation of the head produce reflex movement in the body or its limbs. Microscope sections showed no attempt at regeneration or union of the tissues, but instead, a progressive necrosis, leading more or less quickly to the death, first of the head and later of the trunk. These results followed, even in cases of replantation where the severed head was replaced at once on the body from which it had been detached.

Dr. Finkler may of course be able to point out to us where the German experimenters have gone wrong, but there seems to be a familiar ring in their complaint that he has ignored requests to produce his chimæras, alive or dead, for investigation by others. He cannot ignore the challenge of their final words: " Die Wissenschaft hat angesichts der allen Erfahrungen widersprechenden Angaben des Wiener Autors keine Veranlassung, sich weiter mit ihm und seine Schriften zu beschäftigen."   W. T. CALMAN.
British Museum (Natural History),
South Kensington, S.W.7.

VOL 114, 11 1924

Kammerer revealed himself a fabulist in more areas than one.
He was an associate in Vienna of Eugen Steinach and Serge
Voronoff, the reanimators, by way of testis implants, of vitality
and sexual potency. The reviewer, F. A. E. Crew, director of an
animal breeding institute in Edinburgh, had also experimented
on similar lines and reported success. In time he became
converted to the contrary view.

Rejuvenation.

Rejuvenation and the Prolongation of Human Efficiency :
Experiences with the Steinach-Operation on Man and
Animals. By Dr. Paul Kammerer. Pp. 252.
(London : Methuen and Co., Ltd., 1924.) 8s. 6d. net.

WHEN a new development in science has provided the low comedian with material for jest and the popular author with a plot, when from the biassed misinterpretation of the journalist it appears that a discovery has been made that must revolutionise the life of the individual and of the community,—then it is high time that some one with a training in that particular branch of science and in scientific method, and with a literary ability that enables him to interpret to the man-in-the-train, should give a popular presentation of the facts and the theories based upon them, so that all those who seriously seek information on this subject can glean it. A book in English dealing with the operative methods of achieving rejuvenation and giving in a legitimately popular form a critical review of the work of Steinach, Voronoff, and their followers has, therefore, been expected for some time; and this is attempted in the volume before us.

* * *

As for the methods to be employed, that of injection of such extracts as are available at present is utterly untrustworthy; for the implantation of grafts an adequate supply of human or simian gonad is required, and this is not easily maintained; unilateral ligation of the deferent duct of the male, on the other hand, is a simple operation that has often been performed, though Steinach was the first to do so for the definite purpose of bringing about the rejuvenation effects.

If the Steinach operation is followed by demonstrable improvement in the general well-being of the individual, and there is considerable evidence which seems to show that it is, then this operative procedure must quickly find its way into human and veterinary surgery. For example, it would mean much to the stock-breeder to be able to get another year's crop of offspring from a famous sire.

VOL 114, 530 1924

The term "scientist" was invented by the nineteenth-century scholar William Whewell (of whom the literary clergyman Sydney Smith said that "science is his forte and omniscience is his foible") but did not come into common use until well into the twentieth century. Nature (and scientists) adhered to "man of science" as the preferred usage. Michael Faraday accepted (though never used) the word, but balked at "physicist". ("The equivalent of three separate sounds of i in one word is too much.") Blackwood's Magazine also mocked at "the word physicists, where four sibilant consonants fizz like a squib." The debate suddenly bubbled up again in 1924.

So long ago as 1840, according to the "Oxford Dictionary," Dr. Whewell, eminent as man of letters as well as man of science, wrote: "We need very much a name to describe a cultivator of science in general. I should incline to call him a Scientist." I do not think the objections to the word on merely literary or linguistic grounds can be maintained. It is a hybrid, but the language is full of hybrids: moreover, it may well be argued that *-ist* is naturalised as an English termination.

There is, however, another sort of objection which has to be weighed. Like other words in *-ist*, it has a professional air, as if the man who so described himself were claiming an *ex cathedra* authority for his utterances. Hence its use is not always complimentary. We see this in its derivatives "scientism" and "scientistic." Similarly, when I have been introduced to a public meeting as an "educationist" or "educationalist," I have wished to be saved from my friends: I would rather be called a schoolmaster. My conclusion is that the term scientist is too convenient to be wholly rejected, but that writers would do well to remember its less complimentary use as a label, and not resort to it too frequently.

J. H. Fowler.
16 Canynge Square, Clifton, Bristol.

The voluble Henry E. Armstrong was, as always, to hand with an astringent and conservative opinion.

The Word "Scientist" or its Substitute.

When literary gents, like Sir Clifford Allbutt, Prof. D'Arcy Thompson and Sir Israel Gollancz, come forward in defence of *scientist* and Sir R. A. S. Paget, an expert in vocal sounds, in the most cold-blooded manner possible, says that he would *ist* everybody, it were time that we illiterate sciencers ranged ourselves solidly with Sir Ray Lankester, ever a defender of the faith, proclaiming that we will not have truck with the would-be debasers of lingual beauty.

If I had ever favoured the term—I hate it—I should cease from using it, if only after listening to the High Commissioner for Australia, at the Imperial College of Science (not yet Scientists) dinner, a few days ago. Replying for the guests, at the close of his speech, he referred to the story of two men talking together and one saying—"There will be nothing to laugh at fifty years hence." "What, will there be no scientists!" came the reply. Let us hope there will not be any. The story is a good exemplification of our form in the public eye.

The real men, those who do things—bakers, butchers, builders, boxers, grocers, even green-grocers —all have names ending in *er*. The terminal *ist* is reserved for theosophists, thaumaturgists, even for those who pretend to be but are not chemists, only bits of the same. So far, indeed, is objection taken

to chemist that a wag among them has proposed to substitute *chemor*, not chemor-ist, be it noted. The German *chemiker* was long known as superior to the English *chemist*. Still, *er* has its weak side to some— I am told that, in New York, the undertaker seeks to be known as the moritician. The fact is, none of us likes his name.

The Oxford Dictionary, a mine of inspiration which is too little used, gives *Sciencer* and *Sciential*, both euphonious words. Of late, I have often used sciencer, and like it. Sciential has the authority of Keats and is less committal—it may even be applied not merely to the properly scientific but also to those who neither do nor make anything but merely talk and claim to be of the elect, though I should bar classical telepathists. As to Dr. Jeans, for whom solicitude is properly expressed, he may well be spoken of as a sciencer, if not reckoned with magicians: all will devoutly pray that he be kept away especially from *ists* in the guise of psychists.

I write this without consulting my sons but believe they would all support me, though I have not gifted any one of them with a musical ear—one of them, however, was brought up under Sir Clifford Allbutt in days when he was the boldest of warriors in defence of our English tongue.

We shall do well to take notice that scientist is fast becoming a word of evil import in the public ear—as meaning one of the set of peculiar people who talk a language no fellow can understand. Some day, soon, perhaps, the call may come to label Nature: the Journal of Babel; the Dictionary will then give—Babel, the language of a sect devoted to an obscure practice known as science.

Henry E. Armstrong.

D'Arcy Thompson's classical antecedents inclined him against "scientist", for mixed derivations were generally held to be repugnant. It was said that in the ancient universities the view of television, when it was first presented to the public, was that little good could come of an invention the name of which was half in Latin and half in Greek.

The word "scientist" is, in itself, neither better nor worse than dentist, oculist, socialist, or violinist. It would be pedantic, at the present day, to object to it merely on the ground that it begins in one language and ends in another. If it were a new word, introduced for the sake of brevity and convenience by some respectable writer, I should have little objection to it; I should be reluctant to use it myself, but I should not dream of objecting to its use by others.

It seems to me, however, that the word has already got a sort of taint about it, very much as the word "sophist" did in Greek. It is often used in an equivocal, or even disparaging, sense, by people who have no great respect either for science or the "scientist." Most men of science would surely rather be called so than be dubbed scientist. The widely used term "Christian Scientist" has helped to make matters worse; what that phrase means I do not know, but if I did know I am sure I should not like it any the better.

On the whole, I take it that the word scientist has been in low company, and I should be very slow to introduce it into better.

D'Arcy W. Thompson.

St. Andrews.

Sir Arthur Schuster offers this pleasing anecdote about the rather pompous Samuel Pierpoint Langley.

S. P. LANGLEY (1831–1906).

LANGLEY'S invention of the bolometer, and his pioneer work in the construction of the flying machine, are achievements sufficiently great to ensure a reputation which will outweigh the recollection of defects due to an exaggerated consciousness of dignity, accompanied by a marked inability to see the humorous side of things. I first met Langley on the occasion of the total solar eclipse in August 1878, when he established an observing station on the top of Pike's Peak in order to obtain, if possible, a measure of the thermal radiation of the solar corona. Unfortunately, he suffered severely from mountain sickness, and had to be carried down before the day of the eclipse.

In the following year, Langley visited England and expressed to me the desire to become acquainted with Clerk Maxwell. I was working at the Cavendish Laboratory at the time, and was able to assure him that Maxwell would be interested to meet him as he had, in my presence, referred in very eulogistic terms to a method proposed by Langley to eliminate the personal equation in transit observations. Clerk Maxwell was just then editing Cavendish's scientific manuscripts, and conscientiously repeated every experiment that was described in them. He was specially interested in the method which Cavendish had devised for estimating the relative intensities of two electric currents, by sending the currents through his body and comparing the muscular contraction felt on interrupting the currents : " Every man his own galvanometer," as Maxwell expressed it. When Langley arrived, I took him to the room where Maxwell stood in his shirt sleeves with each hand in a basin filled with water through which the current was laid. Enthusiastic about the unexpected accuracy of the experiment, and assuming that every scientific man was equally interested, he tried to persuade Langley to take off his coat and have a try. This was too much for Langley's dignity ; he did not even make an effort to conceal his anger, and on leaving the laboratory he turned round and said to me : " When an English man of science comes to the United States we do not treat him like that." I explained that, had he only had a little patience and entered into the spirit of Maxwell's experiment, the outcome of his visit would have been more satisfactory.

As an experimenter Langley takes a high rank, though the numerical results he derived were sometimes based on calculations that were not entirely free from defects. This led him occasionally to an optimistic judgment of their accuracy. In sending out an assistant to repeat his measurement of the so-called solar constant, which expresses the total solar radiation in certain units, his final words to him were : " Remember that the nearer your result approaches the number 3, the higher will be my opinion of the accuracy of your observations." The assistant, who since then has himself attained a high position among American men of science, fortunately was a man of independent judgment and skilful both in taking and reducing his observations, with the result that the number 3 is now altogether discredited.

J. J. Thomson, reviewing a biography of his mentor, Lord Rayleigh, adds this reminiscence.

... there never was, I think, a man whose judgment in scientific matters was sounder or more free from prejudice ; if he questioned a conclusion or a result one felt there was a real difficulty which required further investigation. Yet even with these qualities he found prophecy was not free from danger. He says in a letter written at this period : " Yesterday I had an opportunity of seeing the telephone which every one has been talking about . . . it is certainly a wonderful instrument, though I suppose not likely to come to much practical use." Having quoted a bad guess I ought not to omit a good one in his presidential address to the Royal Society in 1908. He says : " We may expect to see flying machines in use before many years are past."

Here is a display of historico-medical knowledge from Karl Pearson, which flattens the anthropologist and champion of Piltdown man (p.143), Sir Arthur Keith.

The Skull of Robert the Bruce.

A LETTER to NATURE is perhaps not the suitable place to discuss the historical evidence for sporadic syphilis being well known in Europe from at least the ninth century, nor do I intend at present to controvert Sir Arthur Keith on several other points in which he disagrees with me in his recent friendly review of my " Bruce " (NATURE, February 28, p. 303). But there is one point at which Sir Arthur seems to me to show less than his usual acumen. He concludes his article with the words : " The writer [Sir Arthur] has searched the pre-medieval graves of England and Scotland for traces of syphilis and found none, and those who know our medical records believe that Robert Bruce had been asleep in Dunfermline Abbey for two centuries before this fell disease appeared in Britain."

Now I strongly suspect that Sir Arthur Keith has overlooked two important considerations : (1) that he has been thinking largely of skeletons dug up from abbeys, and (2) that he has disregarded the fact that the nature of a disease changes with the centuries. Now abbeys in the tenth to the thirteenth centuries were not those seats of vice that a good compatriot of George Buchanan naturally assumes them to have been. Further, the sporadic cases of the disease were undoubtedly confused with leprosy, and it is not in the abbey graveyards, or in the churchyards of the tenth to the fourteenth centuries that I should expect to find evidence of these sporadic cases of syphilis. Sir Arthur should search in the burial-places attached to leper-houses, and even then he must not expect of a certainty to find the osteological appearances identical with those of the post-pandemic times.

St. Hildegard in her " Causae et Curae " of the twelfth century describes under leprosy cases which are certainly not leprosy, but as certainly syphilis. She was quite familiar, as Hildebrand has lately demonstrated, with the general paralysis of the insane. The same writer has also recently proved beyond a doubt that St. Odo of Cluny (*circa* 930) was acquainted with the disease as a result of sexual licence :

Irrepsit vitium—nec jam est simplex neque solum.
Ulcus enim vultum foedans facit esse probrosum,
Et vitium attaminat membrum, cui forsan adheret

At si vulnus abit, de more glabella nitescit,
Membra decore suo renitent vitioque revulso.
Post lapsum sensus terebrat quasi zima libido ;
Illicit, illecebrat, tabo crapulosa saginat.

How shall a man in such state appear on the Day of Resurrection ? the worthy Abbot demands. Even the " saddle nose " so characteristic of the tertiary stages of the disease was recognised in association with sexual licence in the eleventh century. The recent studies of Karl Sudhoff and Philipp Hildebrand seem to me to demonstrate that syphilis existed in Europe, if not in the pandemic form of the fifteenth century, still in not very infrequent cases from the early Middle Ages onwards.

Sir Arthur writes that the " very able medical men who examined the king's skull and bones " had no suspicion in their minds of syphilis. Possibly not ; their report is very inadequate, and they held probably the orthodox view, like Sir Arthur, that syphilis was unknown in Europe before the time of Columbus. But these same medical men did direct attention to the condition of the upper jaw, and to the exfoliated wound on the right side of the sagittal suture, and endeavoured to give explanations of them. These explanations may be correct, but there is the awkward fact that Bruce is said to have died of " leprosy " still unaccounted for.   KARL PEARSON.
Eugenics Laboratory,
University of London.

VOL 115, 571 1925

Similar conclusions to the following have been drawn from more extensive studies in our own time.

PROF. RAYMOND PEARL has endeavoured to estimate statistically the relation, if any, between the number of doctors per unit of population and the death rates observed for the same population (Journ. Amer. Med. Assoc., 1925, vol. 84, p. 1024). It appears that for the thirty-four States of the United States investigated, there is no significant difference in the mortality rate of a community in 1920, whether that community had few or many doctors per unit of population. Two morals suggest themselves from this result. The first is that perhaps the chief social and human value of the physician is in alleviating suffering, rather than in preventing death, at which last task he must in every case ultimately fail. The second is that while there is a great difference between good doctors and poor doctors in respect of the result of their activities, there is no significant difference between a good doctor and no doctor at all !

VOL 116, 217 1925

The irrepressible Henry E. Armstrong warms to a theme that we shall be encountering again:

" THE first epistle of Henry the Chemist to the Uesanians " is the somewhat bizarre title of a characteristic article by Prof. H. E. Armstrong printed in the September issue of the American *Journal of Chemical Education*. Beginning with a very long epigraph, after the manner of a Pauline epistle, the author proceeds to express his opinions on topics connected with the ethics of belief, education, literature, chemistry, and culture. His incisive style reminds one of Shaw and Chesterton, who by wealth of hyperbole and paradox, lame aphorisms, and parodied proverbs, know well how to arrest attention, and also of the late Lord Morley, who, when asked for his opinion on the prose of Carlyle, replied that he preferred the English language. Prof. Armstrong denounces the modern Press, present-day teaching of science, the poor literary value of scientific writing, and the absence of cultural value in the science of to-day ; but he reserves the vials of his wrath for the adherents of the hypothesis of ionic dissociation, who above all others appear to him to sin most against the Pauline injunction of " proving all things, and holding fast that which is good," the freedom to do which, he says, is the only intellectual freedom worth having. Through this excursive " epistle " runs a serious *leitmotif* that will appeal to all, namely, that dogmatism in science is the negation of science, a truth which is of wider applicability than is sometimes assumed.

VOL 116, 827 1925

The "vitality" of various hardy creatures recurs with regularity in the correspondence columns.

Vitality of an Earwig.

AN earwig was found inside a high vacuum pumping set recently. How it managed to get there is not known, but it was possibly in the glass-blower's rubber tubing and was blown in when the apparatus was being modified a few hours previously.

The earwig was not observed until the pumps had been running a quarter of an hour, when it was seen crawling along the glass tubing. The pressure was about 0·001 mm.

In our efforts to extract the earwig it was rather roughly treated, as it fell into the mercury vapour pump (cold) and was eventually poured into a beaker with the mercury, being entirely submerged for a few seconds. It survived all this and actively crawled about on being released.

It is possibly well known that the skin and legs of this creature can withstand such knocking about, but it seems very remarkable that it could be active in a vacuum !   PHYSICIST

VOL 116, 866 1925

Television, pioneered by John Logie Baird, appears on the horizon:

THE art of television is making progress. Mr. J. L. Baird is endeavouring to transmit images of living faces across the Atlantic, and on April 7 the American Telephone and Telegraph Company gave a successful demonstration of television. The president of the company, Mr. Gifford, who was at New York, had a telephone conversation with Secretary Hoover at Washington. The television apparatus also permitted him to see as well as hear Mr. Hoover. By a special device, also, guests assembled at New York could see the expression of his face as he talked and hear him simultaneously on a loud-speaking telephone. The guests at New York afterwards saw individually by television and talked with the guests at Washington. They also saw the face of a clock shown by a member of the staff at the Bell Telephone Laboratories, Whippany, New Jersey. The experiments prove that under favourable atmospheric conditions it is possible to send images over any distance by television. It is impossible, however, to predict as yet when a cheap and trustworthy system of radio-television will be perfected. An entire scene has to be sent in small individual parts in less than the fifteenth of a second. Only apparatus of extreme sensitivity to light can be used, and the exactness of the synchronisation must not exceed about the hundred-thousandth part of a second. Every one must admire the courage of the inventors who have partially overcome these difficulties.

VOL 119, 647 1927

Sir William Arbuthnot Lane (p.130) mounts his hobby-horse again. It is not altogether clear how he relates this view of diet to his advocacy of the virtues of paraffin oil.

THE inter-relationship of food, health, and strength is discussed by Sir Arbuthnot Lane in an article in the *Quarterly Review* for April. While fresh air is undoubtedly important, though animals and primitive natives disregard it, right food seems vastly more important to men and animals. Much of our food nowadays is soft, de-vitaminised, de-mineralised, chemically-coloured and preserved, and highly spiced. Failure to use our strong jaws, powerful jaw muscles, and teeth by the consumption of soft stuff, swallowed without chewing, results in the receding jaw, the narrow nose, weak throat, decaying teeth, toxic gums and tonsils. The capacity of the stomach and bowel clearly indicates that man was intended to live on bulky, not concentrated, food. A 'tablet' diet is an absurdity. Only a well-filled bowel readily empties itself; a sluggish bowel means stagnation and putrefaction of its contents and absorption of poisonous products. A far-reaching reform is necessary; we must go back to sound natural food; and this is one of the objects of the New Health Society founded by the writer.

VOL 117, 733 1926

By 1926 the net was closing in on Paul Kammerer (pp.169 and 184-6). The solitary remaining specimen of his Alytes obstetricans *(the midwife toad) was examined by the American zoologist Kingsley Noble, and fraud was detected.*

* * *

It will be remembered that Kammerer claimed to have produced heritable nuptial pads in a batrachian which normally lacks them. I found the specimen to have its left manus blackened both on its dorsal and ventral surfaces, the extent of the darkened area being fairly well shown in a photograph of the specimen made in Cambridge (Kammerer, 1924, "The Inheritance of Acquired Characteristics," New York, Fig. 9). A slight blackening was also to be seen on part of the right manus. Neither manus had the appearance of possessing nuptial pads, but both seemed to have been injected with a black substance, for the blackening included some of the capillaries.

* * *

The black substance, so irregularly distributed through the muscles, has the appearance of India ink, for under the highest powers the granules are black, not brownish black (or lighter) as most amphibian melanins. However, a critical test as to the nature of this substance is necessarily a chemical one. Oppenheimer (1909, "Handbuch der Biochemie des Menschen und der Tiere," Jena), in describing the properties of melanin, states that it may be changed into a lighter-coloured substance by treating it with concentrated nitric acid. Further, this product is soluble in alkalis. I have carried out a series of experiments with different kinds of amphibian integuments, some injected with India ink, and others merely fixed in alcohol or formol, and have found the test to be critical in distinguishing melanin from India ink. With Dr. Przibram's permission I removed a piece of integument from the palm of Kammerer's specimen at the base of the second finger. A large mass of black substance adhered to the dermal portions of the skin. The piece was cut into three parts and each treated for different periods, first in concentrated nitric acid, and after washing, in concentrated ammonium hydroxide. In spite of this variety of treatment, known to be critical in all cases, the black colouring matter remained intact, while the few small and widely separated melanophores readily observable under the binocular, disappeared. In this resistance to the treatment the colouring matter resembled the India ink masses in our controls. Dr. Przibram has carried these experiments further, and writes : "The black substance has also been subjected to the treatment with antiformin and withstood this reagent, which dissolves all melanins known to now." We may conclude that the substance which gives the dark appearance to the left manus of Kammerer's specimen is not melanin.

VOL 118, 209 1926

A month after this publication, Kammerer, still protesting his innocence, shot himself.

Prof. Paul Kammerer.

I REGRET to have to announce the death of Prof. Dr. Paul Kammerer, who shot himself on the Hochschneeberg, near Vienna, on September 23. In a letter (received after his death) he accuses himself of failures in his personal affairs, but emphasises that he has never committed the scientific tricks hinted at by some of his critics. He deemed the rest of his life too short to be able to take up again the same experiments, and declared himself too weary for this task. Although other than these seem to have been the main causes for his weariness of life, yet this sad end to a precious life may be a warning to those who have impugned the honour of a fellow-worker on unproven grounds. It is in fulfilment of a wish expressed by Kammerer that I beg the editor of NATURE to publish his last word on the much-debated but not solved question of a particular one of his specimens. Having convinced himself of the state it is in now, Kammerer alleges that someone must have manipulated it ; he does not allude to a suspicion whom this might have been.

Need I add that Kammerer's work on the modifiability of animals, especially on poecilogony and adaptation to colour of background in Salamandra and the reappearance of functional eyes in Proteus kept in appropriate light, will secure him a lasting place in the memory of biologists, even if some other of his papers were open to criticism.

HANS PRZIBRAM,
Vienna II. Prater, " Vivarium."

VOL 118, 555 1926

Another posthumous commendation came from the purblind Lamarckian E. W. MacBride. After reproving all those involved in exposure of the fraud, he hints at a conspiracy to discredit Kammerer and other defenders of the Lamarckian doctrine and whips himself into a fine lather, to conclude:

I should like to say to Prof. Huxley that the game of the mutationist opposition to Lamarckism is up. Evidence in favour of Lamarckism is pouring in from all quarters. I direct his attention to the recent work of Metalnikoff in the Institut Pasteur, who, experimenting with the caterpillars of the genus Galleria, showed the inheritability of acquired immunity. This work was continued for nine generations under standardised conditions with adequate controls.   E. W. MacBride.

Imperial College of Science,
    South Kensington, S.W.7.

Vol 119, 160 1927

The correspondence columns of NATURE harbour some unexpected names, such as those of Arthur Conan Doyle — doctor, creator of Sherlock Holmes and fervent devotee of the occult — and of the Roman Catholic historian, reactionary and poet Hilaire Belloc. Belloc had written an ill-informed attack on Darwinism — and particularly, it seems, on the godless H. G. Wells and the Catholic biologist St George Mivart (p.11). The book was harshly dealt with by the reviewer, Sir Arthur Keith. Belloc was temperamentally hostile to science — recall for instance the end of "The Microbe" in his evergreen Bad Child's Book of Beasts:

All these have never yet been seen —
But Scientists, who ought to know,
Assure us they must be so…
Oh! let us never, never doubt
What nobody is sure about!

Belloc's angry response to Keith's review shows that in such circles ignorance is no impediment to strong opinions. He begins—

Is Darwinism Dead?

THE review in NATURE of Jan. 15 of my criticism on Mr. Wells's somewhat antiquated biology has only just been shown to me, hence the delay in my sending this letter. I will make it as brief as possible, for I am only concerned with showing that the distinguished reviewer, Sir Arthur Keith, though he has doubtless been given a few sentences from my book for purposes of quotation, has not read the book itself.

(1) He says: "So adroitly does Mr. Belloc cover his verbal tracks with a smoke screen" that he cannot determine whether I am a 'fundamentalist' or a 'Darwinian.' As a fact, I cannot conceive myself being either, but the point is that no one who had read my book could have imagined that 'Fundamentalism' was the issue. The only issue was whether natural selection were the process whereby the differentiation of species came about.

—and after a series of additional points he comes to the clincher:

(6) I have kept to the last the most damning count in this indictment. The reviewer sets me down as owing my remarks entirely to Mivart, as having merely copied Mivart's work of more than half a century ago:

implying my ignorance of all since. Had he read my book he would have seen that I quoted from authority after authority among the highest names in modern biology from the beginning of the discussion to works which appeared so recently as three years ago. I give their actual words, which prove with what increasing force the old-fashioned Darwinian theory of natural selection has been beaten down. I end by a list of no less than forty such names—I might easily have made it a hundred. No one who had read my book could possibly have missed this continued and repeated citation of authority from every side, which is the principal feature of this section.

I conclude, therefore, that the reviewer has not read my book; for I hope that not even the most violent religious animosity could lead him to deliberate misrepresentation.   H. BELLOC.

Vol 119, 277 1927

Now Keith's rejoinder:

NOT only did I read Mr. Belloc's book with great care, but I also took the trouble of turning up the works of some of the authorities he cites. On p. 12 he mentions, with bated breath as it were, "the great work of Vialleton." This "great work" is a very good elementary treatise on embryology which Prof. L. Vialleton, of the University of Montpellier, wrote for his students, and it stands in much the same relationship to the works of Charles Darwin as do those of Mr. Belloc to Shakespeare's.

Mr. Belloc cites Vialleton as his authority for denying the possibility of birds having been evolved from reptiles. On searching Prof. Vialleton's "Éléments de morphologie des vertébrés"—published in 1911—I found on p. 611 that after citing what Huxley, Owen, Seely, Mivart, and Gadow had to say about the matter, Prof. Vialleton concludes thus: "L'origine des oiseaux reste donc dans le plus complet mystère," which is a very different thing from denying their origin from reptiles. I have collected many other errors of a like kind, enough to convince me that Mr. Belloc's references are untrustworthy. Many of the authorities he cites, such as my friend the late Prof. Dwight, of Harvard, belonged to the generation which never succeeded in assimilating the teachings of Darwin.

ARTHUR KEITH.

Vol 119, 277 1927

AMGUEDDFA GENEDLAETHOL CYMRU

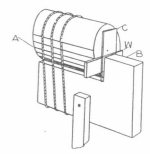

This anecdote from Karl Pearson brings together a surprising number of interesting personages of the nineteenth century. Joanna Baillie, poetess, lived in a large house that still stands opposite the original Medical Research Council laboratory in Hampstead in north London.

The Brain of Laplace.

THE bicentenary of the death of Newton (March 20, 1727) is within a fortnight of the centenary of the death of Laplace (March 5, 1827), and no one acquainted with the work of both can think of one or other except in association. It may, therefore, not be an unfitting occasion to refer to an historical point with regard to the great Frenchman, when we are celebrating the great Englishman.

The physiologist and anatomist Magendie propounded the theory that the intelligence of a human being was in the inverse ratio of the amount of cerebro-spinal fluid contained in the brain case. Writing in 1827, the year of Laplace's death, his " Mémoire physiologique sur le cerveau," he inserted the following words :

" Je me suis trouvé dans la douloureuse nécessité d'examiner le cerveau d'un homme de génie mort dans un âge avancé, mais jouissant encore de la plénitude de ses facultés intellectuelles ; la somme totale du liquide céphalo-spinal ne s'élevait pas à deux onces, et les cavités du cerveau en contenaient à peine un gros " [$= \frac{1}{8}$ once].

I have been unable so far to find any further reference in the writings of Magendie " to the brain of this man of genius who died at an advanced age " and in the fullness of his intellectual powers. Magendie appears to have given no further account of this brain; at least I have found none. Laplace died at the age of seventy-eight in the year Magendie wrote. I have also failed to discover any minute record of Laplace's death which would suggest that an autopsy was made or was a " douloureuse nécessité." I would venture, therefore, to ask those who may be better acquainted than I am with the circumstances of Laplace's death to let me know why his brain came into Magendie's possession and whether a full report on it was ever written. Magendie, indeed, mentions no name, and this might lead one to consider his investigation of the matter was confidential. However, I think the ascription is certain, for quite recently Miss Helen Hunter Baillie—a lady who combines the blood of other famous anatomists with that of a famous author,[2] placed in the hands of Miss Miriam Tildesley a letter of Joanna Baillie to her great niece Miss Sophy Milligan. This letter, dated Hampstead, Monday, 1834, contains the following important paragraph :

" MY DEAR SOPHY. . . . Dr. Somerville told us not long ago a whimsical circumstance regarding the head of La Place, the famous French astronomer. Some Ladies and Gentlemen went one day to the house of Majendie (sic !), the great anatomist, to see the brains of this Philosopher which they conjectured must be of a very ample size, and seeing a preparation on the table answering their expectation they were quite delighted. ' Ah ! see what a superb brain, what organs, what developments ! This accounts completely for all the astonishing power of his intellect, etc.' Majendie, who was behind them and overheard all this, stepped quietly forward and said : ' Yes, that is indeed a large brain, but it belonged to a poor Idiot, who when alive scarcely knew his right hand from his left. This, Ladies and Gentlemen ' (handing to them a preparation of a remarkably small brain), ' this is the brain of La Place.' Dr. Somerville was told this anecdote by Majendie himself. . . .

Your affectionate Aunt, J. BAILLIE."

This Dr. Somerville can scarcely be other than the physician, fellow of the Royal Society, and husband of Mary Somerville, the learned lady who studied Newton's " Principia " in the original, was the correspondent of Laplace, and paraphrased his " Mécanique Céleste." There can thus be no doubt that Magendie was in possession of the brain of Laplace, and very little doubt that the passage in the " Mémoire physiologique sur le cerveau," written 1827, refers to that brain. The questions I would put to the French readers of NATURE are these : What became of Magendie's preparations ? Have they, and with them Laplace's brain, survived until to-day ? If so ; has any one reported on it, or does any account by Magendie other than that I have cited, written or printed, exist ? So few brains of great thinkers have been available for examination, that it would be a real disaster if Laplace's should have had only four lines devoted to it.   KARL PEARSON.

Galton Laboratory,
University College, London,
Mar. 31.

W. L. Bragg's X-ray structure of the sodium chloride crystal upset many chemists; Bragg told of how after a lecture he was approached by a distressed chemist with the entreaty to move each chloride ion just the tiniest bit closer to one sodium than to the others surrounding it. But probably none were as forthright in their objections as the inimitable Henry E. Armstrong.

Poor Common Salt !

" SOME books are lies frae end to end," says Burns. Scientific (save the mark) speculation would seem to be on the way to this state ! Thus on p. 405 of NATURE, of Sept. 17, in a letter on Prof. Lewis's light corpuscles, the statement is made by the writer, that a ' speculation,' by Prof. Lewis, about the quantum, " is repugnant to common sense." Again, on p. 414, Prof. W. L. Bragg asserts that " In sodium chloride there appear to be no molecules represented by NaCl. The equality in number of sodium and chlorine atoms is arrived at by a chess-board pattern of these atoms ; it is a result of geometry and not of a pairing-off of the atoms."

This statement is more than " repugnant to common sense." It is absurd to the n . . .th degree, not chemical cricket. Chemistry is neither chess nor geometry, whatever X-ray physics may be. Such unjustified aspersion of the molecular character of our most necessary condiment must not be allowed any longer to pass unchallenged. A little study of the Apostle Paul may be recommended to Prof. Bragg, as a necessary preliminary even to X-ray work, especially as the doctrine has been insistently advocated at the recent Flat Races at Leeds, that science is the pursuit of truth. It were time that chemists took charge of chemistry once more and protected neophytes against the worship of false gods : at least taught them to ask for something more than chess-board evidence.

HENRY E. ARMSTRONG.

Marie Stopes, lecturer in palaeobotany at the University of Manchester, was chiefly famous as a fearless champion of contraception. Here she receives a sympathetic review of one of her books on the subject.

* * *

Dr. Marie Stopes's book consists in a history of the various varieties of birth-control and especially of contraception. She recalls to our memory the fact emphasised by Prof. Carr-Saunders that birth-control in the form of infanticide and abortion has been practised by all races of men since time immemorial, and that even drugs which prevented conception were known in antiquity. She does a useful service in demolishing the legend, sedulously propagated by a certain coterie, that the legality of birth-control was established by Bradlaugh's fight for it. She shows convincingly that the very pamphlet in connexion with which Bradlaugh was prosecuted had been sold for years without let or hindrance, and that the police only interfered when the publisher added to it several indecent illustrations. Bradlaugh's intervention, so far from helping the cause of birth-control, really hindered it, since it caused it to be associated in the minds of the public with his unpopular atheistic views.

* * *

Dr. Stopes gives an elaborated and detailed account of the physiology of sexual intercourse and of the relation to it of the various methods of birth-control now in use, giving of course prominence to that which she recommends. It would be entirely out of place to discuss these methods in the columns of NATURE, but Dr. Stopes raises several points of considerable physiological interest which may be mentioned here. She maintains that the male sexual discharge, in addition to the fertilising spermatozoa, contains a hormone the absorption of which is most beneficial to a woman's health. *A priori* this seems not unlikely, but it is an exceedingly difficult matter to prove, and when one views the healthy, vigorous unmarried women around one, it is rather difficult to believe. Then Dr. Stopes stresses the fact that in the emotions accompanying coition, there is a female crisis which is reached later than the male crisis, and that when this is not attained the woman is left in an unsatisfied and irritated condition which, often repeated, leads to marital unhappiness and makes a shipwreck of the marriage.

Whilst the normal reader must experience a shock in finding these intimate matters discussed in such detail and in such plain language by a woman, and whilst the reviewer must admit that he sympathises with this feeling of shock, there are certain considerations which give him pause before he indulges in condemnation of the author. Some

years ago, in the common-room of a certain college, the reviewer happened to be the only biologist present when one of Dr. Stopes's books (" Married Love ") was discussed. He was immediately challenged to give his opinion on it, and he replied that while it seemed to him that this book could only have been written by a person entirely devoid of reticence, yet that it would do good in giving information about physiological matters such as the female crisis, which could be obtained nowhere else. To his amazement and stupefaction, two of his questioners, both scientific men and both married, confessed an utter ignorance of the very existence of a female crisis !

VOL 120, 647 1927

A question from Bertrand Russell's errant elder brother, Frank, the "wicked earl" (who was brought to the Bar of the House of Lords, accused of bigamy, but had given Bertie his first course in Euclid):

Selective Association in Kittens.

MY cat has four kittens ; two of them are black and white, and two are black. They are only three weeks old now, but from the beginning they have always been in two pairs according to their colour. Is there any reasonable explanation for this ?

RUSSELL.

VOL 122, 478 1928

From the time of Francis Galton, eugenics became, to some, an inviolable faith. Karl Pearson, in a long review of two books on the subject, comes out with this declaration:

...all racial and ethical obligations are relative to their age, and our conception of moral and national duties will be remoulded step by step as eugenic principles become more widespread. Incest in a family with manic-depressive insanity will remain for ever a crime ; it might actually become a virtue, a national duty in the case of a family of surpassing genius, which had a sound and healthy pedigree.

His concluding lines read:

He reaches the boundary whereat eugenics passes from science to a national faith, or what Galton termed the true religion of the future. Here, at least, we feel entirely in unison with his statements. KARL PEARSON.

VOL 122, 953 1928

John James Waterston's is a sad story. The review acknowledges him as a neglected genius, one whose work prefigured many of the thermodynamic principles enunciated much later by others. The "remarkable paper" to which the reviewer refers was rejected by the Royal Society and return of the manuscript was refused — a considerable misfortune in the days before typewriters and carbon copies. Lord Rayleigh and J. S. Haldane at least did Waterston posthumous justice.

A Neglected Genius.

The Collected Scientific Papers of John James Waterston. Edited, with a Biography, by Dr. J. S. Haldane. Pp. lxviii + 709 + 5 plates. (Edinburgh and London : Oliver and Boyd, 1928.) 25s. net.

IN 1892 the late Lord Rayleigh rescued from oblivion in the archives of the Royal Society a remarkable paper by John James Waterston which had been written in 1845 but had failed to obtain the approval of the Society, and had, therefore, not been printed in the *Proceedings*. So completely has his work been ignored that it will probably come as a surprise to the majority that his writings (published and hitherto unpublished), which have been collected and published by Dr. J. S. Haldane, extend to more than seven hundred pages.

Lord Rayleigh did ample justice to the 1845 paper on the physics of media that consist of perfectly elastic molecules in a state of motion. Concerning it he wrote : " What strikes one most is the marvellous courage with which he attacked questions, some of which even now present serious difficulties. . . . Waterston was the first to introduce into the theory the conception that heat and temperature are to be measured by *vis viva*. . . . In the second section the great feature is the statement that in mixed media the mean square molecular velocity is inversely proportional to the specific weight of the molecules. The proof which Waterston gave is doubtless not satisfactory, but the same may be said of that advanced by Maxwell fifteen years later." Boyle's law, Charles's law, Avogadro's law, and Graham's law of diffusion were all placed on a dynamical footing in this paper. The causes which contributed to it being denied publication in 1845 are difficult to find. At the present time it suffers from having been superseded in style and argument by the work of successors. When written, it apparently suffered from being in advance of its time. Joule's work on the dynamical nature of heat had been in part published, but the theory of conservation of energy was not authoritatively accepted until about six years later. Even so late as 1848, Thomson (Lord Kelvin) wrote : " The conversion of heat (or caloric) into mechanical effect is probably im-

possible, certainly undiscovered. In actual engines for obtaining mechanical effect through the agency of heat, we must consequently look for the source of power, not on any absorption and conversion, but merely in a transmission of heat."

Who was the man whose scientific insight drew from Lord Rayleigh such high praise ? In answer, Dr. Haldane prefaces his collection by a short biography. His grandfather was founder of an important (still existing) firm of manufacturers of sealing-wax and other stationery ; his grandmother was a niece of Robert Sandeman, a well-known religious leader and founder of the body known as Sandemanians—to which Michael Faraday and his blacksmith father belonged—and sister of George Sandeman, who was founder of the well-known firm of port wine merchants.

Vol. 123, 595 1929

Albert Einstein had become a sought-after public figure, though his popular lecture in Nottingham must have been something of an ordeal with serial translation into English. The text, occupying the space between this introduction and the report of cheers at the end of the address, appears formidably demanding. Is the blackboard still extant?

The Concept of Space.

ON Friday, June 6, Nottingham was honoured by a visit from Prof. A. Einstein, who delivered a lecture (in German) in the Great Hall of the University College. After each section an English translation was given by Dr. H. L. Brose. The chair was taken by Prof. H. H. Turner. The lecture was an account of the history of the concept of space, and was addressed to a general audience.

* * *

Amid enthusiastic cheers, Alderman Huntsman, chairman of the College Council, announced that the blackboard used by Prof. Einstein and signed by him would be varnished and preserved in memory of a historic occasion.

H. T. H. PIAGGIO.

Vol. 125, 897 1930

Here J. B. S. Haldane launches himself at the throat of the maddening E. W. MacBride.

Embryology and Evolution.

Four of Prof. MacBride's statements, in Nature of Dec. 6, call for comment. " . . . no one has ever seen 'genes' in a chromosome." Genes cannot generally be seen, because in most organisms they are too small. In *Drosophila* more than 100, probably more than 1000, are contained in a chromosome about $1\,\mu$ in length. They are therefore invisible for exactly the same reasons as molecules. But the evidence for their existence is, to many minds, as cogent. Where the chromosomes are larger, as in monocotyledons, competent microscopists — for example, Belling, in Nature of Jan. 11, 1930—claim to have seen genes. In a case where I (among others) postulated the absence of a gene in certain races of *Matthiola*, my friend Mr. Philp has since detected the absence of a trabant, which is normally present, from a certain chromosome. I shall be glad to show this visible gene to Prof. MacBride.

" . . . if Prof. Gates were a zoologist instead of being a botanist, he would know that the assumption that 'genes' have anything to do with evolution leads to results . . . that can only be described as farcical". I should like to direct Prof. MacBride's attention to the droll fact that in a good many interspecific crosses various characters behave in a Mendelian manner, that is, are due to genes. This is so, for example, with the coat colour of *Cavia rufescens*, which, on crossing with the domestic guinea-pig, behaves as a recessive to the normal coat colour, but a dominant to the black. Hence there has been a change in a gene concerned in its production during the course of evolution. Scores of similar cases could be cited.

" All known chemical actions are inhibited by the accumulation of the products of the reaction. An 'autocatalytic' reaction, in which the products of the reaction accelerated it, must surely be a vitalistic one ! " Autocatalytic reactions are common both in ordinary physical chemistry and in that of enzymes. Thus the acid produced by the hydrolysis of an ester may accelerate its further hydrolysis. As an example of an enzyme action, which for quite simple physico-chemical reasons proceeds with increasing velocity up to 75 per cent completion, I would refer Prof. MacBride to Table 7 of Bamann and Schmeller's [1] paper on liver lipase.

In view of such facts, Prof. MacBride's statement that " The term 'autocatalysis' is a piece of bluff invented by the late Prof. Loeb to cover up a hole in the argument in his book " would seem to be a wholly unfounded attack on a great man who can no longer defend himself. If Prof. MacBride would acquaint himself with the facts of chemistry and genetics, he might be somewhat more careful in his criticism of those who attempt to analyse the phenomena of life. He might also cease to ask the question propounded by him in Nature of Oct. 25, "whether the organs of the adult exist in the egg preformed in miniature and development consists essentially in an unfolding and growing bigger of these rudiments, or whether the egg is at first undifferentiated material which from unknown causes afterwards becomes more and more complicated and development is consequently an 'epigenesis'". The formation of bone in the embryo chick was shown by Fell and Robison [2] to be due to the action of the enzyme phosphatase, which is neither a miniature bone nor an unknown cause. But so long as he does not take cognisance of recent developments in science, Prof. MacBride will no doubt remain a convinced vitalist.

J. B. S. Haldane.

Sir Arthur Conan Doyle was a doctor and maintained an interest in biological advances, as allusions in the Sherlock Holmes stories to fingerprints and other such topical subjects revealed. But his main preoccupation in his declining years was with the spirit world. He was the dupe of fraudulent mediums and was particularly taken with ghostly photographs. Here he hits back at one of his critics.

Mr. Campbell Swinton's account in Nature of August 28 of the incidents connected with the Combermere photograph is both inaccurate and misleading. Since he uses my name so freely perhaps you will permit me to state shortly the true version. The whole story, with the photograph, will be given in the next number of *Psychic Science*—the organ of the Psychic College.

This photograph, which shows plainly the outline of an elderly man seated in an armchair, was sent to me with the endorsement of the Combermere family, who may be expected to know as much about the matter as their relative by marriage. On the back was written that it was taken by a certain lady at the time of the old peer's funeral, and that the shadowy figure was supposed to be the wraith of the deceased man. This I showed (among fifty other psychic photographs) at the Queen's Hall, simply giving the facts as supplied by the family, and making no assertion myself, since I had no personal knowledge of the matter. Shortly afterwards, several violent letters appeared in the press from Mr. Campbell Swinton, in which he used such injurious terms as " photographic fraud." As to the seated figure, he gave in successive letters three different contradictory explanations ; the first that it was a photographic flaw, the second, that the butler had crept into the room and seated himself in the chair; and the third, that plates if kept for some time before development may show strange images. He wound up by challenging me to publish in the *Morning Post* the 'ghost' photograph, alongside of a photograph of the peer taken in life. I at once sent up my photograph without any suggestion whatever that it would not reproduce. That statement is pure invention upon the part of Mr. Campbell Swinton. The editor refused to take the risk of an inferior reproduction, and could only guarantee a good one by touching up, which would be objectionable. A reproduction was afterwards made by the *Daily Sketch*, but whether touched up or not I could not tell.

That is all a technical question with which I had nothing to do. What was, however, strange and rather amusing was that when the photograph of the peer was finally published he proved to be remarkably like the 'ghost,' having a very high forehead and some indication of a short tufted beard. Thus the result of Mr. Swinton's labours was to add one more point to the argument for the authenticity of the picture. There is clear evidence that there was no male visitor or servant in the house who wore a beard.

Arthur Conan Doyle.

September 1.

Fred Allison's observations on the time of onset of the Faraday effect — the rotation of the plane of polarisation of light passing through a liquid in an applied magnetic field — led to the discovery of a plethora of nonexistent isotopes and a whole new physical chemistry, based entirely on self-deception. As with René-Prosper Blondlot's n-rays, the effects were repeated in some laboratories and not in others, and eventually, as if by common consent, all talk of them ceased.

Influence of X-rays upon Time-lags of the Faraday Effect and upon Optical Rotation in Liquids.

DIFFERENCES in the time-lags of the Faraday effect behind the magnetic field in various liquids have been measured by Beams and Allison (*Phys. Rev.*, **29**, 161 ; 1927). Certain considerations have led me to suspect that these time-lag differences might be affected, and even reduced to zero, by the action of X-rays on the liquid. A number of experimental tests very recently carried out demonstrate that the X-rays have such a property. It was found in every case that the time-lag differences of the Faraday effect between any pair of the liquids vanished so long as the liquids were exposed to the X-rays, and that the lags were restored with the screening off of the X-rays. The liquids thus far used are carbon disulphide, carbon tetrachloride, ethyl alcohol, xylene, and chloroform.

The method also affords a means of measuring the absolute time-lags of the Faraday effect, giving values for the various liquids which are consistent with the previously measured time-lag differences.

This work having shown an influence of X-rays upon the lag of the Faraday effect, it was decided to find out whether these rays could produce an effect in rotating the plane of polarisation of light in these same liquids. A preliminary series of tests shows that a beam of X-rays traversing the liquids does impart to them the power of rotating the plane of polarisation, though it is small.

These investigations are being continued, and it is hoped that a detailed report of them will be published in the near future.   FRED ALLISON.

Alabama Polytechnic Institute,
    Auburn, Alabama,
        Oct. 11.

VOL 120, 729 1927

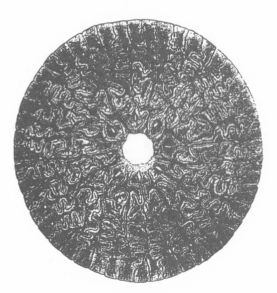

The anthropologist F. G. Parsons considers the evolution of his compatriots in an article with the title "The Englishman of the Future". With many tables, he follows the anthropometric changes in the population since Neolithic times. On head measurements, anno 1927, he finds the following:

IT has been borne in upon me, little by little, that some of the characteristics of the Englishman of to-day do not seem to be hereditary at all, and that in some things we, in our development, are not following any Mendelian laws ; nor are we harking back to Long Barrow, Bronze Age, Celtic, or Saxon types, but that gradually we are building up a new kind of man, differing in certain ways from all of these.

* * *

These records, which run into several thousands, do not give us any reason to think that the Londoner is becoming darker, but do give us reason, though it may need discounting, to believe that he is growing fairer under changing conditions.

The last point to which I wish to direct attention is head shape.

* * *

When we come to measure the educated classes of the community, which have enjoyed a greater share of the modern, improved conditions of environment, the result is still more striking, for we see the members of the British Association with a proportional head height of 0·271, the St. Thomas's Hospital students with 0·272, the Oxford undergraduates with 0·272, a number of British anatomists who met in Dublin in 1898 with 0·275, and the University College staff with 0·278.

And here is his projection for the future of the race:

To sum up, I am left with the belief that the Englishman of the future is, if present conditions persist, making for an average height of 5 ft. 9 in., and the women for one of 5 ft. 6 in. or 5 ft. 7 in. That our people have reached, and are stationary at, a stage in which some 66 per cent. have light eyes and some 34 per cent. dark. That there are no signs whatever that the hair colour has darkened during the last sixty years, though there are signs, which perhaps need discounting, that the hair is lighter than it was sixty years ago. That the head shape is showing unmistakable signs of an increase of its proportional height, with a decrease of its proportional length, and that this increase of proportional height is greater than has been found in any of the stocks from which the modern Englishman is derived. It therefore cannot be looked upon as a harking back to any ancestral form, but must be regarded as an evolutionary process, in harmony with the greatly changed conditions of life which have come about during the last century.

VOL. 120, 482 1927

chapter thirteen
1931-1935

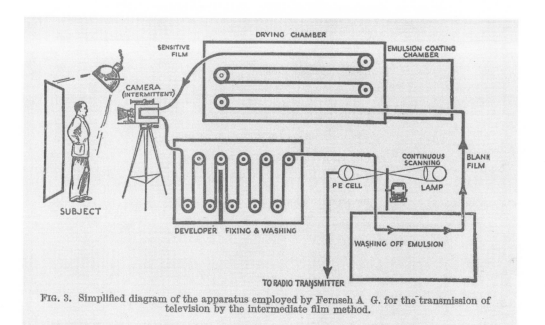

FIG. 3. Simplified diagram of the apparatus employed by Fernseh A. G. for the transmission of television by the intermediate film method.

A. Szent-Györgyi and W. N. Haworth reveal the structure of antiscorbutic factor, vitamin C. C. P. Snow's gaffe ends his career in science, but his novel *The Search* is praised by Joseph and Dorothy Needham. A. V. Hill draws attention to the plight of the Jews in German universities and provokes the champion of Aryan physics, Johannes Stark. The Nazi mathematician Ludwig Bieberbach is derided by G. H. Hardy. Peter Kapitza is detained in Moscow on Stalin's orders. Max Planck remembers James Clerk Maxwell and Sir Charles Vernon Boys's eightieth birthday is celebrated in verse.

1931-1935

Franklin Delano Roosevelt was elected president of the United States, the National Socialists gained control of the Reichstag, and Hitler became German Chancellor and ordered the Reichstag fire. Prohibition was repealed, and in Britain the hunger marchers made their way to London. The treason trials and the terror began in the Soviet Union. J. D. Cockcroft and E. T. S. Walton split the atomic nucleus.

The discovery of vitamin C was one of the great landmarks in biochemistry and medicine, for which the authors of this letter shared a Nobel prize. Albert Szent-Györgyi recognised the antiscorbutic substance that he had isolated from paprika peppers to be a carbohydrate and he proposed the name ignose and then godnose, but these were deemed editorially unacceptable.

'Hexuronic Acid' (Ascorbic Acid) as the Antiscorbutic Factor

IN view of the facts that (1) hexuronic acid is the name of a class of substances rather than that of one individual compound, and that (2) the material described as hexuronic acid isolated from adrenal cortex and now from Paprika contains a molecule of water less than is required for a hexuronic acid, we wish to ascribe the name ascorbic acid to the crystalline substance $C_6H_8O_6$ which has been the subject of earlier communications from our laboratories.

A. SZENT-GYÖRGYI.
W. N. HAWORTH.
Universities of Szeged and Birmingham.
Dec. 19.

VOL 131, 24 1933

Henry E. Armstrong could be relied upon to disagree:

Caution in Christening

THE names of public characters should be chosen circumspectly. To christen the antiscorbutic agent (granted, that we at last have it in hand) *ascorbic acid* is to do it slight justice—even to rob it of the public appreciation it deserves. The name, in fact, is a scurvy one : neither has it obvious significance nor can it well be transposed into either French or German ; in no tongue will it have lilt.

Skorbut, whence scorbutic, apparently is of low German or Dutch origin. Why not call the spade a spade : simply, *antiscorbutic acid*—antiskorbutsäure —acide antiscorbutique ? I hope the distinguished magician who has so deftly conjured one of the great mysteries of our being into tangible form will accept the suggestion.

HENRY E. ARMSTRONG.

VOL 131, 330 1933

Sir Ambrose Fleming, a distinguished figure in telecommunications, may have pinched this trope from Charles Kettering of General Motors.

Sir Ambrose said that the invention of television has gone through the usual three stages. First, when everyone thought it could not be done at all. Second, when leading experts declared that even if it could be done, it was no possible use, and third, when the wiseacres said, we knew it could be done, but it is not a commercial proposition. "It has been my good fortune to be closely and practically connected with the introduction into great Britain of three important inventions, namely, the telephone, the incandescent electric lamp and wireless telegraphy, and I have seen these three stages illustrated in them all.

Sir Ambrose was farsighted:

The Latin poet Horace told us, what every advertiser knows well, namely, that memory of the eye is more tenacious than memory through the ear. We can televise geometrical diagrams, lessons in botany, physics, and zoology, and countless other useful visions. Let us hope that this new weapon science has provided will not be vulgarised or put to base uses, but employed for the instruction, elevation and national entertainment of the public at large."

VOL 131, 539 1933

Adolf von Baeyer was one of the legendary originals of chemistry (p.164). The writer of this article was John Read, professor of chemistry at the University of Aberdeen, author of Humour and Humanism in Chemistry *and himself a student of Baeyer. Bavarian dialect adds spice to this account.* Schmierchemie *was the contemptuous term by which the organic chemists referred to the chemistry of physiological substances.* Melone, *or melon, is the German for a bowler.*

The work was beset with difficulties. At one time, for example, during the intensive search for dihydrophthalic acids, gigantic quantities of sodium amalgam, up to forty kilograms a week, were prepared and used in vain. The author remarks with feeling that the situation became very disagreeable to the assistants. It must have been, indeed, a "schwere, scheussliche und gefährliche Arbeit" ; but no labour was too tedious for the Master and his band of devoted helpers. There was, as Prof. Rupe says, something of the magnificent in this prolonged contest with matter.

Eventually, however, even Baeyer was supersaturated with these hydrogenations ("übersättigt von diesen Hydrierungsarbeiten"), and the sorely tried assistants hailed with deep relief the transference of his interest to succinylosuccinic ester and diketocyclohexane. By means of a 'Kunstgriff' of which Baeyer was very proud (treatment with sodium amalgam in presence of sodium bicarbonate), the diketone was reduced to quinitol. At the first glimpse of the crystals of the new substance Baeyer ceremoniously raised his hat!

It must be explained here that the Master's famous greenish-black hat plays the part of a perpetual epithet in Prof. Rupe's narrative. As the celebrated sword pommel to Paracelsus, so the 'alte Melone' to Baeyer: the former was said to contain the vital mercury of the medieval philosophers; the latter certainly enshrined one of the keenest chemical intellects of the modern world. Hats are not associated as a rule with chemical research, although it is true that Trautschold's illustration (1842) shows the striking variety of headgear which was to be seen in Liebig's original laboratory at Giessen: these choice pieces, although perhaps not including an 'alte Melone', ranged from the postman's cap of Ortigosa the Mexican through the tam-o'-shanter of his unnamed neighbour to the stylish topper favoured by A. W. Hofmann. It now appears that the tradition of laboratory hats descended from Liebig to Baeyer. However that may be, Baeyer's head was normally covered. Only in moments of unusual excitement or elation did 'the Chef' remove his hat: apart from such occasions his shiny pate remained in permanent eclipse.

When, for example, the analysis of the important diacetylquinitol was found to be correct, Baeyer raised his hat in silent exultation. Soon afterwards the first dihydrobenzene was prepared, by heating dibromohexamethylene with quinoline: Baeyer ran excitedly to and fro in the laboratory, flourishing the 'alte Melone' and exclaiming: "Jetzt haben wir das erste Terpen, die Stammsubstanz der Terpene!" Such is the picture from behind the scenes of the dramatic way in which the Master entered upon his famous investigations on terpenes.

Incidents of this kind may appear to be slight, and yet cumulatively they throw a stream of light upon the personality of this great chemist. There is no doubt, for example, that at times 'the Chef' was unduly impulsive. One morning he burst into the private laboratory, and, without having lit his cigar (an indication in itself of unusual emotional disturbance), raised the ancient 'Melone' twice, and exclaimed: "Gentlemen [the audience was composed of Claisen and Brüning], I have just had word from E. Fischer that he has brought off the complete synthesis of glucose. This heralds the end of organic chemistry: let's finish off the terpenes, and only the smears ('Schmieren') will be left!" Prof. Rupe's reminiscences are rich in snapshots of this kind, which are often more revealing than pages of formal description could be.

Baeyer's customary tools were test-tubes, watch-glasses, and glass rods. As an example of his endless patience, Willstätter relates having seen him keep a test-tube in gentle play over a flame for three-quarters of an hour when activating magnesium with iodine. He valued at least three

things which were deemed of fundamental importance by the alchemists; for he impressed upon his students that the essential attributes of the chemist are patience, money, and silence. His lectures were marked by clearness and simplicity of diction, with occasional delicate touches of North German humour or sarcasm. He urged his listeners to learn to think in terms of phenomena; and, like Kekulé, he emphasised the importance of giving occasional rein to the imagination: "so viele Chemiker haben nicht genügend Phantasie".

Although 'the Chef' was often regarded as stiff, unapproachable and severe, he was in reality a kindly man who did much good by stealth. He was free from vanity; and, unlike many men of learning, he was always ready to acknowledge ungrudgingly the merits of others. Baeyer favoured the use of simple apparatus, and the introduction into his laboratory of any device savouring of complexity had to be undertaken with great tact. The first mechanical stirrers, worked by water-turbines, were smuggled in one evening. On the following morning, 'der Alte' beheld them in full working order. For a time he affected to ignore them; then he contemplated them unwillingly, with an air of challenge; next came the first remark, so anxiously awaited: "Geht denn das?" "Jawohl, Herr Professor, ausgezeichnet, die Reduktionen sind schon bald fertig." The Herr Professor was finally so much impressed that he took the exceptional step of summoning the Frau Professor. 'Die Lydia,' as she was called in the laboratory, stood by the merrily clattering apparatus for a while in silent admiration; then she uttered these unforgettable words: "damit müsste man gut Mayonnaise machen können"! What a great deal depends upon one's point of view!

These truly fascinating pages afford glimpses also of the personnel surrounding 'the Chef'. Among them, Herr Leonhard, the old Bavarian factotum inherited by Baeyer from Liebig, stands out by reason of his liberal attitude towards the ethics of lecture demonstrations. Upon occasion he triumphed over his scruples to 'help' his experiments in their fight against the malignity of matter ("Tücke des Objektes")—that bane of the whole race of lecture-assistants: as witness his dry reply to a remark upon the difficulty of making chloroform from alcohol and bleaching powder: "Wissens Herr Doktor, dös Chloroform, dös is schon do herinnen"! There was also the old 'Laboratoriumdiener', Carl Gimmig, a veteran zouave of the war of 1870, with his compelling six-o'clock cry: "Ihr Herre, s'isch Zeit!" If the workers lingered, Carl remorselessly turned off the gas at the main and became 'terribly evident'. There is, too, an instantaneous snapshot of a 'filia hospitalis' of the laboratory, the fair daughter of Herr Inspektor Fehl; but of her it may be said, as of her prototype in the song so beloved of German studentry, that "die Füsschen laufen wie der Wind" off Prof. Rupe's pages.

A narrative of this kind cannot fail to reflect something of the personality of the author, and in the concluding relation of the surreptitious help he once extended to a would-be pharmacist, over-come with 'Examenangst' (expressive word !) in Baeyer's 'Bleikammer', we obtain a glimpse of Prof. Rupe's innate kindness of heart.

"Chemistry has always seemed to me such a *dead* subject—so utterly devoid of human interest!" exclaimed a classicist within the hearing of the present writer the other day, as he stepped some-what reluctantly into a lecture on alchemy. May he and others who share that mistaken view seek a truer orientation in this matter from such writings as these sparkling reminiscences of life and labour in the laboratory of Adolf von Baeyer.

VOL 131, 294 1933

The magic number 137 exerted an almost superstitious power over many physicists. Sir Arthur Stanley Eddington always looked for a peg with that number on which to hang his hat in the cloakroom; Wolfgang Pauli, the great and eccentric Swiss theoretician, died in room 137 in a Zurich hospital and the circumstance was said to have perturbed his spirit during his final days.

A Numerical Coincidence

IN "The Expanding Universe" Sir Arthur Edding-ton refers to the *fine-structure constant*, which appears to be a fundamental in the modern physics of which he treats, as a pure number the value of which is close to, if not exactly, 137, and in a later passage he writes : "nature's curious choice of certain num-bers such as 137 in her scheme."

There is no indication that Sir Arthur includes the following facts in his thesis but it is undoubtedly the case that the number 137 is remarkable because it is the logarithm of itself, or, to be more precise, 13713 is the mantissa to five decimal places of the logarithm of 13713. Also the reciprocal of 13713 used as circular measure is the value of the usual symbol for the rate of rotation of the earth, it being the angle through which the earth turns in one second of mean solar time.

* * *

The resemblance of this salient number to one in a very different connexion may well be taken as an example of Nature's whimsicality.

VOL 131, 550 1933

From the first, NATURE followed with anguish the events in German universities when the Nazis seized power.

Jews in Germany

THE political significance of Nazi revolutionary supremacy under Herr Hitler in Germany is outside our field ; but the treatment of the Jewish learned and professional classes stands condemned in the eyes of the intellectual world. It is a relapse to the crass repression of the Germany of Heine's day and the *Judenhetze* of Prussia fifty years ago. An outstanding case is the resignation by Prof. James Franck, Nobel prizeman with G. Hertz for physics in 1925, of his chair in experi-mental physics in the University of Göttingen. Prof. Franck, it is said, probably would have been spared the forced retirement now operative against Jewish officials, including university pro-

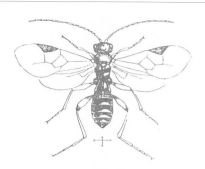

fessors ; but he feels that Germans of Jewish descent are being treated as foreigners and foes of the Father-land, and asks to be released from his office. Prof. Franck served with distinction during the War and received the Iron Cross of the First Class. His action follows fittingly on the retirement of Prof. A. Einstein from Germany, and is the logical, indeed the only, reply for a man of his standing to the acts by which Jews are being excluded from the liberal professions and debarred from the universities. This is the achievement of a movement which, ever since the War, has sought to mould the German people to one pan-Teutonic pattern—in accord with neither the facts nor the conclusions of ethnology.

VOL 131, 612 1933

A. V. Hill, by this time a Nobel laureate for his work on muscle, recognised the plight of German-Jewish and politically "unsound" scientists, driven from their positions by the new regime, and helped found an organisation — the Academic Assistance Council — to afford them succour. The Cambridge biochemist F. Gowland Hopkins was another vigorous activist. Here is the beginning of Hill's stirring appeal to his confrères.

IN 1796, Britain being then at war with France, a French scientific sailor, Chevalier de Rossel, a prisoner of war in England evidently on parole, dined with the Royal Society Club in London on the invitation of Alexander Dalrymple, the Hydrographer to the Admiralty. The Navy, as well as the Royal Society, clearly regarded scientific standing as entitling its holder to civilised and friendly treatment, regardless of the misfortune of a state of war between the two countries.

Among the instructions issued by the Admiralty to the captain of H.M.S. *Rattlesnake*, in which Huxley sailed in 1846 as "a surgeon who knew something about science", was the following :

"You are to refrain from any act of aggression towards a vessel or settlement of any nation with which we may be at war, as expeditions employed on behalf of discovery and science have always been considered by all civilized communities as acting under a general safeguard."

Science and learning have for several centuries been regarded by all civilised communities as entitling those who follow them to a certain immunity from interference or persecution—pro-vided that they keep to the rules.

He describes in due course how matters stand:

The history of science, since the War, has been largely of an effort to break down national barriers of mistrust or lack of understanding. It is quite certain that science cannot progress properly except by the fullest internationalism. Accepting freedom of thought and research as the first postulate, the second is that knowledge, however and wherever won, should be freely available for the use of all.

Up to the beginning of the present year one lived in hopes that reason was being restored. Disillusion, however, has been brought to many by the events of the last nine months. No country has excelled Germany in its contribution to science in the last hundred years, no universities were traditionally freer and more liberal than the German. One felt that the intellectual co-operation of Germany was a necessity in setting science on an international basis. I had intended, in this address, to urge an even closer co-operation. Germany, however, has lately rendered such intellectual co-operation impossible by offending the first and most fundamental rule, that providing freedom of thought and research. It seemed impossible, in a great and highly civilised country, that reasons of race, creed or opinion, any more than the colour of a man's hair, could lead to the drastic elimination of a large number of the most eminent men of science and scholars, many of them men of the highest standing, good citizens, good human beings. Freedom itself is again at stake.

The facts are not in dispute. Apart from thousands of professional men, lawyers, doctors, teachers, who have been prevented from following their profession, apart from tens of thousands of tradesmen and workers whose means of livelihood have been removed, apart from 100,000 in concentration camps, often for no cause beyond independence of thought or speech, something over 1,000 scholars and scientific workers have been dismissed, among them some of the most eminent in Germany. These have committed no fault: many of them are patriotic citizens who fought in the German armies in the War. Many of them are of families which have been in Germany for centuries: not all of them are Jews. It is difficult to believe in progress, at least in decency and commonsense, when this can happen almost in a night in a previously civilised State.

And here is his peroration:

I venture still to think of science and learning, particularly science, which in its experimental method has an absolute means of deciding between opinions, as being the strongest links between the intelligent people of the world. Not many Englishmen, unfortunately, know much about the United States of America. Fortunately with scientific people it is otherwise: they have good reason to know that laborious scientific advances on one hand, or brilliant discoveries on the other, are just as likely to be achieved there

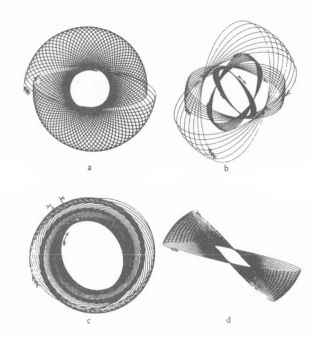

as elsewhere: and they have that close personal contact with the unassuming friendly people who make these contributions to knowledge, which ensures that the scientific community at least would regard as utterly hateful any serious difference between their countries. This friendly rivalry between Britain and the United States, this sense of co-operation, is a stronger link than many may imagine. We scientific people are often poor, and generally without much honour or position: but in the end we exercise more influence than we know—for our fundamental faith is co-operation in the pursuit of an end outside and greater than ourselves.

VOL 132, 952 1933

Nor was it well in Russia:

Prof. Alexis Saposhnikoff.

NEWS has been received that Prof. Alexis Saposhnikoff, the well-known Russian chemist, who, as recorded in NATURE of April 25, 1931, was arrested about a year ago and condemned to ten years' deportation, has had his sentence so far ameliorated that he is transferred daily to the 'Technical Bureau' under guard, where he works without pay, returning to prison in Leningrad at night. An escaped prisoner reports that he was confined with the "old Prof. Saposhnikoff. Frequently he gave us lectures on poison gas. He spoke well but I did not understand much. Prof. Saposhnikoff had been in prison for a year. The Interrogatories of the Ogpu took place between eleven and three o'clock. Prof. Saposhnikoff was often disturbed and returned always very anxious. He and others had their food in the common canteen. Nothing was brought to them from outside, and visits were prohibited. It was like being in a tomb."

VOL 129, 20 1932

A leader addresses the charge of wanton mutilation of the English language, made against scientists by Dr Cyril Norwood, headmaster of Harrow School. Here is the rather histrionic beginning.

Language in the Service of Science

SEEKERS after truth in Nature have always been regarded as disturbers of the peace and associates of the Evil One. It is not surprising, therefore, that throwing stones at men of science is a practice dating from very early times; and it still continues, if in a somewhat different form. Since the powers of evil were let loose during the War, the earlier pebbles of theology have been largely replaced by the heavier stones of sociology, as expressed in a prize-epigram in the *Spectator*:

> Science finds out ingenious ways to kill
> Strong men, and keep alive the weak and ill—
> That these a sickly progeny may breed,
> Too poor to tax, too numerous to feed;

and more recently science has been indicted as a primary cause of unemployment. There is, however, an older missile that critics, chiefly literary men, never cease to hurl: the man of science, they say, cannot make himself intelligible to ordinary people; he abuses the English language, and has little or no literary feeling. Dr. Cyril Norwood, the headmaster of Harrow School, in a recent Friday evening discourse at the Royal Institution on "The Use of the English Language", allies himself with these critics...

Vol 131, 741 1933

This literary contributor has unearthed a scientific, or at least a medical, William McGonagall. Alfred Austin was of course no laggard in this department, for as poet laureate he is supposed to have penned the following deathless lines on the near-fatal illness of the Prince of Wales:

> Across the wires the electric message came:
> "He is no better, he is much the same".

Nature and Science in Poetry

IT is not difficult to imagine how Herculean must have been the effort of the writer of the article in NATURE of August 26 under the above heading, to curb his pen. One realises the strength of mind required to put firmly on one side "La Semaine" of Guillaume de Saluste, Sieur du Bartas (1578); the "De Laudibus Dei" of Blossius Æmilius Dracontius (Fifth Century); and the many references in Dr. Charles Singer's works on oriental science.

There must have occurred to the minds of many readers of NATURE the quatrain of Omar Khayyám:

> Those who have become Oceans of Excellence and Cultivation,
> And from the Collection of their Perfections have become the Lights of their Fellows,
> Have not made a Road out of this Dark Night,
> They have told a Fable, and have gone to Sleep,

magnificently rendered by FitzGerald:

> The Revelations of Devout and Learn'd
> Who rose before us and as Prophets burn'd,
>   Are all but Stories, which, awoke from Sleep,
> They told their Comrades, and to Sleep returned.

The couplet of Alfred Austin reminded me of an amazing poem by a too-little known author—Mr. John Litart—entitled "Death and Disease"—which opens with the lines:

> Oh! happy, happy death, release from earthly cares,
> Think of the thousands who suffer through years and years,
> In such suffering and pain of a hundred kinds,
> And the misery that is borne by all mankind,
> Some which even the skill of the surgeon can't bind,
> Disease of the body, and disease of the mind.

A single verse will give an idea of this masterpiece:

> There is rheumatic *gout* and many will say
> They are scarce free from suffering an hour a day.
> Then there is *cancer* a most terrible disease,
> The pain and tortures it gives will make one's blood freeze.
> Great suffering, a living death, a gnawing pain,
> The surgeon's art no good, aleviation (*sic*) vain.

There are fifty lines, all up to this standard!

This may, I think, be claimed as marking the nadir of scientific 'versification': to my mind the zenith was attained by the exquisite and delicate fantasy of the late Sir Arthur Shipley in his little poem:

> When we were a soft Amœba
>   In ages past and gone,
> Ere you were Queen of Sheba,
>   And I, King Solomon,
> Alone and undivided
>   We lived a life of sloth,
> Whatever you did, I did,
>   One dinner did for both.
> At length came separation
>   By fission and divorce,
> A lonely pseudopodium
>   I wandered on my course.
>   EDWARD HERON-ALLEN.

Vol 132, 446 1933

(This, to be sure, does not quite come up to McGonagall's unique standards, especially when inspired by some feat of technology or science, as in:

> And when Life's prospects may at times appear drear to ye,
> Remember Alois Senefelder, the discoverer of lithography.)

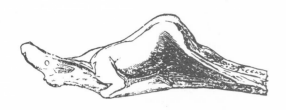

J. B. S. Haldane again takes on E. W. MacBride on Lamarckian inheritance: this extract shows his matchless precision and economy in debate.

I should like emphatically to protest against the view that the contest between Lamarckism and neo-Darwinism has anything to do with that between vitalism and mechanism. It would be quite possible to explain McDougall's results on thoroughly materialistic lines if the increased use by his rats of certain parts of the brain liberated specific substances into the blood, and these substances acted on the genes concerned with the elaboration of cerebral structure. Again, the neo-Darwinian may be a thorough-going vitalist, regarding each gene as an elementary living thing, contributing to the communal life of the cell and the federal life of the organism. He differs from the Lamarckian in believing that the gene has a very high degree of autonomy. There is no more reason to describe him as a mechanist than to dub a man an anarchist if he thinks that Parliament should not reverse the decisions of the London County Council. Similarly, a Lamarckian can be a vitalist, and a Darwinian a mechanist ; or, like myself, he may refuse to take either point of view.

Further, it is often stated that a refusal to admit that conscious purpose has played any important part in evolution is a gloomy and even immoral point of view. McDougall writes of " the Cimmerian darkness in which Neo-Darwinism finds itself ", and I admit that a theory which (like that of stellar structure) is largely based on integral equations is inevitably lacking in popular appeal. Nevertheless, the exact opposite seems to me to be the case. It has taken about a thousand million years for evolution to produce man, and the process has not only been slow but also often cruel. If mind had been an important evolutionary agent, there would be some ground for pessimism. If, on the other hand, it is only in our own time that the purposive guidance of the evolutionary process is becoming possible, then the outlook is altogether brighter. But if this is so, it is all the more our duty to examine as critically as possible the various evolutionary theories and to reject those which do not rest on a solid basis of observation.

From the human point of view the implications of the neo-Darwinian point of view were forcibly stated by the prophet Ezekiel (ch. 18, v. 2) :

" What mean ye, that ye use this proverb concerning the land of Israel, saying, The fathers have eaten sour grapes, and the children's teeth are set on edge ? As I live, saith the Lord God, ye shall not have occasion any more to use this proverb in Israel."

C. P. Snow's scientific career faltered with his publication, (together with P. V. Bowden) of an incorrect and culpably careless paper in Nature *on photochemical conversions in several vitamins. Snow eventually devoted himself to writing novels and — for nothing that happens to a writer is ever entirely without its compensations — used the episode as the basis for his first major success,* The Search. *Bowden and Snow's work was brutally shredded by Sir Ian Heilbron and R. A. Morton.*

Photochemistry of Vitamins A, B, C, D

From their letter in Nature [1] and various reports in the daily Press, it would appear that the work of Drs. F. P. Bowden and C. P. Snow provides a new and powerful technique offering the prospect of an immediate extension of knowledge concerning vitamins. In scrutinising the evidence disclosed in favour of this claim, we find ourselves in difficulties. In the first place, the technique (the study of spectral absorption curves, and irradiation with light of selected wavelengths) is familiar and has already been applied in vitamin studies ; while in the second place, the idea of one experimental method as a key to several vitamins seems to underestimate the differences between organic compounds of widely varying constitution sharing only a capacity to induce unusual (but quite diverse) physiological responses.

* * *

Vitamin D.—Here Drs. Bowden and Snow appear merely to have confirmed earlier observations, except as regards the sentence : " the spectroscopic evidence is, however, already sufficient to indicate that the conversion of ergosterol into calciferol depends on the migration of a double bond *to a position in which its influence on the other double bonds is greatly enhanced* " (our italics). We would inquire what the authors intend to convey by the italicised phrase. So far as the known absorption spectra of calciferol and ergosterol are concerned, the statement conveys no meaning whatever. No indication is given of any other criteria of ' enhanced influence ', neither is it clear what is meant by the *elimination* of an absorbing group from a complex organic molecule by irradiation with light of related frequency. Before such considerations can be of much avail, it is essential to know exactly the nature of the absorbing group—and in most complex organic compounds, the present position of the interpretation of absorption spectra does not make this possible. In this connexion neither the origin of the ergosterol bands nor the calciferol band is yet understood. The possibility of a rearrangement of the double bonds of ergosterol has already been fully studied by Windaus [10] by strictly chemical methods.

In the case of vitamin A, the effective entity is possibly a system of four conjugated double bonds. We can understand a photochemical process breaking up this system, but eliminating the group with " ease and certainty " has no precise meaning.

I. M. Heilbron.
R. A. Morton.

Department of Chemistry,
University, Liverpool,
May 26.

The same year, the centenary of James Clerk Maxwell's birth was celebrated in Cambridge. Max Planck responded in these graceful terms to the toast, proposed by Lord Rutherford.

" Maxwell's centenary is indeed a festive day—a day of honour for the physicists of all countries. In his personality we see not only the erudite savant mastering every domain of physics, but also the imaginative artist, who, like no other, knew how harmoniously to combine the separate material of investigation and thus to further its development. It is pre-eminently with Maxwell that this interweaving of sober intellectual activity with artistic intuition plays a particularly characteristic part.

" In physics, intuition is a peculiar thing. Nowadays it has somewhat fallen into discredit, having really recently done a deal of harm : first in the province of relativity, where the seemingly evident conception of the simultaneousness of two events occurring in different places proved a gross delusion ; then again in the province of the quantum theory, where it has proved absolutely impossible with the common forms of intuition completely to grasp the connexions between corpuscular and undulatory motion.

" And yet those are wrong who wholly reject intuitions by asserting that theoretical physics is only concerned in the mathematical working out of the results of measurements. For he who quite forgoes tentative flights of imagination will never be struck by a new idea which can be made fruitful for experiments. Yet a careful distinction must be made between the operations of the intellect and those of fancy. Here Maxwell sets us a classic example. His theory of electrodynamics was originally founded on quite special ideas of the mechanical nature of the ether, in accordance with the fact that in his time the mechanical conception of Nature was considered nearly a matter of course, having received strong support through the discovery of the principle of conservation of energy. But after, by the aid of such mechanical notions, Maxwell had found his electromagnetic differential equations and had recognised the extent of their efficiency, he did not hesitate in pushing aside as a negligible accessory the mechanical interpretations of the differential equations, in order to make them independent and to give his theory its pure and sublime shape. Perhaps this most clearly characterises Maxwell's physical manner of thinking.

There were reminiscences of the great man, including this one by William Garnett, his first demonstrator at the Cavendish Laboratory:

" It was during the visit of the comet in 1874, when unfortunately the comet's tail was a subject of general conversation, that Maxwell's terrier developed a great fondness for running after his own tail, and though anyone could start him, no one but Maxwell could stop him until he was weary. Maxwell's method of dealing with the case was, by a movement of the hand, to induce the dog to revolve in the opposite direction and after a few turns to reverse him again, and to continue these reversals, reducing the number of revolutions for each, until like a balance wheel on a hair spring with the maintaining power withdrawn, by slow decaying oscillations the body came to rest.

And finally a letter of Maxwell's, making fun of Norman Lockyer:

" How vile are they who quote newspapers, journals, and translations by number of vol. and page, instead of the year of grace, as if one should refer to the standard No. 16240 instead of Oct. 10, 1876. Lockyer always alters a reference to NATURE for Sept. 7, 1876, into vol. ?, p. ?, as if all promoters of natural knowledge counted everything from the epoch when NATURE first began."

Vol. 128, 695 1931

Here Lancelot Hogben, biologist, popularizer and left-wing intellectual, dismisses another scientific novel, now, deservedly it seems, forgotten:

Hunger and Love. By Lionel Britton. Pp. xi + 705. (London and New York : G. P. Putnam's Sons, Ltd., 1931.) 7s. 6d. net.

LIONEL BRITTON first became known to the reading public as the author of " Brain, a Play of the Whole Earth ", sponsored by Bernard Shaw, praised by Hannen Swaffer, and proclaimed a work of genius by St. John Ervine. His new novel has an introduction by Bertrand Russell. One therefore approaches it with high expectations. It is not customary to review novels in this column. After reading the book, the reviewer is not convinced that its contents justify any departure from the usual practice of NATURE. True, there are frequent, almost too frequent, references to protons and protoplasm. These in themselves are not sufficient to justify the claim that the author has written the novel of the machine age. It is a long soliloquy, in which the hazy outline of Arthur Phelps occasionally obtrudes to remind the reader that it is intended to rank as fiction rather than philosophy.

It took Lionel Britton eight years to write the book : the reviewer took six months to read it. Only the prospect of having to write a short notice of its contents sustained him to the end. Britton says a good many penetrating things about contemporary civilisation. The reviewer shares many of his prejudices and most of his opinions, when they are not susceptible of proof. Unfortunately, his way of stating them is prosy, prolix, and disorderly. People who like *Ulysses* may like " Hunger and Love ". Scientific workers, who generally attach importance to intellectual tidiness and coherent expression, will be inevitably repelled. Of the few who start it, still fewer will finish its seven hundred and five closely printed pages. It has captured the temper of contemporary biological realism far less successfully than Charlotte Haldane's " Brother to Bert " or the earlier novels of Mr. Wells.

Vol. 128, 475 1931

A. V. Hill's observations on political events in the German universities drew this response from one of the two eminent physicists who emerged as the apologists for Nazi ideology. Johannes Stark, like his friend and senior Philipp Lenard, was a Nobel laureate (the Stark effect). Both, but more especially Lenard, who was a true zealot, saw a German, Aryan physics polluted by alien forms of thought. These in brief were all that constituted the new physics, relativity and quantum theory in particular — in fact everything that was counter-intuitive and inaccessible to classical approaches. Such physics was held to be a sinister secretion of the Semitic psyche. Lenard and Stark had the ear of the "theorist" of the Nazi movement, Arthur Rosenberg, and published their views in the journal of the SS, Das Schwarze Korps.

International Status and Obligations of Science

IN his Huxley Memorial Lecture, extracts from which were published in NATURE of December 23, Prof. A. V. Hill has made detailed statements regarding the treatment of German scientists by the National-Socialist Government. These statements are not in accordance with the truth. As a scientist, whose duty it is to discover and proclaim the truth, I venture to place on record the following facts as against the inaccurate assertions of Prof. Hill.

The National-Socialist Government has introduced no measure which is directed against the freedom of scientific teaching and research; on the contrary, they wish to restore this freedom of research wherever it has been restricted by preceding governments. Measures brought in by the National-Socialist Government, which have affected Jewish scientists and scholars, are due only to the attempt to curtail the unjustifiable great influence exercised by the Jews. In Germany there were hospitals and scientific institutes in which the Jews had created a monopoly for themselves and in which they had taken possession of almost all academic posts. There were in addition, in all spheres of public life in Germany, Jews who had come into the country after the War from the east. This immigration had been tolerated and even encouraged by the Marxist government of Germany. Only a very small part of the 600,000 Jews who earn their living in Germany has been affected by the National-Socialist measures. No Jewish civil servant was affected who had been in office before August 1, 1914, or had served at the front for Germany or her allies or whose father or son had fallen in the War.

Prof. Hill asserts that something more than a thousand scholars and scientific workers have been dismissed, among them some of the most eminent in Germany. In reality not half this number have left their posts, and among these there are many Jewish and slightly fewer non-Jewish scientists who have voluntarily given up their posts. Examples are the physicists Einstein, Franck, Born, Schrödinger and in addition Landau, Fränkel (mathematician), Fränkel (gynæcologist), Prausnitz (hygienist), and others. Prof. Hill says that there are 100,000 people in concentration camps in Germany and that they are there only because they wished to have freedom of thought and speech. The truth is that there are not even 10,000 in the concentration camps and they have been sent there, not because of their desire for freedom of thought and speech, but because they have been guilty of high treason or of actions directed against the community. It must also be said that no women and children are imprisoned in the concentra-

tion camps in order to bring pressure to bear upon their husbands and fathers.

It would be a good thing to keep political agitation and scientific research apart. This is in the interests of science as well as in the interests of international scientific co-operation. But when a scientist does mix politics with science, he should at any rate fulfil the first duty of a scientist, which is conscientiously to ascertain the facts before coming to a conclusion.

J. STARK.

Physikalisch-Technische Reichanstalt,
  Berlin-Charlottenburg.
    Feb. 2.

VOL 133, 290 1934

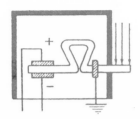

And here is A. V. Hill's response.

WITH Prof. Stark's political Anti-Semitism I need not deal: to an unrepentant Englishman (without any Hebrew ancestry or Marxist allegiance) it appears absurd.

It is a fact, in spite of what he says, that many Jews, or part-Jews, have been dismissed from their posts in universities, although they served in the line in the German armies in the late War. There are dozens of such in the lists of the Academic Assistance Council: whether they were "Beamte" or not is a quibble. Nor is there sense or justice in dismissing persons who were not "Beamte" before August 1, 1914.

Doubtless there are many grades of "dismissal", and in a technical sense certainly some of the persons in our lists were not "entlassen". They have found it impossible, nevertheless, to carry on their work in Germany. Men of high standing do not, without cause, beg their colleagues in foreign countries for help. Whether they were "dismissed", or "retired", or "given leave", or merely forbidden to take pupils or to enter libraries or laboratories is another quibble: the result is the same. It is inconsistent with that "freedom of scientific teaching and research" which the German Government apparently is seeking to restore.

As regards "high treason" and concentration camps, in England we do not call liberalism or even socialism by that name. The statement about women and children is a 'red herring'—I never said or suggested anything of the kind.

No doubt in Germany, after this reply, my works in the *Journal of Physiology* and elsewhere will be burned.

May I take this opportunity of saying that the Academic Assistance Council (Burlington House, W.1) urgently needs funds—for in spite of all the quibbles, scholars and scientists are still being dismissed.

A. V. HILL.

University College,
  Gower Street,
    London, W.C.1.
      Feb. 10.

VOL 133, 290 1934

Ernst Haeckel had died in 1919. E. W. MacBride wrote a reminiscence in NATURE, *which drew from a contributor this little story.*

...one period of the controversy, Haeckel felt that his presence at Jena was jeopardising the good name of his beloved university, so he offered to resign his chair; but the head of the governing body replied: "My dear Haeckel, you are still young, and you will yet come to have more mature views of life. After all, you will do less harm here than elsewhere, so you had better stop here." In point of fact, Jena never forsook Haeckel and Haeckel never forsook Jena, despite the flattering offers he received from the Universities of Vienna, Würzburg, Bonn and Strasbourg...

VOL 133, 331 1934

At about this time, heavy water first became available in sufficient bulk for chemistry and even biology. The great American physical chemist G. N. Lewis made such quantities by electrolysis. E. O. Lawrence was waiting impatiently for the accumulation of enough material to allow him to try deuterons in his particle accelerator, and was disgusted when Lewis, instead of handing over the precious fluid, fed it to a mouse.

IN *Science* of February 16, 1934, Prof. Gilbert N. Lewis summarises the results of certain sporadic attempts to observe the effect of water containing heavy hydrogen, H^2, upon living organisms. Experiments have necessarily been confined to small organisms, though some preliminary observations on mice are included. The first experiments were upon tobacco seeds, the germination of which was completely retarded by pure H_2^2O and slowed up some 50 per cent by water containing 50 per cent H_2^2O. Seeds transferred to normal water after three weeks in pure H_2^2O sprouted in about half the cases but gave unhealthy seedlings. Yeast cultures in an appropriate nutrient medium dissolved in pure heavy water failed to grow, and Pacsu has also shown that the evolution of carbon dioxide by yeast from sugar solution made up with heavy water is much diminished.

In an experiment that was expensive if preparatory in nature, a mouse was supplied in three doses with some 0·66 gm. of pure H_2^2O. The mouse survived, though during the experiment it showed "marked signs of intoxication". The symptoms of distress seemed more marked after each dose but not cumulative, which led Prof. Lewis to conclude that the heavy water was being voided, but no preparation had been made to test this point. Prof. Lewis concludes that H^2 is not toxic in any high degree but that its complete substitution for H^1 leads probably to a complete inhibition of growth, an effect which is to be traced to "the greatly reduced rate of all physicochemical processes when H^2 is substituted for H^1."

VOL 133, 620 1934

Now an example of the sporting inclinations of the Englishman abroad.

Sakai Marksmanship with a Blowpipe

ON looking through some old note-books, I find that I have preserved a record that may be of general interest, one of Sakai marksmanship with a blowpipe. The subject of the record was a Kampar Sakai whom I met on Cameron's Highlands, Malaya, in March, 1926. As I had seen previously some very poor exhibitions by aboriginals with the blowpipe, I asked this man, who was a fine sturdy fellow, to let me see what he could do. The target was a bit of deal-board from the lid of a box, dimensions not noted; but I marked a two-inch bull on it in pencil, stationed the Sakai 50 ft. away, which was the greatest distance possible on account of vegetation, and told him I would give him ten cents for every hit on the target and twenty cents for every bull, the shoot to be limited to ten rounds with unpoisoned darts.

The record was as follows:

1.	$1\frac{1}{2}$ in. from bull.	6.	4 in. from bull.
2.	Bull, side of.	7.	$\frac{1}{4}$ in. from bull.
3.	4 in. from bull.	8.	Missed the target.
4.	1 in. from bull.	9.	1 in. from bull.
5.	Bull, centre.	10.	$1\frac{1}{2}$ in. from bull.

The shoot cost me, in equivalent English currency, 2 shillings and 6·8 pence. I remember that the darts penetrated the target sufficiently to make it difficult to pull them out without injuring them.

68 Chaucer Road,   J. B. SCRIVENOR.
Bedford.
July 27.

VOL 132, 243 1933

An affinity has often been claimed between mathematics and music. Joseph Larmor, mathematician and theoretical physicist, a professor at Cambridge but by then living in retirement in Ireland, contributed this footnote to history.

Psychology of Musical Experience

I HAVE long been impressed by a passage about Lagrange, the prince of mathematicians, in Thomas Young's biographical sketch: "In the midst of the most brilliant societies he was generally absorbed in his own reflections: and especially when there was music, in which he delighted, not so much for any exquisite pleasure that he received from it, as because, after the first three or four bars, it regularly lulled him into a train of abstract thought, and he heard no more of the performance, except as a sort of accompaniment assisting the march of his most difficult investigations, which he thus pursued with comfort and convenience."

I now notice that it correlates rather closely with a remark of Darwin, the prince of naturalists, in the well-known passage in his autobiography where, after describing the atrophy of his tastes for literature and painting and music, he proceeds: "Music generally sets me thinking too energetically on what I have been at work on, instead of giving me pleasure."

JOSEPH LARMOR.

Holywood, Co. Down.
April 5.

VOL 133, 726 1934

In the 1920s and 1930s there was an outbreak of what Irving Langmuir called pathological science, of which n-rays (pp.115-8) afforded a prime example. Among the criteria is the essential irreproducibility of the phenomena, which require special skills to observe— and they remain always at the borderline of detectability. "Mitogenic radiation" was discovered in Russia by A. G. Gurwitsch and occasioned a steady stream of papers, mainly, but by no means exclusively, from the Soviet Union. Here is an analysis of one of the wilder claims.

THIS article is occasioned by recent popular descriptions[1-5] of an apparently well-attested case of luminescence in a human being in Italy, and the references to mitogenetic radiation which accompany them. The subject is a woman suffering from asthma. She is psychologically abnormal—intensely religious and hysterical—and the phenomenon of light emission occurs during light sleep, in circumstances which suggest that it is connected with these abnormalities. It lasts about three seconds, is of sufficiently high intensity to be photographed with an exposure of one sixteenth of a second, and is accompanied by increased respiratory movements, greatly increased pulse rate, and by the utterance of "moaning sounds and expressions".

The phenomenon is certainly unusual. The Italian peasants are said to regard it as a manifestation of holiness ; Signor Protti[5] attributes it, less picturesquely and perhaps less correctly, to the action of "blood radiation" in causing luminescence of certain substances in the skin. Protti's explanation is very unconventional, for bioluminescence is generally supposed to be a type of chemiluminescence, produced during the oxidation of certain substances, the luciferins, in presence of enzymes known as luciferases[6]. Naturally this mechanism has not been demonstrated in the rare cases of luminescence in human beings, but one would hesitate to accept an entirely different kind of explanation without strong positive evidence in its favour. It is possible that some instances of human luminescence are due only to infection by luminous bacteria.

The casual references to "blood radiation" are presumably intended to imply that the existence of such radiation is firmly established and its nature quite generally known. This is not the case. The fundamental experiment of Gurwitsch, claiming to show the emission of radiation from an onion root tip which could stimulate mitoses in a second root placed near it, has been, and continues to be, subjected to severe criticism. Indeed, the state of the subject at present makes a final decision with regard to the validity of this experiment quite impossible. This uncertainty has not, however, deterred Gurwitsch and his pupils from an elaborate development of their ideas, both experimental and speculative ; unfortunately, there are contradictions at almost every stage.

The supposed identity of the radiation with short-wave ultra-violet light, fundamental to the most important later experiments, itself rests on contradiction, the resolution of which should have been the primary object of later research. Thus, although behaving in certain experiments like ultra-violet light (being transmitted by quartz and absorbed by glass, etc.), mitogenetic radiation can pass, without being significantly absorbed, along the interior of an onion root or through a considerable thickness of a suspension of yeast in beer wort. Further, there is no agreement with regard to wave-length. Gurwitsch[7], by experiments with filters and by spectral dispersion of the radiation, found a wave-length 190–250 mμ ; Reiter and Gabor[8], by the same means, found 340 mμ, and both sets of workers were able to confirm fully their own conclusions by experiments with ultra-violet light from artificial sources. Ignoring, or explaining away, these very serious discrepancies, Gurwitsch continues to regard mitogenetic radiation as ultra-violet radiation of wave-length 190–250 mμ.

If this contention is correct, it should be possible to detect mitogenetic radiation by purely physical means, but satisfactory evidence is unfortunately lacking. Positive results obtained with a photosensitive form of the Geiger-Müller electron counter[9,10], the most sensitive apparatus available, are offset by several negative results[11-13], and the latter also demonstrate how easily spurious positive effects can be obtained if experimental conditions are not properly controlled. The most recent experiments[13] suggest that mitogenetic radiation, if it exists, cannot be detected by any known physical method ; its intensity is certainly less than about 300 $h\nu$/cm.[2] sec.

There is no space for a more detailed discussion : some quite characteristic points have already been referred to in NATURE[14] and a detailed review will appear elsewhere[15]. It is only important for the present to note that references to mitogenetic radiation, and with them Protti's reference to blood radiation, should be regarded with scepticism. Even if mitogenetic radiation exists, it is almost certainly too feeble to be capable of causing emission of visible fluorescence. Protti's explanation for his remarkable case of bioluminescence is therefore to be rejected.

Ludwig Bieberbach was another German academic who embraced Nazi doctrine, complete with its bizarre admixture of Teutonic mysticism. G. H. Hardy administers a majestic reproof:

The *J*-type and the *S*-type among Mathematicians

MATHEMATICIANS in England and America have been recently intrigued by reports of a lecture delivered by Prof. L. Bieberbach, of the University of Berlin, to the Verein zur Förderung des mathematisch-naturwissenschaftlichen Unterrichts. They have, however, found difficulty in judging the lecture fairly from secondhand reports. It is now possible to form a more reasoned estimate, Prof. Bieberbach having published a considerable extract, under the title "Persönlichkeitsstruktur und mathematisches Schaffen", in the issue of *Forschungen und Fortschritte* of June 20.

Prof. Bieberbach begins by explaining that his exposition will make clear by examples the influence of nationality, blood and race upon the creative style. For a National Socialist, the importance of this influence requires no proof. Rather is it intuitive that all our actions and thoughts are rooted in blood and race and receive their character from them. Every mathematician can recognise such influences in different mathematical styles. Blood and race determine our choice of problems, and so influence even the assured content of science (den Bestand der Wissenschaften an gesicherten Ergebnissen); but naturally do not go so far as to affect the value of π or the validity of Pythagoras' theorem in Euclidean geometry. . . .

Our nature becomes conscious of itself in the malaise (in dem Unbehagen) produced by alien ways. There is an example in the manly rejection (mannhafte Ablehnung) of a great mathematician, Edmund Landau, by the students of Göttingen. The un-German style of this man in teaching and research proved intolerable to German sensibilities. A people which has understood how alien lust for dominance has gnawed into its vitals . . . must reject teachers of an alien type. . . .

Prof. Bieberbach proceeds to distinguish between the '*J*-type' and the '*S*-type' among mathematicians. Broadly, the *J*-type are Germans, the *S*-type Frenchmen and Jews. The differences of type appear quite clearly in the varying treatments by different mathematicians of the theory of imaginary numbers. For example, in Gauss (an outstanding instance of the *J*-type) one finds above all insistence on the 'anschauliche Bedeutung von $\sqrt{-1}$'. . . . On the other hand, there are expositions of the theory by mathematicians of the *S*-type (for example, Cauchy) which produce a malaise (die Unbehagen verursachen) in one belonging to the *J*-type. . . . Technical virtuosity and juggling with conceptions are signs betraying the *S*-type, hostile to life and inorganic (dem Lebensfeindlichen unorganischen *S*-typus). . . .

Typical of the *J*-type are the 'nordisch-falische' Gauss, the 'nordisch-dinarische' Klein, and the 'ostbaltisch-nordische' Hilbert. . . . One of the crowning achievements of the *J*-type is Hilbert's work on axiomatics, and it is particularly regrettable that abstract Jewish thinkers of the *S*-type should have succeeded in distorting it into an intellectual variety performance (intellektuelles Variété). . . .

But perhaps I have quoted enough; and I feel disposed to add one comment only. It is not reasonable to criticise too closely the utterances, even of men of science, in times of intense political or national excitement. There are many of us, many Englishmen and many Germans, who said things during the War which we scarcely meant and are sorry to remember now. Anxiety for one's own position, dread of falling behind the rising torrent of folly, determination at all costs not to be outdone, may be natural if not particularly heroic excuses. Prof. Bieberbach's reputation excludes such explanations of his utterances; and I find myself driven to the more uncharitable conclusion that he really believes them true.

G. H. HARDY.

New College, Oxford.
July 20.

VOL 134, 250 1934

C. P. Snow's novel (p.214) was reviewed by Joseph and Dorothy Needham, and here is their conclusion. Arthur Miles was Snow's alter ego, who gave up science and entered the world of affairs when the research on which his reputation was based proved wrong. J. B. S. Haldane's criticism was that he could not imagine how any scientist could abandon his vocation for so trivial a reason.

A Crystallographic "Arrowsmith"

The Search. By C. P. Snow. Pp. 429. (London : Victor Gollancz, Ltd., 1934.) 8*s*. 6*d*. net.

* * *

Of criticism, perhaps the most serious that might be made is that Mr. Snow does not sufficiently make clear the nature of Arthur Miles's second enthusiasm. What is this for which an on the whole so successful scientific worker lays down his overall and slide-rule ? We are given to understand that the psychological and political education of mankind comes to seem more important to him than the search for detailed scientific truth. But the outlines of this need greater clarity. Had he turned to propagate a robuster political faith, this would not have been so necessary as it is when his aims seem so mild and moderate, so Lowes-Dickensonian, so L. N. U.

In sum, we have in "The Search" a really important study of human life as it is lived in the world of science. If Mr. Snow can push on along this line, we are not willing to suggest bounds for his possible achievement. But it will need a more definite socio-political outlook, and all the understanding that the closest and most sympathetic observation of human behaviour can give, whether it be of shop-assistant, railwayman, biologist, or parish priest. The results of such a life work are certainly no less valuable than a hundred papers in Proceedings and Transactions.

J. N and D. N.

VOL 134, 890 1934

In the same issue, H. G. Wells was the victim of what he felt to be an unreasonably harsh review. Here is part of his response and creed.

This class-war stuff, this 'dialectic materialism' is essentially un-scientific talk, pseudo-scientific talk; it is literary, pretentious, rhetorical. As sincere, patient and steadfast scientific analysis spreads into human biology we shall begin to get the general concepts of human relationship and social process clear and plain—and then we shall not need to worry about "power"; power will flow to the effective centres of direction. Stalin in our recent conversation accused me of believing in the goodness of human nature. I do at any rate believe in man's ultimate sanity. The political and social imaginations of very many people nowadays seem to me to be obsessed by the transitory triumphs of violence in various countries, and a lot of this talk about the need to organise the illegal seizure of power for direct creative (revolutionary) ends by those masses of the population which presumably have the most unsatisfied desires, is due largely to a lack of perspective in the outlook of the intelligentsia and a want of patience and lucidity in their minds. There is a limit to the concentration of power in human society, beyond which it becomes ineffective and undesirable. The limit has been passed in Germany and Russia to-day.

H. G. WELLS.

VOL 134, 972 1934

And part of a rejoinder by the reviewer, Hyman Levy, professor of mathematics at Imperial College and a prominent and articulate upholder of communism:

(1) He believes in man's ultimate sanity, meaning, I suppose, that he *feels* people in key positions will ultimately accept his solution.
(2) Power will flow to the effective centres of direction, meaning that he *feels* this will be so although he does not see it happening to-day in Russia and Germany.
I do not see why he should expect others to share his sanguine feelings.   H. LEVY.

VOL 134, 972 1934

Sir Oliver Lodge in old age, recording his memoirs for a sound film library being compiled by the Institute of Electrical Engineers, included this lusty blow for the ether, which for 25 years had had a lean time of it.

Before the end of the twentieth century, as I think, or at any rate in the twenty-first, the ether will be recognised as the one means of communication between the atoms, and the whole of physics will become once more luminous and clear, constituting a glorious epoch for our descendants. The ether will come into its own again, not only for practical purposes as the seat of all potential energy, but with a clear understanding of it as the one substance that holds the universe together, in which all matter is embedded, without which even locomotion cannot be properly understood, and which constitutes the physical vehicle for life and mind.   OLIVER LODGE.

VOL 135, 12 1935

The meeting held in Berlin to honour the memory of Fritz Haber — forbidden by the regime — was a brave act of political defiance by Max Planck and Otto Hahn.

Commemoration of Prof. Haber's Death

WHEN Prof. Fritz Haber died in Switzerland a year ago, we were glad to publish in the columns of NATURE an eloquent tribute to his greatness, written by one of his old pupils. By the irony of political circumstances in Germany, the loss of this chemical genius was limited in the journals of that country to a bare announcement, and no obituary notice at all adequate to the influence of his life and work appears to have been published at the time. It is not surprising, therefore, that Haber's scientific friends desired to honour his memory on the anniversary of his death, and that a number of them assembled for this purpose in the Harnackhaus of the Kaiser Wilhelm Gesellschaft on January 29, in spite of the official disapproval of the celebration to which we referred last week (p. 176). The Berlin correspondent of *The Times* reported that the speakers at the meeting laid emphasis on Haber's devotion to his country and his scientific services. Prof. Max Planck, who presided, recalled that Haber's synthetic nitrate process had saved Germany from military and economic collapse in the first months of the War. "We repay loyalty with loyalty," he said, and he laid particular emphasis on the last three words in his closing tribute to "this great scholar, upright man, and fighter for Germany." Prof. Otto Hahn, director of the Kaiser Wilhelm Institute for Chemistry, and other speakers also bore testimony to the debt owed by Germany to Haber for his outstanding contributions to pure and applied chemistry, and in doing so they expressed the feelings of their colleagues throughout the world.

VOL 135, 216 1935

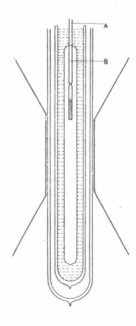

After World War II a participant recalled the atmosphere.

"Re-dedication of Science in Germany"

IN his interesting and noteworthy article "Re-dedication of Science in Germany" in *Nature* of July 13, p. 66, Prof. Polanyi refers to the Kaiser Wilhelm Institut für Physikalische Chemie in Berlin-Dahlem, and to its former director, Prof. Fritz Haber, and to the celebration in his memory on January 29, 1935.

Certain details concerning this impressive ceremony, illuminating the scientific life in Germany under the Nazis, might be perhaps of general interest and might demonstrate that even in 1935, two years after Hitler came to power, there were in Germany men of integrity. I would like to mention that Prof. Max Planck had arranged this ceremony in spite of the greatest difficulties, and in face of the strongest opposition from the Nazi Government, and it was only due to his reputation in the international scientific world that this gathering could be held. These unbelievable conditions are further illustrated by the fact that Prof. K. F. Bonhöffer, as a member of a German university, that is, as a State official in Germany, was forbidden to attend this celebration, and therefore could not deliver his lecture in honour of Haber. It might be of interest to add that Prof. Hahn read Bonhöffer's lecture in his place. At the last moment it became doubtful whether I should be allowed to attend, owing to the following circumstances. As an industrial chemist, I was a member of the Verein deutscher Chemiker. This society was already under Nazi influence, and the members received a strongly worded circular to the effect that they were not allowed to be present at this celebration. Although I was determined to ignore this order, some of my colleagues in the I.G. Farbenindustrie were in doubt as to what course they should adopt, and asked my advice. Acting against this order might have endangered their career, so I asked Prof. Bosch, chairman of the I.G. Farbenindustrie, who together with Haber developed the process for synthetic ammonia, for his advice. He answered that it was the duty of all chemists of the company who were invited to attend this anniversary. Unfortunately, Prof. Bosch died too early to continue his struggle against the Nazi regime.

This gathering of Haber's friends and admirers was, as Polanyi says, really a noteworthy manifestation of independence in German scientific circles.

One act was significant of the Nazis and of Goebbel's Ministry of Propaganda. Placards fixed on the exit doors of the lecture-room ordered the reporters of the newspapers to gather in a special room. Here they were told that no report whatever concerning Haber's anniversary and the celebration was to be published in German papers. So it came about that the German people did not know that there was a Haber celebration in Germany in 1935 under the chairmanship of Planck, and that at this gathering the representatives of many countries, ambassadors and men of science were present. There were, and there still are German men of science of integrity who, if allowed to start research work, could continue their struggle against Nazi ideas with renewed energy, and could try to build up a new German scientific life in a democratic way.

P. MENDELSSOHN BARTHOLDY

9 Addison Crescent,
Kensington,
London, W.14.
July 15.   VOL 158, 170 1946

In October 1934 the Russian physicist Peter Kapitza, who had worked since 1921 at the Cavendish Laboratory in Cambridge under the patronage of Lord Rutherford, was snatched by the Soviet authorities while on a visit to Moscow, on the orders, it was said, of Stalin himself. This comment is by Kapitza's colleague J. D. Cockcroft.

Prof. P. Kapitza and the U.S.S.R.

IT is common knowledge in scientific circles that Prof. P. Kapitza, director of the Royal Society Mond Laboratory at Cambridge, and Messel professor of the Royal Society, has been detained in Russia since last September by order of the Government of the U.S.S.R.

* * *

THROUGHOUT these years of developmental work, Kapitza had visited Russia almost every summer. During these visits he gave lectures and advised on the construction of new institutes, and it was known that he had at one time been offered the directorship of an institute in Russia, but Kapitza himself considered that conditions in the U.S.S.R. were not favourable for the development of his work. It came, therefore, as a shock to his colleagues to learn in October that Kapitza's return passport had been refused, and that he had been ordered to begin the construction of a new laboratory in Russia. The reasons underlying this action may be inferred from the following statement from the Soviet Embassy which appeared in the *News-Chronicle* :—"Peter Kapitza is a citizen of the U.S.S.R., educated and trained at the expense of his country. He was sent to England to continue his studies and research work . . . Now the time has arrived when the Soviet urgently needs all her scientists. So when Prof. Kapitza came home last summer he was appointed as director of an important new research station which is being built at Moscow". This commandeering of Kapitza's services on behalf of the U.S.S.R. ignores the personal and psychological factors involved, as was pointed out by Lord Rutherford in a letter to *The Times* of April 29. A man of Kapitza's highly-strung type must inevitably be profoundly disturbed by a sudden frustration of years of work ; and it comes as no surprise to his friends to learn from reliable sources that his health has already been seriously impaired by anxiety and strain.

VOL 135, 755 1935

The editorial clarity of vision about Hitler's significance and what his rise portended was hardly matched in the daily press; indeed there was still widespread approbation in Britain of the political developments in Germany.

Science and the State in Germany

THE publication in the *Times* of July 24, 25, 27, and 28, of extracts from Herr Hitler's book on the eve of its publication in an abbreviated translation, and the simultaneous account by the *Times* correspondent on July 27 of a Bill approved by the German Government for the compulsory sterilisation of those "considered in the light of medical science as it is understood to-day to be by heredity unfit" give a more illuminating view of the real significance of the Nazi movement than has yet appeared. That the Nazi leader stands self-revealed as ill-balanced, fanatical and otherwise abnormal is immaterial. What is supremely significant is that he has come into power on a wave of popular discontent with present-day social, political and economic conditions which is sufficiently intense to submit to previously unheard of restrictions for which a 'scientific' backing is advanced. Unfortunately, not all the Nazi measures can be supported by argument as sound scientifically as that upon which the sterilisation of the unfit is advocated. Herr Hitler's views of 'Aryans' and Jews and their qualities and character, to accord with which the German race of the future is to be moulded, belong to a 'science' which would be out of date even if it had not failed to justify itself when submitted to the test of scientific analysis. The German people, however, are not alone as victims of the world crisis, and if, as events seem to portend, the world is moving towards a solution of its difficulties through the application of scientific method to its problems by means of a more highly organised form of government than is possible in a pure democracy, it cannot be too widely appreciated outside scientific circles that the science which will provide the solution of our difficulties is not reactionary and does not exclude eminence in any field, whatever its racial origin—in short, it is not the 'science' of Herr Hitler.

VOL 132, 198 1933

The Abbé Georges Lemaître, a young Belgian cleric, propounds here — in 1931 — what is now familiar to all as the Big Bang theory of the origin of the Universe.

The Beginning of the World from the Point of View of Quantum Theory.

SIR ARTHUR EDDINGTON [1] states that, philosophically, the notion of a beginning of the present order of Nature is repugnant to him. I would rather be inclined to think that the present state of quantum theory suggests a beginning of the world very different from the present order of Nature. Thermodynamical principles from the point of view of quantum theory may be stated as follows : (1) Energy of constant total amount is distributed in discrete quanta. (2) The number of distinct quanta is ever increasing. If we go back in the course of time we must find fewer and fewer quanta, until we find all the energy of the universe packed in a few or even in a unique quantum.

Now, in atomic processes, the notions of space and time are no more than statistical notions ; they fade out when applied to individual phenomena involving but a small number of quanta. If the world has begun with a single quantum, the notions of space and time would altogether fail to have any meaning at the beginning ; they would only begin to have a sensible meaning when the original quantum had been divided into a sufficient number of quanta. If this suggestion is correct, the beginning of the world happened a little before the beginning of space and time. I think that such a beginning of the world is far enough from the present order of Nature to be not at all repugnant.

It may be difficult to follow up the idea in detail as we are not yet able to count the quantum packets in every case. For example, it may be that an atomic nucleus must be counted as a unique quantum, the atomic number acting as a kind of quantum number. If the future development of quantum theory happens to turn in that direction, we could conceive the beginning of the universe in the form of a unique atom, the atomic weight of which is the total mass of the universe. This highly unstable atom would divide in smaller and smaller atoms by a kind of super-radioactive process. Some remnant of this process might, according to Sir James Jeans's idea, foster the heat of the stars until our low atomic number atoms allowed life to be possible.

Clearly the initial quantum could not conceal in itself the whole course of evolution ; but, according to the principle of indeterminacy, that is not necessary. Our world is now understood to be a world where something really happens ; the whole story of the world need not have been written down in the first quantum like a song on the disc of a phonograph. The whole matter of the world must have been present at the beginning, but the story it has to tell may be written step by step.   G. LEMAÎTRE.

40 rue de Namur,
Louvain.

Vol 127, 706 1931

When Charles Vernon Boys reached the age of 80, NATURE *printed an affectionate verse tribute, of which these are the last four stanzas.*

To Sir Charles Vernon Boys on his Eightieth Birthday

Why snatch a bullet in its flight,
Lit by a single spark so bright
That on a photographic plate
The fleeting shadow seemed to wait—
With wake and bow-wave primly set—
All posing for their silhouette—
And leave a picture of the *noise ?*
Because, of course, Boys will be Boys !

Why did his bold, untrammelled thoughts
Conceive the scheme of fusing quartz,
Using an arrow, as it fled,
To draw a microscopic thread,
And from the fusion to "unreel"
A gossamer more true than steel,
Which every Physicist enjoys ?
The fact is this : Boys will be Boys.

What made our friend so seeming rash
As to pursue a lightning flash
By lenses rapidly revolved,
And even get the problem solved—
Both of its speed and structure—by
A photograph "which cannot lie" ?
That gave a thrill that never cloys,
And showed us still, Boys will be Boys.

To weigh the earth—to check the Therm—
Explain the logarithmic term—
To build with bubbles, and maintain
The opal colours in their train !
These are his pleasures, these his ploys
(Where skill with mind and Truth alloys)
For which, in Science, as in Toys,
We thank our stars, Boys *will* be Boys !

R. A. S. P.

Vol 135, 984 1935

chapter fourteen
1935-1941

NAZI SCHOLARS SEEK TO EXTIRPATE WORDS OF FOREIGN ETYMOLOGY FROM THE GERMAN LANGUAGE WITH LUDICROUS RESULTS. MAX BORN RUMINATES ON THE MAGIC NUMBER 137. J. B. S. HALDANE RESPONDS WITH SOOTHING WORDS TO THE DEPREDATIONS OF T. D. LYSENKO AND TASTES OXYGEN AND NITROGEN AT HIGH PRESSURES DURING HIS DIVING EXPERIMENTS. J. J. THOMSON DIES. LISE MEITNER AND HER NEPHEW OTTO FRISCH RECOGNISE THE SIGNIFICANCE OF THE NUCLEAR CHAIN REACTION.

1935-1941

The Spanish Civil War began and the Italian army captured Addis Ababa. Germany repossessed the Rhineland. Neville Chamberlain proclaimed peace in our time and the Munich Agreement delivered Czechoslovakia to the Germans. The Molotov-Ribbentrop pact nonplussed the world's communists. Germany and Russia attacked Poland, and France and Britain declared war. Italy entered the war, Winston Churchill succeeded Chamberlain and France capitulated. With the attack on Pearl Harbor, the United States was drawn into the conflict.

Among the many absurdities perpetrated by the Nazi regime was the campaign to purify the German language by expunging all words with foreign stems (Fremdwörter). So the telephone became a "farspeaker" and television a "farseeing apparatus". Here the physicist E. N. da C. Andrade has fun at the expense of the hapless Dr Joos.

New German renderings of 'Foreign' Words

IN NATURE of September 28, p. 495, we published a short notice of the second edition of "Theoretische Physik" by Dr. Georg Joos. Our reviewer commented very favourably upon the book, but animadverted upon the addition of a glossary of "foreign" words (*Erläuterung einiger Fremdwörter*), "in which *Absorption* (*Verschluckung*), *Elastizität* (*Dehnbarkeit*), *Kapillarität* (*Haarröhrchenkraft*) and such-like non-Prussian words are translated into the new German, although these 'foreign' words appear in the articles in Gehler's 'Physikalisches Wörterbuch' of more than a hundred years ago". We have now received an indignant letter from Dr. Joos, containing the following explanation of this glossary. "Many English readers, who according to the reporter are to be impressed comically through these things, will know that the graduates of the 'Oberrealschule' have studied neither Latin nor Greek and that for them an explanation of those words is very desirable", adding, "According to the wording of the report the reader must think that it is my intention to seriously substitute 'Verschluckung' for 'Absorption' or 'Segelstange' for 'Antenne' ". As regards the last sentence our reviewer suggested nothing about Dr. Joos' intentions, but stated the bare fact that the glossary had been added to the book, leaving his English readers free (if we may use the word without offence) to draw any conclusions they wished. Many will, no doubt, share his and our surprise that words which have been in regular use in the German language for four generations and more, a period sufficient, we should have thought, to guarantee their incorporation, should be regarded and named as foreign.

OUR reviewer adds : "We note that graduates of the Oberrealschule have studied neither Latin nor Greek, but then they did not do so at the time of the first edition, which did not contain the glossary ; in fact, they never did. Most students in England and other countries are equally unfamiliar with Latin and Greek, but have no difficulty over 'elasticity' and 'capillarity', while Germans unfamiliar with French have, we believe, no difficulty over '*General*' and '*Soldat*'. Dr. Joos does not seem to realise that, if there is any force in his contention, most German students will not know what his book is about, since both '*Theoretische*' and '*Physik*' are Greek words not explained in the *Erläuterung*. For that matter, *Electricität* (may we suggest *Bernsteinreibungskraft* ?) is not in the glossary, although *Kapazität* (*Fassungsvermögen*) is: *Alkali* (a Semitic word which occurs in the book combined with a Latin word as *Alkalispektren*—shall we suggest *Pflanzenaschenerdelichterscheinung* ?) is missing, although *Kondensator* (*Verdichter*) is included. We hope in the next edition to see the glossary either omitted or properly com-

pleted. We in England find so much in the present-day activities in German universities to move us to tears that Dr. Joos really must allow us a faint smile when we come across something harmlessly amusing from that quarter, and permit us to be our own judges of what is 'comical' ".

German chemistry was similarly taxed, though even the compliant editor here balked at "small-seeing tool" for microscope and "separation craft" for chemistry.

"Purging" Scientific Literature in Germany

THE *Chemiker Zeitung* of November 30, p. 978, prints a notice to German chemists requiring them in future to avoid the use of "foreign" words. It is explained that this can easily be accomplished, and among the illustrations given appear the following :

Förderanlage	instead of	Transportanlage
wirtschaftlich	,,	rationell
für, or je	,,	pro
durchlassig	,,	porös
zusammenpressen	,,	komprimieren
Nachahmung	,,	Imitation
Stück	,,	Exemplar
Ausmasse	,,	Dimensionen
umgrenzen	,,	definieren
Hochstwert	,,	Maximum
Tiefstwert or Niedrigstwert	,,	Minimum.

Vorbild, *Form* or *Muster* are suggested as alternatives to *Schema*, but the Editor, having perhaps seen comments on this subject in the notice of Joos' "Lehrbuch der theoretischen Physik" and in subsequent correspondence in NATURE (September 28, p. 495, and October 26, p. 675), points out that such innovations as *Kleinsehwerkzeug* for *Mikroskop* and *Scheidekunst* for *Chemie* should not be adopted, as they might be regarded as ridiculous by others.

The excesses of the ideologues continued to ravage German universities. By now plainly crazed, Philipp Lenard gave his name to the once illustrious Institute of Physics at Heidelberg. Here is E. N. da C. Andrade again, describing the consecration ceremony.

Philipp-Lenard-Institut at Heidelberg

CEREMONIAL DEDICATION

THE centre of physics teaching and research at the University of Heidelberg, hitherto known simply as the Physikalisches Institut, has recently been solemnly renamed the Philipp-Lenard-Institut. On December 13, at what the local Press justifiably called a unique ceremony, the Minister of Education (*Kultusminister*), Dr. Wacker, deputising for *Reichsminister für Wissenschaft, Erziehung und Volksbildung* Dr. Rust, who was unable to attend owing to illness, formally dedicated the building. His speech may be summarised in a sentence taken from it which, literally translated, reads : "It is, then, very superficial to speak of science 'as such', as a common property of mankind, equally accessible to all peoples and classes and offering them all an equal field of work. The problems of science do not present themselves in the same way to all men. The Negro or the Jew will view the same world in a different way from the German investigator." Prof. J. Stark, the president of the Reichsanstalt, who followed him, was, according to the German report, "particularly zealous against the followers of Einstein and attacked with the greatest frankness the scientific methods of Prof. Planck, who, as is notorious, even to-day stands at the head of a celebrated learned institution !" The ceremony concluded with a *Sieg-Heil* and the Horst-Wessel song.

On the next day, a further ceremony took place, the following account of which is literally translated from the German Press, where it appeared under the heading "A Germ-cell of German Science" (*Eine Keimzelle deutscher Naturwissenschaft*). With one exception, those who took part in this imposing function are, in spite of their high academic positions, comparatively little known in scientific circles in England. The exception is, of course, Prof. Lenard, the student of Hertz, whose papers he edited. A life of Hertz by Prof. Lenard appears in his "Great Men of Science", published before the new régime came into force in Germany and reviewed by Lord Rutherford in NATURE of September 9, 1933 (132, 367).

On December 13 the Minister of Culture and Education, Party-member Dr. Wacker, ceremonially dedicated the Philipp-Lenard Institute of the University of Heidelberg, in the presence of the Reichstatthalter Robert Wagner. On the next day an imposing number of German physicists assembled to make public confession of their union against the Jewish evil (*jüdischer Ungeist*), from which German science must be completely freed.

* * *

Geheimrat Lenard delivered the concluding words, and expressed his approval of these addresses. He exhorted all to continue energetically the fight against the Jewish spirit, which had by no means vanished from the German universities. He recounted many examples of Jewish arrogance (Einstein), supported by Jewish publishing houses (Springer), and expressed his confidence that this movement for German co-operation would embrace all our centres of higher learning.

VOL 137, 93 1936

NATURE *was stirred that the news of King George V's death, borne (so to speak) on the ether, must have reached all corners of the Empire in an instant. In fact, it is now known that his departure was carefully timed, for he was ushered out of the world at midnight by the royal physician, Lord Dawson of Penn, so that the announcement might appear in* The Times *and not, as was feared, the evening papers. The jingle that circulated at the time went:*

Lord Dawson of Penn
Has killed many men,
And that's why we sing:
God save the King.

Broadcast Announcement of King George's Death

ON the occasion of the death of His Majesty King George V on the night of January 20, the organisation of the British Broadcasting Corporation was utilised in communicating the official bulletins to the whole of the British Empire. From 9.30 p.m. onwards, the ordinary broadcasting programmes were stopped, and all the stations of the B.B.C., including those conducting the short-wave Empire service, were linked together, but were kept silent except for the transmission of the official bulletin at 15 minute intervals. At 10 o'clock, a short service of recollection and prayer for the King was broadcast, after which the silent watch between bulletins was resumed. The final announcement of the peaceful death of the King came shortly after midnight. In this way was the great organisation of British broadcasting used in the manner of a gigantic public address system, with literally millions of listeners in all parts of the world constituting the audience. Thus were listeners able to share with the Royal Family the tense anxiety of the last few hours, and to receive simultaneously the news of the passing of our Sovereign. Never before in the history of the world has it been possible for the whole human race to unite in sympathetic response to the messages thus conveyed from Sandringham to listeners everywhere. Truly, "Their sound is gone out into all lands, and their words unto the ends of the world", and the heart of man cannot fail to be touched by this great achievement of science. The imagination of a poet like the late Mr. Rudyard Kipling might well have been stirred by this theme of waves of émotion encompassing the earth to trace the changes which history has seen in methods of proclaiming to the nation the loss of its beloved King.

VOL 137, 141 1936

The fourth Lord Rayleigh followed his father's star and became president of the Physical Society. His presidential address contains a long recollection of Lord Kelvin and Sir James Dewar and of Dewar's remarkable assistant, Robert Lennox, from which this is an extract.

> Those who have had the good fortune to be in personal contact with the great workers of the generation above them should not neglect to record what they think may possibly be valued by posterity. There is no doubt that this duty has been too little regarded in the past. It is hard to realize when one is young that what we see to-day may be gone to-morrow and irretrievably forgotten the day after : and still harder to realize how eager posterity may be to know it."

* * *

Lord Rayleigh spoke also of Dewar's refusal to admit the danger of the work on the liquefaction of gases ; his apparently grudging admission that it was a little "tricky" was no overstatement considering that Lennox and Heath, his two assistants, each lost an eye in the course of the work. He referred also to Dewar's unorthodox engineering, his neglect of proper safety factors, his dislike of unannounced visitors, his secrecy with regard to his laboratory work—he civilly begged to be excused showing his laboratory to Kamerlingh Onnes, saying that then no awkward question could arise—his quarrelsome but essentially kind-hearted nature (was it Scots dourness in an exceptional degree ?), his sleeplessness—he was glad to get even three hours sleep a night—his amateur pharmacology, his generosity to those in need and his refusal to accept thanks, and his curious and almost impenetrable mental processes.

* * *

Lord Rayleigh spoke also of Lennox's stoic courage, his confidence and his cynical wit. Lennox was interested in an acid-resisting alloy, the peculiarities of which had led a puzzled workman to remark "t'aint iron", whereupon the trade name of "Tantiron" was adopted. "It is curious," Lennox remarked afterwards, "that so many analysts should find tantalum in it."

"A serious accident," added Lord Rayleigh, "occurred when he was trying the generation of acetylene under pressure, with a view to storing it liquid in cylinders. There was a violent explosion. Sir Joseph Petavel was in the next room. The first thing he heard after the explosion itself was Lennox's voice saying 'Look out, you men, put out the gas in there'. Next, he saw Lennox, his face streaming with blood and one of his eyes torn to pieces and the fragments spread all over his face, but quite calm. His first thought had been to prevent any further accident from the escaping acetylene."

VOL 137, 419 1936

Here, in volume 137 of NATURE, *is Max Born's cogitation on the magic number 137 (p.211).*

The Mysterious Number 137

IT is a remarkable fact, first made prominent by Sommerfeld's discussion of the fine structure of the hydrogen spectrum, that from three physical constants, h (Planck's constant), c (the velocity of light *in vacuo*) and e (the charge on an electron), a dimensionless pure number can be formed, which usually occurs in the form $hc/2\pi e^2$, with the numerical value 137, or more accurately 137·2.

Dr. Max Born, in a lecture delivered to the South Indian Science Association at Bangalore on November 9, 1935 (*Proc. Indian Acad. Sci.*, 2, 533 ; 1935), declared that the explanation of this number must be the central problem of natural philosophy. Its existence can be ascribed to the fact that there are two different 'natural' units of length, a larger one $\lambda_0 = h/mc$ (the so-called Compton wave-length) taken from quantum theory, and a smaller one $a_0 = e^2/mc^2$ (the so-called radius of the electron). Their ratio is 2π times the mysterious number. After pointing out the great importance of this number in atomic physics, Dr. Born criticised the existing explanations of it.

Sir Arthur Eddington considers that it is associated with the number of degrees of freedom of a pair of electrons, and obtains the value 137. Dr. Born rejects this view, and seeks for an alternative explanation based on the new Born-Infeld-Pryce unitary field theory, which considers matter and field as one and the same. It involves a very large constant called 'the absolute field', which is the magnitude of the field in the centre of the electron. It is suggested that the number 137 is related to the neutralisation frequency of oscillation of a pair of electrons, one positive and the other negative (produced by light quanta passing the field of a nucleus), which approach and finally neutralise each other, emitting light quanta. Apparently the details of the calculation have not been worked out on Born's own theory, but by working on a somewhat similar theory Euler and Kockel (two pupils of Heisenberg) obtain the value 82·4. This differs considerably from the value 137·2, but Dr. Born considers the discrepancy not discouraging in view of the arbitrary assumptions made in the theory.

Dr. Born also uses the new field theory to explain the ratio of the proton and the electron, and obtains the number 2340. The experimental value is 1846·6, and the theory of Sir Arthur Eddington gives the value 1847·6 (*Proc. Roy. Soc.*, A, 134, 524 ; 1931).

VOL 137, 877 1936

In 1936 the storm whipped up by T. D. Lysenko broke over Soviet biology. H. J. Muller, then working in Russia, had disobeyed the Communist Party injunction that human genetics was not to be mentioned at conferences, and at the calamitous meeting that took place in December, Lysenko and his epigone I. I. Prezent denounced all adversaries as enemies of the people. Muller returned to the United States — hustled out in a transport of the International Brigade, heading for the war in Spain — but members of his institute were arrested and even the secretary who had translated his book into Russian was shot. N. I. Vavilov, the most distinguished of Lysenko's victims, died in prison of starvation. Here is the first report in NATURE *of these sombre events, written by the cytogeneticist C. D. Darlington.*

Genetic Theory and Practice in the U.S.S.R.

A REPORT of the fourth session of the Lenin Academy of Agricultural Sciences, which was held in Moscow in December last, has been received from the Soviet Union Year Book Press Service, 623–4 Grand Buildings, Trafalgar Square, London. Apart from seventy-one special papers, there was a general debate on the present position of research in genetics. This apparently took the form of an attack on modern genetic theory by experts who have been engaged in the study of practical problems unrelated to genetics. The grounds of the attack were ostensibly twofold. First, geneticists like Muller and Vavilov were said to have neglected the Marxian principle of the unity of theory and practice in failing to keep their work in touch with the needs of farmers. Secondly, the primary assumptions of genetics were held to be invalid. Presumably, in the absence of other evidence, the second contention was deduced from the first. The attack was reinforced by pointing to work like that of Michurin and Lysenko which, unhampered by academic prejudice, has yielded results of immense practical value by methods of trial and error.

And again:

THE situation revealed by this discussion is astonishing in more than one respect. It is a common-place and as old as Aristotle for scientific men unfamiliar with experimental methods to set forth Lamarckian doctrines in explaining evolution. For experimental workers to set out a Lamarckian method for the advancement of practical breeding is, however, something new and particularly remarkable in the Soviet Union, where the moral bias of this ancient myth might well have made it suspect. Still more astonishing is it to find that these arguments are combined with a charge of doctrinal incompetence against Muller and Vavilov, who, while making

fundamental contributions to genetic theory, have probably done more than any other living workers in this field to show its practical value to mankind.

VOL 139, 185 1937

Some months later, and the issues have become clearer. Darlington again:

Genetic Theory and Practice in the U.S.S.R.

IN a note on genetics in the U.S.S.R. (NATURE, 139, 185 ; Jan. 30, 1937), reference was made to the empirical work of Michurin on the hybridization of fruits, and his published work was said not to have been translated into any foreign language. Our attention had been directed to the fact that a translation, in an abridged form, is available for reference in the Bureau of Plant Genetics at Cambridge. The short published abstracts of the Bureau (*Plant Breeding Abstracts*, 5, 56, 376 and 7, 122) make the character of Michurin's work fairly clear. Like the recent work of Burbank in the United States, it belongs to the period of Kölreuter. It uses the assumptions and deals with the problems that were in favour in the late eighteenth century. Indifference to the refinements of later work has led Michurin, as it did Burbank, to somewhat fantastic conclusions in physiology and genetics. The reason for Michurin's indifference, however, is peculiar and significant. He states that the Mendelian principles are not in accordance with the dialectic of Engels, and must therefore be disregarded. It seems that Aristotelianism is appearing in a new quarter under a new guise.

VOL 139, 1048 1937

J. B. S. Haldane was much reproved for his failure to condemn what was happening in the Soviet Union. The truth was that he did not want to be disloyal to his friends in the Communist Party. This is the kind of equivocation beyond which he was unwilling to go.

The Position of Genetics

THE attacks of Lysenko on Vavilov and other Russian geneticists reported in NATURE of August 21 are not wholly dissimilar to Dr. H. Dingle's attack on Prof. E. A. Milne in a recent issue of this journal. Vavilov was accused of being anti-Darwinian, Milne of going back to Aristotle, in neither case perhaps with full justification. If these attacks have led to a curtailment of Vavilov's work, the situation of genetics in the Soviet Union is indeed serious. If not, hard words break no bones, and the outlook for genetics in Moscow is at any rate no worse than in London, where I understand that the only department of genetics in the University is shortly to come to an end.

J. B. S. HALDANE.

University College,
London, W.C.1.

VOL 140, 428 1937

Here was a novel idea for alleviating the effects of the depression of the 1930s in France.

Aid for Intellectual Unemployed in France

INTELLECTUAL workers, including men of science, writers, artists and others, have suffered no less than industrial workers during the recent years of economic unrest. In 1934 an organization was established in France with the object of providing socially useful work for the unemployed professional men and women.

* * *

The "Confédération des travailleurs intellectuels", consisting of more than 200,000 workers from various professional groups, also had the problem of intellectual unemployment under consideration. "L'Entr'aide des Travailleurs Intellectuels" (E.T.I.) was organized in order to examine the situation and to find ways and means of giving efficient assistance. The poor financial state of France excluded all possibility of help from the Government, and it was impossible to rely upon private donations. A campaign was therefore begun to obtain from the authorities permission to issue special stamps of different values, with a small surcharge, the surcharge being destined for the intellectual unemployed.

* * *

Up to January 1, 1937, the French Post Office paid over to the E.T.I. about a million francs under this scheme. This sum is due mainly to philatelists and stamp-dealers; for the success of the scheme, it is necessary that the public generally should take part in this social and humane work. At present the following stamps have been issued:

FIG. 1.

Vol 140, 612 1937

Lord Kelvin, so Lord Rayleigh remembers, was after all finally induced to come to terms with radioactivity:

In his presidential address to the Physical Society (January 1936), Lord Rayleigh (p. 221) states that Kelvin argued emphatically with Rutherford and himself against the atomic origin of the energy. Rayleigh asked him to make a bet of five shillings that within three (or six) months he would admit that Rutherford was right. Within the allotted period Kelvin came round, and at the British Association he made a public pronouncement in favour of the internal origin of the energy of radium. He also produced the five shillings in settlement of the bet.

Vol 140, 888 1937

Johannes Stark, the voice of Aryan science, was allowed another hearing in NATURE *. This was a shrewd move, for the effect on uncommitted readers must have been devastating.* NATURE *also recorded his vociferations in* Das Schwarze Korps *(p.216), in which he and Philipp Lenard ranted against "White Jews" — those Aryans who compromised with Jewish, that is to say abstract, concepts in physics. Among these was of course Werner Heisenberg, who survived the onslaught with the help of a personal intervention by Heinrich Himmler, a family friend.*

News and Views

The "Jewish" Spirit in Science

IN July of last year, several correspondents sent us copies of an article entitled " 'Weisse Juden' in der Wissenschaft", from the German periodical *Das Schwarze Korps* of July 15. The main theme of the article was that it was not sufficient to exclude all Jews from sharing in the political, cultural and economic life of the nation, but to exterminate the Jewish spirit, which is stated to be most clearly recognizable in the field of physics, and its most significant representative to be Prof. Einstein. "There is one sphere in particular," the article says, "where we meet the spirit of the 'White Jews' in its most intensive form and where what is common between the outlook of the 'White Jews' and Jewish teaching and tradition, can be directly proved, namely, in *Science*. To purge science from this Jewish spirit is our most urgent task. For science represents the key position from which intellectual Judaism can always regain a significant influence on all spheres of national life. Thus it is characteristic that in a time which brings fresh tasks to German medicine and which awaits decisive achievements in the fields of heredity, race-hygiene and public health, our medical journals should, in the space of six months, publish from a total of 2,138 articles, 1,085 from foreign authors, including 116 from Russians of the U.S.S.R. These articles of foreign origin scarcely concern themselves with those problems which seem so urgent to us. Under cover of the term 'exchange of experience' there lurks that doctrine of the internationality of science which the Jewish spirit has always propagated, because it provides the basis for unlimited self-glorification."

Vol 141, 778 1938

William Buckland was the eccentric father of the eccentric biologist Frank Buckland (p.63). These are the first and last of seven stanzas.

Prof. Buckland and Oxford

THE verses which follow are printed by permission of Sir Edmund Phipps, K.C.B., who found them among the papers of his great-grandfather William Foskett of Bath (1763–1843), who was accustomed to preserve current songs, verses and anecdotes, communicated to him in various handwritings. William Buckland (1784–1856) was one of the founders of scientific geology. He began to lecture at Oxford in 1813, and a special readership in geology was founded for him in 1819. He was elected fellow of the Royal Society in 1818, president of the Geological Society of London in 1824 and 1840, president of the British Association at Oxford in 1832, and Dean of Westminster in 1845. His best-known work, "Reliquiae Diluvianae", was published in 1823 ; his Bridgewater Treatise on "Geology and Mineralogy" in 1836.   JOHN L. MYRES.

13 Canterbury Road,
Oxford.

AN INTENDED EPITAPH ON P[ROFESSOR] BUCKLAND AT OXFORD

Mourn, Ammonites, mourn, o'er his funeral urn,
    Whose neck ye must grace no more.
Gniss, Granite, and Slate, he settled your date,
    And his, ye must now deplore.

Then exposed to the drip of some case-hardening
    His carcase let Stalactite cover,   [spring
And to Oxford the petrified Sage let us bring
    When he is incrusted all over.
There with mammoths and crocodiles high on a shelf
Let him stand as a monument raised to himself.

VOL 142, 673 1938

A thought-provoking letter from the aged archaeologist and Egyptologist Sir Flinders Petrie, who had contributed papers on chemical and mathematical topics to NATURE *in the previous century. He expanded his theory of cycles of civilisations in a book entitled* Revolutions of Civilisation, *by which he set great store.*

Science and Mankind

IF we are to deal with the science of man, one of the first steps is to know our own position in the recurring cycles of civilization. Then we can see in what direction we are heading at present.

The cycle of economy and waste covers about 130 years. The waste began at 1535, 1660, 1790, 1920. The more austere periods revived in 1560, 1690, 1820 ; may we therefore look to 1950 or thereabouts ? This is known in the northern saying, "from clogs to clogs in three generations".

The long general cycle of civilization in Egypt bore the best work in 3700, 2600, 1550, 450 B.C., and A.D. 760, an average of 1,115 years, resembling the 1,100 years of the "Great Year" known to the Etruscans. In each of the repetitions there was the same order of development—sculpture and architecture, painting, literature, mechanics, science, and lastly wealth, the stage we seem now to have reached. Following that, other races break in for plunder, and after some centuries of mixture a new dominant rises with a fresh cycle.

A still larger cycle is that of race. The Sumerian, the Semite, and the Perso-Aryan have successively been leaders of the East. We seem now to be nearing the end of Aryan rule, unless we can make recovery.

FLINDERS PETRIE.
Jerusalem.

VOL 142, 620 1938

Here is the outcome of the famous conversation on the winter walk in Sweden between Otto Frisch and his aunt, Lise Meitner, who had found temporary refuge in that country. The perception dawned on them that the results of Otto Hahn and Fritz Strassmann could only mean that nuclear fission had occurred.

Disintegration of Uranium by Neutrons: a New Type of Nuclear Reaction

ON bombarding uranium with neutrons, Fermi and collaborators[1] found that at least four radioactive substances were produced, to two of which atomic numbers larger than 92 were ascribed. Further investigations[2] demonstrated the existence of at least nine radioactive periods, six of which were assigned to elements beyond uranium, and nuclear isomerism had to be assumed in order to account for their chemical behaviour together with their genetic relations.

In making chemical assignments, it was always assumed that these radioactive bodies had atomic numbers near that of the element bombarded, since only particles with one or two charges were known to be emitted from nuclei. A body, for example, with similar properties to those of osmium was assumed to be eka-osmium ($Z = 94$) rather than osmium ($Z = 76$) or ruthenium ($Z = 44$).

Following up an observation of Curie and Savitch[3], Hahn and Strassmann[4] found that a group of at least three radioactive bodies, formed from uranium under neutron bombardment, were chemically similar to barium and, therefore, presumably isotopic with radium. Further investigation[5], however, showed that it was impossible to separate these bodies from barium (although mesothorium, an isotope of radium, was readily separated in the same experiment), so that Hahn and Strassmann were forced to conclude that *isotopes of barium* ($Z = 56$) *are formed as a consequence of the bombardment of uranium* ($Z = 92$) *with neutrons.*

* * *

By bombarding thorium with neutrons, activities are obtained which have been ascribed to radium and actinium isotopes[6]. Some of these periods are approximately equal to periods of barium and lanthanum isotopes[6] resulting from the bombardment of uranium. We should therefore like to suggest that these periods are due to a 'fission' of thorium which is like that of uranium and results partly in the same products. Of course, it would be especially interesting if one could obtain one of these products from a light element, for example, by means of neutron capture.

LISE MEITNER.

Physical Institute,
Academy of Sciences,
Stockholm.

O. R. FRISCH.

Institute of Theoretical Physics,
University,
Copenhagen.
Jan. 16.

VOL 143, 239 1939

Michael Polanyi, one of the outstanding intellects of his time, saw in what is now referred to as "mission-oriented research" a threat to all science.

Prof. Polanyi emphasizes the mutual reactions between science and practical knowledge, and urges that attempts to direct research towards results of possible practical applicability cannot lead to a growth of science that is of much value. A consistent policy on these lines would stop the development of science altogether, turning in effect the efforts now directed to scientific research into attempts to discover empirical solutions for practical problems. Prof. Polanyi condemns, for example, cancer research, and urges that all progress depends on the freedom of the systematic branches of science to pursue their own specific scientific aims. Universal adoption of a policy of endowing research for practical aims would bring science to a standstill and gradually exhaust its practical applications.

Vol 144, 972 1939

In 1940, with the war in progress and the outlook bleak, the notion of bacteriological warfare seems still to have been seen as something of a giggle.

Bacterial Warfare

PROF. TYNDALL will have much to answer for in the results that may be expected from the spread of his "dust and disease" theory. It is stated by the *Athenæum* that a new idea has been broached in a recent lecture by Mr. Bloxam, the lecturer on chemistry to the department of artillery studies. He suggests that the committee on explosives, abandoning gun cotton, should collect the germs of small-pox and similar malignant diseases, in cotton or other dust-collecting substances, and load shells with them. We should then hear of an enemy dislodged from his position by a volley of typhus, or a few rounds of Asiatic cholera. We shall expect to receive the particulars of a new "Sale of Poisons" Act, imposing the strictest regulations on the sale by chemists of packets of "cholera germs" or "small-pox seed". Probably none will be allowed to be sold without bearing the stamp of the Royal Institution, certifying that they have been examined by the microscope and are warranted to be the genuine article.

Vol 145, 523 1940

The death of J. J. Thomson brought a string of eulogies. The astronomer E. A. Milne was aware of Thomson long before he encountered him.

IT is not easy even to recapture any adequate sense of the influence which J. J. Thomson exerted on science in schools in the earlier days of this century. Electrons and J. J. Thomson were spoken of mysteriously by physics masters with the same bated breath. He came into my life, as into those of so many others, long before I came up to Trinity, when as a schoolboy I got hold of a copy of the "Discharge of Electricity through Gases", devoured it, found it much too hard to

understand, resented especially its use of integral signs (which I had not yet "got to"), but gained such thrills from the parts I could understand that I even read it on Sundays, and found it worth the rebuke I received at home for violating the Sabbath. It sent me to his early *Phil. Mag.* papers of 1880 or so, on the stability of rings of charged corpuscles inside a sphere of the opposite sign of charge.

Vol 146, 356 1940

Max Born was one of the others who ruminated on Thomson's dedication to science.

IT was Prof. J. J. Thomson's name which took me to Cambridge in 1906.

* * *

More than fifteen years later, on a visit to Cambridge, I met Thomson's son, who took me to the Cavendish and into the basement room where "J. J." was working, surrounded by the usual complicated structures of apparatus, glass tubes and wires. I was introduced: "Father, here is an old pupil of yours who studied with you years ago. . . ." The grey head, bent over a glowing vacuum tube, was lifted for a minute: "How do you do. Now, look here, this is the spectrum of . . .", and we were in the midst of the realm of research, forgetting the chasm of years, war and after-war, which lay between this rencontre and the days of our first acquaintance. This was Thomson in the Cavendish: science personified.

Vol 146, 356 1940

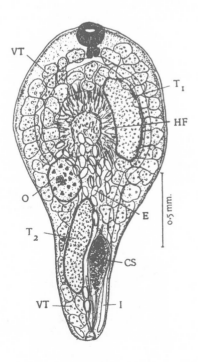

There has been much debate about whether Albert Einstein was religious and what exactly he meant by "God does not play dice". The article from which these two short passages are taken shows that he was without religion in any meaningful sense.

SCIENCE AND RELIGION

By Prof. Albert Einstein, For.Mem.R.S.

IT would not be difficult to come to an agreement as to what we understand by science. Science is the century-old endeavour to bring together by means of systematic thought the perceptible phenomena of this world into as thorough-going an association as possible. To put it boldly, it is the attempt at the posterior reconstruction of existence by the process of conceptualization. But when asking myself what religion is, I cannot think of the answer so easily. Even after finding an answer which may satisfy me at this particular moment, I still remain convinced that I can never in any circumstances bring together, even to a slight extent, all those who have given this question serious consideration.

* * *

The more a man is imbued with the ordered regularity of all events, the firmer becomes his conviction that there is no room left by the side of this ordered regularity for causes of a different nature. For him neither the rule of human nor the rule of Divine Will exists as an independent cause of natural events. To be sure, the doctrine of a personal God interfering with natural events could never be *refuted*, in the real sense, by science, for this doctrine can always take refuge in those domains in which scientific knowledge has not yet been able to set foot.

But I am persuaded that such behaviour on the part of the representatives of religion would not only be unworthy but also fatal. For a doctrine which is able to maintain itself not in clear light but only in the dark will of necessity lose its effect on mankind, with incalculable harm to human progress. In their struggle for the ethical good, teachers of religion must have the stature to give up the doctrine of a personal God, that is, give up that source of fear and hope which in the past placed such vast power in the hands of priests.

Vol 146, 605 1940

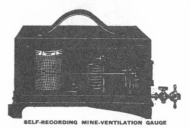

SELF-RECORDING MINE-VENTILATION GAUGE

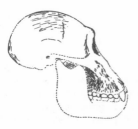

Frederick Soddy reviewed G. H. Hardy's A Mathematician's Apology, *a work of crystalline elegance. His review has a pleasant postscript.*

The reviewer may have laid himself open to the retort that he has given no indication of what he considers real mathematics really is. As Huxley remarked, when you want to test the strength of a gymnast you do not ask him to lift a 200-lb. pair of dumbbells (such, for example, as is being dealt with in the hieroglyphics on the dust-sheet of this book, from "Asymptotic Formulæ in Combinatory Analysis" by G. H. Hardy and S. Ramanujan, which the publishers have thought the best way of frightening the public into buying it), but a 10-lb. Indian club *to see what he can do with it.*

$$\sum_{m=1}^{\infty} \sum_{n=1}^{\infty} (2n+1)^{-2m} = \sum_{n=1}^{\infty} \frac{(2n+1)^{-2}}{1-(2n+1)^{-2}}$$

$$= \sum_{n=1}^{\infty} \frac{1}{(2n+1)^2 - 1} = \sum_{n=1}^{\infty} \frac{1}{4n(n+1)}$$

$$= \frac{1}{4} \sum_{n=1}^{\infty} \left(\frac{1}{n} - \frac{1}{n+1} \right) = \frac{1}{4}$$

On a glorious summer afternoon in 1931, misspent in a way for which even the author could scarcely apologize, attending a faculty meeting, to improve the anything but shining hour of imprisonment the reviewer slipped across a note to the author, demanding the sum of all the reciprocals of each of all the odd integers, except unity, raised to the power of each of all the even integers, a question which the veriest tyro must admit is scarcely up to scholarship standard. After a surprised and concentrated stare of disgust at this common and probably poisonous insect, he started jabbing it with a pencil, viciously but cautiously at first, then with lightning ferocity as he chased it down the paper to the kill, and *in an incredibly short space of time* flung back the dead body here reproduced.

F. S.

Vol 147, 5 1941

E. W. M. (presumably E. W. MacBride) is admonished by Joseph Needham for peddling the nastier prescriptions of eugenics.

IT is impossible to let the article "Cultivation of the Unfit", signed by E. W. M. in NATURE of January 11, pass without comment. With remarkable inconsistency, the author felicitates Sir Arthur Keith's apologia for war as being "the result of increasing population and race pressure" and "the means by which Nature decides which race shall 'inherit the earth' "; yet admits in an adjacent paragraph that the spread of birth-control will bring to an end the "cultivation of the unfit". If birth-control can be made to do this, why can it not also be made to bring to an end the existence of War, with all its horrors ? Population-control by a world authority is the obvious goal towards which all our efforts should be tending.

E. W. M. then describes the weeding out of the (physically) unfit in animal communities, and goes on to refer to the "elaborate and costly social services" which keep alive the "morally, mentally, and physically" human unfit. This apparently guileless transition seems to involve several non sequiturs. Thus we are given no evidence that the deformed specimens of chamois or red deer were also morally and mentally unfit, or if so, unfit for what ? The application of purely biological concepts to sociological phenomena is surely inadmissible.

Finally, the compulsory and *punitive* sterilisation of parents who "have to resort to public assistance in order to support their children" is offered as a remedy. Are we to assume that E. W. M. includes shipowners, beet-sugar shareholders, and other persons receiving financial benefit other than wages from industries subsidised by the State, though privately owned, in this category ? And can he even be serious in suggesting biological "punishment" for the two million unemployed ?

It is difficult to express the dismay experienced in seeing these doctrines, so dangerous for humanity, receiving the imprimatur of what is perhaps the most famous scientific weekly in the world.

JOSEPH NEEDHAM.

Caius College,
Cambridge.

Vol 137, 188 1936

NATURE made no concessions to the rigours of the Blitz, at this time in full swing.

Though situated in an area which has certainly had a share in the wanton destruction resulting from indiscriminate bombing, we are glad to say that, so far, the offices of NATURE have suffered nothing but a few broken windows and occasional interruption of work caused by the proximity of delayed-action bombs. In any event, arrangements have been in existence for some considerable time for NATURE to carry on elsewhere in the unhappy event of its present home becoming uninhabitable for any length of time.

Vol 147, 321 1941

J. B. S. Haldane, who was completely impervious to physical danger (and one of the few combatants who owned actually to enjoying World War I), was engaged in highly hazardous underwater experiments aimed at maximising the chances of survival of submariners trying to escape from a sunken craft. Haldane and a group of volunteers, consisting mainly of Spanish Civil War veterans from the International Brigade and also Juan Negrin, physiologist and prime minister of the Spanish Republic, exposed themselves for long periods to pressures of ten atmospheres and extreme cold. Haldane suffered an attack of the divers' bends and a gas bubble lodged in his spine, causing him pain for the rest of his life. He and Martin Case, the co-author of this letter to NATURE, with a few others, then went on to study the effects on themselves of almost lethal concentrations of carbon dioxide, which induced hallucinations and convulsions. Haldane called his father's gas-analysis apparatus, which he was attempting to manipulate in these conditions, "a purple bitch", and was with difficulty deterred from smashing it to pieces. He also insisted (and continued afterwards to insist) that the aged and distinguished physician Sir Leonard Hill, who was an adviser on the project, had tried to murder him during the experiments. The incidental observation recorded below evidently gave Haldane inordinate pleasure, mainly, it seems, because it contradicted the textbook assertion that oxygen was a tasteless gas; it also reinforced a tenet of Marxist dialectic, which states that the quality of a thing is changed by the quantity.

Tastes of Oxygen and Nitrogen at High Pressures

WHILST carrying out experiments on behalf of Admiral Sir M. E. Dunbar-Nasmith's Physiological Sub-Committee for saving life from sunken submarines, we and other subjects have had occasion to breathe oxygen, air and other gas mixtures at high pressures.

When oxygen was breathed at 6 atmospheres, several subjects noticed a peculiar taste, which was enhanced at 7 atmospheres. None of them noticed it at 3 atmospheres. The taste is both acid and sweet. Two subjects described it as "like dilute ginger beer", and "like dilute ink with a little sugar". It was felt unevenly, by one subject mainly on the back of the tongue, by another beneath it. In one case it persisted for some minutes after ceasing to breathe oxygen. It may be remarked that although oxygen is a convulsant at such high pressures, it can be breathed with complete safety for long enough to taste it.

In air at 10 atmospheres, and sometimes even at 8 atmospheres, a number of subjects reported a taste which is variously described as harsh, metallic, and indefinable. It is certainly not due to oxygen, and one subject who tasted it regularly in air did not do so when mixtures in which the nitrogen of air had been replaced by helium or hydrogen were breathed at 10 atmospheres. We therefore attribute it to nitrogen.

Not all subjects reported these tastes. This was probably often due to the fact that other sensations were distracting them, and to the narcotic effect of nitrogen at high pressures. However, one subject who was repeatedly on the look-out for both tastes has never tasted nitrogen, and only tasted oxygen very faintly at 7 atmospheres. His sense of taste is, however, poor as a result of cerebral concussion.

Vol 148, 84 1941

Sir James Jeans and Sir Arthur Stanley Eddington, two titans of cosmology, engaged in a running argument that was ventilated in Nature. *Here is how it was seen by J. T. C. Moore-Brabazon (later Lord Brabazon of Tara), a pioneer of motor racing and of aviation (and holder of the first pilot's licence to be granted in Britain).*

A correspondence between two men of such astronomical mental calibre as Jeans and Eddington, firing long-range shots at each other, should, I suppose, be read in silence and with respect by the ordinary man in the street. It is indeed very enjoyable in these days to have such a discussion, but first of all we must thank Jeans for the very human confession that he had been "re-reading" Eddington's book. It is comforting to think it was not just our fault we did not get it all the first time.

Jeans sums up Eddington's contention in paragraph 2 by saying that "all those laws of Nature that are usually classed as fundamental, as well as the values of the constants of Nature, can be foreseen 'from epistemological considerations, so that we can have *a priori* knowledge of them' ". "*A priori*" knowledge is given, quoting Eddington again, as "knowledge which we have of the physical universe prior to actual observation of it". From this it is fair to say that Eddington claims that fundamental laws are objective, yet in his answer towards the end he states that there is no such thing as a truly objective law. Eddington should, therefore, challenge Jeans's summary of his main contention, yet he does no such thing.

I have always regretted the Michelson-Morley experiment. Things were perfectly satisfactory before their distressing negative results. I feel we are not at the end of this story, just as we are not at the end of the story that the red shift in the spectrum means receding speed, unless we tie ourselves to the Hilaire Belloc creed and "never, never let us doubt, what nobody is sure about". To say that without the Michelson-Morley experiment we should find ourselves "faced with a universe far more complicated than we have lately imagined" can only be agreed to by those who can dart with such facility from physics to metaphysics.

Finally, a protest against 'plugging' the word 'epistemological'. It is neither pronounceable nor understandable.

J. T. C. Moore-Brabazon.
81 Albert Hall Mansions,
S.W.7.

E. W. Scripture was the foremost authority of his time on the science of phonetics.

Experimental Phonetics and Ancient Greek Verse

Analysis of a macrophonic registration[1] of "Devon to Me!" spoken by John Galsworthy[2] showed that 15 of the 40 lines of the poem had a rhythmic form indicated by $\smile\ \cup\ \smile\ \cup\ \smile$; an example is the line "Where my fathers stood". This is the form known in ancient Greek metrics as the hypodochmius (for example, "Oed. rex", 1208). Twenty lines had the form $\smile\ \cup\ \cup\ \smile$; an example is "Watching the sea". This is the choriambus, a very common ancient form. Five lines had the form $\smile\ \cup\ \cup\ \smile$ as in "Taste of the cream pan!" This is the form termed adoneus. Every line of the poem was spoken with some form of ancient Greek rhythm. The registration of the first stanza of "Drake's Spirit" by Galsworthy him-

self showed that the line "I, Francis Drake" had the rhythm of the third epitrite $\smile\ \smile\ \cup\ \smile$ as in "Medea", 628, and that the line "When the land needs" had the rhythm of the rising ionic $\cup\ \cup\ \smile\ \smile$ as in "Phoen.", 1539.

A registration of "Hickory, dickory, dock" showed a rhythm of $\smile\ \cup\ \cup\ \smile\ \cup\ \cup\ \smile$; this is a hemiepes as in "Medea", 412. The line "Gems of a master's art" (Bridges, "Buch der Lieder") was spoken with the rhythm $\smile\ \cup\ \cup\ \smile\ \cup\ \smile$; this is one form of the dochmius as in "Agamem.", 1166. "Here a little child I stand" (Herrick, "Grace for a Child") registered with the rhythm $\smile\ \cup\ \smile\ \cup\ \smile$; this is a lekytheion as in "Phoen.", 642. "Tell me thou star whose wings of night" (Shelley, "The World's Wanderers") was spoken with the rhythm $\smile\ \cup\ \cup\ \smile\ \cup\ \smile\ \smile$; this is the choriambic dimeter as in "Antig.", 332. "Fear no more the heat of the sun" (Shakespeare, "Cymbeline", 2, 4, 249) had the rhythm $\smile\ \cup\ \cup\ \smile\ \cup\ \smile$; this is the choriambic dimeter as in "Antig.", 107. "A wet sheet and a flowing sea" (Cunningham, "A Wet Sheet . . ."), yielded the rhythm $\cup\ \smile\ \smile\ \cup\ \cup\ \smile\ \cup\ \smile$; this is a form of the glykoneus as in "Iphig. Taur.", 1097. The line "Ein schönes, wohlgewachsnes Buch" (Ginzkey, "Das Buch"), spoken by the poet himself[3], showed the rhythm $\cup\ \smile\ \cup\ \smile\ \cup\ \smile$; this is the iambic dimeter, a common ancient form. "Die Damen im schönen Kranz" (Schiller, "Der Handschuh"), spoken by a German, registered with the form $\cup\ \smile\ \cup\ \cup\ \smile\ \cup\ \smile$; this is the ancient telesillion.

The lengths of the vowels in ancient Greek verse are known. The examples given here seem to justify the conclusion that, wherever the lengths of the vowels in ancient Greek verse correspond with the lengths in English and German verse as determined by registration and measurement, the rhythmic scheme is the same. Many—perhaps all—the ancient forms may be established on this principle. Greek metrics thus becomes an exact science consisting of numbers obtained by measurements.

E. W. Scripture.
Phonetic Laboratory,
62 Leytonstone Road,
London, E.15.
June 28.

[1] See Nature, **132**, 138 ; 1933.
[2] Janvrin, "Analyse von zwei von John Galsworthy gesprochenen Gedichten", *Z. Exper.-Phon.*, **32**, I, 147 ; 1930.
[3] Scripture, "Anwendung d. graphischen Methode auf Sprache u. Gesang", 73, Leipzig, 1927.

Another correspondent enlarges on William Buckland and the verses (p.230). He reproduces, besides these anecdotes, a long and convoluted poem with the title "Ode to a Professor's Hammer", which readers will wish to be spared.

Buckland's own comment upon it in a letter which he wrote to Miss Jane Talbot at Penrice Castle in Gower on December 11, 1820. Buckland had just returned from a long Continental trip which, he said, had been "less adventurous in the line of imprisonments and banditti than the last, but had not been lacking in curious incidents". He then went on : "In contemplation of the possibility of my remaining underground upon the Continent for ever, I found upon my return that Mr. Whateley of Oriel had composed for me an elegy which I am happy to have it in my power to forward you a copy."

* * *

"I received last week a poetical epistle from my friend Mr. Shuttleworth containing one of my lectures done into rhyme. It is extremely neat and full of humour, and will be highly entertaining to you."

* * *

It being the object of the versifier to produce at present merely a specimen of his intended work, he has omitted the following fifty verses, exclusively geological, concluding with :

> Those Bones I brought from Germany myself
> You'll find fresh specimens on yonder shelf,

and also a digression of 2,300 lines of which the concluding couplet is :

> So, curl the tails of puppies and of hogs
> From left to right the pigs, from right to left the Dogs,

and also for the same reason, the subsequent and still more digressive digression, which is terminated by the following admirable reflexion—the whole passage consisting of 5,700 fresh lines :

> And not wild, but *tame* cats only, teaze their prey.

The concluding couplet which is given without any alteration from the mouth of the learned Lecturer is here subjoined solely because it serves as an additional proof, if such were wanting, of the close connexion which subsists between geological speculations and not the ideas only, but also the language, of complete poetry. It will be observed that tho' intended only as a common sentence of adjournment, it has all the fluency and grace of the most perfect rhythm, and of its own accord 'glides into sense and hitches in a rhyme' :

> Of this enough—on Secondary Rock
> To-morrow, Gentlemen—at 2 o'clock."

VOL 142, 1040 1938

An anonymous correspondent relates how the depredations of the Nazi regime have brought low a proud institution, the ancient University of Heidelberg.

THE general character of the changes in the German universities was expounded by Herr Rust, Reichsminister for Education, at the recent celebration of the 'jubilee' of the University of Heidelberg. His address is translated in the November issue of the *Universities Review*. Herr Rust believes the New Germany to be the true heir of Sparta and suggests that these changes are as though Sparta had triumphed over Athens. Had that calamity befallen the world, all that we could have inherited from Greece would have been 'discipline', for Sparta had no other gifts to bestow.

But the new "Weltanschauung", Herr Rust assures us, "is the life-blood of a new science . . . National Socialism has provided science with new principles from which she can derive the strength of self-confidence. . . . The old idea of science based on the belief in the supremacy of the intellect is finished."

The article then describes the indignities heaped on the academic faculty, enumerates the distinguished scholars dismissed on grounds of race or political sympathies and ends with this resounding peroration:

Faced with a series of unprecedented attacks on the interest of learning as well as on their own rights, no member of the staff has had the temerity to utter any word of public protest. There are several retired or very senior members of independent standing from whom one might have reasonably hoped for some sign of courage. But Germany has openly reinstated serfdom, the essence of which is to restrain the worker from marketing his labour as he will and to place him at the disposal of an overlord. It is, therefore, appropriate that the minds of Germans should be enslaved along with their bodies, for slavery mercifully provides its own spiritual anæsthesia. This, in truth, is the 'new principle' for science which Reichsminister Rust has discovered. But, to speak yet more plainly, science has been abolished in the German universities and its spirit has abdicated from the Reich.

VOL 139, 98 1937

J. C. (afterwards Sir Jack) Drummond was the leader of a group of nutritional biochemists who planned the British diet during the privations of World War II. So successful were they that the health of the nation was better during this ordeal than ever before. Not long after the end of the war, Drummond, camping in France, was murdered along with his family by a Breton peasant. Here is the beginning of an article by Drummond on the value of margarine; after analysing the nutritive value of the dismal substance, he finishes up with the following:

EARLY HISTORY

IT is unlike the French to be unmindful of a son of their country who has rendered notable service in so important a matter as food, yet I am not aware that the achievement of Mège-Mouries is commemorated in France by a public monument or inscription or even by a street name. Equally surprising is the scanty information supplied by biographical works of reference about that most ingenious inventor. Perhaps we can attribute this to the lack of sympathy one would expect from a nation which appreciates above all other things good food and good cooks, towards one who hoped to pass off the greasy and rather unpalatable products of his laboratory as butter. Whatever the reason for such neglect, the fact remains that the French might well be proud to claim the inventor who made possible one of the greatest food industries of our time.

It is often erroneously believed that Mège-Mouries set out to concoct a substitute for butter. It is true that he was stimulated by the desire of the Victualling Department of the French Navy to find such a substitute ; but he himself was more ambitious. He attempted to produce butter itself, and that by a series of laboratory treatments that he thought would reproduce exactly the changes by which he believed body fat to be converted into milk fat in the animal body. His laborious process, in which beef fat was digested at blood heat, first with macerated sheep's stomach and then with chopped cow's udder and milk, actually gave him nothing more unusual than the softer portion of the original fat, carrying, presumably, some traces of a milky flavour. Nevertheless, in the early days he was quite satisfied that he had made butter. Others, less optimistic, were content to believe that he had solved the problem of preparing an edible substitute for butter, and for this achievement he was honoured in 1870 by Napoleon III.

Had it not been for the Franco-Prussian War, it is probable that France would have led the world in the technical development of the new process. As it happened, the opportunity was seized in the United States and in Holland. In a remarkably short time factories were springing up in these countries for the manufacture of what was then called 'butterine'. It was soon found that Mège-Mouries's complicated digestion was quite unnecessary, and that all it was necessary to do was to mix as thoroughly as possible with skimmed milk, preferably slightly sour, a fat of the appropriate melting point. On separating the fat again it was found to have acquired a buttery taste. It could then be 'worked', salted and coloured to taste.

Manufacture in Great Britain lagged far behind in the seventies. So far as can be judged from the contemporary press, the home-produced material was, moreover, of very inferior quality. Most of it was made in small, dirty, 'back-street' premises under highly insanitary conditions. To a large extent it was knowledge of these facts that gave 'butterine' so bad a reputation at that time. The English public was ready to believe any fantastic yarn about the doubtful nature of the fats used by the manufacturers, and they also suspected, often with more reason, that a good deal of the despised stuff was employed as an adulterant for butter. The passing of the Margarine Act in 1887 did something to allay uneasiness ; nevertheless, the reputation of the new food remained an unsavoury one for a good many years.

* * *

The product Monsted turned out was made from that fraction of beef fat which is known as 'oleo' or 'premier jus'. It was, in fact, a highly digestible fat possessing a calorific value equivalent to that of ordinary butter and, what was of great importance although unknown at the time, its vitamin A, and probably also its vitamin D, content was by no means negligible.

* * *

With rationing of butter in force, and having in mind the limited purchasing power of the poorer sections of the community, the desirability of enriching all margarines with vitamins A and D supplements must appear obvious.

At the present time a large proportion of the margarine sold over retail counters is so enriched. Unfortunately, there does not appear to be uniformity between the various brands, although for the purpose of general estimates such as have been made in this article, the content of an 'average' vitaminized product can be taken as 450 units of A and 30 units of D per ounce. A weekly supplement to the butter ration of 6 oz. of reinforced margarine could be regarded, therefore, as adding some 2,700 units of A and 180 units of D to the diet. It is certainly nothing to boast about ; but one must remember how valuable is every unit where poverty diets well below the marginal nutritional level are concerned.

It will not be possible to plead ignorance this time if vitamin deficiency is allowed to be responsible for a decline in general health such as occurred during the years of 1914–18.

VOL 145, 53 1940

chapter fifteen
1942-1953

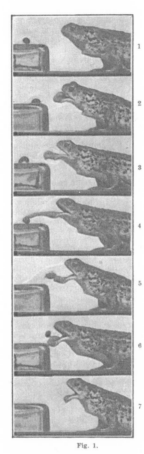

Fig. 1.

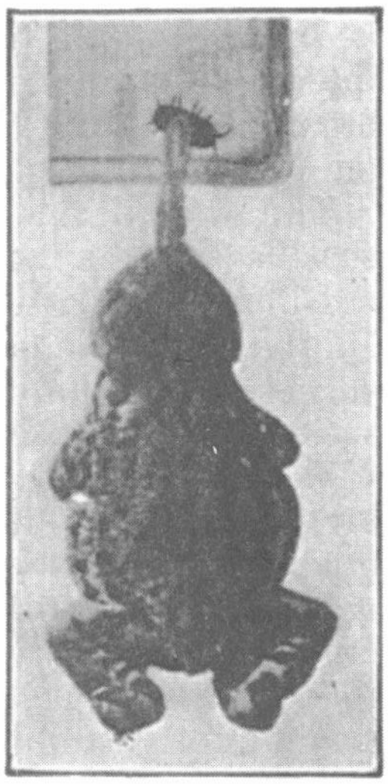

Fig. 2.
TOAD FEEDING : FROM ABOVE.
TONGUE FULLY EXTENDED.

JOSEPH NEEDHAM SURVEYS CHINESE SCIENCE IN ADVERSITY. GEOLOGISTS CONDUCT RESEARCH IN A PRISONER-OF-WAR CAMP. EUROPEAN SCIENCE COUNTS THE COST OF ITS LOSSES IN THE WAR. A. V. HILL DEPLORES THE RIFT BETWEEN THE SCIENTISTS OF RUSSIA AND THE WEST. JULIAN HUXLEY DISSECTS MARXIST DOCTRINE ON GENETICS. SIR CHARLES VERNON BOYS AND THE EARL OF BERKELEY ARE TREATED TO FULSOME OBITUARIES. THE STRUCTURE OF THE GENETIC MATERIAL IS SOLVED AND THE NEW AGE OF BIOLOGY DAWNS.

1942-1953

The tide of war turned: the Germans were defeated in North Africa and at Stalingrad. The Western Allies landed in Normandy. The armies met in Germany and the war in Europe ended. The Hiroshima and Nagasaki atomic bombs brought an abrupt end to Japanese resistance. The League of Nations was succeeded by the United Nations and the Iron Curtain descended on Europe. India became independent and was partitioned. The Chinese communists ousted Chiang Kai-shek. The Korean War was fought and Senator Joseph McCarthy's House UnAmerican Activities Committee did its worst. The nuclear arms race began.

The Earl of Berkeley was a rare product of the aristocracy, more interested in science than hunting. There was supposedly a component of Berkeley in Aldous Huxley's Lord Edward in Point Counterpoint, *but J. B. S. Haldane's father, J. S., also saw himself (to his annoyance) in the same character. This obituary describes Berkeley's background as well as his work on osmotic pressure, for which he is remembered.*

The Earl of Berkeley, F.R.S.

RANDALL THOMAS MOWBRAY BERKELEY, eighth Earl, was born on January 31, 1865, and died on January 15. He was head of the historic house which acquired the Berkeley lands in the twelfth century, and is among the very few that can rightly claim a pre-Conquest pedigree ; even the critical genealogist Horace Round allows the Berkeley descent from Eadnoth the Staller, an officer of the household of Edward the Confessor.

In feudal days the Castle of Berkeley was held directly of the King, so that its owner was a baron by tenure, a status which in 1421 was merged in a summons by writ and a hereditary peerage. Since such baronies can be held and transmitted by females, on the death of the sixth Earl in 1882 it passed to his niece, wife of General Milman, who thus became Baroness Berkeley, while the earldom, created in 1679, went to a cousin as heir-male. For some seventy years this title had not been assumed, while there was doubt about its inheritance, leading to a famous lawsuit. Meanwhile the lands were in possession of the Lords Fitz-Hardinge. But in 1891 Randall Berkeley established his claim to the earldom before the Committee of Privileges.

Lord Berkeley began his active career in the Royal Navy and reached the rank of lieutenant. But being drawn to scientific research, he left the Service and built a laboratory at Foxcombe, on Boar's Hill near Oxford, a house which he much enlarged, adding a really fine stone hall. It is now the Theological College known as Ripon Hall. There he carried out his experiments, the success of which was shown by his election to the Royal Society in 1908.

Berkeley's work centred around the idea of measuring the physical properties of concentrated solutions, in order to test the possibility of applying to them equations of the type of that used by van der Waals for gases, vapours and liquids. The fundamental determination required is obviously that of osmotic pressure, either measured directly, or deduced by the principles of thermodynamics from the measured vapour pressures. To obtain accuracy in the results, many and great experimental difficulties had to be overcome. The original method of directly measuring osmotic pressure, invented by Pfeffer, is quite unsuitable when concentrated solutions are used. Instead of allowing solvent to enter a porous porcelain cell closed by a semipermeable membrane of copper ferrocyanide until equilibrium is obtained, Berkeley and his colleague E. G. J. Hartley deposited the membrane in the walls of a porous tube with its ends sealed, and varied the hydrostatic pressure on the solution until no further movement, either in or out, occurred. From 1904 until 1919 a series of papers, most of them in the *Proceedings* or *Transactions of the Royal Society*, appeared from the Foxcombe laboratory, gradually improving the methods and results, and dealing with other allied subjects, such as the vapour pressures and densities of solutions of varying temperature and concentration.

A remarkable galaxy of academic stars inveighs against the public school system, a fight for which today's socialists have no stomach.

The system of public school education is undemocratic. It segregates those children who, by inheritance, proceed to leading positions in industry, politics. the civil services, the armed forces, and by its training strengthens their social privileges. It is socially injurious, since it divorces this section of our people from the rest, from the life of the main part of the community, especially from all those engaged in the productive processes and manual and technical labour. Men brought up in the public school tradition are out of touch with modern social realities and often incapable of grasping modern social problems.

The prospect of endowing these schools with Treasury grants is usually linked with a proposal to admit into these schools a limited number of the more gifted members of the poorer sections of society. This proposal is born as much from the financial crisis of the public schools as from an awareness of their social responsibilities. If it were adopted, the main results would be :

(*a*) a direct subsidy given by the taxpayer to the schools of the wealthy ;

(*b*) the psychological dislocation of the gifted poor boys, who would thus be in the position of poor relation in the family of the rich ;

(*c*) the stultification of the gifts of these hostages, since in this atmosphere they would lose contact with the life and problems of the mass of the people ;

(*d*) the intellectual impoverishment of the State-aided secondary schools which would thus lose their most gifted pupils.

* * *

We call, therefore, for the full incorporation of the public schools into a unified, State-aided system of secondary education as a step towards democracy in our educational system and towards greater health in our national life.

J. D. BERNAL, P. G. H. BOSWELL, S. BRODETSKY, J. CHADWICK, W. E. LE GROS CLARK, F. A. E. CREW, C. LOVATT EVANS, B. FARRINGTON, C. B. FAWCETT, P. SARGANT FLORENCE, P. F. FRANKLAND, LANCELOT HOGBEN, JULIAN HUXLEY, R. K. KELSALL, R. D. LAURIE, JOSEPH NEEDHAM, R. PASCAL, J. A. RYLE, F. SODDY, H. G. STEAD, GEO. THOMSON, F. E. WEISS, F. WOOD-JONES.

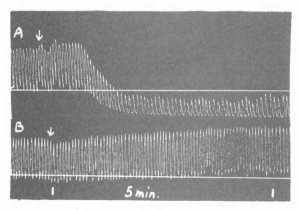

A, CLOVER JUICE : *B*, GRASS JUICE

This from the exuberant Sir Charles Vernon Boys at 87:

"Why does Sir C. V. Boys elect
To do the things we least expect ?"
Sir Richard gave the explanation*.
But how ? Well, here's an illustration—
"A photograph which cannot lie"
Shows clearly that he does it by
A flight of pure imagination.

VOL 150, 464 1942

Boys's inventiveness did not flag:

Mechanical Lighters

WHEN about a year ago we found it difficult to get matches I thought of a plan for producing a light especially of value where gas is available, which if brought into use would mean economy in chlorate of potash and elimination of match-sticks and match-boxes—surely a desirable end. But then the definition of a mechanical lighter in the 1928 Act of Parliament was shown to me, which would impose a 2s. 6d. duty on an article which otherwise could be sold for a penny. I concluded, and the recent Bow Street Police Court tinder box case has shown that I was right, that the Excise would not hesitate to impose the duty on any lighter however advantageous it might be.

My plan was to make a paste of chlorate of potash with so much diluted silicate of soda as would, when dry, make a firm button which could be glued at the end of a rod and used to strike a spark against a safety match-box, or against a sheath coated with the same red phosphorus mixture. Hung up by a gas burner, it would light the gas with the expenditure of perhaps one tenth or one twentieth of the chlorate in a safety match head. To my surprise, I found it would light a methylated spirit lamp the only time I ever tried.

C. V. BOYS.

St. Marybourne,
Andover.

VOL 151, 336 1943

In 1945, Boys died in his ninetieth year. He was an institution and a physicist of a now probably extinct breed. C. T. R. Wilson (of the cloud chamber) wrote his obituary.

In 1897 Boys became one of the Metropolitan gas referees. He greatly improved the methods of gas calorimetry, and the calorimeter described by him in the *Proceedings of the Royal Society* in 1903 was adopted as the standard instrument for testing London gas ; it came into general use in gas-works throughout Great Britain. He devoted much thought during many years to the planning of a still better gas calorimeter ; but it was not until 1934 that he finally arrived at a design which completely satisfied him. This was described in his Guthrie Lecture of that year. Boys tells in this lecture that the idea underlying one important part of the mechanism came to him in a dream. "I was sufficiently impressed by it to get up at six and go to Victoria Street, where I blew in glass the bulb and tube you now see". He was then in his eightieth year.

* * *

In spite of the handicap of the loss of one eye and very defective vision in the other, Boys continued his varied scientific activities until the end of his long life ; when he was eighty he published little books on the natural logarithm and on weeds. It was in this year that he received his knighthood.

Boys does not appear to have been greatly interested in theoretical physics. His delight was in designing, constructing and manipulating apparatus for physical measurements of the highest accuracy, and in overcoming experimental difficulties which to most would have seemed insuperable. He was a really great experimenter, and his methods of working were original and often unconventional. He appears to have been equally original and unconventional in ordinary life.

VOL 155, 40 1945

Irving Langmuir's presidential address to the American Association for the Advancement of Science, with the title "Science, Common Sense and Decency", appeared as an article in NATURE, *whence this little moral tale.*

The formation of crystals on cooling a liquid involves the formation of nuclei or crystallization centres that must originate from discrete, atomic phenomena. The spontaneous formation of these nuclei often depends upon chance. At a camp at Lake George, in winter, I have often found that a pail of water is unfrozen in the morning after being in a room far below freezing point, but it suddenly turns to slush upon being lifted from the floor.

Glycerine is commonly known as a viscous liquid, even at low temperatures. Yet if crystals are once formed they melt only at 64° F. If a minute crystal of this kind is introduced into pure glycerine at temperatures below 64°, the entire liquid gradually solidifies. During a whole winter in Schenectady I left several small bottles of glycerine outdoors and I kept the lower ends of test tubes containing glycerine in liquid air for days, but in no case did crystals form. My brother, A. C. Langmuir, visited a glycerine refinery in Canada which had operated for many years without ever having any experience with crystalline glycerine. But suddenly one winter, without exceptionally low temperatures, the pipes carrying the glycerine from one piece of apparatus to another froze up. The whole plant and even the dust on the ground became contaminated with nuclei, and although any part of the plant could be temporarily freed from crystals by heating above 64°, it was found that whenever the temperature anywhere fell below 64°, crystals would begin forming. The whole plant had to be shut down for months until outdoor temperatures rose above 64°. Here we have an example of an inherently unpredictable divergent phenomenon that profoundly affected human lives.

VOL 151, 268 1943

Joseph Needham's acquaintance with Chinese science was furthered by his travels in the war. He wrote a series of articles describing how Chinese scientists were faring in different parts of their country during a time of extreme deprivation. Here is a sample.

SCIENCE IN SOUTH-WEST CHINA
I. THE PHYSICO-CHEMICAL SCIENCES
By DR. JOSEPH NEEDHAM, F.R.S.
British Council Cultural Scientific Mission in China

OWING to the course of the War and world conditions, Free China has now been isolated from the rest of the world for such a long time that an account of what our Chinese scientific colleagues have been doing will surely be of interest to readers of NATURE both in Britain and the United States.

* * *

The scientific institutions are well scattered at different distances from the city, and transport is by horse-cart, charcoal-burning bus, 'jeep', lorry or car.

* * *

All the departments are housed in 'hutments' built of mud brick, and roofed very simply with tiles or tin sheets, though some have curving roofs in the great tradition of Chinese architecture. Inside, the floors are beaten earth, with a little cement, and extreme ingenuity has been used in fitting up laboratories for research and teaching under these conditions. For example, since no gas is available, all the heating has to be done with electricity, and hence when the supply of element wire for heaters (home-made out of clay) ran out some time ago, work was at a standstill until it was found that gun lathe shavings from one of the Yunnan arsenals would do very well. When hæmatoxylin became unobtainable, it was found that a dye something very like it could be obtained from an orange-coloured wood native to Yunnan, *Cæsalpinia sappan*. When microscope slides could not be had, window-panes broken by air raids were cut up, and the unobtainable cover-slips were replaced by local mica. For glass-blowing, the blowpipes are fed with the vapour of power alcohol (derived from molasses) passed through electric furnaces. Instances could be multiplied at length.

VOL 152, 9 1943

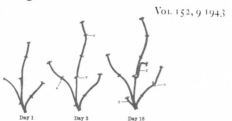

Day 1   Day 3   Day 18

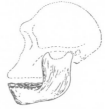

Professorial chairs in the Scottish universities were still for the most part not subject to age limits.

Sir D'Arcy W. Thompson, C.B., F.R.S.:
A Professorial Record

"YOU will never live to my age, without you keep yourselves in breath with exercise, and in heart with joyfulness"—and so successfully has Sir D'Arcy Thompson fulfilled the injunction of Sir Philip Sidney that ere Christmas Day he will have completed sixty years as professor of natural history. On December 22, 1884, at the age of twenty-four, he was elected, as its first incumbent, to the chair of natural history in the newly opened University College of Dundee.

* * *

His predecessor retired in his seventy-ninth year; in his eighty-fourth Sir D'Arcy continues to teach with vigour and to take part in many activities outside the University.

VOL 154, 761 1944

With the end of the war, the uneasy alliance with the Soviet Union began to crumble and relations with scientists too were affected.

Cancelled Visit of British Men of Science to the Academy of Sciences of the U.S.S.R.

MANY readers of *Nature* will have been astonished and repelled by the studied discourtesy with which eight of the intending guests of the Soviet Academy of Sciences were prevented by His Majesty's Government last week from going to Moscow. Not only were they put to gross inconvenience and annoyance by the refusal, without warning and at the last moment, of permission to travel, but also the explanation given was as incredible as the real reason was insulting.

In this prohibited group were those whose talents and devotion have rendered priceless service to the nation during the War. But let us remember the words of the "Preacher":

"There was a little city and few men within it; and there came a great king against it and besieged it and built great bulwarks against it:

"Now there was found in it a poor wise man and he by his wisdom delivered the city: yet no man remembered that same poor man.

"Then said I, Wisdom is better than strength: nevertheless the poor man's wisdom is despised, and his words are not heard".

The offensive treatment of our scientific colleagues, inconceivable towards members of most other professions, is a sufficient comment on the patronizing Ministerial praise with which science and scientific men are occasionally favoured. When they, and others, are offered reparation later on in 'awards' or 'honours', let them recall the words of T. H. Huxley:

"The sole order of nobility which, in my judgment,

becomes a philosopher, is the rank which he holds in the estimation of his fellow-workers, who are the only competent judges in such matters".

A. V. HILL.

16 Bishopswood Road,
London, N.6.

Vol. 155, 753 1945

Sir Henry Dale (the discoverer of acetylcholine), as president of the Royal Society, contributed this historical morsel to NATURE.

A HITHERTO UNPUBLISHED LETTER OF ISAAC NEWTON

THE Royal Society recently received, on indefinite loan from its present owner, a hitherto unrecorded autograph letter of Isaac Newton. The letter is a long one, it has an intrinsic interest for the history of science, it is in a remarkably good state of preservation and its authenticity is established with unusual completeness. It was written in 1677 from Cambridge to the Hon. and Rev. Dr. John North, then living in London, but later Master of Trinity College, Cambridge. The present owner of the letter, Mr. Roger North, of Rougham, King's Lynn, Norfolk, who has placed it in the custody of the Royal Society, makes the very probable suggestion that the "new Treatise of Musick", with which the letter deals, was "A Philosophical Essay on Music", by Francis North, Lord Guilford, to whose brother the letter is addressed.

H. H. DALE.

Here is a passage from Newton's letter.

Pag 5 lin 14 & pag 6 lin 30, the Author asserts y^t sound is produced in ye Torricellian vacuum, & thence seems to collect y^t ye medium of sound is not ye grosser Air but some subtiler aerial fluid of a middle nature between y^t Air & AEther, wch can penetrate glass & other gross bodies. But it is to be suspected y^t this experim^t of ye Torricellian Vacuū holds only when ye glas is not well emptied of Air. For Mr Boyle (Exp^t 27) repeating it wth a watch hung in his Receiver, found that as ye receiver was more & more emptied, ye noise made by ye Ballance wheel grew fainter & fainter till at last it was not heard at all though ye handle & wheels of ye watch were still seen to continue their motion as freely as at first. Yet ye louder sound of a Bell continued audible when ye air was drawn out, though perhaps it would not have done so could ye Receiver have been fully exhausted of Air, & ye Bell have been susteined in ye Receiver by something wth might not touch ye glas. To ye best of my remembrance I have also some where read of an Alarm Watch whose Alarm being made to go, ye sound of ye Bell grew faint by drawing out ye air.

Pag 5 lin 18 The Author asserts y^t a sound seems to come in streight lines to ye ear though an obstacle be situated between ye ear & ye sounding body so y^t it cannot come in streight lines. But this I doubt of if, he mean y^t it seems to come in streight lines from ye sounding body. If indeed ye interposed obstacle be not too gross & compact, suppose a glass window or thin wall of wood or mortar, ye air may by shaking it propagate ye sound through it, & then ye sound will be heard in a streight line from ye sounding body: but if ye obstacle be massy, suppose a very steep & high hill or a sollid high wall of brick or stone, I am apt to think ye sound will seem to come from ye top of ye Hill or Wall rather then in a direct line from ye sounding body behind it. And some such diverting of sounds I have observed occasionally, in walking on a street close by a single house whilst bells were ringing on ye other side. The house to ye best of my remembrance had no windows on that side next me & ye sound of ye bells seemed to come from y^t end of ye house wch I was nearest to, though ye bells were directly behind it so in a Room of stone walls wth but one window, the sound will seem to come from ye window though ye sounding body without lye not that way.

To conclude:

I am much obliged to you for giving me notice of the objection made against my notion about colours. But ye experiment succeeds otherwise then tis reported. If you place your eye where ye blew light falls on ye wall so y^t a by-stander see your eye of a blew colour you will at ye same time see ye Prism of a blew colour; and so if you place you eye in ye red light you will see ye Prism red. What colour a By-stander sees fall on yor eye you will see at ye Prism: as I can affirm by iterated experience.

The last week I called at yor Lodgings & hope ere long to have an opportunity to wait on you again. In ye meane time I rest wth my thanks to you for yor kind acceptance of my former Letters

Yo^r humble Servant
& honourer
Is. Newton

Vol. 156, 193 1945

The war over, scientists were beginning to count the cost. The appalling losses, especially in Eastern Europe, make sombre reading. Similar roll-calls of the dead appeared for other countries.

Losses among Polish Zoologists during the War

Among more prominent names may be mentioned : Prof. K. Białaszewicz (University of Warsaw, physiology, died 1943), Prof. T. Garbowski (University of Cracow, animal psychology, died 1940 in concentration camp, Oranienburg), Prof. E. Godlewski, jun. (University of Cracow, embryology, died 1944), Prof. S. Kopeć (University of Warsaw, general biology, executed 1941), Dr. Z. Koźmiński (hydrobiology, killed in action 1939), Prof. R. Kuntze (Warsaw School of Agriculture, economic zoology, executed 1944), Dr. S. Minkiewicz (economic entomology, died 1944), Prof. W. Roszkowski (University of Warsaw, general and systematic zoology, executed 1944), Prof. M. Siedlecki (University of Cracow, general zoology, died 1940 in concentration camp, Oranienburg), Dr. J. Wiszniewski (hydrobiology, killed in action 1944). It is feared that the above data are still not quite complete, as the fate of several persons who left the country in 1939 or later, or were forcibly displaced by the Germans, is not yet certain.

Vol. 157, 258 1946

Nor did German scientists escape unharmed. The ethologist Konrad Lorenz, it became known, had aligned himself with an association of German biologists for Hitler. In prison he did not waste his time, and wrote from memory on scraps of paper a textbook on animal behaviour, rediscovered only after his death and just published.

Fate of German Ornithologists

Mrs. Margaret Nice, the well-known American ornithologist, has heard from Dr. Stresemann that Dr. Konrad Lorenz, one of the pioneers in the study of behaviour, notably in birds, whose probable death was reported in *Nature* of November 10, 1945, p. 578, is alive. He is a prisoner in Russian hands, unwounded, and employed as a camp physician. The same letter also contains news of other German ornithologists : Dr. Steinfatt is in Denmark, but will shortly be returning to Germany ; Dr. Meise is alive, but still in Russian hands.

Vol. 157, 510 1946

Here by way of contrast is an instance of humane actions in hard times.

Japanese Men of Science in Malaya during Japanese Occupation

The circumstances of the publication of C. F. Symington's "Foresters' Manual of Dipterocarps", recently reviewed in *Nature*[1], are known to very few persons, but they are interesting enough to be recorded in detail. The Manual was issued from Raffles' Museum, Singapore (Syonan Hakubutukan), towards the end of 1943 and was on sale solely for *bona fide* men of science. At that time, the Japanese Military Administration and the Syonan (Singapore) Municipality were endeavouring harder than ever to stamp out all traces of the British, even their language. That the Manual was published, and that there was a stock of some 280 copies for the British in September 1945, we owe to the far-sightedness, influence and discretion of a few Japanese men of science.

The acting director at Raffles' Museum in 1942 was Prof. Hidezo Tanakadate, of Tohoku Imperial University, Sendai. He obtained the temporary release of Mr. H. E. Desch, of the Malayan Forestry Service, from the Changi Military Camp and, at the end of June, took Mr. Desch to Kuala Lumpur, where he found the galley-proof of the Manual. It was decided to publish the work (500 copies) on the ground that it would be more likely to survive the War in that way than as a single galley-proof, for the whereabouts of Mr. Symington and his manuscript were unknown. The cost of printing was met personally by Prof. Tanakadate and by Marquis Yositika Tokugawa, who acted as president of the Museum and Library. It was insisted by Prof. Tanakadate that the book should conform exactly with the previous series of *Malayan Forest Records*, of which it is No. 16, so that it should stand the test of time, as a scientific work, regardless of hostilities and racial prejudice. He therefore added a brief preface, as a single page of romanized Japanese, and he issued the Manual from the Museum to give it official standing and to prevent pilfering of the stock by what he called 'common people'.

The proofs were read mainly by Mr. Desch, even after his return to the Military Camp in January 1943. The Japanese officers who succeeded Prof. Tanakadate, namely, Prof. Kwan Koriba and Dr. Y. Haneda, took the proofs personally to the camp and fetched them again on correction. As the printing was continued by the Caxton Press in Kuala Lumpur, great care had to be taken in sending the proofs from Singapore, for there was a very strict censorship and the post was unreliable. Japanese staff officers travelling to and fro carried them personally, while duplicates were kept at the Singapore Botanic Gardens. The co-operation of military officers was possible only because they were known personally to the professors as students or colleagues.

Similar action was taken by Dr. Koga, the director of the Tokyo Zoological Gardens, in publishing M. W. F. Tweedie's "Poisonous Animals of Malaya", which was rescued from the broken and looted premises of the Methodist Publishing House in Singapore. A large remainder is also at Raffles' Museum.

In the interest of science, one must distinguish carefully between the 'Japanese' of popular conception and the Japanese men of science, who in Malaya, at least, endeavoured to serve science with impartiality.

E. J. H. Corner

15 The Park,
Great Shelford,
Cambridge.
May 28.

[1] *Nature*, **157**, 671 (1946).

The death of Philipp Lenard, old, alone and reviled, was marked by a penetrating obituary by E. N. da C. Andrade. Here is its conclusion.

Lenard was personally a difficult man, whose character contained many contradictions. An intimate friend of mine who knew him well once wrote to me : "Was Lenard betrifft, so ist er so klug und so dumm wie immer". He was profoundly disappointed not to have discovered the Röntgen rays, which he had almost under his hand and would have in all probability found within a year or so of the actual date if Röntgen had not anticipated him. He never used Röntgen's name in referring to the rays. He took as a personal affront any inadequate acknowledgment of his work and was incapable of any generosity or even justice towards anyone who, in his opinion, had failed to appreciate any part of his services to science. Although he owed much of his success to Jews, for example, Hertz and Könisberger, and at one time freely acknowledged the debt, he became a bitter anti-Semite and even treated Einstein as not far from an impostor. He refused to fly the Institute flag at half-mast when Rathenau was murdered and was with difficulty saved from popular indignation. He became a whole-hearted supporter of the Nazi regime and of the 'German physics' movement : in fact, he wrote a book called "Deutsche Physik". Yet he had a kindly side to his nature and was often a pathetic rather than a menacing figure. He possibly felt a deep personal need of friendship which he was unable to win or, if he could win it, to retain. His lack of trust in others, his failure to awaken the self-reliance or sympathy of those working under him were the cause that he did not found a great school of physics. It has been suggested, probably correctly, that the clue to his character was that he was a weak personality that sought to protect itself by a hard shell.

As an experimental physicist, Lenard was certainly one of the greatest figures of his time. His work on the physics of the electron was distinguished by a masterly experimental technique and his discoveries had a profound influence on the course of physics, in particular his work on the release of electrons by electron impact and by light. Yet he seemed fated never to achieve supreme greatness. He missed the discovery of Röntgen rays ; he came near to the discovery of the true structure of the atom, but just went astray, and his work on light emission was the first to indicate the important part which the release and return of the electron played, but left to Bohr the great advance. He was a whole-hearted enthusiast for experimental physics, whose appreciation of the great men of science of the past times was generous and informed, as can be seen from his book "Grosse Naturforscher" (translated into English under the title "Great Men of Science"). He was a dark genius beclouded by strong personal fears, doubts and envies, but undoubtedly a genius and one who has left an abiding impression in physics.

E. N. da C. Andrade

Vol. 160, 896 1947

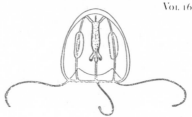

In Stalin's Russia, denunciations of "idealist" or "cosmopolitan" science were all too closely reminiscent of Philipp Lenard's and Johannes Stark's caperings in Nazi Germany only a decade before.

The Crisis in Soviet Science

LAST August scientific workers all over the world heard with deep disappointment that the Soviet Union had officially adopted an isolationist attitude on certain branches of biology. For the first time in the U.S.S.R. there was established a 'party line' in one of the natural sciences. Since then there has been speculation as to whether this attitude might extend to other natural sciences, and a recent broadcast from Moscow gives point to these speculations. On January 26, 1949, the philosopher Alexander Alexandrovitch Maximov, who is a corresponding member of the Academy of Sciences, and who belongs to the staff of its Institute of Philosophy, gave a broadcast on the Moscow Radio Home Service. The theme of his talk was the correct Bolshevik attitude to natural science. He attacks those foreign physicists who "regard as synonymous the philosophical definition of matter and the objective idea of reality", and who are responsible for other "idealistic misinterpretations" in relativity and quantum theory. He indicts by name Einstein, Niels Bohr and Heisenberg. He warns his listeners against the "Kantian acrobatics of modern bourgeois atomic physicists". He contrasts the ideology of these "social traitors" in capitalist countries with the scholars in capitalist countries who "raise their voice in support of genuine science, of a scientific materialist outlook"; and he cites with approval Langevin, Joliot-Curie, Blackett, Haldane and Levy. The purpose of the broadcast was twofold: (a) to emphasize the importance of a correct philosophical approach to physics, based on Lenin's famous "Materialism and Empiriocriticism", and (b) to encourage an attitude of "militant intransigence towards bourgeois idealistic philosophy and sociology".

VOL 163, 354 1949

A detailed article in 1952 set out what the reasoning was. Linus Pauling's resonance theory of the chemical bond came in for particular opprobrium, and the deadly epithet "idealistic" was applied to it, to Pauling himself and to several of Russia's best physical chemists, such as Syrkin, Dyatkina and Volkenshtein, who were made to suffer for their errors.

It should be noted that, throughout the discussions, the terms 'resonance', 'resonance structure', 'resonance energy', etc., are used almost entirely to indicate the objective concepts to which exception is taken. Thus, at the Conference, the mathematical physicist E. I. Adirovich said[6]: "The Heitler–London method, since it reflects an objective reality, cannot be regarded as idealistic. It is not on quantum mechanics or on quantum-mechanical calculations that idealism in physics and chemistry rests, but on scientific and philosophical perversions of these. The theory of resonance may be taken as an example".

The view was expressed that the adoption of the idealistic theory of resonance has led to the replacement of investigation into real factors determining molecular structure and properties by fruitless investigation into separate 'resonance structures', none of which has any physical reality; it is considered, therefore, that the theory has impeded the progress of chemical science.

* * *

Finally, it is objected that the theory of resonance is mechanistic. It has attempted to reduce chemical phenomena to the mechanics of the electron, and this violates the principle, developed, for example, by Engels, that a higher form of movement cannot be reduced to a lower form of movement. It is here that a parallel is drawn between resonance theory and Mendelian theories in biology.

VOL. 169, 862 1952

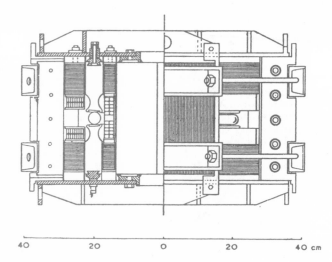

40   20   0   20   40 cm

In the biological sciences, and especially in genetics, matters were of course far more devastating. The sinister jargon of totalitarianism made rational discourse impossible. W. H. Auden had put it like this:

> The ogre does, as ogres can,
> Deeds quite impossible for man.
> But one prize is beyond his reach:
> The ogre cannot master speech.
> Across the dark and ravaged plain,
> Among the desperate and the slain,
> The ogre strides with hands on hips,
> With drivel gushing from his lips.

Julian Huxley wrote a long and masterly analysis of the politics of Soviet biology. It begins:

SOVIET GENETICS: THE REAL ISSUE

BY DR. JULIAN HUXLEY, F.R.S.

NOW that the long-drawn-out dispute over genetics in the U.S.S.R. has come to a close, with the complete defeat of the neo-Mendelians at the hands of Lysenko, it is time for men of science outside the U.S.S.R. to take stock of the situation and to see what implications and consequences it has for them. I believe that the situation is very grave. There is now a party line in genetics, which means that the basic scientific principle of the appeal to fact has been overridden by ideological considerations. A great scientific nation has repudiated certain basic elements of scientific method, and in so doing has repudiated the universal and supranational character of science.

Huxley then considers the most important resolutions passed by the Praesidium of the Soviet Academy of Sciences. For example:

The Bureau of the Division of Biological Sciences shall revise the syllabuses at biological institutes, *bearing in mind the interests of Michurinism.*'

An explanatory statement follows, including the following remarks: 'At a number of Academy institutes *formal genetics has not been combated with sufficient vigour.* For this the Praesidium of the Academy takes the blame. The Bureau of the Division of Biological Sciences and its head L. A. Orbeli (who was released from his duties as Academician-Secretary under Resolution 1) have failed to give a correct orientation to the biologists of the Academy.

'The report by Lysenko (1948 ; and ref. 3), which has been approved by the Central Committee of the Communist Party, *has exposed the scientific inconsistency of the reactionary idealist theories of the followers of Weismannism*—Schmalhausen, Dubinin, Zhebrak, Navashin and others.'

A letter to Comrade Stalin is summarized as follows : 'A pledge is here given by the Praesidium of the U.S.S.R. Academy of Sciences to further Michurin's biology *and to root out unpatriotic, idealist, Weismannite-Morganist ideology*'.

A further statement by the Praesidium ("To the prosperity of our progressive science") is finally summarized : 'Michurin's materialist direction in biology *is the only acceptable form of science, because it is based on dialectical materialism and on the revolutionary principle of changing Nature for the benefit of the people. Weismannite-Morganist idealist teaching is pseudo-scientific, because it is founded on the notion of the divine origin of the world and assumes eternal and unalterable scientific laws. The struggle between the two ideas has taken the form of the ideological class-struggle between socialism and capitalism on the international scale, and between the majority of Soviet scientists and a few remaining Russian scientists who have retained traces of bourgeois ideology, on a smaller scale. There is no place for compromise. Michurinism and Morgano-Weismannism cannot be reconciled.*'

*   *   *

Lysenko in his "Report" says that a major error of Darwin lay in his "transferring into his *teaching Malthus' preposterous reactionary ideas* on population". Later he writes "Reactionary biologists everywhere have . . . done everything in their power to empty Darwinism of its materialist elements"—a somewhat strange accusation to bring against the neo-Mendelians who have discovered and elucidated the material basis of heredity in the chromosomes, and in so doing appear to be on the track of the essential material basis of life itself.

This is a very sore spot with the Michurinites. Thus when Nemchinov was saying that he could not agree that "the chromosomes have no relation with the mechanism of heredity", there was a disturbance in the hall, and one delegate shouted "There are no such mechanisms". More important, Prezent, the philosophic interpreter of Lysenko, says that "nobody will be led astray by the Morganists' false analogies concerning the invisible atom and the invisible gene. *Far closer would be an analogy between the invisible gene and the invisible spirit.*"

V*ol.* 163, 935 1949

And he comments on the terms in which the academy attacked the ideologically unsound:

The Academy, in its official replies, has descended to extraordinary depths for a scientific body. It calls both Dale and Muller "tools in the hand of reactionaries", and implies that they could have had to expect persecution in their respective countries if they had not resigned. It states that Sir Henry Dale has shown himself a partisan "of the theories which were in vogue in Hitlerite Germany, which served as pretext for so many sanguinary horrors, and are still defended by the upholders of slavery and racial discrimination such as the Americans, who wish to impose their hegemony on the world". It adds that all believers in progress will approve the measures taken "and to be taken" by the Academy "to destroy for ever the continuance of this criminal obscurantism".

V*ol.* 163, 935 1949

Huxley sought to explain the official Soviet doctrine of Lamarckian inheritance to a mystified public by means of a simple analogy: if the theory were correct, it would follow that male Jewish babies are all born without foreskins. Shakespeare, he noted, had rejected any such foolishness:

There's a divinity that shapes our ends,
Rough-hew them how we will.

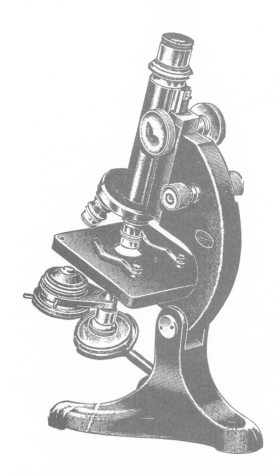

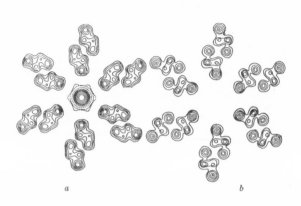

a   *b*

Next a warming example of what can be achieved in adversity by dedicated enthusiasts.

PRISON CAMP GEOLOGY

THE fascinating memoir, referred to below*, is worthy of notice, not only because it is a major contribution to 'front' petrology and deep-seated tectonics, but also because of the extraordinary conditions under which the research it records was carried out. The University of Edelbach was founded by French prisoners-of-war in Oflag XVII A (1940–45). Not content with lectures alone, the geologists made a thorough investigation of the area—only 400 metres square—enclosed within the barbed wire. No stone was left unturned, and trenches and secret tunnels provided many critical exposures. A microscope was constructed in the camp and equipped with polarizers improvised from piled cover glasses. Thin sections were mounted with a mixture of violin wax and edible fat. Only the determination of certain untwinned feldspars remained to be completed on the return to France.

The greater part of the memoir is devoted to the crystalline rocks of the Waldviertel complex and the syntectonic granitization phenomena displayed by them. The country rocks are tectonites with pronounced linear structures. They include biotite- and graphite-schists, and types ranging from plagioclasite to amphibolite, all originally poor in quartz. Some excellent examples of micro-tectonic analysis are given, and it is shown that in spite of the intense deformation undergone, lattice discontinuities have been largely healed by granoblastic 're-cooking'. Movement and recrystallization were probably simultaneous, rather than alternating, phenomena. Granitization and what was formerly styled 'injection', due to geochemical migrations, consisted of quartzification, followed by development of perthitic orthoclase at the expense of both the original rock material and the newly crystallized quartz. The evidence suggests that while diffusion through the lattices took place locally, the main transport was by way of intergranular boundaries. Complementary to the addition of silicon and potassium, the cafemic elements calcium, iron and magnesium were expelled from the granitization zone to form basic fronts in the surrounding rocks. The amphibolitic aureoles of the Moldanubian 'orthogneiss' are believed to be large-scale results of the same process.

*Métamorphisme, silicifications et pédogénèse en Bohême Méridionale. By F. Ellenberger in collaboration with R. Dézavelle, M. Fischer, A. Guilleux, V. Host, A. Moyse and P. Pérault. Pp. 169. (Besançon: Annales Scientifiques de Franche-Comté, 1948.)

It is shown that quartz and orthoclase were remarkably plastic during the physico-chemical conditions that attended their formation, and that, in consequence, the granite formed by the transformation of pre-existing rocks could readily become intrusive. It follows that to prove a granite intrusive does not prove that it has ever been in a liquid condition.

The memoir is full of important observations and stimulating suggestions, and should be read by all workers in the field of plutonic geology.

VOL. 163, 967 1949

When the Russian army entered Berlin in 1945, Otto Warburg, who had sat out the war in his institute in Dahlem, decamped to his summer house on the island of Rügen. Since the Russians had crated up the entire contents of the laboratory and despatched them to Russia, there was nothing for Warburg to do but write. Among the results was his treatise on metals in biology. He refused initially to allow it to be translated into English, for he feared that one of his many enemies, such as David Keilin, whom he had traduced in the book, might review it. Warburg was approached by Paul Rosbaud, a physicist and publisher — an opponent of the Nazis (code-named "The Griffin"), who had braved many dangers to take information about scientific progress in the Third Reich, especially concerning the atomic bomb project, to the Allies by way of neutral Sweden, where in the line of business he was allowed to travel. The Griffin finally succeeded in talking Warburg into permitting a translation, but could not persuade him to take out for the English version a few paragraphs containing the most disobliging passages about Keilin and others. The book was indeed reviewed by Keilin, who comported himself with beatific restraint. This is how he signs off.

The reader, in his attempt to understand better the aim of the author, will possibly turn his attention to a quotation from Pasteur used as the 'motto' to this book. The misleading appearance of this motto, which has recently been mistaken for a quotation from a letter of Pasteur to Berthelot, is due to its faulty transcription. In fact, it should read as follows, the missing words again being given in italics : "Comment *M. Berthelot* n'a-t-il pas senti que le temps est *le seul juge en cette matière et* le juge souverain? Comment n'a-t-il pas reconnu que du verdict du temps je n'ai pas à me plaindre ?" These were the words of challenge of Pasteur to Berthelot in the course of one of their dramatic discussions which took place in the Paris Academy of Sciences on February 10, 1879. This discussion, which concluded their twenty-year-old controversy, was revived following the unfortunate publication by Berthelot of Claude Bernard's incomplete and inconclusive notes, which appeared to throw doubt upon the whole of Pasteur's work on fermentation.

One can scarcely escape the impression that had Prof. Warburg full confidence in the meaning of Pasteur's words, this book would have been written in a different spirit : as it is, this 'motto' reflects a certain feeling of disappointment. Yet the author has no reason to complain that his work has not received its due recognition.   D. KEILIN

VOL. 165, 4 1950

Sir Lawrence Bragg quarried the Cavendish Laboratory archives for the substance of a lecture and article. Here are two extracts.

Much of this material has until recently been stored in the Cavendish without any attempt at review or classification. During the past eighteen months, it has fortunately been possible to arrange that Mr. Derek Price, who is a historian of science, should arrange it and make a list of what exists. He has unearthed many treasures, and I propose to give some examples here.

* * *

Another note of Maxwell's is headed "Concerning Demons" and starts "Who gave them this name ? Thomson". All lecturers on thermodynamics quote 'Maxwell's Demons', who can defeat the second law by opening a door when a fast molecule approaches and closing it against a slow one. Apparently Kelvin was actually their parent.

Maxwell and Tait were indefatigable correspondents. They had been school-fellows at the Edinburgh Academy. The postcards and letters which they interchanged were a running commentary on scientific events of the day, largely in a kind of punning mathematical shorthand which they had invented. Some two hundred of these are among the Maxwell papers. One, clearly referring to a lecture Tait is to give at the Cavendish, starts "Ο Τ΄, θαγξ φορ Αλλες" and goes on to say that Prof. Liveing will provide the lime-light, the gases will go for half an hour, and that if he is particular about his lantern he had better bring his own like Guy Fawkes or the Man in the Moon. . . . "Yr's $\delta p/\delta t$" (Τ΄ = Tait, $\epsilon p/\delta t$ = J. C. Maxwell).

The Cambridge period is initiated by the letter inviting him to be a candidate for the Cavendish chair at a salary of £500 a year, and encouraging him by an assurance that Kelvin did not wish to be considered. A great contribution which Maxwell made was the design of a laboratory for physical investigations which is recognizably of a modern type. The idea of such a laboratory was quite new in Britain ; when at King's College, for example, Maxwell carried out his experiments in the garret of his home. Maxwell's note-book at that period contains references to details of the building which have quite a modern ring ; care that the lecture-room can be quickly darkened by shutters, positions of taps and leads on the lecture bench, and facilities for showing diagrams. The rooms were definitely planned for different types of experiment. An interesting relic which came to light recently in a drawer of an old desk is a small screw of paper which had apparently been overlooked in tidying operations ; it had on it the names of "Gentlemen at Practical Work, Lent Term, 1877", including Schuster, Shaw (Sir Napier Shaw), Glazebrook, Fenton, McAllister. Another interesting letter is one to Maxwell from Stokes about one of Maxwell's papers submitted for publication by the Royal Society. It is dated 1878, and is typewritten ! Stokes had the first typewriter in Cambridge, and remarks in his letter that he is going to London to get a new ribbon for it, as the old one is worn out.

* * *

The other example is taken from the remarkable collection of letters from colleagues about radioactivity. One file contains the letters written by my father to Rutherford between the years 1904 and 1909. In 1904 my father was at the University of Adelaide. He had had a very full and busy life organizing his department, and indeed much of the University, and it was not until the age of forty-two that he attempted to do any original research. He was asked to read a paper on radioactivity to the Australian Association for the Advancement of Science. This aroused his interest in α-rays, and he purchased some radium bromide and made experiments on the variation of ionizing power along the path of the α-particles. He was able to prove that all α-particles from each type of radioactive atom have a definite energy, that they go through gas or solid without deviation, and have a definite range. He had no colleague in Australia with whom he could discuss his results ; they are set out in a series of long letters to Rutherford at McGill in Montreal. The first letter starts, "I have lately obtained some curious experimental results in connection with the absorption of the α rays which are, I think, new". In the second letter he has had a reply, and is thrilled to know that Rutherford thinks the ideas are new and really important.

Vol. 169, 684 1952

In 1945, as the fog of war cleared, the scientists of Europe emerged from the rubble, blinking in the sunlight. Reports from ravaged countries of who had survived and who had perished began to appear in the pages of Nature *(p.242). This fragment is from a letter to Julian Huxley from Nikolaas Tinbergen, the Dutch ethologist and future Nobel laureate.*

A new star in the sky of animal psychology is A. Kortlandt, a young fellow, very clever, very original, very 'cormorant-minded' ; for years he has lived among the Lekkerkerk cormorants, day and night, most of the time in a hide in the nest-trees, where he often stayed for one or more weeks, day and night continuously, receiving his food by a 'téléférique' (*Drahtseilbahn*).

Kluyver, the starling man, is all right ; he had to leave Wageningen in September 1944, and his house was badly damaged, but he managed to save all the notes of his extensive study on the great tit, and most of his books, and has recently returned. The Government Phytopathological Service, where Kluyver was the ornithologist, has been absolutely ruined by the Germans. All records and files have been burned, as well as the library, one of the many instances of the enormous setback caused by the War. Sirks, Hazelhoff, Dijkgraaf (nephew and pupil of Von Frisch), Raven, Krijgsman, my brother L. Tinbergen, Ihle, De Beaufort are all right. Ihle very nearly died of starvation. Owing to the very limited possibilities of travelling and writing (no trains; cycling only possible during some hours of the night, to avoid the German slave-hunters), we heard such things mostly too late to help each other. De Burlet, much to our regret, has been co-operating with the Germans. He has fled and nobody knows where he is. Hirsch, who was a German, seems to be in Germany.

You may wonder whether we will have time to resume 'pure' scientific research, now that all kinds of reconstruction work will demand so much of our energy. I am sure that we will succeed in keeping part of our time for research.

Vol. 156, 576 1945

In the years immediately following the war, a rancorous debate sprang up between Werner Heisenberg and the Dutch-American physicist Samuel Goudsmit. Goudsmit had led an expedition, code-named Alsos, that followed the Allied armies through Germany. Its mission was to investigate the activities of German laboratories in the furtherance of Hitler's war. There was in particular still a fear that an atomic bomb was being built. Heisenberg, who had led this project, sought to exculpate himself by claiming that he and his colleagues were aiming to construct an atomic reactor and he insinuated that moral scruples alone had deterred them from developing a bomb. A group of the most prominent German physicists, Heisenberg included, were rounded up and brought to England. There they were confined in a country house near Cambridge and their conversations were bugged. The so-called Farm Hall transcripts revealed that Heisenberg was above all mortified, when news came of the Hiroshima bomb, that the scientists of the Manhattan Project had so far outstripped their German competitors. In an extended account in NATURE, *based on an even longer article in the German journal* Naturwissenschaften, *Heisenberg presents the German wartime nuclear physics effort in a highly favourable light. His account terminates with this rather disingenuous explanation of why his project had not in the end prospered.*

We have often been asked, not only by Germans but also by Britons and Americans, why Germany made no attempt to produce atomic bombs. The simplest answer one can give to this question is this: because the project could not have succeeded under German war conditions. It could not have succeeded on technical grounds alone: for even in America, with its much greater resources in scientific men, technicians and industrial potential, and with an economy undisturbed by enemy action, the bomb was not ready until after the conclusion of the war with Germany. In particular, a German atomic bomb project could not have succeeded because of the military situation. In 1942, German industry was already stretched to the limit, the German Army had suffered serious reverses in Russia in the winter of 1941–42, and enemy air superiority was beginning to make itself felt. The immediate production of armaments could be robbed neither of personnel nor of raw materials,

nor could the enormous plants required have been effectively protected against air attack. Finally—and this is a most important fact—the undertaking could not even be initiated against the psychological background of the men responsible for German war policy. These men expected an early decision of the War, even in 1942, and any major project which did not promise quick returns was specifically forbidden. To obtain the necessary support, the experts would have been obliged to promise early results, knowing that these promises could not be kept. Faced with this situation, the experts did not attempt to advocate with the supreme command a great industrial effort for the production of atomic bombs.

From the very beginning, German physicists had consciously striven to keep control of the project, and had used their influence as experts to direct the work into the channels which have been mapped in the foregoing report. In the upshot they were spared the decision as to whether or not they should aim at producing atomic bombs. The circumstances shaping policy in the critical year of 1942 guided their work automatically towards the problem of the utilization of nuclear energy in prime movers. To a German physicist, this task seemed important enough. The mere possibility of solving the problem had been rendered possible by the discovery of the German scientific workers Hahn and Strassmann; and so we could feel satisfied with the hope that the important technical developments, with a peace-time application, which must eventually grow out of their discovery, would likewise find their beginning in Germany, and in due course bear fruit there.

Vol. 160, 214 1947

The year 1953 could be said to mark, in biology at least, the end of history. Here is James Watson and Francis Crick's paper on the structure of DNA, which ushered in the new era with the celebrated understatement near the end.

MOLECULAR STRUCTURE OF NUCLEIC ACIDS

A Structure for Deoxyribose Nucleic Acid

WE wish to suggest a structure for the salt of deoxyribose nucleic acid (D.N.A.). This structure has novel features which are of considerable biological interest.

A structure for nucleic acid has already been proposed by Pauling and Corey[1]. They kindly made their manuscript available to us in advance of publication. Their model consists of three intertwined chains, with the phosphates near the fibre axis, and the bases on the outside. In our opinion, this structure is unsatisfactory for two reasons: (1) We believe that the material which gives the X-ray diagrams is the salt, not the free acid. Without the acidic hydrogen atoms it is not clear what forces would hold the structure together, especially as the negatively charged phosphates near the axis will repel each other. (2) Some of the van der Waals distances appear to be too small.

Another three-chain structure has also been suggested by Fraser (in the press). In his model the phosphates are on the outside and the bases on the inside, linked together by hydrogen bonds. This structure as described is rather ill-defined, and for

This figure is purely diagrammatic. The two ribbons symbolize the two phosphate—sugar chains, and the horizontal rods the pairs of bases holding the chains together. The vertical line marks the fibre axis

this reason we shall not comment on it.

We wish to put forward a radically different structure for the salt of deoxyribose nucleic acid. This structure has two helical chains each coiled round the same axis (see diagram). We have made the usual chemical assumptions, namely, that each chain consists of phosphate diester groups joining β-D-deoxyribofuranose residues with 3′,5′ linkages. The two chains (but not their bases) are related by a dyad perpendicular to the fibre axis. Both chains follow right-handed helices, but owing to the dyad the sequences of the atoms in the two chains run in opposite directions. Each chain loosely resembles Furberg's[2] model No. 1; that is, the bases are on the inside of the helix and the phosphates on the outside. The configuration of the sugar and the atoms near it is close to Furberg's 'standard configuration', the sugar being roughly perpendicular to the attached base. There is a residue on each chain every 3·4 A. in the z-direction. We have assumed an angle of 36° between adjacent residues in the same chain, so that the structure repeats after 10 residues on each chain, that is, after 34 A. The distance of a phosphorus atom from the fibre axis is 10 A. As the phosphates are on the outside, cations have easy access to them.

The structure is an open one, and its water content is rather high. At lower water contents we would expect the bases to tilt so that the structure could become more compact.

The novel feature of the structure is the manner in which the two chains are held together by the purine and pyrimidine bases. The planes of the bases are perpendicular to the fibre axis. They are joined together in pairs, a single base from one chain being hydrogen-bonded to a single base from the other chain, so that the two lie side by side with identical z-co-ordinates. One of the pair must be a purine and the other a pyrimidine for bonding to occur. The hydrogen bonds are made as follows: purine position 1 to pyrimidine position 1; purine position 6 to pyrimidine position 6.

If it is assumed that the bases only occur in the structure in the most plausible tautomeric forms (that is, with the keto rather than the enol configurations) it is found that only specific pairs of bases can bond together. These pairs are: adenine (purine) with thymine (pyrimidine), and guanine (purine) with cytosine (pyrimidine).

In other words, if an adenine forms one member of a pair, on either chain, then on these assumptions the other member must be thymine; similarly for guanine and cytosine. The sequence of bases on a single chain does not appear to be restricted in any way. However, if only specific pairs of bases can be formed, it follows that if the sequence of bases on one chain is given, then the sequence on the other chain is automatically determined.

It has been found experimentally[3,4] that the ratio of the amounts of adenine to thymine, and the ratio of guanine to cytosine, are always very close to unity for deoxyribose nucleic acid.

It is probably impossible to build this structure with a ribose sugar in place of the deoxyribose, as the extra oxygen atom would make too close a van der Waals contact.

The previously published X-ray data[5,6] on deoxyribose nucleic acid are insufficient for a rigorous test of our structure. So far as we can tell, it is roughly compatible with the experimental data, but it must be regarded as unproved until it has been checked against more exact results. Some of these are given in the following communications. We were not aware of the details of the results presented there when we devised our structure, which rests mainly though not entirely on published experimental data and stereochemical arguments.

It has not escaped our notice that the specific pairing we have postulated immediately suggests a possible copying mechanism for the genetic material.

Full details of the structure, including the conditions assumed in building it, together with a set of co-ordinates for the atoms, will be published elsewhere.

We are much indebted to Dr. Jerry Donohue for constant advice and criticism, especially on interatomic distances. We have also been stimulated by a knowledge of the general nature of the unpublished experimental results and ideas of Dr. M. H. F. Wilkins, Dr. R. E. Franklin and their co-workers at King's College, London. One of us (J. D. W.) has been aided by a fellowship from the National Foundation for Infantile Paralysis.

J. D. Watson
F. H. C. Crick
Medical Research Council Unit for the
Study of the Molecular Structure of
Biological Systems,
Cavendish Laboratory, Cambridge.
April 2.

[1] Pauling, L., and Corey, R. B., *Nature*, 171, 346 (1953); *Proc. U.S. Nat. Acad. Sci.*, 39, 84 (1953).

[2] Furberg, S., *Acta Chem. Scand.*, 6, 634 (1952).

[3] Chargaff, E., for references see Zamenhof, S., Brawerman, G., and Chargaff, E., *Biochim. et Biophys. Acta*, 9, 402 (1952).

[4] Wyatt, G. R., *J. Gen. Physiol.*, 36, 201 (1952).

[5] Astbury, W. T., Symp. Soc. Exp. Biol. 1, Nucleic Acid, 66 (Camb. Univ. Press, 1947).

[6] Wilkins, M. H. F., and Randall, J. T., *Biochim. et Biophys. Acta*, 10, 192 (1953).

index

Numbers in roman type indicate articles in which a person is mentioned or is the subject of discussion. Italics indicate that the article is written by the person named.